Infrastructure Asset Management with Power System Applications

Infrastructure Asset Management with Power System Applications

Lina Bertling Tjernberg

CRC Press
Taylor & Francis Group
Boca Raton London New York

CRC Press is an imprint of the
Taylor & Francis Group, an **informa** business

MATLAB® is a trademark of The MathWorks, Inc. and is used with permission. The MathWorks does not warrant the accuracy of the text or exercises in this book. This book's use or discussion of MATLAB® software or related products does not constitute endorsement or sponsorship by The MathWorks of a particular pedagogical approach or particular use of the MATLAB® software.

CRC Press
Taylor & Francis Group
6000 Broken Sound Parkway NW, Suite 300
Boca Raton, FL 33487-2742

Printed on acid-free paper

International Standard Book Number-13: 978-1-4987-0867-8 (Hardback)

Visit the Taylor & Francis Web site at
http://www.taylorandfrancis.com

and the CRC Press Web site at
http://www.crcpress.com

To my parents Jan and Margareta
for your endless love and support

&

To my husband Oscar and
our two sons Gunnar and Walter
for providing the greatest
love, challenge, and inspiration of life.

Contents

List of Tables

List of Figures

Preface

This book is about infrastructure asset management, which can be expressed as the combination of management, financial, economic, engineering, and applied to physical assets, with the objective of providing the required level of service in the most cost-effective manner. It includes management of the whole life cycle of a physical asset from design, construction, commission, operation, maintenance, modification, decommissioning, and disposal. Moreover, there are typically constraints, for example, in budget, which result in the need of a strategy for performing asset management.

There is a long history of applying infrastructure asset management, for example, in the military and aviation sectors. This book focuses on asset management of an infrastructure for energy, that is, the electric power system.

Electricity has been appointed as one of the greatest inventions. The launch of the light bulb in 1879, which is listed as one of ten most important inventions of all time, revolutionized society by the use of electricity. Electricity provides energy in applications for transport, heating, lightning, communications, and computation. The electric power system is the infrastructure that manages the electricity from generation, delivery to consumption.

The electric power system is currently undergoing a major reconstruction, captured in the concept of Smart Grid. Main drivers for this change are the climate and energy goals towards a sustainable energy system. There are, however, also crucial drivers for cost efficiency and efficient use of resources—in this context the tools of infrastructure asset management becomes one of the important solutions. Moreover, new technology developments open up for new solutions for asset management of the electric power system. A general trend is in new possibilities for control and operation, for example, with Phasor Measurements Units, generally located in the transmission grid providing measurements of voltage and current with accuracy up to 30–120 times per second, or smart meters placed at the end consumer, which enables integration of small-scale electricity generation, for example, from solar cells, electrical vehicles, and energy storage, and a distributed control of energy use. Another trend is the development of diagnostic measurement techniques for assessing the insulation condition and prediction of lifetime of physical assets, and new methods for condition monitoring, for example, using sensor networks. These different trends have in common an overall development towards access to large volumes of data for handling and analysis, and another concept growing in interest is referred to as Big Data—which provides new means for infrastructure asset management.

Acknowledgments

This book is the result of approximately 20 years of dedicated work in the field of applied reliability theory and numerous applications for the electric power system. I would like to express my deep gratitude to the many people who have inspired me, who have supported me, and who have contributed to the work contained in this book. It is not possible to name you all who have been with me over the course of the years but you are all thought of with true gratitude. In the following paragraphs, I have tried to capture the main contributions to this work from three overall parts consisting of academic mentors, industrial collaborators and mentors as well as students and researchers who have worked within my research groups.

The dream of publishing this book came with the start of my own PhD studies on reliability assessment of electrical distribution systems at KTH (Kungl Tekniska Högskolan) Royal Institute of Technology. I was recruited by Professor Roland Eriksson and Professor Göran Andersson (later professor at ETH), and I am truly grateful for your faith in me. Your valuable advice and great leadership continues to guide and inspire me. The main books that inspired me were written by Professor Ron Allan, UK (emeritus professor at the University of Manchester), Professor Roy Billinton, Canada (emeritus professor at University of Saskatchewan) and by Professor Arne Holen (emeritus professor at Norwegian University of Science and Technology [NTNU]), Norway. I was honored to have Professor Allan as the main advisor of my PhD student project, and visits with Professor Billinton and Professor Holen during my studies and continued work. For my post-doc, I had the privilege of visiting with Dr. John Endrenyi and Dr. George Anders, then at Kinetrics and also associated with the University of Toronto in Canada. The inspiration and knowledge I gained from them and what they taught me about the process of performing research constitutes the basis for this book.

The Competence Centre of Electric Power Engineering (EKC) at KTH sponsored my PhD studies. Later, I became a program manager further developing the research program on maintenance management, and funded the research group on reliability centered asset management (RCAM). EKC involved several industrial partners and the Swedish energy authority. The core partners were ABB, EON (formerly, SydKraft), Vattenfall, National Swedish Grid, and Ellevio (formerly, Stockholm Energy and later, a part of Fortum) and STRI (High-Voltage Testing and Consulting). A core input for the research presented in this book was the access to reliability data, which was gained from the utilities and I am truly grateful for the access to data and for the sharing of expertise and the collegial friendship. I like to acknowledge a few persons to whom I am specifically grateful are Sven Jansson (at the former Sveriges Distribution Utilities, now a part of Energiforsk), Olle Hansson (a key person for the maintenance program within EKC and a mentor for my work), Anders Björk at the then Energiforsk and now OX2, and Sven-Erik Thor at Vattenfall who supported several projects in wind power, and also Ying He my former PhD student colleague and since then at Vattenfall R&D. The international colleague friendship and volunteer work for IEEE Power and Energy Society has been a huge inspiration for my work during all this period and I am truly grateful for friendship, and sharing and exchange of knowledge. In 2007, I was recruited to the Swedish National Grid (SvK) by the current technical director, Sture Larsson. I am truly grateful for your faith in me and for sharing your expertise and knowledge in the electric power system. Other former colleagues at SvK who I like to extend special thanks are: Stefan Arnborg, (whom I known since my PhD studies at KTH and we kept the same interest in power system reliability), Per Kvarnefalk and Lars Marketeg (for introducing me to maintenance aspects at SvK), and Göran Ericsson (who took over my role to coordinate the R&D). In 2009, I obtained a position at Chalmers University and I am truly grateful for the collegial friendship shown by my colleagues during the intensive years that followed. I would especially like to mention Professor Ola Carlson at Chalmers, Per Norberg at Vattenfall for having trusted me in this position and being a mentor during my time at Chalmers, and my closest colleague, Assistant Professor Tuan Le. Another key person I like to acknowledge is Professor Michael Patriksson at Chalmers who has made large contributions with his expertise in applied optimization, and Professor Math Bollen at Luleå Technical University for numerous inspiring discussions. The main industrial collaborators during this period were Vattenfall, ABB, EON, Göteborg Energi, and SKF. I am truly grateful for the access to data that they provided and which have made it possible and several interesting. In summary, I am truly grateful for the sharing of expertise, for the collegial friendship, and for the necessary input data performing the different studies which are the basis for the case studies in this book.

I would like to express my sincere thanks to the many master theses students, PhD students, and postdocs, who have worked within, and visit with, my research groups during this time. You are the core of my professional family and it is your dedicated work that has been the force forward. It has been a privilege being your mentor and supporting you on your path forward. The different case studies selected for this book are a form of "best-of" selection of publications within the core topic of RCAM (Reliability Centered Asset Maintenance). Work from the following master theses students are included in the case studies: Erik Persson, Otto Wilhelmsson, Johan Ribrant, Sabina Karlsson, and Gloria Puglia. Work from the following former PhD students (and the last postdoc) are included in the case studies (or are related to the development of the computer tool RADPOW): Patrik Hilber (now at KTH), Tommie Lindquist (now at Swedish National Grid), Carl Johan Wallnerström (now at the Energy Market Inspectorate), Johan Setréus (now at Swedish National Grid), Julia Nilsson (Stockholm University), Anna Franzen, Francois Besnard (now at OX2), Pramod Bangalore (now at Greenbyte), Katharina Fischer (now at Fraunhofer Institute for Wind Energy Systems—IWES). I would wish to specially thank Dr. Bangalore, who has also contributed the material for the Appendix, and is currently assisting the supervision of the next exciting case study.

For the accomplishment of this book I would like to thank a few additional persons. Dr. Richard Brown (exponent) for encouraging me to submit my book proposal to the CRC Press Taylor & Francis Group. I like to express my thanks to Nora Konopka, editor at CRC Press/Taylor & Francis Group for taking on this project with support from Kyra Lindholm, editorial assistant. I also like to thank Professor Rajeev Thottappillil, the head of department at KTH, for being positive and supportive of me taking on and finalizing this task. Thanks also to Mr. Himanshu Gautam, a former master student, for taking on the job of assisting me in the editing and formatting the draft of this manuscript. Without the joy of working with you and your support, I am not sure if I would have managed to complete this project. I also like to thank my colleague Docent Nathaniel Taylor for reading and commenting on the manuscript despite your very tough schedule. Finally, I owe gratitude to my dear friends and wonderful family who have stood by me and provided the required support.

I am solely responsible for any error in this book. This material is necessarily a work in progress and suggestions from readers for improving its accuracy and clarity will be warmly welcomed.

Lina Bertling Tjernberg
Stockholm, Sweden, February 1, 2018

Author

Dr. Lina Bertling Tjernberg is a full professor in power grid technology at KTH (Kungliga Tekniska Högskolan)—the Royal Institute of Technology since 2013. She did her research study and undergraduate education from KTH. She was away from KTH for a period of seven years. During 2007–2009 she was with the Swedish National Grid as director of Research and Development. During 2009–2013 she was professor in sustainable electric power systems at Chalmers University of Technology, and during 2009–2012 she was the head of the Electric Power Engineering Division. She has spent two research periods in Canada; during 2000 at the University of Saskatchewan and during 2012–2013 at the University of Toronto and Kinectrics. She has spent a sabbatical period during 2014 as visiting professor at Stanford University at the Department of Civil and Environmental Engineering with Stanford Sustainable Systems Laboratory. Her research area is within applied mathematics for reliability evaluation, life time modelling, and maintenance optimization. Her application areas have been focusing on the electrical power system including distribution, transmission, high-voltage components (such as circuit breakers, disconnectors, gearbox, transformers), and hydro, nuclear and wind power, and on new developments for a sustainable energy system, including; large energy storage, electrical vehicles and micro grids. Current research also includes electrification of transportation with the overall objective to minimize the use of fossil fuels and reduce energy usage.

Dr. Bertling Tjernberg is a senior member of IEEE (Institute of Electrical and Electronics Engineers), is the chair of the Swedish PE/PEL Chapter and is an appointed distinguished lecturer (DLP) of the IEEE Power & Energy Society (PES). She has served on the Governing Board of IEEE PES, the Board of IEEE Sweden, the Board of the IEEE PES Subcommittee on Risk, Reliability and Probability Applications (RRPA) and she has served as associate editor for the *IEEE Transactions on Smart Grid Technologies*. She is an expert for the EU commission within Energy, ICT, and Security. The Swedish Government has appointed her as a member of the advisory council for the Energy Markets Inspectorate and she is a member of the National Strategic Council for Wind Power as well as the Scientific Board of the Swedish Civil Contingencies Agency. She has published over 100 papers and several book chapters.

Abbreviations

AC	Alternating current
ABAO	As Bad As Old
ACER	Agency for the Cooperation of Energy Regulators
AENS	Average energy not supplied per customer served
AGAN	As Good As New
ASAI	Average service availability index
CA	Criticality Analysis
CAIDI	Customer average interruption duration index
CAIFI	Customer average interruption frequency index
CBA	Cost Benefit Analysis
CIGRE	International council on large electric systems
CM	Corrective maintenance
CTAIDI	Customer total average interruption duration index
DC	Direct current
DSR	Demand Side Response
EC	European Commission
ELFORSK	Swedish Power Utilities Research and Development Organization
EPR	Ethylene propylene rubber
ENTSOE	European Network of Transmission System Operators
EPRI	Electric power research institute
FMEA	Failure Mode and Effect Analysis
FMECA	failure mode, effects, and criticality analysis
GTC	Grid Transfer Capacity
HD	Hogalid power station
HV	High voltage
IEA	International Energy Agency
int.	Interruption of voltage
LH	Liljeholmen power station
LV	Low voltage
MATLAB	Matrix laboratory
MLE	Maximum Likelihood Estimation
MLS	Method of Least Squares
mth	Months
MV	Medium voltage
NKR	Norwegian crowns
no.	number
PCI	Project of Common Interest
PD	Partial discharging

PE	Polyethylene
PM	Preventive maintenance
pp	Polypropylene
PVC	Polyvinyl chloride
RCM	Reliability-centered maintenance
RADPOW	Reliability assessment program of power distribution systems
RBD	Reliability block diagram
RES	Renewable Energy Sources
rms	Root mean square value
rpm	Revolutions per minute
SAIDI	System average interruption duration index
SAIFI	System average interruption frequency index
SEK	Swedish crowns
SINTEF	Foundation for scientific and industrial research at the Norwegian Institute of Technology
SJ	Statens jarnvag
SEW	Socio-Economic Welfare
SoS	Security of Supply
Svel	Association of Swedish electrical power utilities (included in Swedenergy AB)
Svk	Svenska Kraftnat (the Swedish National Grid)
TSO	Transmission System Operator
TYNDP	Ten Years Network Development Plan
UMIST	The Manchester Centre for Electrical Energy
USASK	The University of Saskatchewan
XLPE	Cross-linked polyethylene

1

Introduction

1.1 Infrastructure Asset Management

The value of making smart decisions gives a reason for adopting Asset Management (AM). AM is a broadly used concept typically used either in the financial or engineering sector. "**Asset Management** is defined as a coordinated activity of an organization to realize value from assets" [1]. The scope of this book is on AM of engineering systems. The first step of AM is always the motivation. This book is therefore to a wide extent a collection of case studies presenting application studies using AM.

The case studies in this book are all related to the infrastructure of the electric power system, which is a technical system that has been identified as one of the most important inventions of the past century [2]. Access to electricity for lighting and transportation was fundamental for the industrial revolution. Today electricity is fundamental to the digital society. A future trend is the returning to use electricity for transportation making it possible to use renewable fuels. This topic is further discussed in Section 1.2 introducing the electric power system as an enabler for the transition into a sustainable energy system, also today referred to as Smart Grid (SG).

In several parts of the world, this energy infrastructure was developed in large scale during the last century typically in Europe and the USA. The developments started with commercial electric power systems in 1880s [3]. The trend for the development of the electric power systems was to larger and interconnected systems expanding geographically into large synchronous grids. This trend was motivated by the economics of scale motivating large-scale electricity production and transport of energy over long distances using the energy efficient system of electric power. In other parts of the world, large developments can be seen for the current century like in Asia with huge developments in China and India building the most modern electric power systems today.

In summary for the coming years, there is an expected major need of investments in the electric power systems worldwide, either as a result of an aging infrastructure, from the need of expansions or/and a result from transition into a sustainable energy system. In the USA, the cumulative investment gap between 2016 and 2025 has been estimated to be 177 billion dollars [4]. A 10-year strategy plan in Europe foresees around in total 150 billion euros of investments in electric power infrastructure [5].

AM is a tool helping to make these investments smart. All the case studies presented in this book have the joint overall goal in supporting decision making for AM. Moreover, the decision making is supported by numerical variables, which could be used for showing on the benefit resulting in selecting different alternatives in the planning.

1.2 Towards a Sustainable Energy System

The energy system is in a global transition towards a sustainable energy system. Resource efficiency and environmental concerns push towards change into the use of renewable energy resources and to optimize the energy usage. This transition has been motivated by climate and energy goals and a growth in energy needs, the latter specifically in the Asian region. The United Nations adapted a resolution for a sustainable development with 16 goals until 2030 [6]. Recently, the European Commission (EC) launched the new targets towards 2030 [7,8]. In these targets, EC states "renewable energy will play a key role in the transition towards a competitive, secure and sustainable energy system." The EU has a legally binding target to 2030 of at least 27% renewable energy in final energy consumption at European level. This target translates into 46%–49% electricity generated by renewable energy resources according to the EC [9]. Moreover, wind energy should take at least 21% of that share [9]. Worldwide investments in renewable energy rose sixfold between 2004 and 2015, from USD 46.6 to 285.8 billion [10].

The electric power system is being modernized to enable a sustainable energy system. New developments include possibilities and challenges with generation, delivery, and usage of electricity as an integrated part of the energy system. This involves new forms of usage of electricity, for example, for transportation and demand response, and to the updating of existing electricity infrastructures. For electricity generation, the trend is toward new large-scale developments like offshore wind farms, as well as small-scale developments like rooftop solar energy. At the same time, digitalization of society is creating new opportunities for control and automation as well as new business models and energy related services. The overall trend for technology developments is new possibilities for measurement and control. An example is Phasor Measurements Units (PMUs), generally located in the transmission network, which provide measurements of voltage and current up to 30–120 times per second. Smart Meters placed with the end consumer, which enable integration of private small-scale electricity production from solar cells or energy storage from electric vehicles and general distributed control of energy use, are others. Another trend is the development of diagnostic measurement techniques for assessing the insulation condition and prediction of lifetime of physical assets, and new methods for condition monitoring, for example, using sensor networks. These different trends have in common an overall development towards access to large volumes of data for handling and analysis, and another concept growing in interest is referred to as Big Data, which provides new means for infrastructure asset management.

1.3 Reliability Centered Maintenance

The Reliability Centered Maintenance (RCM) method is a structured approach that focuses on reliability aspects when determining maintenance plans. The concept RCM was established in 1978 by Nowlan and Heap [11]. They were active in the commercial aviation industry. The need to improve reliability while containing the cost of maintenance was the driving factor behind this work [12]. Maintenance and reliability are important because of the large costs associated with maintenance tasks and costs due to the loss of production and breakdowns. Breakdowns can also lead to consequences that affect the environment or personal safety. These aspects could also be taken into consideration when performing an RCM analysis.

There are different versions of the RCM method in use. This book outgoes from the original RCM method and presents the Reliability Centered Asset Maintenance (RCAM) method. The RCAM method combines the proven method of RCM with quantitative maintenance optimization techniques. The RCAM method was originally developed for the application on electric power distribution systems [13,14] and later shown promising for the maintenance strategy selection and optimization of wind turbines as was described in e.g. [15].

While sole RCM as a qualitative method is limited in determining which maintenance strategy is the most cost-effective option, mathematical maintenance optimization techniques alone do not ensure that maintenance efforts focus on the right components. By merging these two approaches, the RCAM method provides an instrument for the quantitative assessment and comparison of maintenance strategies.

1.4 Outline and Scope of the Book

This book has two main parts where one is generic related to infrastructure asset management including reliability and maintenance theories and tools, and the other is specific with applications for the electric power system. This combination is not commonly found in the literature where typically books focus either on generic approaches or on specific applications. Two examples of related books on AM are given in [16,17]. The first gives a presentation of AM in the role of "whole-life management of physical assets." It presents AM as "a powerful new logic for building value and generating better returns from asset-intensive businesses." The second provides a selection of case studies and shows how a variety of organizations operating in different sectors have embraced asset management thinking and how this has affected them. Another example of books in related topic is given by [18]. The authors underline maintenance, replacement and reliability for optimization of AM. This book provides both description of the software tools, underlying methods, and includes cases from a variety of sectors.

Techniques and theories for reliability evaluation and maintenance planning in power systems have been well developed separately (for example, [15–28]). The earliest publication appearing in the bibliographies [15–21], relating to this field comes from Calabrese in 1933, and the first major group of papers from Lyman in 1947. However, there is a lack of techniques in use that relate to both of these two issues. This may indicate that the techniques currently available are not good enough. One probable reason for this is the gap between theory and practice, where sophisticated theoretical methods are too complicated to be used effectively. Advanced theoretical optimization theories do not automatically solve the practical problems of planning maintenance. Another reason could be that the data input required for the models are not available and that there has been a lack of incentive to solve this. However, there is a belief that this will change.

The applications presented in this book include a wide selection of different case studies. The case studies include examples from both system studies and component studies. The system studies include different parts of a power system including electrical distribution systems, wind power systems, and hydropower systems. The component studies, for example, breakers, transformers, gearbox, show examples of proposed reliability and maintenance models. There is also a specific chapter with examples on optimization.

The case studies show examples of different input data, theories and models and solution techniques. It is shown that performing these types of studies there would always be a need of both systematic models based on underlying theories, and on specific experience and knowledge of the actual system and its equipment. A fundamental challenge in the AM is that these types of needs are typically not in the same place in an organization and there might be a lack of communication and dialogue to overbridge such a gap. Possible standards like the ISO 55000, presented in the next chapter, would help in overcoming such a gap and contribute to a sound AM planning.

1.5 Goal of This Book

The main goal of this book is to provide the reader with knowledge of the fundamental terminology and theory used for infrastructure asset management and different analysis tools, and to give examples on how these can be used for different electric power system applications. Specifically, this book shows how systematic techniques, such as RCM and RCAM, can be used to achieve efficient strategies for infrastructure asset management.

1.6 Audience

This book is meant for practitioners or students within the different areas of infrastructure asset management with interest in applications for the electric power system. Infrastructure asset management is a wide field and covers different expertise areas, for example, management, economics, and engineering. The target audience for this book includes readers from different expertise areas providing a common understanding ground for the joint work on asset management.

A vision with the book is to encourage the collaboration and exchange of knowledge between different actors such as industry, academia, regulators, and decision makers. It is believed that the collaboration leads to better decisions and higher quality in the results. In a report presented by power industry and academia, it is concluded that the university education for the emerging workforce should be closely coordinated with industry whenever feasible to make sure the hands-on learning process is based on the most advanced instructional delivery tools used in the industry [29].

The background required for the reader is a basic knowledge of mathematical analysis and fundamentals in electrical engineering. For readers without this background knowledge, some explanation is provided in the appendix, and background references are provided to recommended literature, and it is hoped that the reader still can get all the essential ideas and important points.

2

Maintenance as a Strategic Tool for Asset Management

2.1 Introduction

The **aim** of asset management is to handle physical assets in an optimal way in order to fulfill an organization's goal whilst considering risk, where the goal could be maximum asset value, maximum benefit, or minimal life cycle cost, and the risk could be defined by the probability of failure occurrence and its consequence, for example, unavailability in power supply to customers.

There are different possible actions to handle these assets: acquire, maintain, replace, or redesign. AM implies consequently to make **right** decisions on:

- What assets to apply actions for
- What actions to apply
- How to apply the actions
- When to apply the actions

To make the right decisions, there is a need of:

- Condition data
- Failure statistics
- Reliability modeling techniques
- Reliability assessment tools
- Maintenance planning tools
- Systematic techniques for maintenance planning, for example, the RCM and RCAM methods

This book shows an example of how maintenance can be used as a tool to answer these questions in the support of a systematic asset management process.

2.2 Standards for Asset Management

"Navigating a world in transition" is the title of the ISO Annual Report 2016 [34]. The ISO Secretary-General expresses "Our world is in transition. We are being pulled in different directions by the forces of globalization, climate change, and the unprecedented speed of technological breakthroughs. We stand on the brink of a fourth industrial revolution driven by increased automation, advanced technologies, and robotics. These developments carry enormous potential for positive transformation, but our success will depend on our ability to seize and navigate the waves of change. We can only get there if we work together. Global supply chains, interoperability, safety, and traceability are just some of the issues that need concerted international effort." This citation does not only give a context to the value and purpose of using standards. It also well captures the needs of and fundamentals of asset management. This book uses the standard ISO 50000 as a basis for a description of asset management as a tool for infrastructure maintenance management with learnings from the electric power sector.

2.2.1 About ISO and ISO 55000

The ISO is an independent nongovernmental international organization for standardization. The ISO was created in 1946 with the objective to facilitate the international coordination and unification of industrial standards. Currently, the ISO comprises 163 national members [34]. The ISO 55000, Asset Management—Overview, principles, and terminology, was first launched in 2014. It provides an overview of asset management and management systems for the management of assets. It also provides the context for ISO 55001: Asset management—Management systems—Requirements and ISO 55002: Asset management—Management systems—Guidelines for the application of ISO 55000.

The target audience of these standards is [35]:

- Those considering how to improve the realization of value for their organization from their asset base
- Those involved in the establishment, implementation, maintenance, and improvement of an asset management system
- Those involved in the planning, design, implementation, and review of asset management activities

The adoption of these standards is believed to enable an organization to achieve its objectives through the effective and efficient management of its assets. The application of an asset management system helps to assure that those objectives can be achieved consistently and sustainably over time [35].

The ISO 55000 family of standards is the first set of international standards for asset management. They are emerging from the Publicly Available Specification (PAS) 55. The PAS55 was launched by the British Standards Institute (BSI), in 2004, resulting from an effort led by the Institute of Asset Management (IAM). It is considered as the first internationally recognized specification for Asset Management.

The ISO 55000 family of standards aligns with other major management systems. This includes ISO 9001 for quality management, ISO 14001 for environmental management, and ISO 31000 for risk management. The ISO 55000 family of standards provides the first management standard to implement the ISO Annex SL. The Annex SL is a high-level structure to provide a universal structure, identical core text, and common terms and definitions for all management standards.

The migration to ISO 55000 from PAS55 is considered as fairly straightforward with similar elements. A major difference is that the PAS55 is focused on the optimal management of physical assets, but the ISO 55000 is a standard for any asset type.

2.2.2 Definitions and Terminology

In the following, a selection of definitions from the ISO 55000 [1] is presented.

- *Capability*: Measure of capacity and the ability of an entity (system, person, or organization) to achieve its objectives
- *Competence*: Ability to apply knowledge and skills to achieve intended results
- *Continual improvement*: Recurring activity to enhance performance
- *Effectiveness*: Extent to which planned activities are realized and planned results achieved
- *Monitoring*: Determining the status of a system, a process, or an activity
- *Measurement*: Process to determine a value
- *Policy*: Intentions and direction of an organization as formally expressed by its top management
- *Objective*: Result to be achieved. An objective can be strategic, tactical, or operational
- *Requirement*: Need or expectation that is stated, generally implied or obligatory
- *Risk*: Effect of uncertainty on objectives
- *Asset*: Item, thing, or entity that has potential or actual value to an organization
- *Asset life*: Period from asset creation to asset end-of-life

- *Life cycle*: Stages involved in the management of an asset
- *Asset management*: Coordinated activity of an organization to realize value from assets
- *Strategic Asset Management Plan (SAMP)*: Documented information that specifies how organizational objectives are to be converted into asset management objectives, the approach for developing asset management plans, and the role of the asset management system in supporting achievement of the asset management objectives
- *Asset management plan*: Documented information that specifies the activities, resources, and timescales required for an individual asset, or a grouping of assets, to achieve the organization's asset management objectives
- *Preventive action*: Action to eliminate the cause of a potential nonconformity or other undesirable potential situation
- *Predictive action*: Action to monitor the condition of an asset and predict the need for preventive action or corrective action
- *Corrective action*: Action to eliminate the cause of a nonconformity and to prevent recurrence
- *Management system*: Set of interrelated or interacting elements of an organization to establish policies and objectives and processes to achieve those objectives
- *Asset management system*: Management system for asset management whose function is to establish the asset management policy and asset management objectives

2.3 Generally about Maintenance

2.3.1 Maintenance as a Tool for AM

Maintenance is one of the main tools of asset management. It can be defined as an activity of restoration where an unfailed device has its deterioration arrested, reduced, or eliminated [26]. Its goal is to increase the duration of useful component life and postpone failures that typically would require expensive repairs. Consequently, the task of maintenance is to slow the deterioration process. This is shown in Figure 2.1 where increasing deterioration is expressed in terms of decreasing asset value [36]. The asset value curves in the diagram are here referred as life curves.

The curves in Figure 2.1 illustrate how maintenance can be used as a tool for AM by showing the benefits of different maintenance policies. The figure illustrates conditions for two maintenance policies and for the case where

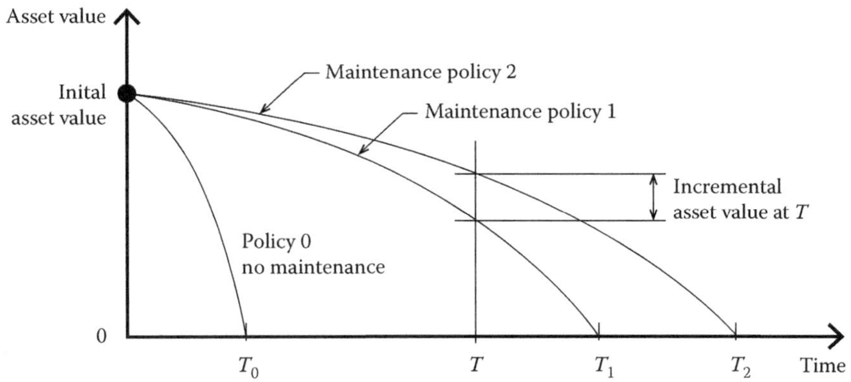

FIGURE 2.1
Illustration of the impact of maintenance policies on life curves (Adapted from Endrenyi, J. et al., *Comparison of Two Methods for Evaluating the Effects of Maintenance,* the 8th International Conference on Probabilistic Methods Applied to Power Systems (PMAPS), Ames, Iowa, September 2004.) [36].

no maintenance is carried out at all. If failure is identified with the condition of zero asset value and lifetime is defined as the mean time to failure, the life extensions T_0 to T, when Policy 1 is applied instead of neglecting maintenance, and T_1 to T_2, when Policy 1 is replaced by Policy 2, are clearly indicated in the figure. As far as reliability is concerned, Policy 2 is superior to Policy 1. Maintenance has its own costs which has to be taken into account when comparing policies and the most cost-effective policy chosen. The costs of maintenance should be balanced against the gains resulting from increased reliability. When costs are considered, Policy 2 may or may not be superior to Policy 1.

2.3.2 Maintenance Planning

Maintenance procedures are an integrated part of the planning, construction, and operation of a system. Moreover, they are of central and crucial importance to the effective use of available equipment. The aim of maintenance activities is to continuously meet performance, reliability, and economic requirements, while also adhering to the constraints set by system and customer requirements.

The concept maintenance refers to all actions undertaken to keep or restore equipment to a desired state. For example, the electrical power system must abide by the regulations and norms for heavy current and maintenance. The cost of maintenance must be taken into consideration when handling system assets to minimize the lifetime costs of the system. However, some maintenance activities must be undertaken even if they are not profitable. For an electrical system this could, for example, be inspections and earth-plate-metering inspections stipulated in the power regulations.

2.3.3 Definitions for Maintenance

In the following section, some main definitions for failures are presented. There are several books presenting main definitions. The main source used for this book is [37].

Functions are what an asset is expected to perform and can also be anything an asset has to comply with, such as a color or shape. The function of an asset does not have to be exact. In some cases, it is sufficient if the performance of an asset lies within some interval, for example, for an oven to provide a temperature. It is possible to divide functions into two subcategories, namely,

- Primary functions describe the main purpose of the asset
- Secondary functions describe additional features the asset should meet such as color or safety aspects

Functional failures is the inability of an item of equipment to fulfill one or more of its functions.

Failure modes are events that cause functional failures.

Failure effects are what happen when a failure mode occurs. The effects include evidence of failures, safety, and environmental hazards or production effects.

Failure consequences The consequences of all failures can be classified as being either Hidden, Safety, Environmental, Operational, or Nonoperational.

Potential failures are a state of an asset which indicates that a functional failure will occur.

2.3.3.1 P–F Interval

The P–F curve in Figure 2.2 shows the relationship between operation time and the condition of an asset. At point (A), a failure of the asset starts to occur. At the point (P), it is possible to detect that a failure will occur. This is called a potential failure. Point (F) corresponds to the time when a functional failure has occurred. The P–F interval is the time between it is possible to detect that a functional failure will take place and the time when it is expected to occur. This interval is also referred to as the warning period.

If an asset is checked at regular intervals and a potential failure is found, the time left until a functional failure occurs is the net P–F interval (Figure 2.3). This is the worst-case time available to prevent the functional failure from occurring after detecting the potential failure.

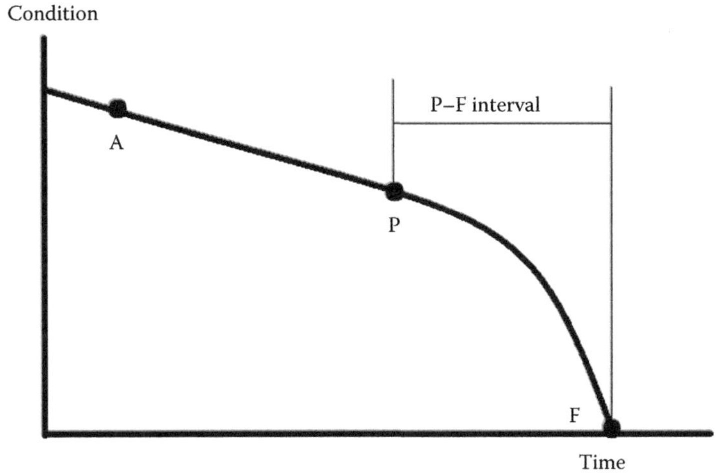

FIGURE 2.2
The P–F curve. The graph shows the events of: A, the failure starts to occur; P, the potential failure; and F, the functional failure.

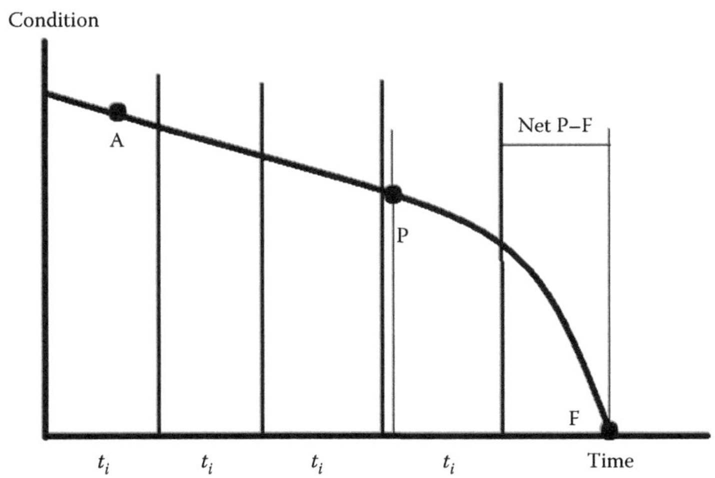

FIGURE 2.3
The net P–F interval. The graph shows the events of: A, the failure starts to occur; P, the potential failure; F, the functional failure; and the Inspection interval t_i.

2.3.3.2 Failures

2.3.3.2.1 Age-Related Failures

Many failures can be classified as age related. These kinds of failures are often found on mechanical equipment and are, for example, due to fatigue, corrosion, oxidation, and evaporation [37]. To be classified as an age-related failure, the probability of failure should increase at some point in time. The increase of failure probability can, for example, be constant in time or start at a specific age when the component usually wears out.

2.3.3.2.2 Non-Age-Related Failures

Often failures are not due to age but rather to surrounding factors such as a nonuniform stress or handling errors which may occur at random. For these failures, there is no certain point in time when the likelihood of failure increases [37].

2.3.3.2.3 Hidden Failures

If a failure is not noticed under normal working conditions, then the failure is a hidden failure. These failures alone are often of minor issue but in case of multiple failures they can be disastrous. Failures of safety equipment are often classified as hidden failures and will not become evident until there is another failure [37].

2.3.3.3 Maintenance

Maintenance is a combination of all technical, administrative, and managerial actions during the life cycle of an item intended to retain it in, or restore it to, a state in which it can perform the required function [38]. Maintenance can be carried out in different ways. One way to illustrate the different types of maintenance is shown in Figure 2.4.

2.3.3.3.1 Preventive Maintenance

Preventive maintenance is carried out at predetermined intervals or according to prescribed criteria and intended to reduce the probability of failure or the degradation of the functioning of an item [38].

2.3.3.3.2 Predetermined Maintenance

Predetermined, or scheduled, maintenance is a preventive maintenance carried out in accordance with established intervals of time or number of units of use but without previous condition investigation [38]. Predetermined maintenance is an option if a failure is age related and the probabilities of failure in time can be established. Depending on the consequences of a failure, different maintenance intervals can be chosen. If the consequences of a failure are not too severe and the cost for predetermined tasks are high, one might choose to allow the intervals between tasks to be longer than if the functional failure leads to a safety hazard which has consequences that cannot be tolerated. Examples of predetermined maintenance tasks are scheduled restoration and discard tasks. These tasks involve restoring or replacing the asset at determined intervals of time [37].

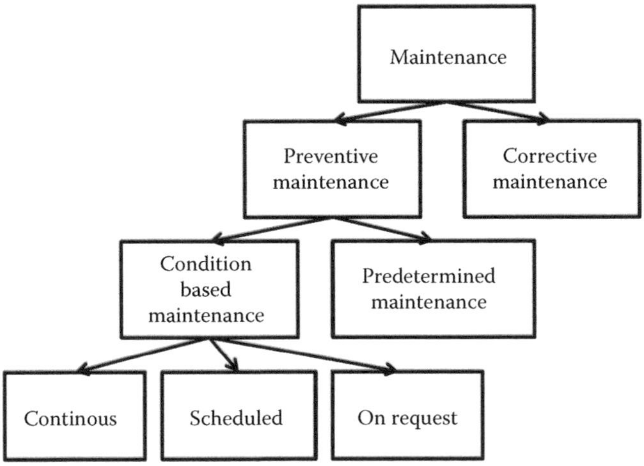

FIGURE 2.4
Overview of maintenance concepts. Outgoing from standard EN13306 (*Maintenance Terminology.* European standard: EN 13306:2010.) [38].

2.3.3.3.3 Condition-Based Maintenance (CBM)

Condition-based maintenance is a preventive maintenance based on performance and/or parameter monitoring and subsequent actions [38]. When dealing with non-age-related failures, CBM tasks are often used. CBM or on-condition tasks are performed by inspecting assets to determine if any potential failures have occurred. Some common methods include using human senses such as listening for unusual noise or condition monitoring of the asset. These tasks do not help postpone a failure but can detect that a failure will occur and thereby make it possible to act in order to avoid the consequences of a failure. CBM can be continuous, scheduled, or on request. If CBM is scheduled, intervals for conducting CBM tasks could be chosen with the P–F and net P–F intervals previously described. To make sure that a CBM task will detect a potential failure before a functional failure occurs, the interval between inspections must be shorter than the P–F interval. When a potential failure is detected, it is often desirable to have some additional time to correct the fault before a functional failure occurs. To accomplish this, the maintenance interval must be shortened further. Equation 2.1 shows the relationship among the inspection interval, the P–F interval, and the net P–F interval.

$$(P\text{–}F \text{ interval}) - (\text{Net } P\text{–}F \text{ interval}) = \text{Inspection interval} \tag{2.1}$$

The time needed after a potential failure is detected to plan and repair the asset is called action time. If this time and the P–F interval is known, Equation 2.1 gives you the longest allowed inspection interval if net P–F time is switched to action time [37].

2.3.3.3.4 *Corrective Maintenance*

Corrective maintenance is carried out after fault recognition and intended to put an item into a state in which it can perform a required function [38]. Corrective maintenance can be used if there is no way to detect or prevent a failure or it is not worth doing. It simply means that the asset is run till a failure occurs and then the system function is restored. This may of course not always be an option if the consequences of a breakdown are severe.

2.3.3.3.5 *Failure-Finding Tasks*

To find out whether a hidden failure has occurred, failure-finding tasks are used. This is done by checking if equipment is working in the way it should. These tasks can for example include switching on and checking if spare equipment is functioning properly [12].

2.3.3.3.6 *Redesign*

Redesign is used if there is no possibility to conduct maintenance on an asset and the probability and consequences of the failure are too high. Redesign might also be advantageous if maintenance is too expensive compared to redesigning [37]. Redesign is not always considered to be part of maintenance but is included here.

2.4 Preventive Maintenance

The aim of PM is to extend the lifetime of the components. Component lifetimes can be prolonged when the causes of component failure and therefore component failure rates themselves reduce. This is of course dependent on the nature of the causes. One way of understanding the relationship between maintenance and reliability would be to study these causes. Moreover, the cost-effective planning of a system implies that the correct components are maintained, at the correct time and with the correct maintenance activity. In this context, this means focusing maintenance on the critical components that is those that have a significant impact on system reliability and to reduce the dominant causes.

2.4.1 Maintenance Planning

Essentially, there are four tasks in a scheduled maintenance program [11]:

1. Inspection of an item to detect a potential failure
2. Reworking an item before a maximum tolerable age is exceeded
3. Discarding an item before a maximum tolerable age is exceeded

4. Inspecting an item and finding failures that have occurred but were unknown

For each failure type, an RCM program should be able to determine the type of maintenance task, if any, that should be applied and how frequently. Decision diagrams assist in this process.

The following are some typical maintenance questions.

- What efforts are required for this maintenance?
- Is the maintenance performed too little or incorrect?
- Is preventive maintenance done on the correct items?
- What is the relationship between failure and maintenance costs?

Applying the RCM method would be one way of providing answers to these maintenance questions.

2.4.2 Design for Preventive Maintenance

PM should be a prime consideration for any new equipment installation [39]. Effective preventive maintenance begins with good design with a conscious effort toward maintainability. Quality, installation, configuration, and application are fundamental prerequisites in attaining a satisfactory preventive maintenance program. Installation cost without regard for performing efficient and economic maintenance influences system design. In many instances, the additional cost of performing maintenance plus lost production from outages due to a lack of maintenance more than offsets the savings in initial cost. A system that is not adequately engineered, designed, and constructed will not provide reliable service, regardless of how good or how much preventive maintenance is accomplished.

Experience shows that equipment lasts longer and performs better when covered under a PM program [39]. An effective PM program could reduce accidents and operator error, minimize costly breakdowns and unscheduled outages by identifying and solving problems early avoiding major problems.

2.4.3 Equipment Deterioration

Equipment begins to deteriorate with installation. This is normal and if unchecked, the deterioration can progress and cause equipment malfunction or failure. Harsh environmental conditions and system stresses such as overload, severe duty cycle, load increases, circuit alterations, and changing voltage conditions can accelerate the deterioration process. An effective PM program can detect and mitigate these conditions. Equipment PM procedures should be developed to accomplish four basic functions [39]:

1. To keep the equipment clean
2. To keep the equipment dry

3. To keep the equipment sealed tight
4. To minimize the friction

Water, dust, high or low ambient temperature, high humidity, vibration, component quality, and countless other conditions can affect proper operation of equipment. Without an effective PM program, the risk of a serious failure increases.

2.5 Data Analytics, Data Analysis, and Big Data

Statistics is a branch of mathematics including the collection, analysis, interpretation, presentation, and organization of data [40]. Data is a set of values being either qualitative or quantitative variables. Qualitative properties are observed and can generally not be measured. Quantitative properties on the other hand have numerical values. The methods presented in this book have its basis in the analysis of quantitative data. It is, however, believed that a sound maintenance management requires both qualitative and quantitative data since the experience of the specific component or system being analyzed not always can be expressed in numerical values.

Two main statistical methods are used in data analysis: descriptive and inferential statistics. The further summarize data from a sample using indexes. The data analytics term for analysis of this data is exploratory data analytics (EDA). The EDA aims to find patterns and relationships in data. The latter draw conclusions from the data that are subject to random variation and are commonly referred to as probability theory. The data analytics term for the analysis of this data is confirmatory data analysis (CDA). The CDA applies statistical techniques to determine whether hypotheses about a data set are true or false.

Data analytics (DA) is the process of examining data sets. Advanced types of DA include typically the different methods of:

- *Data mining* involving the sorting of large data, identifying trends, patterns, and relationships
- *Predictive analytics* seeking to predict customer behavior, equipment failure, future events
- *Machine learning and artificial intelligence technique* that uses automated algorithms, making possible to solve larger problems faster compared to traditional analytical models

Big Data is a term for a data set that is so large or complex that traditional data analysis methods and tools are not possible to use to get results within a tolerable elapsed time. Challenges typically include capturing, storage, analysis, searching, sharing, visualizing, querying, and updating. These challenges might result in that, for example, data mining is not possible to apply. Below is a definition from the Gartner IT Glossary:

"Big Data is high-volume, high-velocity and/or high-variety information assets that demand cost-effective, innovative forms of information processing that enable enhanced insight, decision making, and process automation" [41].

3

Reliability Evaluation and Lifetime Evaluation

3.1 Introduction

Dependability is the collective term used to describe the availability performance and its influencing factors: reliability performance, maintainability performance, and maintenance support performance [42]. This chapter introduces basic evaluation methods and techniques for reliability modeling and analysis. Reliability theory is well documented and this section presents some fundamental aspects directly related to the analysis performed within the different case studies. The main sources of the material contained in this section include [43–48].

Reliability assessments of a technical system are performed in order to determine where and when new investments, maintenance planning, and operation are going to be made. This assessment can be performed either by deterministic or probabilistic methods, each with its own advantages. The key advantages of the probabilistic technology are its ability to quantify the stochastic nature.

3.2 Basic Concepts of Reliability

3.2.1 Reliability and Availability

The concept of reliability is widely used for engineering systems meaning the ability of a system or component to perform its required functions under stated conditions for a specified time. Reliability is closely related to availability, which is typically described as the ability of a component or system to function at a specified moment or interval of time.

Reliability engineering can be considered as a subdiscipline of engineering system emphasizing dependability in the life cycle management of a product [49]. This includes designing in the ability to maintain, test, and support the product throughout its total life cycle [39].

3.2.2 Definition of Basic Reliability Measures

The international vocabulary using IEC Standard [75] formulates three-level stages for *dependability of service*; outgoing from reliability, availability, and dependability. This section gives an overview of the commonly used definitions in industry for reliability and availability. Note that the models are applicable only based on several assumptions. The models presented in the following section are more general.

3.2.2.1 Availability

The concept of availability has several different definitions. According to Reference 45 the definition is: *The probability that the component, system is capable of functioning at a time t.* Different types of availability are [75] given by: asymptotic availability, asymptotic average availability, average availability, and instantaneous availability. According to IEC standard [75], the definition for asymptotic availability is

$$A(\infty) = \lim_{t \to \infty} A(t) \tag{3.1}$$

where A describes availability, or

$$A(\infty) = \frac{m_s}{m_s + \text{MDT}} \tag{3.2}$$

where m_s denotes the average service time and MDT the mean down time.

This concept can be used if failure intensity and repair intensity are constant. When these conditions are valid the definitions for availability and asymptotic availability are the same, and are commonly called availability.

According to Reference 45, the asymptotic availability is identical to asymptotic average availability when the threshold value for the asymptotic availability exists and this is then called availability A. The availability A is defined as

$$A = \frac{E(T)}{E(T) + E(D)} = \frac{\text{MTTF}}{\text{MTTF} + \text{MTTR}} \tag{3.3}$$

where T if the life time and D the repair time.

3.2.2.2 Mean Time between Failure

Mean time between failure (MTBF), is the most common way to determine a maintenance interval. It indicates the time from a component failure to its next failure. This time includes mean down time (MDT) and mean time to failure (MTTF).

$$\text{MTBF} = \text{MDT} + \text{MTTF} \tag{3.4}$$

According with Holen [45], MTTF can be written as

$$\text{MTTF} = E(T) = \int_0^\infty tf(t)\,dt \tag{3.5}$$

MTBF can with help of historical data be estimated to get an approximated value. *The large numbers law* is valid at estimation of values and says that when the number of observations (N) goes to infinity the value of the estimated parameter goes to the parameter's true value [46].

$$\text{MTTR} = \frac{\sum_{n=1}^{N} t_n}{N} \tag{3.6}$$

where t_n = time between failure for failure n and N is a large population.

If the component is not in operation during 100% of the time this must be taken into consideration.

3.2.2.3 Mean Time to Recovery

MTTR together with the MTW is the MDT:

$$\text{MDT} = \text{MTTR} + \text{MTW} \tag{3.7}$$

where MTW is the mean waiting time and is also estimated with help of historical data. MTTR is estimated with the help of historical data in the same way as MTBF:

$$\text{MDT} = \frac{\sum_{n=1}^{N} r_n}{N} \tag{3.8}$$

r_n = repair time between failure for failure n and N is a large population.

3.2.2.4 Failure Rate

Failure rate, λ, indicates how many times that a component fails. MTBF indicates how long time passes between each failure. The relation between these is therefore

$$\lambda = \frac{1}{\text{MTBF}} \tag{3.9}$$

3.3 Modeling of Life Distribution Functions

A common application of probability theory, as well as the fundamental issue for this book, is to predict the lifetime of a unit using probability theory. The lifetime of a unit can be described by a random variable X. The distribution function describes the probability that the lifetime is less or equal to t.

The fundamental functions for a one-dimensional, continuous, stochastic variable is defined as follows.

One way to describe the characteristics of a one-dimensional continuous random variable (X) is to use its distribution function. For a given outcome of X, (x) is the probability $P(X \leq x)$ that X is smaller or equal to x. If this is made for all of x, a function $F_x(x) = P(X \leq x)$ is obtained, which is defined for all values in the interval $-\infty < x < \infty$.

Definition 3.1

The distribution function for the continuous one-dimensional random variable X is defined by

$$F_x(x) = P(X \leq x), \ -\infty < x < \infty \tag{3.10}$$

The distribution function is evaluated as follows:

$$F_x(x) = \int_{-\infty}^{x} fx(t)\mathrm{d}t \tag{3.11}$$

If a function $F_x(x)$ exists so that Equation 3.11 applies, then x is said to be a continuous random variable. The function $F_x(x)$ is called the density function for x.

Definition 3.2

The density function for the continuous one-dimensional random variable X is defined by

$$f_X(x) = F'_X(x) \tag{3.12}$$

in every point x where $f_X(x)$ is continuous.

The density function describes how the total probability is distributed over the infinite number of possible x values. Based on the two functions above introduced ($F_X(x) = F(t)$ and $f_X(x)$), the lifetime evaluation uses the two functions defined below.

Definition 3.3

The **reliability function** (or the survival probability function) $R_x(t)$ is defined by

$$R_X(t) = 1 - F_X(t) \tag{3.13}$$

Definition 3.4

The failure rate function (or hazard function), $\lambda(t)$, is defined by

$$\lambda(t) = \frac{f_X(t)}{R_X(t)} \tag{3.14}$$

An unavailability function that is the inverse of availability is consequently $1 - R_X(t) = F_X(t)$. In addition, there are two measurements that are of specific interest here, being the expected lifetime and the variance $V(X)$. The variance is commonly denoted by σ^2. The variance measures the distribution of different outcomes for the random variable, and the standard deviation is defined by $D(X) = \sqrt{V}(X)$. The result is a measure of the distribution that has the same units as the random variable. The standard deviation is commonly denoted by σ.

Let X be the random variable for the length of life of a component or system of components. Table 3.1 summarizes the different functions and measurements for this, as well as presents their relationships defined by the distribution of the variable X.

These characteristics are general and could be used to define any continuous random variable X. However, there are several distribution functions that are widely used for modeling, for example, uniform distribution, normal distribution, log normal, exponential distribution, Weibull distribution, and more. One way of modeling the lifetime of a component is consequently to assume that it can be described by the characteristics of a known distribution, and then select parameter values that fit the specific purpose.

TABLE 3.1

Basic Functions and Measurements for Life Distribution Modeling

Notation	Name	Definitions
$F_X(x)$	Distribution function	$F_X(x) = P(X \leq x)$
$R_X(x)$	Reliability function (or survival probability)	$P(X \geq x) = 1 - F(x)$
$f_X(x)$	Probability density function	$F_{X'}(x)$
$\lambda(x)$	Failure rate (hazard) function	$\lambda(x) = f_X(x)/R(x)$ $= F_{X'}(x)/(1 - F_X(x))$
m	Expected lifetime	$m = \int_0^\infty x f_X(x)\mathrm{d}x$ $= \int_0^\infty R_X(x)\mathrm{d}x$
σ^2	Variance	$\sigma^2 = \int_0^\infty (x - \mu)^2 f_X(x)\mathrm{d}x$

3.3.1 Exponentially Distributed Random Variable

The assumption that the lifetime follows an exponential function is widely adopted. The reason for this is that the resulting failure rate functions imply a constant failure rate. The characteristic functions are presented later.

An exponentially distributed random variable $X \in Exp(m)$ has the following characteristics for the density and distribution functions, respectively:

$$f_X(x) = \begin{cases} 1/m \cdot e^{-x/m} & \text{if } x \geq 0 \\ 0 & \text{if } x < 0 \end{cases} \tag{3.15}$$

$$F_X(x) = \begin{cases} 0 & \text{if } x < 0 \\ 1 - e^{-x/m} & \text{if } x \geq 0 \end{cases} \tag{3.16}$$

It can be easily deduced that the resulting failure rate function (defined in Table 3.1) is constant, meaning that another notation form is therefore λ.

Figure 3.1 shows the characteristic functions for the exponentially distributed lifetime with $m = 2$. The first graph shows the density function. The larger the value of m, the more widespread the probability mass over the interval $(0, \infty)$ becomes.

The most essential characteristic of the exponential distribution, which is the common pdf in availability analyses, is that the failure rate is constant over time. It should not be assumed that all components exhibit this characteristic. Most do not. Its popularity is a function of the fact that it is the best pdf given the data that supports the reliability metrics of most power components. Essentially, the exponential requires only the MTBF, which can be easily determined by a total component run time and a total of component failure events.

3.3.2 Weibull-Distributed Random Variable

Accurate modeling of the system components requires distribution functions that allow different characteristics of the failure rate functions. One of the widely known distribution functions that includes several characteristics for the failure rate function is the Weibull function.

A Weibull-distributed random variable X follows a Weibull (a, b) has the following characteristics for the density and distribution functions, respectively:

$$f_X(x) = \begin{cases} \frac{b}{a}(x/a)^{b-1} e^{-(x/a)^b} & \text{if } x \geq 0 \\ 0 & \text{if } x < 0 \end{cases} \tag{3.17}$$

where a and b are positive numbers, and

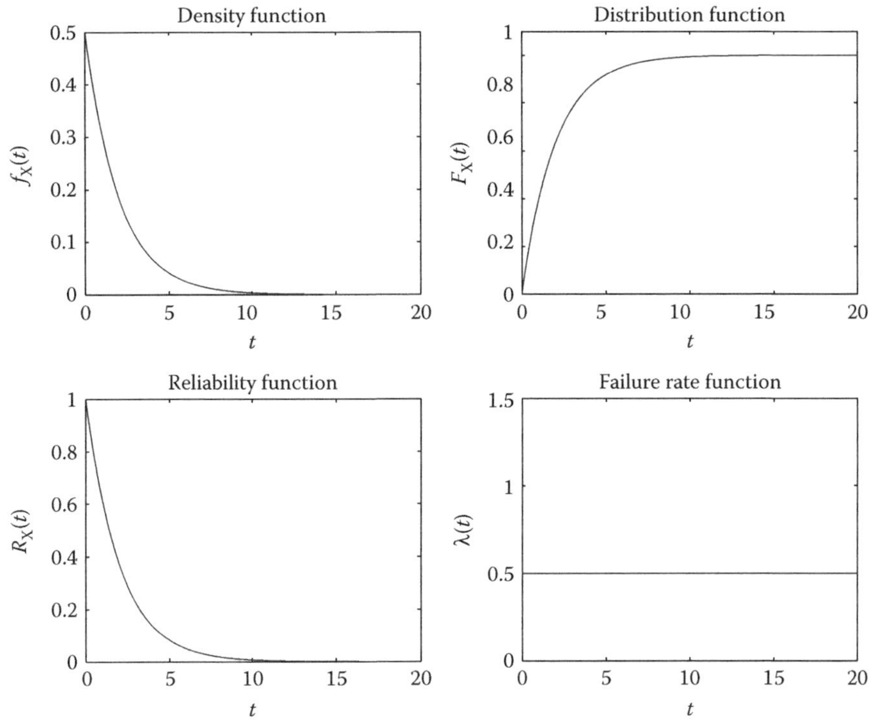

FIGURE 3.1
Functions for an exponentially distributed random variable $X(m=2)$.

$$F_X(x) = \begin{cases} 0 & \text{if } x \geq 0 \\ 1 - e^{-(x/m)^b} & \text{if } x < 0 \end{cases} \tag{3.18}$$

This function has been shown to be useful, for example, with breakdown strength data. Different values of the parameters a and b provide a band of distributions, and some special cases are: $b=1$ which gives the exponential distribution, and $b=2$ which gives the Rayleigh distribution.

Figure 3.2 presents results for the Weibull distribution.

The Weibull distribution of the lifetime variable, as presented in Figure 3.2, is shown for a collection of functions with multiple characteristics. Moreover, it can be seen that the failure rate function (the bottom right graph) is shown for the following three characteristics: (i) increasing failure rate ($b>1$), (ii) decreasing failure rate, when $b<1$, and (iii) constant failure rate, when $b=1$.

The last of these characteristics is the special case when the Weibull distribution equals the exponential distribution $Exp(m)$.

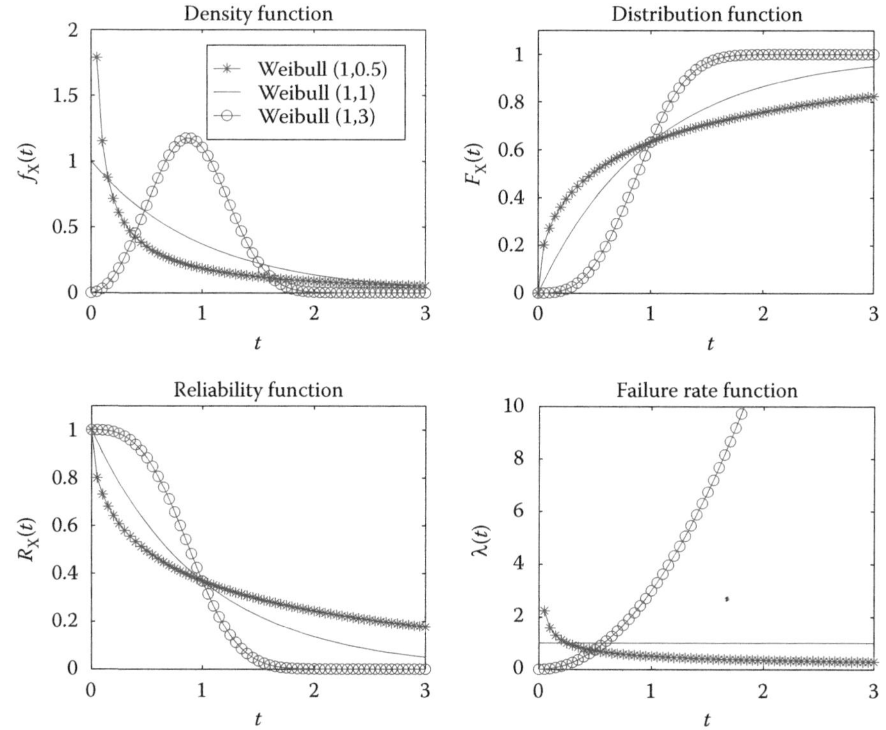

FIGURE 3.2
Functions for the Weibull-distributed random variable X for three cases.

3.3.3 System of Components

3.3.3.1 Series Reliability

Consider the system represented by the reliability block diagram (RBD) in Figure 3.3.

The number below each block in Figure 3.3 is the failure rate, in failures per million hours. The inverse of the failure rate is the MTTF (exponential failure rate assumed).

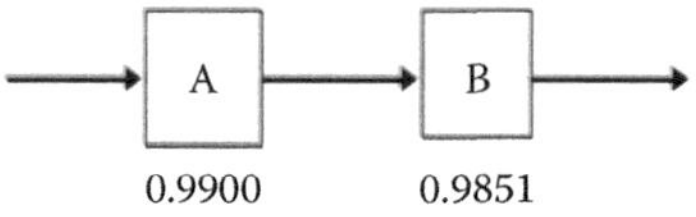

FIGURE 3.3
Example reliability block diagram.

3.3.3.2 Series Configuration-Weakest Link

Components A and B in Figure 3.3 are said to be in series, which means all must operate for the system to operate. Since the system can be no more reliable than the least reliable component, this configuration is often referred to as the weakest link configuration. An analogy would be a chain; the strength of the chain is determined by its weakest link.

3.3.3.3 Series Calculation Method 1

Since the components are in series, the system reliability can be found by adding together the failure rates of the components. The system failure rate is $0.001000 + 0.001500 = 0.002500$. The reliability is shown in Equation 3.19:

$$R(t) = e - 0.0025 \cdot 10 = 0.9753 \tag{3.19}$$

3.3.3.4 Series Calculation Method 2

Alternatively, the system reliability could be evaluated by multiplying the reliabilities of the two components as follows: $0.9900 \cdot 0.9851 = 0.9753$.

3.3.3.5 Reliability with Redundancy

Consider the RBD shown in Figure 3.4.

The number above each block in Figure 3.4 is the failure rate in failures per million hours. The inverse of the failure rate is the MTTF (exponential failure rate assumed). The number below each block is the reliability.

The system represented by the block diagram in Figure 3.4 has the same components (A and B) used in Figure 3.3, but two of each component are used in a configuration referred to as redundant or parallel. Two paths of operation are possible. The paths are top A–B or bottom A–B. If either of two paths is intact, the system can operate. The reliability of the system is most easily calculated by finding the probability of failure for each

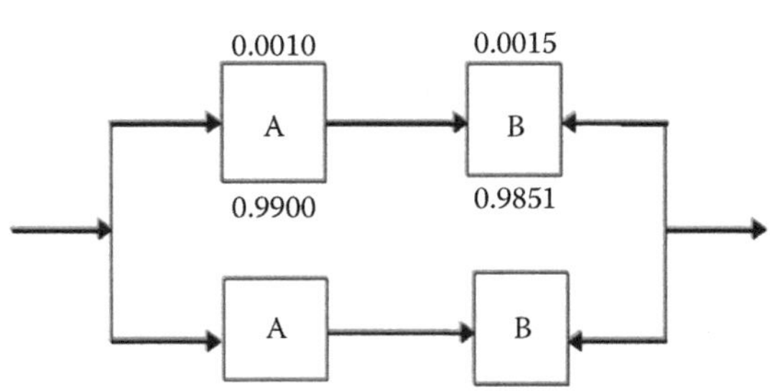

FIGURE 3.4
RBD of a system with redundant components.

path, multiplying the probabilities of failure (which gives the probability of both paths failing), and then subtracting the result from 1. The reliability of each path was found in the previous example. Next, the probability of a path failing is found by subtracting its reliability from 1. Thus the probability of either path failing is 10.9753 = 0.0247. The probability that both paths will fail is $0.0247 \cdot 0.0247 = 0.0006$. Finally, the reliability of the system is $1 - 0.0006 = 0.9994$, a significant improvement over the series-configured system, which had a reliability of 0.9753.

3.3.3.6 *M* of *N* Calculations for Reliability

Equation 3.20 can be used for calculating the reliability of an m of n system for any arbitrary m or n:

$$R(t) = \sum_{k=m}^{n} \frac{n!}{k!(n-k)!}(e^{-\lambda k})^{k}(1 - e^{-\lambda k})^{n-k} \tag{3.20}$$

where n is the total number of components and m is the required components.

3.4 Comparison of Evaluation Techniques and Tools

3.4.1 Analytical versus Simulation Method

Like all mathematical analyses, reliability analysis concentrates first on modeling the mathematical problem and second on finding solutions to the problem using the model. Furthermore, the model can be used to solve the problem directly and mathematically (analytical method), or indirectly by numerical experiments (simulation method). A comparative study of these two fundamental methods for assessing reliability is discussed later on in this section and summarized in Table 3.2.

3.4.2 Analytical Method

In the analytical method, reliability evaluation techniques are used directly on the model by solving the mathematical equations. Moreover, there are two fundamental techniques used for reliability evaluation, namely network modeling and Markovian modeling. The most commonly used out of these two is the network modeling technique.

TABLE 3.2

Comparable Features of Computational Methods

Analytical Method	Simulation Method
The model often needs to be simplified	Every phenomena which can be modeled can be simulated
Relatively short solution time needed	A large amount of computing time is needed for finding solutions
Always the same results for a given model and input data	Results dependent on the actual random number of repeated simulations
Output limited to average values	Probability distributions associated with the reliability indices as output

3.4.2.1 Network Modeling

A technical system can always be thought of as being composed of components. In the network modeling technique, the structural relationships between a system and its components are considered. The physical system is transferred into a reliability network using the system operational logic and knowledge about the physical behavior and requirements of the system. Details of these techniques have been presented in [45,50].

One of the central problems in reliability analysis is modeling the failure behavior of the system. This can be done by deducing possible failure modes as the minimal cut set does.

Definition 3.5

A **cut set** is a set of components which upon failure cause a failure of the system. A cut set is minimal when it cannot be reduced any further and still remain a cut set.

There are a number of different approaches to obtaining minimal cut sets. A distribution system can be treated as a system composed of load points the subsystems. One approach, referred to as *the load-point driven* technique, deduces minimal cut sets for each load point by identifying each event that leads to a failure of that load point. Furthermore, reliability indices are deduced for each load point and these indices are then combined to provide the system indices. An algorithm for the deduction of minimal cut sets by using the *minimal paths* technique has been specified in [51].

Definition 3.6

A path is a set of components that when operating guarantees the operation of the system. A path is minimal when it cannot be reduced any further and still remain a path.

Another approach to obtaining cut sets not based on reliability networks is the MOCUS algorithm [45]. This algorithm uses *fault trees* for evaluating the failure probability of a subsystem. A fault tree is a logical diagram that displays the relationship between an undesired event in the system and the cause of that event. The aim is to identify events via logical conditions that lead to the event that is under investigation the top event.

In a comparative study [52], a load point-driven approach has been compared with an event-driven approach. The principle of the event-driven approach is to treat each failure event separately and see the effect of the failure on the whole system by identifying the affected load points. Consequently, the major difference between the two approaches is that in the load point-driven approach all failure events for each load point are considered in turn, but in the event-driven approach, all load points affected by one failure event are considered in turn.

3.4.2.2 Markovian Modeling

In the Markovian model, it is assumed that in an interval of given length, the probabilities of a working component failing (or a failing component being repaired) are only dependent on the state of the system at the beginning of this small interval and the length of the interval [53]. This property of lack of memory means that information on how the system entered the current state or when it entered the state is not needed. For a detailed description, see [45,53].

The major drawback with this technique is the large number of states needed to model a system. This is due to the fact that the number of states increases exponentially with the number of factors studied. Therefore, many simplifying assumptions have to be made to limit the Markov model to a manageable size.

Markov models can be used in reliability assessments to model the stochastic processes of systems. Markov modeling implies the assumption of a system event process with no memory. This means that the future states for the system are independent of all past states except the immediately preceding one [54]. The stochastic process of the system also needs to be stationary or time homogeneous, meaning that the behavior of the system must be the same at all points of time. This leads to the basic characteristic of Markov modeling that the probability of making a transition between two specific states is constant at all times in the model.

A Markov model can be applied to systems whose random behavior varies either discretely or continuously with time. The discrete case of Markov modeling is referred to as Markov chain and the continuous as a Markov process.

3.4.3 Simulation Method

In the simulation method, an actualization of the process is simulated and after having observed the simulation process for a given period, estimates are made of the unknown parameters. The simulation is consequently treated as a series of real experiments. There are several types of simulation processes. In reliability analysis, simulation often concerns stochastic processes that is the simulation of random events. These simulation methods are commonly referred to as Monte Carlo simulations. Stochastic simulation can be used in a random or a sequential way. In the random approach, the simulation creates randomly chosen intervals, and in the sequential approach the intervals are chosen in chronological order. This implies that if a system model is simulated, where events for the system model in one interval depend on the previous interval, then the sequential approach is appropriate, which is often the case in reliability studies.

The simulation process is intended for examining and predicting the stochastic behavior of a system in simulated time. Therefore during the simulation period, events are made to occur at randomly determined times obeying predetermined probability distributions. One of the central problems in simulation methods is therefore how to generate random numbers from the different probability distributions. By using random numbers and converting them into distribution functions that represent the behavior of the system, the actual behavior of the system can be realized.

3.5 Bayesian Modeling

This chapter has introduced different techniques for reliability modeling of components. These models are probabilistic and allow the user to predict the likely future behavior of the component.

A main challenge in reliability modeling is however the lack of input data. For example, failure data or other forms of operational information that is not usually stored and can therefore not easily be used for new equipment. When analyzing failure data, it is common to consider samples in which not all components have failed at time t. These lifetimes are then said to be right censored. If, on the other hand, the lifetime is interval censored it means that the exact lifetimes are not known; only an interval of time in which the failure occurred is recorded. Bayesian modeling is a technique that can be used if there is a lack of initial input data for solving an analytical model.

An example is presented in order to understand the concept. The example is based on a case study presented in Reference [55]. A method to model the reliability of power system components is proposed using development test data from the manufacturer. The method provides a means of updating the model as new information becomes available. Figure 3.5 illustrates the method. Steps 3–8 are performed for every subcomponent. The two input data boxes in the figure represents input data to the reliability model and the output data box shows the resulting reliability model.

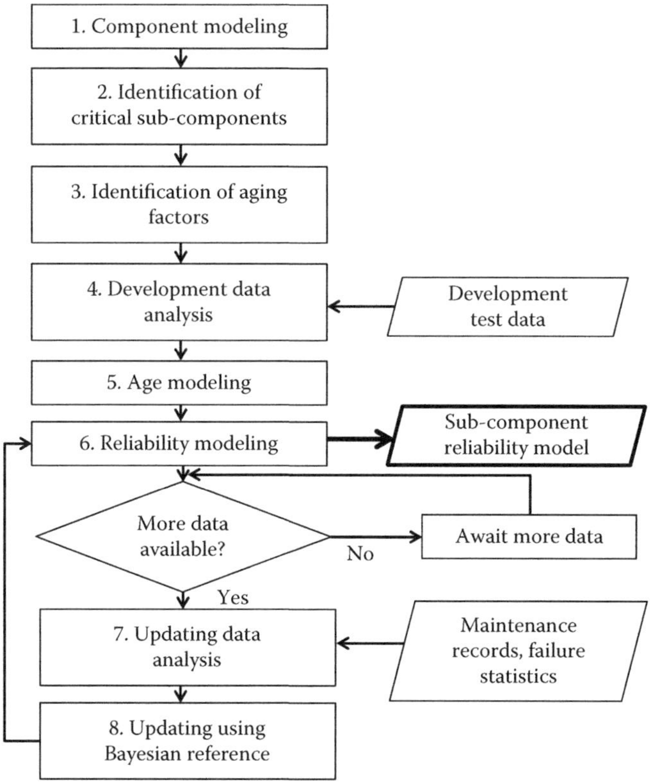

FIGURE 3.5
Illustration of the process developing a sub-component reliability model using Bayesian reference (Adapted from Lindquist, T., Bertling L., and Eriksson, R., *A Method for Age Modeling of Power System Components based on Experiences from the Design Process with the purpose of Maintenance Optimization,* Reliability and Maintainability Symposium (RAMS), IEEE, 2005 [55].)

- *Stage 1—Component Modeling.* The definition of a subcomponent is the smallest replaceable item in the power system component. All subcomponents in the component model are considered to be non-repairable and statistically independent. Consider a component comprising m non-repairable subcomponents. Each subcomponent, i, has a lifetime Xi, where Xi is an independent random variable (r.v.) with a probability density function (p.d.f.) $fi(x)$, where x is an observation of X. Using the proposed method, a power system component is modeled as a serial system comprising m subcomponents, each with a lifetime Xi.
- *Stage 2—Identification of Critical Subcomponents.* The second step is to collect and analyze failure statistics in order to determine which subcomponents are critical to the component. To be useful, the failure

statistics need to be very detailed. If the component to be modeled is of a new design and therefore has not yet failed, failure data or experiences from previous similar designs, or experiences from the development process may be used. In this example, the critical subcomponents are defined as those subcomponents that are most critical to the component with respect to some reliability importance measure. The reliability importance measures will be different depending on the component to be studied and the purpose of the study. Examples of reliability importance measures are hazard rate, availability, repair costs, etc. The number of critical subcomponents will differ depending on the component and the accepted accuracy of the model.

- *Stage 3—Identification of Aging Factors.* The next step is to identify what factors affect the aging of the different critical subcomponents. A critical subcomponent may have more than one aging factor. It is expected that the dominating age factor can be established by relying on engineering judgments and experiences from the development process. However, in cases when it is not so obvious which aging factor is dominant the method of lifetime coefficient of variation described in References 56 and 57, may be used. For the two critical subcomponents in the CB example the aging factors represent the number of operations for the operating rod and accumulated short-circuit current for the interrupter unit. This example will from here on only focus on the wear of the operating rod.
- *Stage 4—Development of Data Analysis.* This step involves the collection and analysis of development test data in order to support age models for every factor affecting the relative age of the critical subcomponents. These data typically result from tests and calculations carried out by the manufacturer during the design and development process, such as reliability studies, FMEA and different types of mechanical strength tests and calculations. In the CB example, a reliability study of a CB was used as input data. In the study several strength and endurance tests were carried out as well as dynamic simulations.
- *Stage 5—Age Modeling.* This step involves the modeling of the relative age of the critical subcomponents for each factor critically affecting its age. The relative age is typically a value between 0 to 1, where 0 means that the subcomponent is new and 1 means that it has reached the accumulated stress, for which it was designed. Note that a subcomponent may have a relative age, $A(y)$, larger than 1,

$$A(y) = \frac{y}{c}, c \neq 0 \tag{3.21}$$

where y is the accumulated stress and c is the set accumulated stress limit the subcomponent was designed to withstand.The advantage of using relative age is that it is easy to compare different subcomponents with respect to their relative age, even though they might have different failure mechanisms. In the example, the CB shall be able to withstand at least 10,000 operations according to the manufacturer and therefore, $c = 10{,}000$.

- *Stage 6—Reliability Modeling*. This is the last stage before the updating loop starts for the reliability modeling as shown in Figure 3.5. In this example, the hazard function is used as a measure of reliability. The hazard function is modeled by first fitting a p.d.f. to the data, acquired in stage 4. Then, for each critical subcomponent i, the hazard function, $h_i(x)$, is calculated as

$$h_i(x) = \frac{f_i(x)}{1 - \int_0^x f_i(\tau)\, d\tau} \tag{3.22}$$

where the hazard function for the serial system, comprising m subcomponents, is then

$$h_{system}(x) = \sum_{i=1}^{m} \binom{n}{k} h_i(x) \tag{3.23}$$

Applied to the CB example, a prior p.d.f. for X_i is formulated based on the data from the development tests. The lifetimes of the operating rod is assumed to be Weibull distributed. The Weibull distribution was earlier presented in Section 3.3. The *two parameters of the Weibull distribution* are estimated from the observations of X, using the method of least squares (MLS). The resulting p.d.f. and hazard rate represents models for the operating rod until new information becomes available.

- *Stage 7—Updating Data Analysis*. In this stage the reliability model is updated and improved based on new available data. These data may be failure statistics and/or maintenance records. Maintenance records could, for example, contribute with important information in the form of right censored failure data. In the CB example both failure data and maintenance history for the operating rod were found.
- *Stage 8—Updating Using Bayesian Inference*. Consider a quantity θ that we wish to estimate. In classical statistics this parameter is treated as fixed but unknown. In Bayesian statistics the parameter θ is considered to be a result of an r.v. with some p.d.f. f, called the prior p.d.f. This p.d.f. may reflect any prior knowledge we may have regarding the values of θ. The greatest advantage of using Bayesian methods is the way prior knowledge is combined with collected data into the resulting posterior distribution that can be used for inference. The computation of the posterior distribution uses Bayes' theorem

[40,58,60]. A general form of Bayes' theorem involving data x and the parameter θ is as follows:

$$f(\theta|x) = \frac{f(x|\theta)\cdot f(\theta)}{\int f(x|\theta)\cdot f(\theta)\,d\theta} \tag{3.24}$$

where the conditional density $f(\theta|x)$ is the posterior distribution of θ, given X, and $f(x|\theta)$ is a model for the observed data. In this example the posterior expectations are evaluated using Markov chain Monte Carlo methods (MCMC) [59]. Details are presented in Reference 55.

- *Stage 6—Reliability Modeling (with new input data).* The predictive distribution in this stage now is used to study the distribution for a new observation of X_0. After observing $D = \{X_1 X_2, \ldots X_n\}$ the predictive density of X_0 given D, is defined as [60]

$$f(X_0|D) = \int_0^\infty f(x|\theta)\cdot f(\theta|D)\,d\theta \tag{3.25}$$

When the predictive distributions, and subsequently the hazard functions, are calculated for all critical subcomponents in the serial system the component hazard function is obtained from Equation 3.23. In the CB example, the posterior p.d.f. and the predictive p.d.f. were simulated, that is, not calculated analytically using Equation 3.16 [55].

In summary, the example in this section shows on a method for modeling the reliability of a component, constituted of subcomponents, in absence of or with limited access to failure statistics. Further examples on using Bayesian methods are presented in Chapter 8. For further reading of Bayesian methods the following reference is recommended [58].

3.6 Risk Assessment

Risk is often expressed in terms of a combination of the consequences of an event and the associated likelihood of occurrence. Thus risk may be quantified as the product of the probability of a detrimental event occurring by the consequence of the event (i.e., Risk probability consequences) [62].

$$\text{Risk} = \text{probability} \cdot \text{consequences} \tag{3.26}$$

The definitions above should also clarify the distinction between risk and probability. Probability is the likelihood of an event occurring. Risk is the end result, that is, likelihood cost or consequence of occurrence.

An example of a Risk Assessment is provided in Chapter 10 and Figure 10.1. The figure shows a typical Risk matrix with consequences and the probability of the event.

3.7 Life Cycle Cost (LCC) Analysis

3.7.1 LCC Analysis

The LCC for a technical system is its total cost during its lifetime. The goal is to minimize the total lifetime cost. The total cost includes all costs associated with planning, purchasing, operation and maintenance, and liquidation of the system. Power plant financial concerns could typically be investment, maintenance, production loss, and rest value.

It is realized that the LCC is different for different technical systems. The case studies included in this book show different examples of performing LCC analyses, including electrical distribution systems, wind power plants, components in nuclear power plant and hydropower plant.

In the following section, a brief introduction of some fundamental concepts and methods used for the economic analysis performing the LCC analysis is given.

3.7.2 Fundamentals of Economic Analysis

In this section, some fundamental techniques for cost analysis are introduced that will be used in the modeling of cost functions for solving the maintenance question.

An economic analysis is crucial for evaluating the effect of different maintenance strategies and maintenance methods. These analyses relate to long-term profitable assessment, that is, investment calculations. Moreover, PM could be considered as an investment with the purpose of reducing the number of failures or improving component condition.

In general terms, an investment refers to the use of capital, which results in long-term payment consequences [63]. For an electric power system, long term would typically refer to an expected lifetime of about 30 years.

Investments are taken up as assets in the balance sheet, and commonly as building and construction assets. The cost of the investment is then divided by depreciations over the economic lifetime of the investment.

Investment decisions are often times difficult. This is mainly due to their characteristic uncertainty. For example, it is difficult to decide the future payments and the economic lifetime of the investments. Investment calculations aim to simplify and support investment decisions. The most fundamental concepts are the in-and-out payments. The investment calculation is primarily about deciding whether or not a resource requirement (in the form of a base investment) is justifiable with respect to size and distribution over the period of payment surplus (of in-and-out payments). However, payments at different points in time are not comparable. The money that is available today is of greater value than the same amount of money after a period of time (for instance a year). This is due to the fact that the amount of money that is available today could be increased by a favorable interest rate. Put simply, a one-hundred US dollar note today has a current value of USD 100, however, USD 100 after 20 years will have a value that is appreciably lower than 100 USD today. Therefore, to make different payments comparable it is necessary to re-evaluate them to the same point in time.

Another fundamental concept is the interest rate or the owners earning requirement on income from capital. There is no simple way to decide on this, however, it should be defined in respect to the following: the rate on which loans are made, the alternative uses of the investment resources, or the risk that is associated with the investment.

Economic calculations are for reasons of simplicity, normally made on a yearly basis. This implies the assumption that all payments are considered as transferred at the same time each year.

Three commonly used methods for investment calculations are presented below.

1. *Payback Method*: The payback method defines how long it will take for the investment to recover initial outlay. This is the payback period. In this simple form of the method, this time is defined as the time when the invested capital without the rate equals the sum of the yearly surplus expressed by a simple ratio. This method is suitable for comparing investments with short lifetimes. This is, for example, because it does not consider when the payments are made and therefore would ignore the benefits from early payments.
2. *Present Value Method*: The present value method means that all the payments are compared at one point in time, being the *present time*. The present value (PV) means the amount of money that should be put into the bank **now** at a certain rate (d) to pay for an outlay (C) after n years. This means that all future payments are recalculated to the equivalent value for the present time. This is made by discounting the payment surplus for each year by the use of a capital rate, for example, an interest rate.

   The present value of one outlay (C) to be payed after n years is gained by multiplying this with the *present value factor* ($PV_f(n,d)$) as follows:

$$PV = C \cdot PV_f(n,d) \Leftrightarrow C \cdot (1+d)^{-n} \quad \text{[USD]} \tag{3.27}$$

where C is the outlay in [USD], n is the number of years from the present to the date of outlay, and d is the discount rate where for example an interest rate of $7\% \Rightarrow d = 0.07$.

When discounting with the present value method, real or nominal interest rates could be used. The nominal interest rate takes inflation into account. The discount rate r depends on the real interest r_1 and the inflation r_2, according to:

$$1 + r = (1 + r_1)(1 + r_2) \tag{3.28}$$

The interest is decided by the company management and indicates the return that is required for making an investment. The choice of interest rate is not obvious or trivial. The choice of interest value depends on, for example, the length of the investment. A long investment can include larger risks as the future is unknown. Investments where a higher risk is taken require a higher interest.

A common situation is where there are equal annual payments. In this case, the total present value is the sum of present values for these yearly payments. For this case, a specific formula for the present value can be evaluated by using the *present value sum factor* ($PV_{sf}(n, d)$), which is defined as follows:

$$PV = C \cdot PV_{sf}(n, r) \Leftrightarrow C \cdot \frac{(1 + d)^n - 1}{d \cdot (1 + d)^n} \quad \text{[USD]} \tag{3.29}$$

This method is a powerful tool and is widely used. However, the main disadvantage with it is that the discount rate, which to a great extent decides the result, is somewhat subjective. One way of getting around this is by using the internal rate of return. The *internal rate of return* is defined by the discount rate that results in a zero net present value, which is when the present value of the costs equals the revenues. However in this analysis here, only the costs are being analyzed and this alternative method will therefore not be investigated further.

3. *Annual Cost Method*: In this method, the costs are evaluated as annual payments. This method is useful when comparing alternative investments with different lifetimes. The present value is then spread over the lifetime, for example, with a yearly equal sum of the depreciation and the interest rate over an asset lifetime. This constant annual payment is usually referred to as the *annuitized value* (a). The annuity that defines equal yearly payments of a present value can be evaluated by using the previously defined present value sum factor as follows:

$$a = \frac{PV}{PV_{sf}(n, r)} \Leftrightarrow PV \cdot \frac{d \cdot (1 + d)^n}{(1 + d)^n - 1} \quad \text{[USD]} \tag{3.30}$$

The following simple calculations have been used to provide examples of the above methods. (For these examples, 1 MSEK is approximately 0.1 USD or Euro.)

EXAMPLE 3.1

Evaluate the present value of an investment in PM performed on cables after 9 years of service, for a cost of MSEK 2/km. Assume that 30% of the cables with average length of 8.016 km are effected by the PM measure. Furthermore, assume that this amount of money could be put into the bank with a rate of 2%.

Solution

The cost of PM is a one time cost of: $2 \cdot 0.3 \cdot 8.016 \approx MSEK4.864$, using Equation 3.27 $\Rightarrow PV = C \cdot (1+d)^{-n} = 4.864 \cdot (1.02)^{-9} \approx MSEK4.07$.

EXAMPLE 3.2

Assume the same situation as in Example 3.1, but with PM applied at years 11 and 12, respectively. Now, evaluate the total present value for the three investment costs.

Solution

Equation 3.27 $\Rightarrow PV = C \cdot (1+d)^{-n} + C \cdot (1+d/100)^{-11} + C \cdot (1+d)^{-12} = 4.864 \cdot (0.837 + 0.804 + 0.788) \approx MSEK11.82$. If there was no rate, the cost would instead be $4.864 \cdot 3 \approx MSEK14.59$.

These two examples illustrate the benefit of, for example, a utility having a contract with a fixed price for future PM activities. However, if the future investments were payed with loans instead, the economic evaluation should take into account the rate of the loan that would increase the required cost of the investment.

EXAMPLE 3.3

Based on Example 3.2, use present value analysis to show which of the three alternative time points for PM (9, 11 or 12) is the most cost effective.

Solution

Evaluate the present values for the three investment alternatives and choose the solution with the lowest present value. Equation 3.27 $\Rightarrow PV = C \cdot (1+d)^{-n}$

$$\text{Alternative 1:} \Rightarrow PV = 4.864 \cdot (1.07/100)^{-9} \approx MSEK4.07$$
$$\text{Alternative 2:} \Rightarrow PV = 4.864 \cdot (1.07/100)^{-11} \approx MSEK3.91$$
$$\text{Alternative 3:} \Rightarrow PV = 4.864 \cdot (1.07/100)^{-12} \approx MSEK3.835$$

$\Rightarrow$ Alternative 3.

This example demonstrates that the lowest present value is gained with the investment as late as possible. Consequently, it has been shown that it is most cost effective to perform the PM measure as late as possible.

EXAMPLE 3.4

Evaluate the annual constant payment of the depreciation and the interest rate. Assume the investment cost for PM by replacement of cables is $MSEK4.864$. Furthermore, assume a remaining lifetime of 21 years and an interest rate of 7%.

Solution

Equation 3.30

$$\Rightarrow a = PV \cdot \frac{d \cdot (1+d)^n}{(1+d)^n - 1} = PV \cdot \frac{0.07 \cdot (1.07)^{21}}{(1.07)^{21} - 1} \approx 0.092 \Rightarrow$$

Annual payment is $4.864 \cdot 0.092 \approx MSEK0.449$.

EXAMPLE 3.5

Evaluate the present value for a yearly payment of $MSEK0.449$ for 21 years, and with an interest rate of 7%.

Solution

Equation 3.29

$$\Rightarrow PV = C \cdot PV_{sf}(21,7) = C \cdot \frac{(1.07)^{21} - 1}{0.07 \cdot (1.07)^{21}} = 0.449 \cdot 10.8335 \approx MSEK4.865$$

It can be seen from these simple examples that an economic analysis could be very extensive and complicated. However, making a comprehensive cost analysis lies outside the scope of this book and the cost functions that are being modeling require quite simple cost considerations, though these are by no means trivial.

3.8 Computer Tools for Reliability Evaluation

There is in general a need to have computer tools for reliability assessment. There are reliability tool available on the market. Typically, the drawback is that the underlying methods and assumptions are not fully understood. Therefore, it is common to develop your own tool, especially for research.

Example on a commercially available tool is NEPLAN [64]. NEPLAN AG was founded in 1988 and develops software tools for the electrical, gas, water, and district heating networks as well as specialized services related to these fields. Other examples of tools developed at universities are given by the RELNET program from University of Manchester in the UK [65] and RELRAD from EFI in Norway [66] both based on the networking modeling technique. The Markovian modeling technique has been used for the R-RADS program at the University of Washington [67]. One reason for the popularity of network modeling techniques may stem from the simplicity of the method and the natural similarities between the network model and the distribution system topology. A more comprehensive overview of different tools is provided in the CIGRE Study Committee c4 Report [62].

For the RCAM Case Study on Electrical Distribution System, an in-house tool Reliability Assessment for Electrical Distribution Systems (RADPOW) was used. The analytical method has been used as the basic approach for the reliability assessment in RADPOW. It uses the network modeling technique and the minimal cut set (load point-driven) approach for the deduction of failure modes, and the Monte Carlo simulation technique has been incorporated with the basic approach to provide the probability distributions associated with the reliability indices of interest.

The RADPOW program can compute basic indices for reliability (load point and system-performance indices), determine the impact of new technologies, determine the impact of protection and control in the system [39,68], analyze general network configurations (radial, parallel, and meshed), and is flexible and easy to develop further (object-oriented programming). The RADPOW tool has been used for the RCAM Case Distribution System presented in Chapter 6.

4

Reliability-Centered Asset Management Method

4.1 Introduction

This chapter presents the generic method of reliability-centered maintenance (RCM) and its extension into a quantitative approach, that is, Reliability-Centered Asset Management (RCAM) method. It presents the different stages of performing an RCM and RCAM analysis including performing a Failure Mode and Effect Analysis (FMEA) of critical components. Finally, it discusses the different needs of input data.

4.2 RCM in General

4.2.1 RCM Concepts

The term RCM identifies the role of focusing maintenance activities on reliability aspects. The RCM methodology provides a framework for developing optimally scheduled maintenance programs. The aim of RCM is to optimize the maintenance achievements (efforts, performance) in a systematic way. This method requires maintenance plans and leads to a systematic maintenance effort. Central to this approach is identifying the items that are significant for system function. The aim is to achieve cost effectiveness by controlling the maintenance performance, which implies a trade-off between corrective and preventive maintenance and the use of optimal methods.

4.2.2 The Emergence of RCM

The RCM concept originated in the civil aircraft industry in the 1960s with the creation of the Boeing 747 series of aircraft (the Jumbo). One prerequisite for obtaining a license for this aircraft was having in place an approved plan of preventive maintenance (PM). However, this aircraft type was much larger and more complex than any previous aircraft type, thus PM was expected to be very expensive. Therefore, it was necessary to develop a new PM strategy. United Airlines led the developments and a new strategy was created. This was primarily concerned with identifying maintenance tasks that would eliminate the cost of unnecessary maintenance without decreasing safety or operating performance. The resulting method included an understanding of the time aspects of reliability (aging) and identifying critical maintenance actions for system functions. The maintenance program was a success. The good outcome raised interest and the program spread. It was further improved, and in 1975 the US Department of Commerce defined the concept as RCM and declared that all major military systems should apply RCM. The first full description was published in 1978 [11], and in the 1980s the Electric Power Research Institute (EPRI) introduced RCM to the nuclear power industry. Today RCM is under consideration by or has already been implemented by many electrical power utilities for managing maintenance planning.

4.2.3 Different Versions of RCM

There are different versions and concepts of the RCM approach in use. First, there is the MSG-3 concept used by the civil aviation industry, which is a development of the initial MSG-1 version. Second, there is the original version of RCM defined in [11]. In the 1980s, environmental questions became important issues, which led to a change in the RCM decision diagram, and a separate treatment of environmental aspects of failures was defined. This addition changed the decision diagram, and according to Moubray [12] the change was enough to warrant adopting a new name that is RCM2. Attempts to develop the RCM methodology further have been made by EPRIGEN (a subsidiary of EPRI), with the adoption of an RCM process called streamlined RCM (SRCM), which is a simplified version of RCM aiming at lowering the cost of performing RCM. The reason behind this development was EPRI's concern that the classical RCM resource requirements were too great for an average system [68]. However, the concept and fundamental principles for these different versions of RCM remain true to the original definitions. Finally, an RCM with the objective of realizing the inherent reliability characteristics of the equipment at a minimum cost [11] is the RCM concept of today. Its implementation must reflect current needs and resource capacity.

4.2.4 The RCM Method

RCM provides a formal framework for handling the complexity of the maintenance issues but does not add anything new in a strictly technical sense. RCM principles and procedures can be expressed in different ways [11,12, 68,69], however, the concept and fundamental principles of RCM remain the same.

The following features originate from the first definition of RCM [69], and define and characterize the RCM method. The RCM method facilitates the:

- Preservation of system function
- Identification of failure modes
- Prioritizing of function needs
- Selection of applicable and effective maintenance tasks

There are several different formulations of the process of creating an RCM program and achieving an optimally scheduled maintenance program found in the literature. Three formulations have been addressed in this section. The first two have both been derived from the original RCM definition, and the third is an approach based on a set of questions rather than steps.

4.2.4.1 Smith

Smith defined a systematic process for RCM by implementing the following features that have been defined earlier [69].

1. System selection and information collection
2. System boundary definition
3. System description and functional block diagrams
4. System functions and functional failures
5. Failure mode and effects analysis (FMEA)
6. Logic decision tree analysis (LTA)
7. Selection of maintenance tasks

4.2.4.2 Nowlan

The process of developing an initial RCM program when the information required is lacking consists of the following steps [11]:

- Partitioning the equipment into object categories in order to identify those items that require intensive study.
- Identifying significant items that are those that have essential safety or economic consequences and hidden functions that require scheduled maintenance.
- Evaluating the maintenance requirements for each significant item and hidden function in terms of the failure consequences and selecting only those tasks that will satisfy these requirements.
- Identifying items for which no applicable or effective task can be found, then either recommending design changes if safety is

involved or assigning no scheduled maintenance tasks to these items until further information becomes available.

- Selecting conservative initial intervals for each of the included tasks and grouping the tasks in maintenance packages for application.
- Establishing an age exploration program to provide the factual information necessary to revise initial decisions.The first step is primarily an activity for reducing the problem to a manageable size. The following three steps are the essence of the RCM analysis, constituting the decision questions, as the seven questions listed below.

4.2.4.3 Moubray

To analyze the maintenance aspects of a system and its components, the first step is to identify the system items and which of these ought to be analyzed. Thereafter, the RCM process can be formulated into seven questions for each of the selected items [12]. The seven general questions are:

1. What are the functions and performances required?
2. In what ways can each function fail?
3. What causes each functional failure?
4. What are the effects of each failure?
5. What are the consequences of each failure?
6. How can each failure be prevented?
7. How does one proceed if no preventive activity is possible?

4.3 The RCAM Method

The RCAM has been formulated based on an understanding of RCM concepts and experience gained from RCM application studies. The RCAM identifies the central role of defining the relationship between component behavior and system reliability that is made through the evaluation of the causes of failures. The RCAM method was first published by Bertling [14] and has since then been applied for several different power system applications. A selection of case studies is included in the following chapters. In [36], the methods of RCAM are compared with the methods developed by Kinectrics, engineering consultants in Toronto, Canada, focusing on the evaluation of the effect of various maintenance policies on the reliability of a device.

Table 4.1 presents the RCAM that includes the main procedures for developing RCM plans, and consequently is also the first result in the development process. Table 4.1 also identifies the following issues: (i) the logical order of

TABLE 4.1
The RCAM Approach

Step	Procedure	Level	Data Required	Results
1	Reliability analysis	Syst.	Comp. data	Reliability indices
2	Sensitivity analysis	Syst.	Comp. data	Critical comp.
3	Analysis of critical components	Comp.	Failure modes	Critical comp. affected by maint.
4	Analysis of failure modes	Comp.	Failure modes causes of failures, and so on	Frequency of maint.
5	Estimation of composite failure rate	Comp.	Maint. Frequency	Composite failure rate
6	Sensitivity analysis	Syst.	Frequency of maint.	Relationship between indices and PM schedules
7	Cost/benefit analysis	Syst.	Costs	RCAM

Abbreviations: Syst., system; comp., component; maint., maintenance.

the different procedures required, (ii) the need for interaction between the system and the component levels, and (iii) an indication of the different input data needed.

4.3.1 The Three Stages of the RCAM Method

The following three main stages can be identified for the procedures.

Stage 1 System reliability analysis (system level analysis) defines the system and evaluates critical components for system reliability (Steps 1 and 2).

Stage 2 Evaluation of PM and component behavior (component level analysis) analyzes the components in detail and with the support of necessary input data, a quantitative relation between reliability and PM measures can be defined (Steps 3–5).

Stage 3 System reliability and cost/benefit analysis (system level analysis) puts the understanding of the component behavior gained in a system perspective. The effect of PM on components is analyzed with respect to system reliability and benefit in cost for different PM strategies and methods (Steps 6 and 7).

Figure 4.1 illustrates the three stages of the RCAM giving an overall picture of the approach.

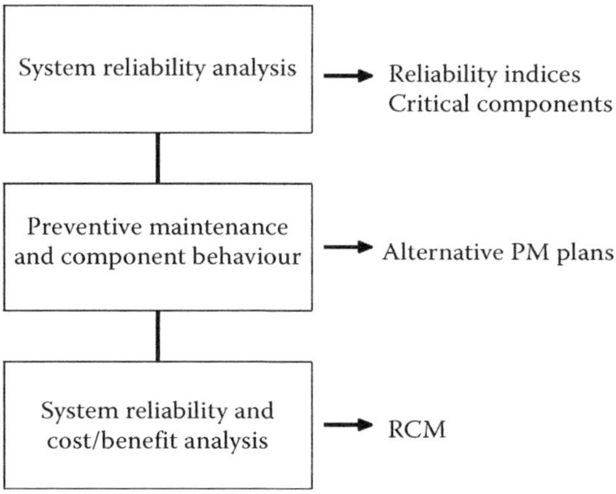

FIGURE 4.1
The three main steps in the RCAM method.

4.3.2 Relationship between Reliability and Maintenance

The central part of the RCAM, and the greatest challenge, is the definition of a relationship between reliability and preventive maintenance, that is, Stage 2 in the RCAM analysis discussed previously. This could be done through a generic approach involving the steps to be processed or a generic theory involving mathematical relations to be solved. Below are two alternative approaches presented for Stage 2 in the analysis. The availability of data and recourses in time would be the main arguments for the selection of approach. The alternatives are summarized as follows.

Definition 4.1

Approach I implies that a PM activity results in a percentage reduction in the causes of failures for affected components. Furthermore, it assumes that the failure rate is also reduced by the equivalent percentage. The resulting model of the relationship between failure rate and PM is referred to as $\lambda(PM)$.

Definition 4.2

Approach II implies that a functional relationship is established between failure rate and PM activities. This must be based on detailed knowledge and understanding of the condition of the components as well as the failure rate characteristics. The resulting model of the relationship between failure rate and PM is referred to as $\lambda(t, PM)$.

The following two general remarks can be made about the failure rate modeling. First of all, that determining the functional relationship with time for each failure cause implies a subdivision of the overall failure rate into rates for each failure cause. This requires an understanding of the physical processes involved that relates component behavior to lifetime. It may not be possible to do this for all the failure causes. It should also be kept in mind that the present failure rates are a consequence of existing maintenance procedures and the effect is not known or easily extracted. Furthermore, the estimation of the effect of maintenance on each of these functions must be made with knowledge of the physical processes involved. Second, that the combinations of the life functions associated with each cause of failure used for obtaining a new failure rate function for each component must be made for specific systems because the location and use of a specific component determines whether a particular failure may exist.

4.3.3 Implementation of the RCAM Method

An overall conclusion from the implementation of the RCAM method for the Electric Distribution System Case Study, which is described in detail in Chapter 6, is that it is an extensive effort required to accomplish such an analysis. To enable the carrying out of an RCAM analysis, comprehensive input data and knowledge were required, specifically about the cable component behavior. Furthermore, it can be identified that input data for the cable component behavior were obtained primarily from two different types of sources: (i) data based on experience, which were used for predicting the failure rate characteristic, and (ii) knowledge about the cable condition, by modeling the component behavior. Furthermore, it can be seen that some of the steps in the approach are the same as those used as input data in the framework. This is an interesting result since it reflects that one of the greatest challenges and difficulties when applying RCM is to define the data required to support the model, not only because of difficulties in obtaining these, but because it is simply part of the analysis to define suitable data (e.g., choosing maintenance frequency).

It is important to recognize that first, a model for the component failure rate characteristic is defined and then second, this is implemented for a PM strategy. The system analysis is consequently an analysis of the effect of implementing the model developed for different PM strategies.

Figure 4.2 illustrates the different steps in performing an RCAM analysis. These steps have been implemented and validated for the RCAM Case Study presented in Chapter 6. This figure includes the different procedures in the RCAM approach as well as the systematic process for analyzing the included components and causes of failures in the system. Furthermore, this figure shows the logic of the model for the functional relationship between failure rate and PM. The choice of input variables that was used for the results presented is shown on the left-hand side of this figure. Figure 4.2 illustrates the resulting logic implementing the RCAM method. This figure includes the different procedures in the RCAM method as well as the systematic process for analyzing the components and causes of failures included in the system.

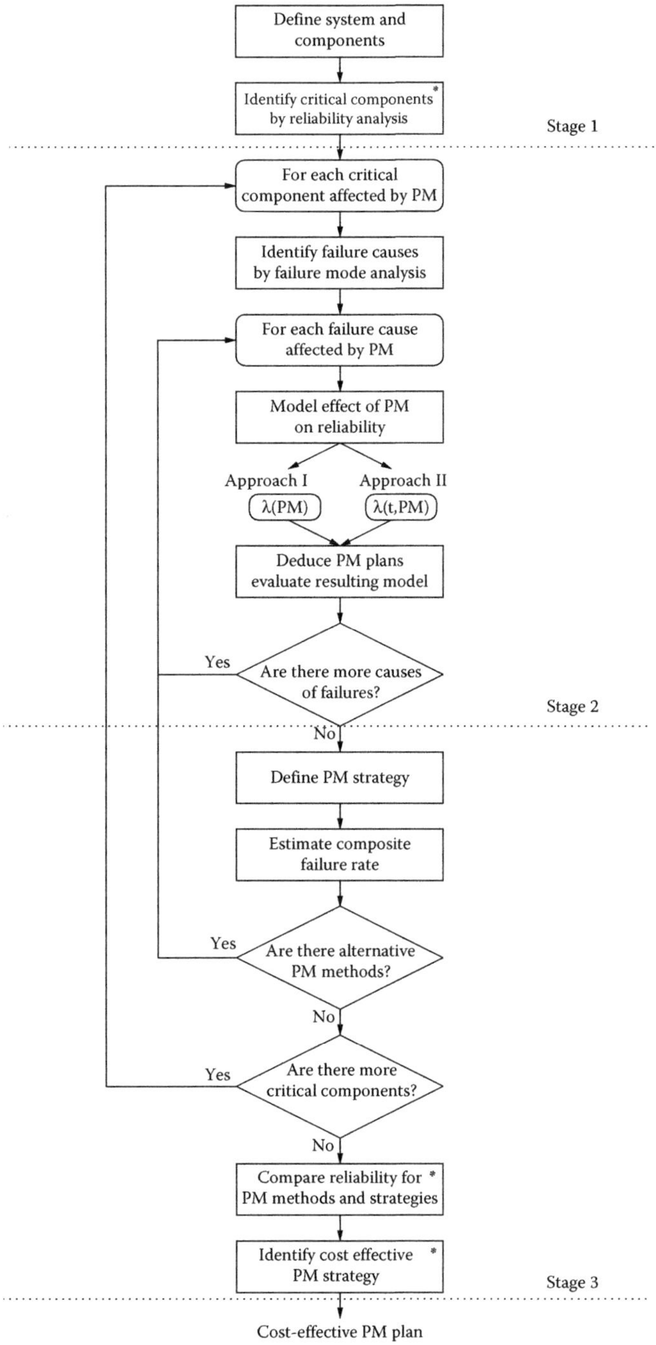

FIGURE 4.2
Implemented RCAM method for Electrical Distribution System Case. The steps that feature the asterisk (*) would typically need a software tool for the reliability analysis. For the RCAM Case Study for Electrical Distribution System the tool RADPOW was used.

4.3.4 Remarks about the Meaning of PM Strategy

The PM strategy provides a recommendation for performing an activity in the future. However, on the actual day for the PM measure, the decision would be based on whether it would be beneficial to do the same at that time, or would it be more profitable to postpone it, or even to do something else. Also, the policy or strategy may be to perform PM on a percentage of the components spread over a period of time, maybe several years. This policy is set up several years in advance of the actual time PM is to be done and therefore does not determine which specific components will be maintained. The actual components chosen for PM would be those that appear to be performing worst at the time PM is to be done.

The RCAM method implies the development of a policy based on predicted behavior that should be used to set a policy for determining when and how to use resources for maintenance. The resulting logic can be illustrated as follows:

Failure rate → predicted result → policy.
Implement policy → measure component condition → apply PM.

4.4 Failure Mode and Effects Analysis

Failure mode and effects analysis (FMEA) is a useful tool when performing an RCM analysis. FMEA is a way to evaluate potential failure modes and their effects and causes in a systematic and structured manner. Failure modes mean the ways in which something could fail. *Effects analysis* refers to studying the consequences of those failures. The purpose of the FMEA is to take actions to eliminate or reduce failures, starting with the highest-priority ones. By itself, an FMEA is not a problem solver, it should be used in combination with other problem-solving tools. The analysis can be done either in a qualitative or quantitative way. The basic steps in performing an FMEA could be:

1. Define the system to be analyzed. Complete system definition includes defining of system boundaries, identification of internal and interface functions, expected performance, and failure definitions.
2. Identify failure modes associated with system failures. For each function, identify all the ways failure could happen. These are potential failure modes.
3. Identify potential effects of failure modes. For each failure mode, identify all the consequences on the system. "What happens when the failure occurs?"

4. Determine and rank how serious each effect is. The most critical pieces of equipment which affected the overall function of the system need to be identified and determined.
5. For each failure mode, determine all the potential root causes.
6. For each cause, identify available detection methods.
7. Identify recommended actions for each cause that can reduce the severity of each failure.

The FMEA classifies each potential failure according to the severity of the mission success and personnel/equipment safety [39]. The FMEA could be extended with a criticality analysis (CA). The CA will provide the estimates of system critical failure rates based on past history and current information. The resulting FMECA (failure mode, effects, and criticality analysis) is a reliability evaluation and design technique that examines the potential failure modes within a system in order to determine the effects of the overall system and the equipment within the system [39]. The FMECA should be initiated as soon as preliminary design information is available [39]. The FMECA is a living document that is not only beneficial when used in the design phase but also during system use. As more information on the system is available, the analysis should be updated in order to provide the most benefit.

Generic maintenance data is a valuable tool when historical information is not available or when the engineering is establishing a maintenance-based line for a new system. This type of data is extremely rare but important to the establishment of a good RCM program.

4.5 Systematic Asset Management Process with RCM and RCAM

The overall goal is to reach a systematic AM process with RCM and RCAM. This does not only mean to implement the RCM and RCAM, but most important to find a process for updating the maintenance programs resulting from performing RCM-analysis. It is worth underlining that experience shows that it is challenging to reach this goal. The last chapter in this book gives an example from early adoption of RCM for Hydro Power Systems which might give more insight in such an experience.

The RCM process starts in the design phase and continues for the life of the system as shown in Figure 4.3. There are several major tasks required to implement the RCM concept.

In order for the RCM or RCAM analysis to be possible, there is a need for comprehensive knowledge about the system, its components, and other suitable input data to support a quantitative analysis. In the following section, these needs are further discussed.

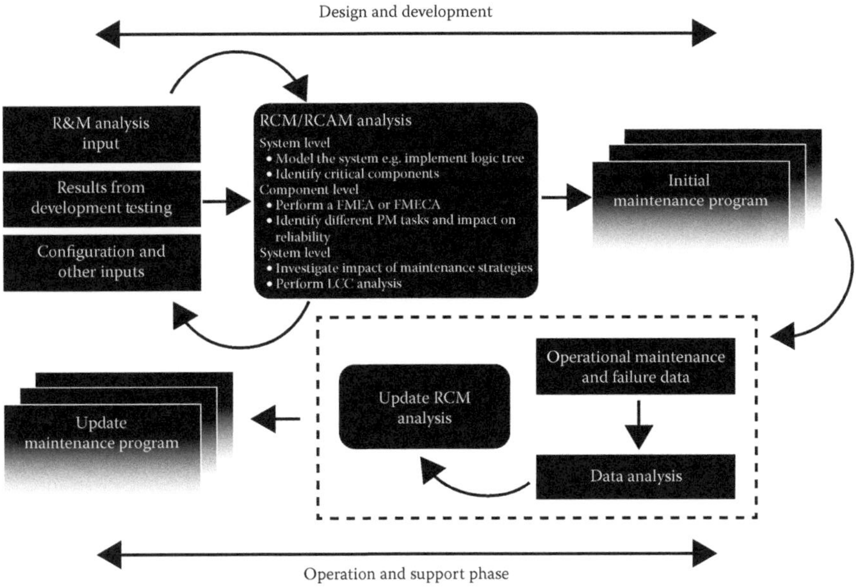

FIGURE 4.3
An overview of steps of the RCM process. This figure is an extension of the RCM analysis presented in [39] extending into the RCAM method for the RCM analysis.

According to [39], these tasks can be grouped into two main tasks as follows.

1. *Conduct supporting analyses:* RCM is an information-intensive process. Supporting analyses providing these data include the FMEA, fault tree analysis, functional analysis, and others.
2. *Conduct the RCM analysis:* The RCM analysis consists of using a logic tree to identify effective, economical, and, when safety is concerned, required preventive maintenance.

Planning to implement an RCM approach to defining the preventive maintenance for a system or product must address each of the tasks noted in the preceding paragraph. The plan must address the supporting design phase analyses needed to conduct the RCM analysis. Based on the analysis, an initial maintenance plan, consisting of the identified preventive maintenance with all other maintenance being corrective, by default, is developed. This initial plan should be updated through life exploration during which initial analytical results concerning frequency of failure occurrence, effects of failure, costs of repair, and so on, are modified based on actual operating and maintenance experience. Thus the RCM process is iterative, with field experience being used to improve upon analytical projections.

4.5.1 Data Requirements

A central part of RCM analysis is how to collect, select, and define these different input data needs. A summary of the different data requirements is given below [39].

At the system level, the following data are required:

- System descriptions
- System drawings
- PM and control programs
- Commitments or requirements for existing programs

At the component level, the following data are required:

- A list of components
- Component maintenance history data

For the cost/benefit analysis, the following data are required:

- The cost of reliability, that is, investment cost and outage cost
- The cost of undertaking maintenance, that is, the cost of manpower, materials, components, and so on
- The cost of not undertaking maintenance, that is, outage costs for the utility and for the customers

4.5.2 Input Data from the Practice

The input data for performing the RCM and RCAM analysis also involves the experience from practice. The case studies presented in this book give examples of different approaches in accessing this type of data.

The guidelines [39] summarized the different input needs as system or component level input as follows.

At the system level, the main areas for input are from two areas. First, the understanding of the overall procedures involved in maintenance planning. Second, a dialogue is needed with the utilities to get to know the relationship between maintenance and reliability.

At the component level, the following issues need to be discussed:

- Which modes can be affected by maintenance
- The relationships between failures and lifetimes
- Whether maintenance can affect "constant" failure rates or wearing out

For the cost/benefit analysis, it is critical that the following issue is discussed:

- What are the factors that need to be balanced

4.6 Input Data Required for the RCM and RCAM Analysis

4.6.1 Data Categories

The RCM analysis requires an extensive amount of information. Much of this information is not available early in the design phase and therefore the RCM analysis for a new product cannot be completed until just prior to production.

The data needs fall into four categories: failure characteristics, failure effects, costs, and maintenance capabilities and procedures. Below are some discussions presented for these categories [39]:

- *Failure Characteristics:* Preventive maintenance is effective only for certain underlying probability distributions. Components and items, for example, for which a constant failure rate applies (e.g., the underlying probability distribution is the exponential), do not benefit from preventive maintenance. Only when there is an increasing probability of failure should preventive maintenance be considered. This is why RCM is performed on components in failure mode.
- *Failure Effects:* The effects of failure of some items are minor or even insignificant. The decision whether or not to use preventive maintenance for such items is based purely on costs. If it is less expensive to allow the item to fail (and then perform corrective maintenance) than to perform preventive maintenance, the item is allowed to fail. As stated earlier, allowing an item to fail is called run to failure.
- *Costs:* The costs that must be considered are the costs of performing a preventive maintenance task(s) for a given item, the cost of performing corrective maintenance for that item, and the economic penalties, if any, when an operational failure occurs.
- *Maintenance Capabilities and Procedures:* Before selecting certain maintenance tasks, the analyst needs to understand what the capabilities are, or are planned, for the system. In other words, what is or will be the available skill levels, what maintenance tools are available or are planned, and what are the diagnostics being designed into or for the system.

TABLE 4.2

Data Sources for the RCM and RCAM Analysis

Data Source [39]	Comment
Lubrication requirement	Component level data. Determined by designer
Repair manual	Component level. Manufacture recommendations
Engineering drawings	Component or system level. These could be updated after construction
Repair part list	Component level
Quality deficiency reports	Component and system level. From operation stage
Recorded observations	Component and system level. From test of new equipment and from field of use
Hardware block diagrams	Component and system level
Bill of materials	Component level
Functional block diagrams	Component and system level
Existing maintenance plans	Component level
Maintenance orders or manuals	Component and system level
Discussion with field personnel and field operators	Component and system level
Results from the FMEA, FMECA, and other reliability analyses	Component level
Results of maintenance task analysis	Component and system level

4.6.2 Sources of Data

Table 4.2 lists some of the sources of data for the RCM analysis. Note that when RCM is being applied to a technical system in use, the historical maintenance and failure data will be inputs for the analysis. However, when historical data is not available or during the design phases of a system, generic data is an invaluable source for establishing a base line and making comparison analysis on the system. An effective failure reporting and corrective action system (FRACAS) is an invaluable source of data.

4.6.3 Data Collection

Five categories of information contain the necessary data for reliability modeling [39]: site identification, site one-line drawings, nameplate information, critical equipment designation and sparing, and maintenance data. When combined, this information gives the analyst all the necessary data to populate a reliability model. Data collection for facilities is not intended to be done in a single setting nor in a single month. This is an ongoing activity that should be completed in as timely a manner as possible without impacting the readiness of the facility. Once completed, updates to the information are only necessary as maintenance is performed on the equipment.

4.6.4 Maintenance Data

Maintenance data provides the reliability analyst with time to failure data as well as insight into the level of periodic maintenance performed on a piece of equipment. Time to failure data provides data for the calculation of time to failure metrics while periodic maintenance data allows a validation of manufacturer-supplied failure data. This data contains both scheduled and unscheduled maintenance actions. Scheduled maintenance lists periodic maintenance while unscheduled maintenance lists equipment failures and repairs. Maintenance data typically consists of both handwritten log books or record, and computerized maintenance records.

5

Electric Power System Reliability and Asset Management

5.1 Introduction

The electric power system is by its invention a smart system to support energy on demand. This means availability of supply for customers with a certain reliability performance.

The electric power system is one of the fundamental infrastructures in the modern society. The classical electric power system involves generation, transmission and distribution, and consumption under requirements from owners, customers, authorities, and regulators. The electric power system is an extremely complex system involving a large number of components, operated in real time, with a constant requirement on a balance between electricity generation and consumption, which is spread over a wide geographic area possible including several nations and markets.

The ongoing transition of the electric power system for a sustainable energy system, today commonly referred as smart grid, extends the electric power system with additional features. Examples are given by the customers being a so-called procumers that are electrical customers that also could be small-scale electricity producers. Moreover, the procumers could have home storage for electricity and electrical vehicles which could be charged or connected to the grid for feeding in. Another example, on the larger scale, is the large-scale offshore wind farms connected with a long distance meshed high voltage dc grid, including energy storage with large numbers of batteries.

In summary, the electric power system is a complex technical system, which is becoming more complex with the ongoing transition for a sustainable energy system. A nicely definition of a complex system is given by "a complex system is one where no individual can, at any time, understand the entire thing" [71]. Consequently, it is important to be accurate in the modeling and humble in making conclusions of the understanding of the results.

These different developments for a sustainable energy system result in several challenges for the existing and future electric power systems. The challenges are, however, different involving different solutions and actors.

A summary capturing different challenges for current developments of the electric power system is as follows (resulting from the EU project ERA-Net):

1. Enabling an increased flexibility of the power system to cope with the growing share of intermittent, variable, and decentralized renewable generation, and managing the complex interactions.
2. Increase network capacity to support increased generation and transmission resulting from renewables and in support of the internal energy market.
3. Provide information, services, market architectures, and privacy guarantees to support open markets for energy products and services, whilst facilitating the active participation of customers.

These three different challenges well capture different needs for developments of the electric power system in different areas of the world. The first and second challenges can be solved with technologies. The first, for example, by introducing new solutions with storage and ways of controlling the load. The second involves infrastructure projects building new lines. The third is, however, related to new areas of the power system focusing on the active customer provided with new possibilities being active in the production, operation, and consumption of electricity as part of the local energy system.

5.2 Background

5.2.1 Historical Note

Until the 1870s, electricity was a matter of interest only to engineers and researchers [72]. Experiments were conducted to learn more about electrical phenomena, and batteries were the main source of power. Then the Belgian researcher Zenobe Gramme invented the generator, which could provide greater electrical currents than the battery. This saw the beginning of the breakthrough for electricity. Electrical feeders were then built from small and large power plants to provide light and run electrical machines for primary and secondary industries. The first incandescent lamp came into being around 1880, invented simultaneously by the American Thomas Alva Edison and the Englishman Joseph Swan. It was then that the benefits of electricity for daily (especially nightly) life became clear, and with it came the birth of institutions such as public libraries. Electricity had finally reached its customers providing the demand and rational for electrical power delivery systems.

5.2.2 Developments of the Electric Power System

The electric power system includes electrical transmission and distribution systems (T&D systems) for the transport of electrical power from producers to customers. These consist of thousands of overhead lines, substations, transformers and other equipment spread over large geographical areas, and interconnected to deliver power on demand to customers. The major design of a T&D system is based on two physical and economic constraints.

The first constraint being that it is more economical to transport power at high voltages because of the reduction in losses, but higher voltage transmission requires equipment with a greater capacity which is in turn itself more expensive. However, the voltage level that European customers utilize is 250 V/416 V (three-phase). This is not an economical level for transmission, therefore costly voltage transformers are required.

The second constraint is that power in past have been more economical to produce in large amounts, but it must be delivered for the usage in small quantities at low voltage levels (120–250 V). Although with the introduction of the market in the late 1990s, it became economically feasible also with distributed generation. The current shift of the electric power system for the sustainable energy system involves several changes focusing on meeting the needs of the electricity customer, who also could be a small-scale producer of electricity and having both home batteries, electrical vehicles and solar panels integrated with the local power system called micro grid.

The traditional T&D system is designed to transport power from a few large generating plants to many smaller sites (the customers). Furthermore, this has to be done using hierarchical voltage levels, from high to low.

As a consequence of this hierarchical structure, the T&D system can be thought of as being composed of several distinct levels of equipment, denoted by the following: transmission, switching, sub-transmission, substations, feeders, service/secondary and customer equipment [73].

5.3 Terminology and Definitions

5.3.1 Main Concepts

In this section, selected main terms are presented for the electric power system. These are generally used terms in the power industry [74].

- The customer is the purchaser of electricity from a supplier.
- The supplier is the party who provides electricity via a public distribution system (referred to here as an electricity utility).

- The supply-terminals are the points of connection to the public system used by the customer, for example, the electricity metering point or the point of common coupling (referred to here as load points).
- The supply voltage is the root mean square (rms) value of the voltage at a given time at the supply-terminals, measured over a given interval.
- The nominal voltage (Un) is the voltage by which a system is designated or identified and to which certain operational characteristics are referred.
- The declared supply voltage (Uc) is the voltage applied to the supply-terminal, which is normally the nominal voltage (Un) of the system, but by agreement between the supplier and the customer, a different voltage (Uc) can be applied.
- The supply interruption is a condition where the voltage at the supply-terminals is lower than 1% of the declared voltage. A supply interruption can be classified as:
    - Prearranged (or planned) when consumers are informed in advance, or
    - accidental, when caused by failures. Failure can, for example, be related to external events or to equipment.

Accidental interruptions are classified as

- Long interruptions (interruptions > 3 min) or
- Short interruptions (interruptions < 3 min)

5.3.2 Failures

The general definitions of failure for the T&D system that is used in this book can be structured according to Figure 5.1. Failures can be divided primarily into damaging faults and nondamaging faults. Outages caused by damaging faults are usually called permanent forced outages, while outages caused by nondamaging faults are categorized again after the action of restoration into: (i) transient forced outages when the system is restored by automatic switching and the outage time is negligible, and (ii) temporary forced outages when the system is restored by manual switching or fuse replacement. Long interruptions are often caused by damaging faults (permanent faults) and short interruptions are often caused by transient faults.

A structure for the different types of outages, capturing the models of restoration, is defined in [75] with a corresponding presentation in [76].

Furthermore, damaging faults can be separated into two models of failure: passive failure and active failure defined as follows in [75].

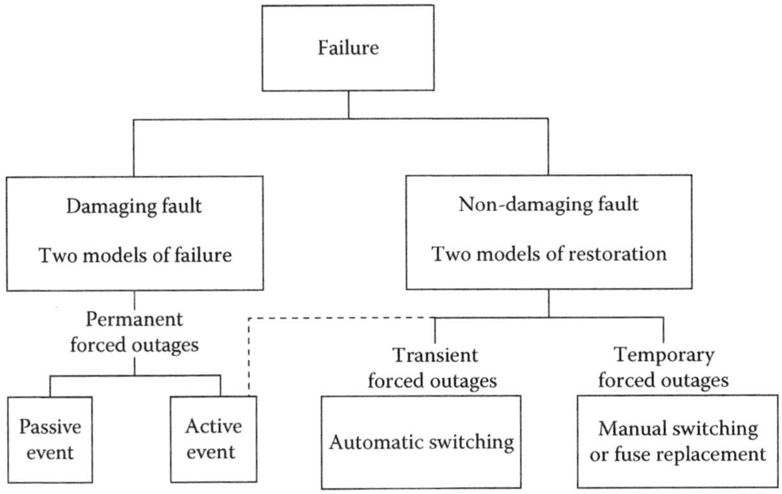

FIGURE 5.1
Definitions of failure for T&D systems.

Definition 5.1

An active failure is a failure of an item which causes the operation of the protection devices around it and results in the opening of one or more fuses.

Definition 5.2

A passive failure is a failure that is not an active failure.

The failed item (component) by an active failure is consequently isolated and the protection breakers are re-closed. This leads to service restoration to some or all of the load points. However, for the passive failure service is restored by repairing or replacing the failed component (or by re-closing a disconnecter and using another feeder for supply).

The outage time of a failure is made up of various items depending on the cause. Figure 5.2 shows two different time sequences following active and passive failures. As can be seen in the figure, the active failures can be restored by either repair or replacement, or by switching. The dotted line in Figures 5.1 and 5.2 indicates active failures that are restored by switching and not caused by a damaging fault. These are referred to as additional active failures.

Definition 5.3

An additional active failure is a failure mode that occurs when a component fails actively and causes interruption through its impact on other components.

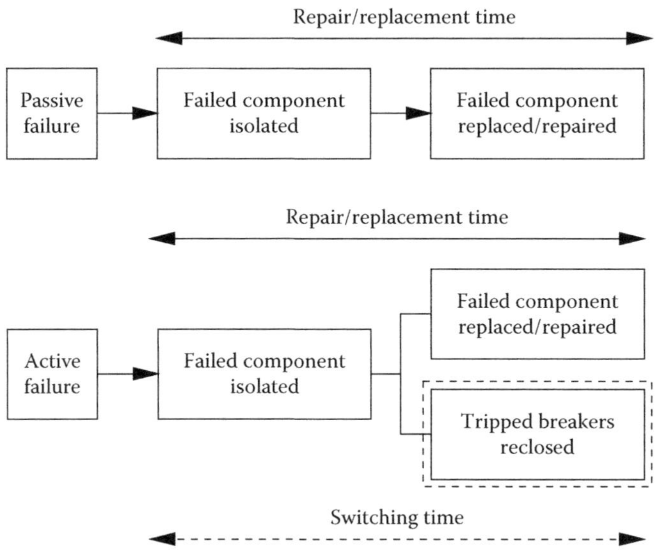

FIGURE 5.2
Outage time sequences for T&D systems.

5.3.3 Definitions for Electric Power System Operation

Recently, the EU presented a new regulation on the operation of the electric power systems in Europe [77]. A few selected definitions are presented as follows:

- *Operational security* means the transmission system's capability to retain a normal state or to return to a normal state as soon as possible, and which is characterized by operational security limits.
- *Constraint* means a situation in which there is a need to prepare and activate a remedial action in order to respect operational security limits.
- *Alert state* means the system state in which the system is within operational security limits, but a contingency from the contingency list has been detected and in case of its occurrence the available remedial actions are not sufficient to keep the normal state.
- *Blackout state* means the system state in which the operation of part or all of the transmission system is terminated.
- *Internal contingency* means a contingency within the TSO's control area, including interconnectors.
- *External contingency* means a contingency outside the TSO's control area and excluding interconnectors, with an influence factor higher than the contingency influence threshold.

- *Influence factor* means the numerical value used to quantify the greatest effect of the outage of a transmission system element located outside of the TSO's control area excluding interconnectors, in terms of a change in power flows or voltage caused by that outage, on any transmission system element. The higher is the value the greater the effect.
- *Contingency influence threshold* means a numerical limit value against which the influence factors are checked and the occurrence of a contingency located outside of the TSO's control area with an influence factor higher than the contingency influence threshold is considered to have a significant impact on the TSO's control area including interconnectors.
- *Contingency analysis* means a computer-based simulation of contingencies from the contingency list.
- *Critical fault clearing time* means the maximum fault duration for which the transmission system retains stability of operation.
- *Fault* means all types of short circuits (single-, double-, and triple phase, with and without earth contact), a broken conductor, interrupted circuit, or an intermittent connection, resulting in the permanent nonavailability of the affected transmission system element.
- *Transmission system element* means any component of the transmission system.
- *Operational security analysis* means the entire scope of the computer-based, manual, and automatic activities performed in order to assess the operational security of the transmission system and to evaluate the remedial actions needed to maintain operational security.
- *Operational security indicators* means indicators used by TSOs to monitor the operational security in terms of system states as well as faults and disturbances influencing operational security.
- *Operational security ranking* means the ranking used by TSOs to monitor the operational security on the basis of the operational security indicators.
- *Operational tests* means the tests carried out by a TSO or DSO for maintenance, development of system operation practices and training and to acquire information on transmission system behavior under abnormal system conditions and the tests carried out by significant grid users for similar purposes on their facilities.
- *Forced outage* means the unplanned removal from service of a relevant asset for any urgent reason that is not under the operational control of the operator of the concerned relevant asset.
- *Outage planning incompatibility* means the state in which a combination of the availability status of one or more relevant grid elements,

relevant power generating modules, and/or relevant demand facilities and the best estimate of the forecasted electricity grid situation leads to the violation of operational security limits taking into account remedial actions without costs which are at the TSO's disposal.

- *Outage planning agent* means an entity with the task of planning the availability status of a relevant power generating module, a relevant demand facility, or a relevant grid element.

5.4 Power System Reliability

5.4.1 Overall Problem Statement

Figure 5.3 captures the fundamentals of the problem with any engineering system. They are all exposed to unknown events with a probability of failure. The risk and cost of unexpected events are items that typically any organization wants to minimize. In order to do this, models are formulated. The models are created with enough complexity to capture the main features of the risk and cost. Reliability theory can be used for such purpose. These models can be solved directly with analytical techniques. The problems could, however, be too complex and large in order to achieve results solving the analytical models in a reasonable time and with available computational resources.

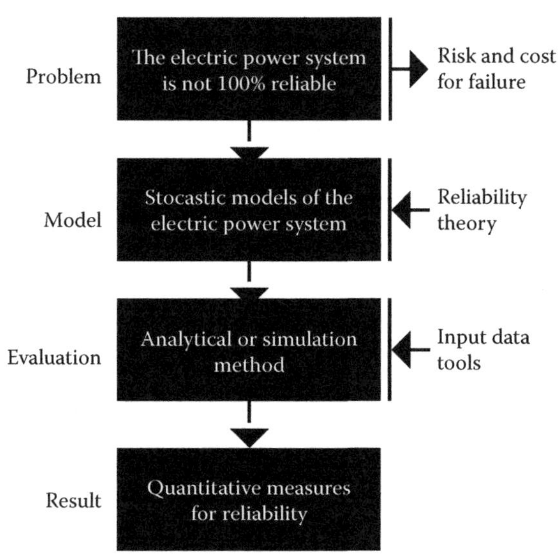

FIGURE 5.3
Fundamentals of problem.

In those cases, simulation methods are used. The simulation is by definition an experiment, and therefore will not have the exact numerical value each time. The numerical result can be used for decision support in the planning and operation of the engineering system. For most engineering systems, the final decisions are made by human beings, who not only have quantitative input data from this type of processes but also have access to qualitative input based on own and others experience. There could also be, for example, political or regulatory inputs that complement the input from the engineering system. The focus of this book is to show examples and underlying theory for giving decision support for an engineering system using electric power system examples. Moreover, this book shows how maintenance can be used as a tool for decision support considering the impact on system reliability.

5.4.2 Introduction to Electric Power System Reliability

The electric power system has its overall function to deliver electricity to customers on demand with a certain quality. The NERC (North American Electric Reliability Corporation) defines reliability as "the degree to which the performance of the elements of the electrical system results in power being delivered to consumers within accepted standards and in the amount desired" [78].

The electric power system is an extremely complex system, even before the current developments into smart grid. Some persons claim that Internet was firstly launched by the electric power system. Models have in past therefore typically divided the power system into different parts such as distribution, transmission, and generation, or into planning and operation. The latter relates to the key factor of the different the time horizon. Figure 5.4 shows the concept of operation time for an electric power system. It is obviously suitable with different models depending on time perspective.

In summary, the power system reliability has two fundamental different concepts.

1. *Security*: "The ability of the system to withstand sudden disturbances." This concerns short-term operations and is addressed by ancillary services which include: Voltage support, Congestion relief, Regulation capacity, Spinning reserves, Nonspinning reserves, etc.
2. *Adequacy*: "The ability of the system to supply the aggregate electric power and energy requirements of the consumers at all times." This concerns planning and investment and is addressed by Planning reserves, Installed capacity, Operable capacity, or Available capacity.

An important reflection of the two concepts as presented above is that the ongoing developments of the electric power system will result in that these two areas might merge. The planning in future will be more based on real-time measurements. In summary, power system reliability concerns the

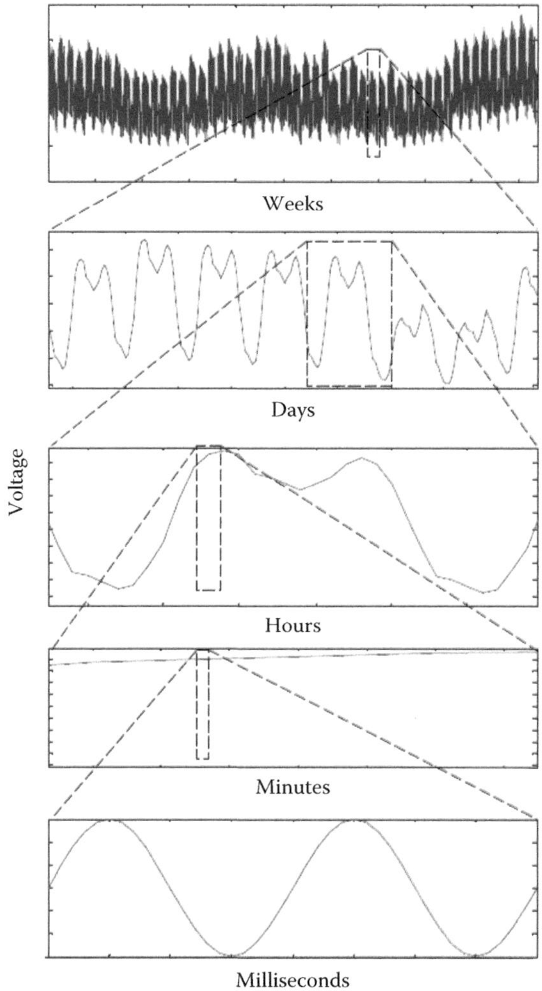

FIGURE 5.4
Illustrative example of operation time for electric power systems.

development of methods to stochastically predict the number and duration of interruptions of the electricity supply. However, no reliability analysis will ever be able to predict that an interruption will take place in a certain area, in a certain day and time, and last for a certain time.

5.4.3 Electric Power System Reliability Assessment

An Electrical Power System consists of a large number of interconnected subsystems and components each of which interact with and influence the overall system reliability. One definition of reliability is the ability of a component or system to perform required functions under stated conditions for a stated period of time [79]. Reliability assessment of the electrical power system is often divided into two functional aspects: system adequacy and security. Adequacy is the ability of the power system to supply the aggregate electric power and energy requirements of the customer at all times, taking into account scheduled and unscheduled outages of system components. Security is the ability of the power system to withstand sudden disturbances such as electric short circuits or nonanticipated loss of system components [79].

A reliability model that includes the whole complexity of the electrical power system would be impossible to implement. The analysis would be far too complex and the results would be very difficult to interpret. Instead, it is preferable to model the system as separated into different parts. The traditional way for electrical power system reliability modeling is by using the three hierarchal levels: generation (HL1), generation and transmission (HL2), and distribution (HL3). Each level can then be modeled and evaluated individually. A study of HL2 is also referred to as a composite system reliability assessment and this can include both adequacy and security analysis. Models for this type of modeling are fully described in [79].

The Gold book is an IEEE Recommended Practice for the Design of Reliable Industrial and Commercial Power Systems [39]. The objective of the recommended practice is to present the fundamentals of reliability analysis applied to the planning and design of industrial and commercial electric power distribution systems. It motivate the reliability analysis of the power system by the high costs associated with power outages, the necessity to consider the cost of power outages when making design decisions for new power distribution system, and to have ability to make quantitative "cost-versus-reliability" trade-off studies [39]. Moreover, the recommended practice underlines that the lack of credible data concerning equipment reliability and the cost of power outages have hindered engineers in making reliability studies in past. The new edition of the recommendation practice overcomes this by including an extensive mechanical and electrical equipment reliability data from the U.S. Army Corp of Engineers Power Reliability Enhancement Program (PREP) database.

5.4.4 Equipment Data for Electric Power Systems

A knowledge of the reliability of electrical equipment is an important consideration in the design and operation of industrial and commercial power distribution systems. The failure characteristics of individual pieces of electrical equipment (i.e., components) can be partially described by the following basic reliability statistics:

- Failure rate, often expressed as failures per year per component (failures per unit-year).

- Downtime to repair or replace a component after it has failed in service, expressed in hours (or minutes) per failure.
- In some special cases, the probability of starting (or operating) is used. Reliability data on the pertinent factors (e.g., cause and type of failures, maintenance procedures, repair method, etc.) is also required to practically characterize the performance of electrical equipment in service [39].

Reliability data collected from equipment reliability surveys and a data collection program over a period of 35 years or more are presented in [39]. According to [39], this data is the most comprehensive database publicly available on electrical and mechanical equipment reliability in the world. The data provide historical experience to those who have not been able to collect their own data. The data can be an aid in analyzing, designing, or redesigning an industrial or commercial system and can provide a basis for the quantitative comparison of alternate designs.

5.4.5 Electric Power System Reliability Indices

This section presents some basic reliability indices that are commonly used for quantitative measure of reliability of electric power systems. These indices correspond to general used indices as defined in the literature [80,81] and the focus is on measures for T&D systems. For electric power generation systems like; wind, nuclear and hydro power main reliability measures are availability which was introduced in Chapter 3.

5.4.5.1 Definitions and Terminology

Commonly used terms in system reliability analyses are defined here. These definitions are from the guidelines [39] and are commonly used for power system reliability assessment.

- *Availability:* The ability of an item under combined aspects of its reliability, maintainability, and maintenance support to perform its required function at a stated instant of time or over a stated period of time.
- *Component:* A piece of electrical or mechanical equipment viewed as an entity for the purpose of reliability evaluation.
- *Failure (f):* The termination of the ability of a component or system to perform a required function.
- *Failure rate (λ):* The mean (arithmetic average) is the number of failures of a component and/or system per unit exposure time. The most common unit in reliability analyses is hours (h) or years (y).

Therefore, the failure rate is expressed in failures per hour (f/h) or failures per year (f/y). Syn: forced outage rate.

- *Inherent availability (Ai):* The instantaneous probability that a component or system will be up or down. Ai considers only downtime for repair to failures. No logistics time, preventative maintenance, etc., are included.
- *Maintenance downtime (Mdt):* The total downtime for scheduled maintenance (including logistics time, spare parts availability, crew availability, etc.) for a given time period (Tp) (hours).
- *Mean downtime (MDT):* The average downtime caused by scheduled and unscheduled maintenance, including any logistics time. Syn: mean time to restore system (MTTRS).
- *Mean time between failures (MTBF):* The mean exposure time between consecutive failures of a component.
- *Mean time between maintenance (MTBM):* The average time between all maintenance events, scheduled and unscheduled, and also includes any associated logistics time.
- *Mean time to failure (MTTF):* The mean exposure time between consecutive repairs (or installations) of a component and the next failure of that component. MTTF is commonly found for nonrepairable items such as fuses or bulbs.
- *Mean time to maintain (MTTM):* The average time it takes to maintain a component, including logistics time. MTTM is primarily a measure of the preventative maintenance frequency and durations.
- *Mean time to repair (MTTR or simply r):* The mean time to replace or repair a failed component. Logistics time associated with the repair, such as parts acquisitions, crew mobilization, are not included. It can be estimated by dividing the summation of repair times by the number of repairs and, therefore, is practically the average repair time. The most common unit in reliability analyses is hours (h/f).
- *Operational availability (Ao):* The instantaneous probability that a component or system will be up or down, but differs from Ai in that it includes all downtime. Included is downtime for unscheduled (repair due to failures) and scheduled maintenance, including any logistics time.
- *Reliability:* The ability of a component or system to perform required functions under stated conditions for a stated period of time. The term reliability is also used as a reliability characteristic (metric) denoting a probability of success or a success ratio. In general usage, reliability refers to system performance over time.
- *Repair downtime (Rdt):* The total downtime for unscheduled maintenance (excluding logistics time) for a given Tp (hours).

- *Repair logistics time (Rlt):* The total logistics time for unscheduled maintenance for a given Tp (hours).
- *System:* A group of components connected or associated in a fixed configuration to perform a specified function.
- *Total downtime events (Tde):* The total number of downtime events (including scheduled maintenance and failures) during the Tp (previously referred to as all actions, maintenance, and repair).
- *Total failures (Tf):* The total number failures during the Tp.
- *Total maintenance actions (Tma):* The total number of scheduled maintenance actions during the Tp.
- *Total period (Tp):* The calendar time over which data for the item was collected (hours).
- *Year (y):* The unit of time measurement approximately equal to 8765.81277 hours (h). Any rounding of this value will have adverse effects on analyses depending on the magnitude of that rounding; 8766 is used commonly as it is the result of rounding to $365.25 \cdot 24$ (which accounts for a leap year every 4th year); 8760, which is $365 \cdot 24$, is the most commonly used value in power reliability field. By convention, 8760 will be used throughout this recommended practice.

5.4.5.2 Load Point Indices

Basic reliability indices for each specified load point in an electric distribution system are as follows:

- Expected failure rate, λ [f/yr]
- Average outage duration, r [h/f]
- Annual expected outage time, U [h/yr]
- Average loss of energy, LOE [kWh/yr]

$$LOE_s = U_s \cdot pnotsup$$

and *pnotsup* is the product of total number of customers and the total active power per customers [kW/cust].

As discussed previously in Chapter 3, network modelling and Markovian technique are different approaches used in the modelling and evaluation of reliability. For a detailed study of larger and more complicated systems, the network modelling provides an over-simplified model, and the Markovian model gets too complicated. Therefore, an alternative method based on the Markovian approach but with a set of approximate equations has been proposed [79] and has been implemented for the RADPOW tool [13]. For a detailed presentation of the method the reader is referred to literature e.g. [13,44,79].

5.4.5.3 System Performance Indices

The basic problem with trying to measure reliability is how to relate the two quantities: frequency and duration [73]. One way of solving this is to use the following system performance indices classified according to [82] and [79].

- System average interruption frequency index (SAIFI) [int/yr, cust]:

$$\frac{\text{total number of customer interruptions}}{\text{total number of customers served}} = \frac{\sum \lambda_i N_i}{\sum N_i} \tag{5.1}$$

  where λ_i, the expected failure rate of the load point i, and N_i, the number of customers for load point i.

- System average interruption duration index (SAIDI) [h/yr, cust]:

$$\frac{\text{sum of customer interruption durations}}{\text{total number of customers served}} = \frac{\sum U_i N_i}{\sum N_i} \tag{5.2}$$

  where U_i, the annual expected outage time of load point i.

- Customer average interruption frequency index (CAIFI) [int/yr, cust]:

$$\frac{\text{total number of customer interruptions}}{\text{total number of customers interrupted}} = \frac{\sum \lambda_i N_i}{\sum N_a i} \tag{5.3}$$

  where $N_a i$, the number of customers affected at load point i.

- Customer average interruption duration index (CAIDI) [h/int]:

$$\frac{\text{sum of customer interruption durations}}{\text{total number of customers interruptions}} = \frac{\sum U_i N_i}{\sum \lambda_i N_i} \tag{5.4}$$

- Customer total average interruption duration index (CTAIDI) [h/yr, cust]:

$$\frac{\text{sum of customer interruption durations}}{\text{total number of customers interrupted}} = \frac{\sum U_i N_i}{\sum N_a i} \tag{5.5}$$

- Average energy not supplied per customer served (AENS) [kWh/yr, cust]

$$\frac{\text{total energy not supplied}}{\text{total number of customers served}} = \frac{\sum LOE_i}{\sum N_i} \tag{5.6}$$

  where LOE_i, average loss of energy of load point i.

- Average service availability index (ASAI):

$$\frac{\text{customer hours of available service}}{\text{customers hours of service demand}} = \frac{\sum N_i \cdot 8760 - U_i N_i}{\sum N_i \cdot 8760} \tag{5.7}$$

Note that CTADI and CAIFI include the total customers that are interrupted, which implies that each individual customer is only counted once regardless of the number of times their supply is interrupted. This, however, does not apply to CAIDI where all interruptions for each customer are counted. Moreover, the indices are related according to the following:

$$\text{CAIDI} = \frac{\text{SAIDI}}{\text{SAIFI}}$$

$$\text{CTAIDI} = \frac{\text{CAIDI}}{\text{CAIFI}}$$

5.4.6 Need of Probabilistic Assessment

The electric power system planning and operation have historically been made based on deterministic criteria. More advanced analytical techniques are, however, used for the assessment of power system security. In addition to advances in computational methods, this includes the development of emerging approaches such as risk-based security assessment and the application of intelligent technologies [61].

Probability-based reliability assessment approaches for electric power system have been well developed for use in planning-related decision making. The standard approaches on which most commercial grade programs are, however, typically made for planning, and need further refinement for use in operational decision making. The main reason for is that the nature of the decision is quite different. Long-term planners identify the facilities necessary to satisfy reliability criteria for the range of expected operating conditions. The decision is then whether or not to reinforce the facilities. In contrast, operational planners identify the operating limits within which reliability criteria are satisfied. The decision is whether or not to take action to modify the operating conditions. This fundamental difference has two important implications: one on the loading conditions to study and another on the type of indices to calculate.

5.4.6.1 Deterministic versus Probabilistic Criteria

This section gives a brief description of typically deterministic criteria for electric power system planning and shows their main weaknesses.

Examples on deterministic criteria used to handle the probability of failure

1. *Generation capacity* = expected maximum load + fixed percentage of expected load
2. *Operation capacity* = expected load + reserve (=one/more of the largest units)
3. *Network capacity n–k* criteria where the system should be in function although *k* units are in failure

The following weaknesses in the deterministic criteria summarized above can be seen:

- A fixed percentage of generation capacity do not guarantee enough reserve capacity
- All planning of the electric power system is based on prognosis for expected load which always included uncertainty
- Failure rate, for example, for lines depends on several issues such as length, design, geographical, environment, and consequently the (n–1) criteria cannot guarantee a certain level of disturbances

5.4.6.2 Method for Deterministic Operation

A deterministic method has traditionally been used to guarantee a high operational security of power transmission systems. Under the deterministic method, the stability of the system is evaluated for a list of predefined events. This list contains all single contingencies (one fault, loss of one generation unit) that may occur starting from the existing operational state. The method is normally referred to as the $(N-1)$ criterion. The consistent use of the $(N-1)$ criterion has resulted in an extremely good track record for the transmission system reliability in almost all industrial countries. There are, however, a number of reasons for developing probabilistic methods for use next to the $(N-1)$ criterion [83].

1. The $(N-1)$ criterion neither considers the likelihood nor the consequences of the events on the list.
2. A number of recent blackouts have shown that the $(N-1)$ criterion is not sufficient to guarantee the high reliability that is required for the transmission system.
3. During periods of adverse weather, single redundancy is often insufficient. The failure rate of, especially, overhead lines increases several orders of magnitude during periods of adverse weather. The probability of losing two or more components within a short period of time is no longer negligible.
4. Open markets for electricity over large geographical areas have resulted in less controllable and less predictable power flows at transmission level.

5. The large-scale introduction of weather-dependent sources like wind power and of nondispatchable small-scale generation (distributed generation) will introduce even more uncertainty in transmission system operation.

Next to these technical reasons, the reliability demands on power transmission systems are still increasing. Also have developments in information technology made it possible to perform large amount of calculations in a short period of time and to move beyond ($N-1$).

5.4.6.3 Operational Risk Assessment

Operational risk assessment aims at determining (or predicting) the performance of the system during a short period of time, starting from a well-defined and known operational state and considering weather and load predictions available during the operation. There is difference between reliability calculations and operational risk assessment, which is in the amount of time one looks into the future. Reliability calculations are a design tool, where the future stretches over several years, thus covering many operational states and all kinds of weather. Operational risk assessment covers a time scale of hours (the so-called lead time) with the (initial) operational state known and weather observations and predictions available.

There are two steps in calculation of operational risk indices:

1. Determining which combinations of component outages (contingencies) result in interruptions or severe system disturbances
2. Calculating the probability of these contingencies occurring

Like with composite system reliability analysis, the former is the main difficulty. For individual contingencies, accurate and strong methods for stability analysis are available [84]. However, for a large transmission system the number of possible states is extremely large. There are several methods for addressing this and under development in the literature on composite system reliability. The literature on power system stability (i.e., determining if a contingency will result in an interruption or not) has very much been emphasized on accurate detailed models; however, for probabilistic operational risk assessment, simplified models and/or direct methods for stability assessment may be more appropriate [67]. For operational risk assessment, contrary to reliability analysis, there is no need to consider a large number of initial operational states, load levels, and types of weather, so that the amount of calculations to be performed is actually less.

However, for operational-risk assessment, it is no longer sufficient to consider failure rates obtained from long-term observation of large numbers of identical components. To consider the influence of the weather, a major impact factor for operational risk, the momentary failure rate during the period being considered should be used. In the same way, component repair times and switching times appropriate for the actual operation and weather conditions should be used.

5.5 Asset Management Trends

5.5.1 Changing Requirements for Maintenance Methods

The developments of the electric power system push towards increased complexity and greater requirements for maintenance procedures. For example, maintenance decisions have been traditionally based on experiences and measurements which could be supported by diagnostics methods. The increase in the expectations of maintenance has kept pace with the increasing knowledge about the dynamic characteristics of the power system. These higher expectations are due to the increasingly complex systems and higher demands on the cost-effective use of resources. The increasing knowledge about the system has been gained primarily by an understanding of the relationships among failure frequency, reliability, and maintenance, and by methods and techniques provided by continuous measurements.

5.5.2 Maintenance Specifications and Performance

In the perspective of a state-owned transmission system operator (TSO), the following factors have been given as reason for performing maintenance [85]:

- To maintain a high level of safety for personnel
- To maintain a good level of availability (required by customers)
- To prevent the grid from aging too quickly
- To postpone reinvestments or new investments
- To preserve the environment
- To maintain good knowledge about the status of the grid
- If possible, to assist personnel (the contractors) maintain a high level of competence
- To respond to changing environmental aspects (for example, tree growth)

As one effect of the de-regulation, many utilities made a major change in their handling of maintenance going from in-house resources to contract resources. This solution was typically cost-efficient and also lead to less experience and competence developments in the own organization. The following factors were summarized as constituting problems in maintenance:

- Maintenance is a long-term process, and consequences caused by too little or incorrect maintenance do not show up until up to 3–6 years later
- Lack of available methods for evaluating the efficiency of maintenance
- Lack of available methods for describing the status of the grid
- Human factors causing faults increase with increased maintenance
- Reduced maintenance and low investment lead to loss of competence

There were further problems identified including loss of expertise as mentioned earlier, against a background of aging personnel, the majority of the whom are middle-aged or older and there is a low influx of younger people. The situation where contractors are responsible for performing maintenance and the trend towards the rapid turnover of personnel result in knowledge reduction.

5.5.3 Discussion about Probabilistic AM and RCM

Electrical power utilities are changing their activities to increase their competitiveness and reduce costs. Tools are needed that can be used to quantify costs associated with the whole life cycle of their equipment. However, to enable taking full advantage of techniques like RCM, it is necessary to predict the remaining life cycle of an item and to quantify the effect of changes in maintenance or operating policies on the remaining lifetime and associated costs.

One methodology for doing this has been developed by Ontario Hydro Technologies [86]. A generic technology called Asset Management Planner is used, which was originally applied to rotating machines. This method is described in a paper entitled Probabilistic Asset Management and RCM at substations [86]. This method enables decisions to be made based on quantitative analyses, which is desirable in this context. The results demonstrate the importance and merit of modeling maintenance policies for power system equipment.

One problem identified is that numerous pieces of equipment in electrical distribution systems are not well documented. There are no international standards, and these are needed to form a basis for specifications.

It is probable that a lack of data is one reason why a heuristic approach to RCM is commonly used.

6

RCAM Case for Electrical Distribution Systems

6.1 Introduction

A complete application of the RCAM (Reliability-Centered Asset Maintenance) method is presented in this chapter. The application studies have been made on two electrical different distribution systems: (i) a rural system of overhead power lines (the Flymen system) and (ii) an urban underground cable system (the Birka system). These application studies have been performed using the simple approach called Approach I and a comprehensive approach called Approach II. The Birka system has been run through the complete RCAM method in Approach II, which is the failure rate modeling approach ($\lambda(t, PM)$). This study is referred to as the Cable Application Study. Consequently, both Approaches I and II have been applied to the Birka system. The difference between these two lies in the way they deal with the issue of relating reliability and the effect of PM activities.

6.2 Distribution System Components

This section introduces the different components in the distribution system and focuses on the underground cable component. It also presents different sources and results for component statistics and in particular the causes of failures. Results presented in this chapter provide input data for the analysis of the Flymen system (see Section 6.5.2) and for comparing this with the resulting data from the Birka system analysis.

6.2.1 Components in General

The main components in a distribution system include:

- Lines (lines, poles, and related items)

- Cables (cables, junctions, and related items)
- Breakers
- Transformers
- Disconnectors
- Load disconnectors
- Fuses
- Bus bars

For further presentation of the individual components, the reader is referred to the literature in the field. For the purpose of this study, one component is analyzed in greater detail that is the cable component.

6.2.2 The Underground Cable Component

The main reference for the general description of the cable component in this section is the Electrical Engineer's Reference Book [87].

6.2.2.1 Introduction to the Cable Component System

Cable component systems contain two essential components: (i) the metallic conductor with low resistance to carry the current and (ii) the insulation to provide a dielectric medium for isolating conductors from one another and from their surroundings.

In order to specify suitable insulation and construction for the service performance required, the design voltages are quoted in the form U_0/U, that is, voltage-to-earth (phase-to-ground voltage)/voltage between phases [87]. For example, power distribution to end users that is usually distributed in three phase lines with a zero conductor [88] provides voltage levels of 230/400 V. Moreover, the voltage range extends from automobile cables of 6–12 V to the highest transmission voltages reaching around 800 kV [87]. Cables are normally grouped in either of two ways: (i) by design voltage or (ii) by usage. The first way typically uses the traditional categories of low, medium, and high voltages, while the second typically uses, for example: (i) wiring and general, (ii) power distribution, and (iii) transmission. Traditionally, these cables had different insulation types and production places, which are no longer true for new cable types [87]. It is, however, important to bear in mind that system analysis as undertaken in this book includes system components that have been in operation for long time. Many systems are more than 30 years old and due to deregulation are expected to operate close to maximum length [89]. These demarcations may, however, vary over years and between countries, and are therefore in themselves not a significant source of differentiation.

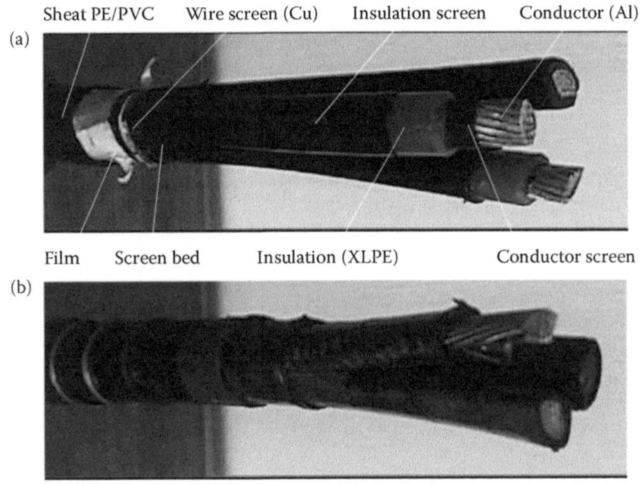

FIGURE 6.1
Two different types of underground cables which are in use for the 11 kV distribution in Stockholm City network, that is, the Birka system.

This Case Study focuses on electrical distribution system analysis. Furthermore, the focus of the analysis is on the underground cable component, which is presented further in Sections 6.4 and 6.6, and relates to failures in the cable insulation. Therefore, the insulation part of the cable is presented in greater detail in the following section. There are two main types of insulation material Figure 6.1 shows a piece of cable length of the both types. For a thorough description of the cable component, please refer to specific literature. A comprehensive literature reference is provided in [87].

6.2.2.2 Cable Insulation

Underground cable insulation can be either polymeric material (thermoplastic and elastomeric materials) or impregnated paper. These types of insulations are discussed further below. The latter is representative of the first types of cables and the former of more modern material. From about the 1960s onwards, polymeric materials have become increasingly preferred in insulation because of their potentially higher operating temperature.

6.2.2.2.1 Polymeric Material

The properties of good insulation material are flexibility and the ability to withstand high temperatures. Natural rubber fulfills the first but not the second, and therefore other materials have been used. This shortcoming of rubber can be solved by using, for example, synthetic polymeric materials.

Elastomeric material returns rapidly to approximately its initial dimensions and shape after deformation at room temperature caused by a weak stress. Under such conditions, thermoplastic material shows permanent deformation. Conventional elastomeric compounds need to be cross-linked by vulcanization generally by chemical methods to provide them with characteristics akin to rubber compounds. Two most common elastomeric materials are ethylene propylene rubber (EPR) and cross-linked polyethylene (XLPE). These combine the flexibility and electrical properties of natural rubber with a higher operating temperature and easier strippability. Examples of thermoplastic materials are polyvinyl chloride (PVC), polyethylene (PE), and polypropylene (PP) [87].

One of the disadvantages with PVC in cabling relates to its use in buildings or tunnels. Although PVC is basically flame retardant, if a serious fire does develop it can transmit flame and decompose generating noxious acidic fumes and dense smoke [87].

Compared with paper insulation, XLPE has the main advantages of not needing metallic sheathing, and being much cleaner and easier to handle in laying and jointing [87].

The most important single problem with XPLE insulation is that like paper insulation, internal partial discharges occur at voltages of 5 kV upwards at any irregularities within or at the surface of the insulation.

While it was recognized that the XLPE insulation must be extremely clean and free from voids, and that screening at both surfaces of the insulation was necessary, many cables were put into service without adequate testing to ensure freedom from discharging. It was not until the mid-1970s that ideal forms of screening were developed that could be readily removed for jointing and could adequately deal with thermal expansion and contraction. In the years that followed, the final problem was to identify and find solutions to problems caused by the effects of water insulation contact. Tree-like structures were found in the insulation and it was eventually established that these could lead to electrical breakdown [87]. This phenomenon is called water treeing and is further discussed in Section 6.4. Many of the early cable failures were due to imperfections resulting from the use of semiconducting fabric tapes for conductor screens. A thin layer of extruded semiconducting polymeric material is now mandatory, and to ensure a clean interface it is normally extruded in tandem with the main insulation and cured with it.

6.2.2.2.2 Impregnated Paper

Impregnated paper insulation as the name implies consists of layers of paper tapes overlapping the conductor. The cable is dried and impregnated before application of a metal sheath, which is required to keep the insulation dry and undamaged. The impregnation of the paper by a mass-impregnating process is carried out before the application of the metallic sheath. Moreover, the paper consists of a felted matting of long cellulose fibers derived from wood pulp. An important factor in the control of the properties of the paper cables is the washing of the fibers. Large quantities of water are used for this, and with paper intended for the highest voltage cables this water has to be deionized to ensure a minimum power factor [87].

From its introduction at the end of the 1800s, impregnated cables have provided excellent service to the cable industry. Under normal conditions, users have been able to install cables and then simply forget about them. Ultimate lifetimes of 50–60 years have been common, and the majority of cables have been replaced only because they became too small for their loads. While the basic dielectrics have changed little throughout the years, there have been considerable improvements in the quality of materials and manufacturing techniques, and these have led to successive reductions over the years [87].

6.2.2.2.3 The Change of Use of Insulation Material

The excellent dielectric properties of PE and XLPE materials came into prominence in the early 1960s in higher voltage applications (11–45 kV) and an increasing efforts have gone into them ever since [87]. From the results of a survey of XLPE cables effected by water trees, it was concluded that XLPE cables that are produced now are estimated to have the same life expectancy as paper insulated cables [94], which is discussed further in Section 6.4.

6.2.3 Review of Component Statistics

6.2.3.1 Swedish National Outage Data Bank

6.2.3.1.1 Introduction to the Reporting System

National disturbance statistics for the Swedish distribution system have been summarized annually by Swedenergy AB (formerly the Association of Swedish electrical power utilities (Svel), from the DAR reporting system, and its precursor FAR (see, e.g., [95]). DAR is a system for compiling and analyzing information about disturbances, supply interruptions, and failures in the power distribution system. Statistics are collected for the voltage levels 22–0.4 kV [96]. DAR was developed to provide the utilities with statistics about complications due to disturbances and failures in their equipment, and to provide the authority with reliability indices to compare the utilities and illustrate levels of power availability to customers. In the future, the need for collecting such data is expected to increase. The authority is introducing strategies to control and reward utilities in respect to supply availability and outage costs suffered by customers [97]. To meet these new requirements, a new generation of reporting systems called DARWIN was developed [98].

The output from DAR is presented in relative values to facilitate the comparison of statistics between different utilities. The following list summarizes some of its features.

- Data is provided at the individual voltage levels

- Faults are allocated by specific types of components
- Fault type is identified, along with how and when the fault is cleared
- How and when the failed component is isolated and identified
- The number and types of causes of faults are provided and classified as either:
  - Weather
  - Damage
  - Material/method
  - Personnel
  - Other

Within each group, a more specific cause can be stated. Note that only one cause is reported for a failure, this being the primary and initial cause of the failure.

An understanding of the fundamental concepts in the reporting system is required for the correct reporting and analyzing of the disturbance data. These are as presented in [96] and the glossary [99].

6.2.3.1.2 Analysis of Disturbance Data from the DAR Reporting System

Distribution system statistics from the DAR system for the years 1995–1997 are summarized in Table 6.1 [95]. This includes a survey of the interruption of supply incidents for different voltage levels and geographical locations. The survey includes the following voltage levels: 22, 11, less than 10, and 0.4 kV. The results show that the 11 kV distribution system contributes some 35%–40% of all the interruptions at the voltage levels studied here.

Table 6.2 presents an outline and overview of interruptions to supply (not forced outages) for these years. The interruptions are shown as both the total number of reported interruptions (voltage interruptions) and those that lead to supply interruptions. The results illustrate that the 11 kV range is responsible for most of the unavailability of supply for the voltage levels studied here.

The report [95] provides a comparison of causes of failure between two rural networks and two urban networks for the period 1990–1997. These rural networks are located in the south of Sweden (RuralS) and the north of Sweden (RuralN). The urban networks are located in the middle of Sweden (UrbanM) and in the north of Sweden (UrbanN). Table 6.3 presents the results from the 11 kV level. For the rural networks, the most marked difference is in the failure caused by lightning. This may be due to the higher frequency of lightning in the south due to regional meteorological effects. For the urban networks, the most marked difference is seen in other weather. However since the unknown causes in the UrbanM network are as high as 30%, no relevant conclusions can be made about this difference. The major reason

TABLE 6.1

Causes of Failures Leading to Interruption of Supply in Distribution Systems in Sweden 1995–1997

	Estimated Average	
Cause of Failure	**Number**	**Percentage**
Weather		
Lightning	343	12.3
Other weather	1199	43.1
Wind	242	8.7
Snow	213	7.7
Rainwater	36	1.3
Salt	5	0.2
Tree (wind)	485	17.4
Tree (snow)	218	7.8
Damage	283	10.2
Material/method	453	16.3
Personnel	62	2.3
Overloading	62	2.2
Returning load	5	0.2
Unknown	371	13.4
Total	2778	100

Source: Jansson, S., *Availability in Power Supply in the Swedish Electrical Distribution Systems 1995–1997*, Technical Report, Swedenergy AB (Former Association of Swedish Electric Power Utilities), (Original title: *Rapport om leveranssäkerhet i de svenska elnäten 1995 till 1997*, in Swedish), 1999. [95]

for registering an unknown cause is because the interruption has not been analyzed correctly [100]. By comparing the rural and urban networks, it can be seen that other weather shows the greatest diversity. This is probably an effect of the more frequent use of overhead lines in rural networks and underground cables in urban networks. It can be concluded from the statistical analysis that the cause of failures are closely connected to where the distribution system is situated, in regards to both geographical and environmental conditions. It has also been identified that there is a problem with reporting, evident by the significant number of unknown causes reported.

Table 6.3 presents average values for the causes of failures for the complete study performed in [95]. These values are based on interruptions to supply for the period 1995–1997. Other weather makes up about 40% of the causes of failures in all the networks together, which is the same level as for the rural networks.

It has been shown that the 11 kV voltage level contributes significantly to failures in the distribution systems, and that the causes of these

TABLE 6.2

Disturbance Statistics for Swedish Distribution Systems

Year	Voltage Levels	Total Interruptions	Supply Interruptions	Energy not Supplied [MWh]
1995	All	6364	6220	6506
	11 kV	2632 (41%)	2565 (41%)	5561 (85%)
1996	All	7342	7209	9519
	11 kV	2755 (38%)	2698 (37%)	7655 (80%)
1997	All	11,078	10,9390	5574
	11 kV	3745 (34%)	3668 (34%)	4432 (80%)

Source: Jansson, S., *Availability in Power Supply in the Swedish Electrical Distribution Systems 1995–1997*, Technical Report, Swedenergy AB (Former Association of Swedish Electric Power Utilities), (Original title: *Rapport om leveranssäkerhet i de svenska elnäten 1995 till 1997*, in Swedish), 1999. [95]

TABLE 6.3

Causes of Failures Leading to the Interruption of Voltage for Different Distribution Systems in Sweden 1995–1997

Causes of Failures	RuralS [%]	RuralN [%]	UrbanM [%]	UrbanN [%]
Lightning	23	5	0	3
Other weather	45	50	1	29
Willful damage	7	11	13	24
Material/method	11	12	37	21
Personnel	1	0	19	9
Overloading	1	5	1	3
Unknown	13	18	30	11

Source: Jansson, S., *Availability in Power Supply in the Swedish Electrical Distribution Systems 1995–1997*, Technical Report, Swedenergy AB (Former Association of Swedish Electric Power Utilities), (Original title: *Rapport om leveranssäkerhet i de svenska elnäten 1995 till 1997*, in Swedish), 1999. [95]

failures depend on geographic location and impacts from the surrounding environmental. Therefore, it would be of great interest to analyze distribution systems that are spread across the country and represent both rural and urban systems. Section 6.5 presents results from such an analysis of two different systems, one rural network and one urban network.

6.2.3.1.3 Statistics for Underground Cables from the DAR and FAR Systems

Table 6.4 presents statistics from the DAR and FAR systems, and primarily for the Stockholm City network (Birka Nat AB). These data were prepared in connection with the study reported in [101].

TABLE 6.4
Statistics for Underground Cables from the DAR and FAR Systems

Component Types	Voltage Level [kV]	Total Length [km]	Failures [Year and km]
DAR—National Statistics 1990–1996			
Total	132	2341	0.00564
Total	33	582	0.01279
Total	11	16,704	0.00902
Total	0.4	132,315	0.00128
DAR—Internal Statistics, Birka 1990–1994			
Total	11	2298.8	0.01201
XLPE	11	819.9	0.01638
Impregnated paper	11	1478.9	0.00427
FAR—Internal Statistics, Birka 1982–1990			
Total	110	126	0.00088
Total	33	594	0.00937
Total	11	2069	0.03057
Total	0.4	5173	0.05475

6.2.3.2 Data from the National Fault and Interruptions Reporting System (NAFIRS) in the UK

Data based on the NAFIRS from the UK are presented in Table 6.5. This table presents results for high voltage (11 kV) distribution systems and separates the information according to different component types. Note that in this table, "Line" refers to overhead lines and "Cables" refers to underground cables.

Maintenance can have an impact on some of these causes but not others. The cause of failures in Table 6.5 that are assumed to be influenced by maintenance are: wind-borne materials (detected by inspection and then resolved by, for example, removal), corrosion, vibration, trees, vermin, farm animals, willful damage, faulty construction and installation, aging and wear, faulty manufacture and design. Other weather includes: lightning, rain, snow, ice, and gales. In practice, however, maintenance may overcome none, some, or all of these causes of failures.

6.2.3.2.1 *Window Rule for the Collection of Data*

When using statistics to predict the future, it is important to consider which time interval is relevant to the specific analysis. For example, if the aim is to predict the lifetime function of a component type for which significant changes in design have recently been introduced, then when predicting the behavior of this component, use only a short interval, or question whether these statistics should be used for the purpose of prediction at all. Generally,

TABLE 6.5

Faults per Cause in the UK, from NAFIRS

Cause of Failure	Line [%]	Cable [%]
Weather	52.3	3.6
Wind-borne materials	2.6	0.3
Condensation	0	0.1
Corrosion	1.3	0.7
Vibration	0.5	0.3
Fires	0.1	0.1
Trees	10.6	0.1
Birds	3.5	0.1
Vermin	0.3	0.3
Farm animals	0.7	0
Willful damage	1.7	0.7
Other utilities	0.2	21.8
Farming activities	1.1	0.7
Individuals	1.3	11.6
Builders and developers	5.3	20.3
Utility personnel	0.4	2.9
Faulty construction and installation	1.3	4.4
Aging and wear	13.1	29
Operational restrictions	3.5	1.5
Faulty manufacture and design	0.2	1.5

Source: Allan, R., Professor of Electrical Energy Systems. The Manchester Centre for Electrical Energy (UMIST), 2002. [102]

however, changes happen over time, both with the introduction of new techniques as well as resulting trends in the statistics. Therefore, the data used should cover a reasonable length of time to achieve adequate statistical significance, but not be too long, as the aging data would not be relevant to new data. A reasonable data collection time of 10 years has been used as guideline in the UK [102].

6.2.3.3 Other Sources Available

This section presents a selection of different sources of reliability component and system data, and indicates what types of information these include.

6.2.3.3.1 Nordel

Fault statistics covering the Scandinavian countries (Denmark, Finland, Norway, Sweden, and Island) are compiled within the Nordel reports. These statistics refer to faults causing disturbances in 100–400 kV power systems.

The total number of faults in 1999 came to 2204, and the loss of energy was 10.9 GWh [103]. The Nordel data include:

- Component type
- Number of components
- Number of disturbances
- Percentage breakdown of the causes of failures for the period 1990–1999 into lightning, other weather, damage, personnel, technical equipment and other

New guidelines for the classification of disturbances and failures that provide greater detail and are common for all these countries are currently being prepared [104].

Svenska Kraftnät owns and operates the Swedish transmission system and provides Nordel with the Swedish statistics. Other information is available but not published yet in the Nordel reports and covers component types reporting on: (i) the failure frequency (from 1997) and (ii) the unavailability of supply per disturbance and assorted probabilities for failure of function [105]. Moreover, Statnett provides Nordel with the Norwegian statistics, for example, those presented in [106].

6.2.3.3.2 *Canadian Electricity Association*

The Canadian Electricity Association (CEA) was founded in 1891 and is a national forum for the evolution of electricity enterprises in Canada. The CEA mission is to provide a comprehensive database for component and system reliability and performance data to assist its member utilities. The first stage of this system was implemented in 1977, covering reliability data from the generation system. These data are still being presented annually (see, e.g., from 1998 [107]) and include both generating units and outage data. Moreover, transmission equipment data are published, for example, for lines, transformers with failure causes, and more [108]. The only part missing is the statistics for the distribution system.

6.2.3.3.3 *CIGRE*

CIGRE, as earlier introduced, is an international association, registered in France and founded in 1921. Technical work is undertaken within its various study committees, and whose activities have resulted in some very interesting reports, including disturbance statistics. For example, a comprehensive survey of circuit-breaker failures can be found in [109], and data from service experience in high voltage gas-insulated substations (GIS) can be found in [110].

6.2.3.3.4 T-Book

The T-book [111] provides a comprehensive presentation of reliability data on components in Nordic nuclear power plants. This collection of statistics started in the mid-1970s and the first edition of the book was released in 1982 and included operational statistics from 21 reactor years. The most recent edition of the book includes data from a total of 108 reactor years. Reliability data are presented for different failure modes and the failure rates are presented for different confidence intervals. The long tradition of careful collection and management of statistics by the nuclear power plants in the generation of power is exemplary in the power transmission and distribution industry.

6.2.3.3.5 A Concluding Remark

As can be seen from the data sources presented earlier, there is a lack of distribution system data. However, some data from Sweden were presented in Section 6.2.3.1 and from the UK in Section 6.2.3.2 that relate to the distribution system level. Nonetheless, the conclusion is that future work requires international efforts to generate statistics for distribution-level component and system data. One of the reasons for not compiling and publishing data may be the competitive nature of the relationships between power utilities. However, it would in the common interest of all the parties to provide support for improving their systems.

6.3 Causes of Failures in Underground Cables

This chapter summarizes results from a survey of the causes of failures based on knowledge from both disturbance reporting system and practical experience from maintenance. This information focuses on the underground cable component and relates experience from one Swedish electrical power utility that owns and operates the Stockholm City network, namely Birka Nat AB. The results provide input data for the analysis of the Birka system following in Sections 6.5.3 through 6.7.

6.3.1 Assumptions and Identified Weaknesses

6.3.1.1 Introduction

One observation from the RCM framework was that in order to make an RCM analysis possible, comprehensive knowledge about the system and its components is required together with suitable input data to support a quantitative analysis. This issue has been analyzed in detail in this section. The results of a survey have been presented which identify the knowledge and data required, should they be or become available. The results from the survey also provide input data for the RCM analysis of the Cable Application Study that follows.

The material used here comes mainly from the report [112], which includes a comprehensive presentation of the statistical survey. Results have also been published in an RCM perspective in [113].

6.3.1.2 Cooperation with an Electrical Power Utility

In the first stage, different possible sources of data and contacts were investigated. The aim was to find data that could provide detailed knowledge about the system and its components, and to support the objective of defining a relationship between reliability and PM. It was identified that the best approach was to focus the communication on one utility, which would capture deep rather than broad knowledge, though also specific to the actual utility. The utility chosen was the owner of the Stockholm City network, Birka Nat AB (which also includes the former Stockholm Energi). From 2015, the electrical distribution system of Stockholm City is owned and operated by a consortium named Ellevio AB.

Cooperation with Birka Nat was a good choice for several reasons, including: (i) the availability of good long-term disturbance data, (ii) the previous contacts were good, (iii) close proximity to both head office and the network, facilitating visits to the actual distribution systems and simplified meetings, and (iv) good contact possibilities with maintenance personnel (Birka Service was now contracted to undertake the maintenance of the network as a separate business unit within the same company (Birka Nat), however, they share a history with Stockholm Energi which implies close relationships between the interests). The utility (Birka Nat), on the other hand, has an interest in evaluating methods for improving the management of maintenance with the aim of achieving more cost-effective PM.

6.3.1.3 Approach to and Summary of the Survey

1. System statistics analysis
   a. Disturbance reports for the Stockholm City network were analyzed.
   b. The conclusion was that for system reliability the critical voltage level is the 11 kV, and the critical component type is the underground cable.
2. Cable component analysis
   a. Analysis of specific cable system data
      i. Disturbance reports for a specific 33/11 kV substation were analyzed.
      ii. Faults due to the 11 kV underground cable were identified and analyzed in great detail.

   iii. The conclusion was that causes of failures denoted material and method is predominant in cable system failure, and that more information is needed to relate causes of failures to PM activities.

  b. Discussions with maintenance personnel
   i. Several meetings were held with people from both the utility (Birka Nat) and the PM contractor (Birka Service). The focus of the discussion was on the causes of failures of the 11 kV underground cables and how these could be prevented by maintenance.
   ii. Outcomes from the discussions were presented in a report that has been used in the discussion and other sections that follow.

3. The results were put into the RCM framework for the failure mode and effect analysis that is the deduction of the causes of the failures. This provides input data for the following RCM analysis.All the analyzed disturbance reports imply voltage interruptions and not forced outages.

6.3.2 System Statistics Analysis

Failure reports within the period 1982–1990 and also from the reporting system FAR have been analyzed for the whole Stockholm City distribution system. These reports refer to voltage interruptions for which the supply interruptions have also been defined. A total of 3980 reports have been investigated.

6.3.2.1 The Stockholm City Network

Figure 6.2 provides an overview of the distribution system in Stockholm City. This distribution system includes different types of stations, from the complex constructions for the transmission voltage at 220 kV, down to the simpler designs for households at 400 V. The end customers are defined by the number of annual subscriptions, being 450,000 at 400 V. Some of the end customers are, however, so-called high-voltage customers, which in this system include: 17 customers at the 33 kV level and 329 customers at the 11 kV level. This group consists mainly of the underground railway system (33 kV), industries, and hospitals. The customer data is further discussed in detail for the system analysis presented in Section 6.5.3.3.

Table 6.6 summarizes the different feeder lengths in the system, which are laid out in tunnels, the ground, or overhead. It can be seen that underground cables constitute 90% of the Stockholm City system. About 70% of these are 11 kV feeders. Consequently, the Stockholm City distribution feeders are dominated by the 11 kV underground cable.

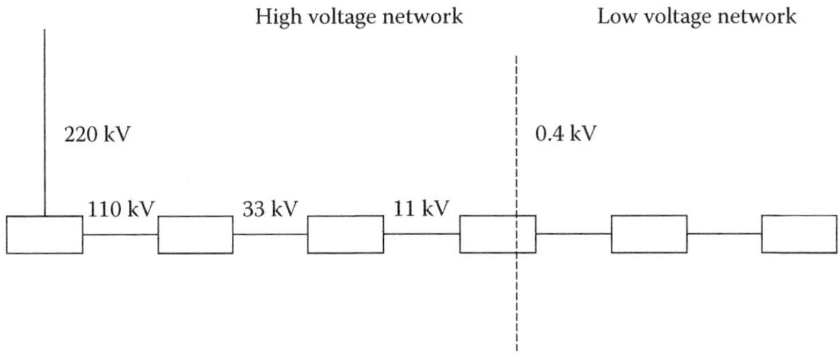

FIGURE 6.2
Illustration of the distribution system in the Stockholm City network. The boxes represent substations.

TABLE 6.6
Feeder Lengths Used in the Distribution System in Stockholm City

Feeders [kV]	Tunnel [km]	Underground [km]	Overhead [km]	Total [km]
220	20.3	5.4	0	25.7
110	30.6	103.6	0	134.2
33	67.3	465.0	0	532.3
11	199.1	2101.6	14.2	2315.0
Total	317.4	2675.6	14.2	3007.2

Source: Birka Nät statistics 1998-12-31.

6.3.2.2 System Level Statistics

Failure reports state the system voltage at the actual failure point. In this survey, the reports were sorted into three voltage groups: HV (220, 110, 33 kV), 11 kV, and other. It should be mentioned that the distinctions between voltage levels are not completely systematic in the reporting system, for example, auxiliary voltages are included in the one category but belong to different voltage levels.

Table 6.7 compares the three categories of reports, referred to as HV (220, 110, 33 kV), 11 kV, and other. It can be seen that few interruptions occur for the first category (HV), however, they do affect a large number of customers with an average outage time of about half an hour. Furthermore, 35% of the failures occur at the 11 kV level. The number of customers affected is somewhat lower than for the higher voltages, but the average outage time is about 1 h. The highest number of failures (about 58%) occurs within the voltage group referred to as other voltage levels, which consists mainly of lower voltages. However of these, some 48% are related to auxiliary voltages which are not considered to be relevant component types in this context [114]. The conclusion is that the 11 kV voltage level is critical to the system regarding the number of interruptions and the effect on customers.

TABLE 6.7

Failure Reports 1982–1990 in the Stockholm City Area

Class	Quantity of Reports	Custom. Affected	Total Cust. Outage Time [h]
HV	$262 \approx 7\%$	455,501	208,955
11 kV	$1392 \approx 35\%$	419,602	420,134
Other	$2326 \approx 58\%$	386,762	243,044
Total	$3980 = 100\%$		

6.3.2.3 Component Level Statistics

In the next step of the survey, the reports were sorted by failed component type and cause of failure, for both the entire Stockholm City system and the different voltage classes.

Tables 6.8 and 6.9 present the results for the whole Stockholm City network. The tables show the different components (failure causes) and their individual impacts on failures of the power delivery system. It can be seen that the cable component contributes about 36% of the total supply interruptions at all voltage levels. Another significant result is that the causes of failures reported as unknown were almost 50%.

The disturbance statistics were further analyzed for the different voltage levels separately. Tables 6.10 and 6.11 show the results for the 11 kV level. These tables confirm the hypothesis that the dominant contribution to failure is made by the cable system components. The cable system consists of the cable itself (including insulation and conductor), terminations and joints. The cable system causes about 30% of the failure reports, 60% of the interruptions, and 80% of the affected customers. Moreover, these affected customers experienced an average outage time of about 1 h due to cable system failure. The dominant failure causes, as shown in Table 7.6, are attributable to "manufacturing and material," which contributes about 20% of total failures. However, "other equipment" has been reported as causing 40% of failures.

6.3.3 Cable Component System Analysis

6.3.3.1 Statistics from the 11 kV Station Liljeholmen (LH11)

The previous analysis of statistics for the complete Stockholm City system could not provide enough detailed information about the causes of failures. Therefore, the survey continued by studying one particular station in the system, the Liljeholmen station (LH11) to see whether additional data could be provided for this specific station.

A system for the case study has been defined as shown in Figure 6.3. This system is based on the Liljeholmen-Bredang area (220–11 kV) and is

TABLE 6.8

Failed Component Types for Voltage Disturbances in the Stockholm City Network 1982–1990

	Numbers of Custom.			
Failed Component	**Voltage Inter.**	**Supply Inter.**	**Affected Custom.**	**Outage Time[a] [h]**
Bus bar	21	14	5404	7590
Circuit breaker	211	76	63,453	12,622
Load disconnector	255	51	19,750	14,102
Disconnector	62	16	130,704	112,236
Earthing switch	12	9	15,760	3712
Power transformer	519	144	36,129	26,956
Reactor	3	0	0	0
Power capacitor	81	69	0	0
Measuring transformer	15	8	92,600	88,530
Fuse				
Fuse apparatus	20	12	1523	2791
Incorrect fuse	46	18	22	59
Incorrect automatic fuse	12	1	0	0
Overhead line				
Pole	1	1	75	304
Insulator	13	10	1750	1078
Other material	4	4	550	395
Clear line path[b]	5	5	552	1265
Cable system	≈14%	≈36%	≈35%	
Conductor	175	166	101,719	163,609
Conductor insulation	262	236	152,916	139,214
Joint	60	57	163,960	90,721
Terminations nonprotected	6	6	9050	2764
Terminations protected	20	14	11,262	12,729
Surge diverter	1	1	0	0
Charging rectifier	75	14	0	0
Power rectifier	298	192	0	0
Battery	18	4	0	0
Other current supply	76	19	11,200	3275
Control equipment				
Incorrect relay protection	140	60	212,561	108,180
Incorrect relay	97	21	130,850	50,277
Control equipment	79	16	9500	9317
Computer equipment	50	10	11,100	6660
Other control equipment	462	35	69,800	5204
Compressed air	200	13	0	0
Building, other	414	32	13,600	14,633
Total	3714	1335	1,265,790	878,223

[a] $\sum N_a \cdot r$, where N_a, number of affected customers, and r, outage time.

[b] For example, cleared from close trees.

TABLE 6.9

Causes of Failures in Voltage Disturbances in the Stockholm City Network 1982–1990

	Numbers of Custom.			
Causes of Failures	**Voltage Inter.**	**Supply Inter.**	**Affected Custom.**	**Outage Time[a] [h]**
Overvoltage	170	108	12,958	17,507
Returning load	1	1	0	0
Dirt	11	1	0	0
Lightning	85	27	12,060	21,058
Wind	16	14	2202	8853
Snow, ice	25	12	3801	5471
Rain	102	13	2250	5690
Other weather	30	6	3050	4533
Digging	88	83	34,177	51,540
Falling tree	2	2	650	440
Animal	11	11	2532	1718
Traffic injury	5	3	201	71
Fire	2	2	0	0
Other damage	68	46	22,881	19,603
Incorrect operation	25	21	132,311	20,322
Incorrect handling	39	16	78,300	11,606
Incorrectly adjusted protection	49	18	26,850	18,442
Lack of surveillance	18	4	0	0
Incorrect installation	62	21	77,851	77,724
Contractor	6	3	3000	1700
Other personnel	122	51	67,304	27,100
Manufact./material	565	196	340,719	263,020
Lack of PM	204	35	12,506	8133
Dimension fault	31	12	2000	900
Falling trees (lack of PM)	1	1	200	470
Other equipments	1976	628	427,987	312,322
	≈53%	≈47%		
Total	3714	1335	1,265,790	878,223

[a] $\sum N_a \cdot r$, where N_a, number of affected customers, and r, outage time.

considered to be sufficiently delimited to enable an adequate RCM analysis. The system is referred for further studies as the Birka system.

Disturbance reports on interruptions to supply during the period 1982–1999 have been studied. These reports come from the two reporting systems FAR (1982–1990) and DAR (1991–1999).

FAR provided 23 interruptions and DAR 35 interruptions at Liljeholmen. These led to the disconnection of the 11 kV main breaker, which implies a

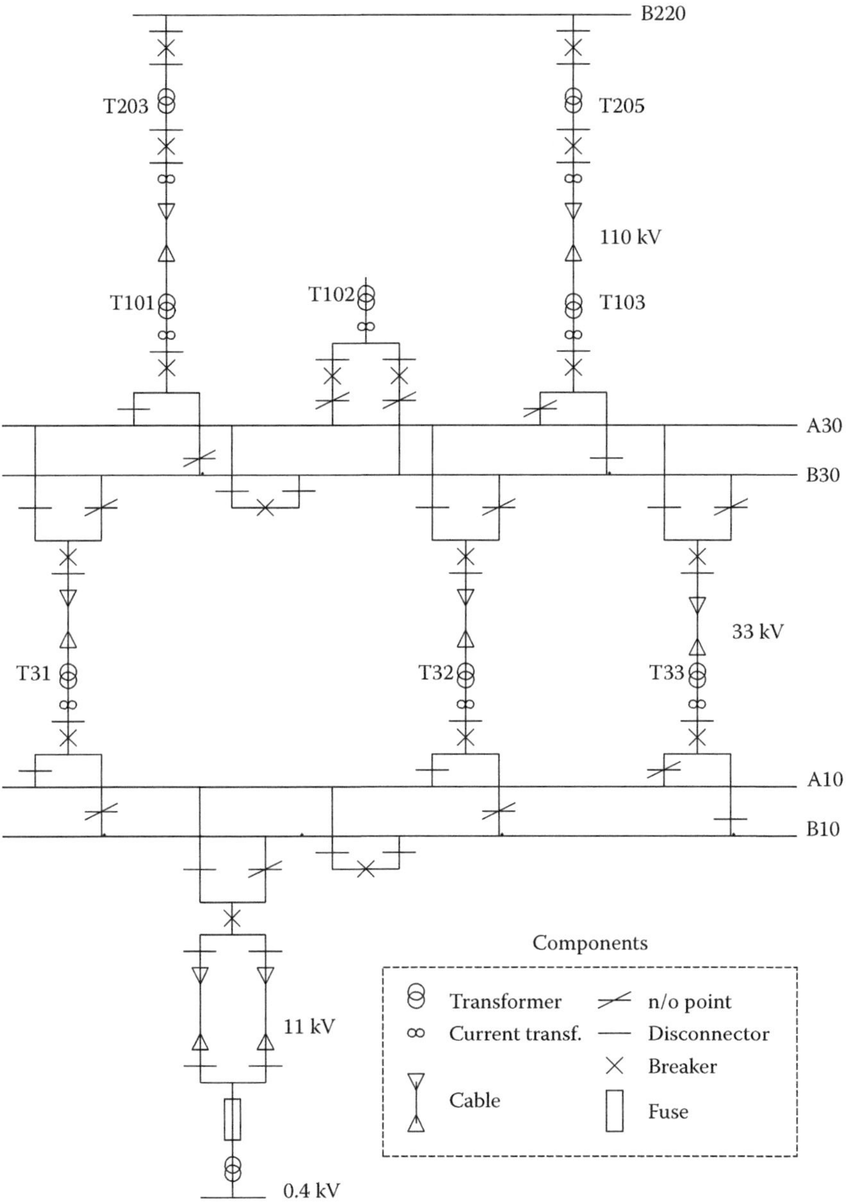

FIGURE 6.3
Illustration of the case system LH11 used in the RCM study at Birka Nat. Later referred as the Birka system.

TABLE 6.10

Failed Component Type in Voltage Disturbances in the Stockholm City Network at the 11 kV Level during the Period 1982–1990

	Numbers of Custom.			
Failed Components	**Voltage Inter.**	**Supply Inter.**	**Affected Custom.**	**Outage Time[a] [h]**
Bus bar	16	10	4604	5616
Circuit breaker	79	21	34,453	8988
Load disconnector	244	46	19,700	14,052
	≈22%	≈7%	≈5%	
Disconnector	19	7	5304	9853
Earthing switch	7	7	5760	2045
Power transformer	174	69	16,379	17,138
Reactor	2	0	0	0
Power capacitor	60	52	0	0
Measuring transformer	3	0	0	0
Fuse				
Fuse apparatus	3	2	23	41
Incorrect fuse	8	8	22	59
Incorrect automatic fuse	1	0	75	304
Overhead line				
Pole	1	1	75	304
Insulator	10	9	1750	1078
Other material	4	4	550	395
Clear line path	5	5	552	1265
Cable system	≈31%	≈62%	≈78%	
Conductor	147	143	99,718	161,094
Conductor insulation	212	207	132,115	124,653
Joint	54	51	87,560	60,161
Terminations nonprotected	3	3	50	64
Terminations protected	19	13	6262	9562
Surge diverter	1	1	0	0
Charging rectifier	1	0	0	0
Control equipment				
Incorrect relay protection	3	3	3700	2303
Incorrect relay	5	5	750	1013
Control equipment	1	1	0	0
Other control equipment	4	0	0	0
Power rectifier	298	0	0	0
Building, other	8	1	200	450
Total	1093	669	419,527	420,134

[a] $\sum N_a \cdot r$, where N_a, number of affected customers, and r, outage time.

TABLE 6.11

Causes of Failures in Voltage Disturbances in the Stockholm City Network 1982–1990

		Numbers of Custom.		
Causes of Failures	**Voltage Inter.**	**Supply Inter.**	**Affected Custom.**	**Outage Time[a] [h]**
Overvoltage	38	33	11,057	14,860
Dirt	1	0	0	0
Lightning	11	10	5760	16,228
Wind	14	13	2202	8853
Snow, ice	8	7	3801	5471
Rain	3	1	600	280
Other weather	13	6	3050	4533
Digging	73	70	34,177	51,540
Falling tree	2	2	650	440
Animal	7	7	2532	1718
Traffic injury	1	1	1	8
Other damage	29	26	19,570	16,066
Incorrect operation	15	14	7311	3689
Incorrect handling	8	5	4200	1504
Incorrectly adjusted protection	7	7	4600	3191
Lack of surveillance	8	0	0	0
Incorrect installation	30	12	18,851	4191
Contractor	2	1	3000	1700
Other personnel	32	21	20,604	14,863
Manufact./material	312	131	121,318	75,961
	≈29%	≈20%	≈29%	
Lack of PM	91	23	11,706	8083
Dimension fault	9	7	2000	900
Falling trees (lack of PM)	1	1	200	470
Other equipment	378	271	142,337	185,585
	≈35%	≈40%	≈34%	
Total	1093	669	419,527	420,134

[a] $\sum N_a \cdot r$, where N_a, number of affected customers, and r, outage time.

fault on the redundant 11 kV underground cable system. This redundancy system features two cables in parallel and loop feeding. Consequently, about 60 interruptions occurred during a 20-year period. There are about 30 similar stations in the Stockholm City area. This gives $30 \cdot 60/20 = 90$ int./yr, or 2 int./wk, which is considered to be in good agreement with real supply disturbances (which could also be validated by comparison with the $669/9 \approx 75$ int./yr in Table 6.10 from the FAR system). Furthermore, the network consists of about 2000 km of underground cable, which is $90/2000 \approx 0.05$ int./yr and km. This value can be compared with the average

TABLE 6.12

Supply Interruptions Caused by Cable System Failures at Substation LH11 in the Stockholm City Network from the Two Periods from the Reporting Systems FAR and DAR

	Supply Interruptions		Total
Causes of Failures	**1982–1990**	**1991–1999**	**[%]**
Damage			
Digging	2	3	8.6
Sabotage	–	1	1.7
Other damage	–	3	5.2
			≈16
Personnel			
Incorrect manoeuvre	–	3	5.2
Incorrect installation/laying	–	3	5.2
Other personnel	1	–	1.7
			≈12
Material/method			
Manufac./material	5	3	15.5
Lack of PM	1	2	13.8
Other material/method	14	9	5.2
			≈35
Unknown	–	8	13.8
Total	23	35	100

value for cable faults, being about 0.051 int./yr and km, in Norway for the period 1967–1983 [65]. This shows that the cable fault data is significant. Furthermore, in [95] an average value is given for the failure rate of 11 kV cables for 1995–1997, being about 0.03 int./yr and km. Finally, [115] shows that the failure rate for mass impregnated cables is about 0.02 int./yr and km. The conclusion is that the failures reported in the cable system give a realistic value for interruptions to the Stockholm City system and therefore validate the data used in the survey.

6.3.3.1.1 Causes of Failures

Table 6.12 shows the results of the failure and disturbance reports from FAR and DAR. These results show the interruptions due to cable faults for the LH11 system. Materials or methods cause 59% of these faults. Furthermore, it can be seen that within the Material/Method cause, failures by manufacture and material predominate with about 14%. No failures were caused by weather, which is a typical behavior for underground cables. The conclusion from these statistics is that materials and methods are the most significant causes of failures for cables. However, the statistics do not provide enough details about the failure causes, to relate these to PM measures.

6.3.3.1.2 Cable-Specific Information

Information reported about failed cables has been studied in more detail here relating to cable-specific information (e.g., manufacturer or size) and the parts of the failed components (e.g., joint, conductor, or other). In both reporting systems (DAR and FAR), a detailed classification of the failed component can be reported. In FAR, the identity of the components can be included which is not the case for DAR. However, the reporting of this information is imperfect. In DAR, failure reports were made in 31% of cases, and in FAR, only 3 of 23 cases were reported with detailed cable-specific information, but information about which part of the component that caused the interruption was provided in 13 of the 23 cases.

The cable-specific data could provide information to illustrate the trend of failures in a certain type of cable. The DAR data shows that the types of cables exposed to failure are equally divided between the two main cable types, XLPE insulation and mass impregnated cable. The FAR data was too poor to support any conclusions.

The reporting frequency supporting the question about which part of the cable system fails is quite poor. Such reports have been made in only 24 of the 58 cases (41%). Results from both reporting systems indicate that the cable itself is the major contributor to failure rather than joints and terminations. This result does not correspond with the expected outcome for Birka Nat, which estimates that the joints provide the greatest contribution to failure [116]. The result is, however, emphasized by [115], which indicates that mass impregnated cables when divided into their constituent parts (being the cable itself, the terminations, and the joints) contribute 0.014, 0.005, and 0.003 int./yr and km, respectively.

6.3.3.1.3 Failure Location

The cable failures were studied in regard to failure location. All the failures were pinpointed on a geographical map showing the cable layout. It could be deduced that the failures were geographically scattered over the area, and that some critical areas featured repeated failures between stations. These critical areas contained three different stations with a total of 16 failures, which implies about 28% of the failures during the period. In all of these cases, repetitive faults were overcome by reconnections, which implied that another cable was used as a feeder. In addition, one of the stations was taken out of service.

6.3.3.2 Maintenance Personnel Experience

6.3.3.2.1 Introduction

In this part of the survey, the statistics have been compared with practice the experiences from the 11 kV underground cable faults in the Stockholm City system. The aim here has been to identify what additional information is available and what the present reporting systems do not tell, plus to provide recommendations for how the reporting systems and the maintenance routines could be improved.

The information in Table 6.12 shows the causes of failures for the 11 kV underground cable system at LH11. The table shows what has happened but not why or how. This consequently leads to questions such as: "What further information can be identified?" and "What additional details can be reported in the present and future reporting systems?" From these, a discussion document was prepared for interviews with maintenance personnel. The focus of these discussions can be summarized as "Experience of the causes of failures and PM in 11 kV underground cables."

The first meeting resulted in a working document that was discussed further in two subsequent meetings. Participants came from Birka Service Operation Planning Group and also involved the actual electricians that performed the recouplings, repairs, and similar work. This section provides a brief summary of the outcomes from these discussions that in turn resulted in recommendations. It is worth noting that the survey was presented in a 59 page report, and that the material here is very much reduced.

6.3.3.2.2 General Aspects

The power delivery system in the Stockholm City area is a mixed network where the feeders consist of several types of cables and installation years, and are joined together in many different combinations. Working and failure-free cables exist that were put into operation in 1902. On the other hand, cable joints from 1970 to 1991 have been shown to be prone to failure.

In general, no direct relationship exists between commissioning year and failure occurrences, but experience shows that when work is performed near to cables, the probability of cable failure increases. However, these particular failures are likely to be primarily caused by digging, which in some areas and time periods could be the dominant cause of failures. In this case study, the main focus is on the cause of failure due to material or method. These have been illustrated in Table 6.12 and can be seen to be a significant cause of failure and may be preventable by maintenance.

6.3.3.2.3 "Materials and Method" as a Cause of Failure

Material and method has been reported as the cause of failure in more than half the cable system faults. It is difficult to confirm whether this value is correct or reasonable. These failures are more complicated to deduce than failures caused for instance, by digging. Nonetheless the following can generally be said about the failures caused by material and method: (i) the classifications are not sufficiently developed to allow individual judgments about the cause of failure and the cause of aging. However, aging could in turn be related to different causes and would therefore probably not be a suitable cause for a reporting system.

One main focus for the discussions was to identify what the cause denoted as material and method means. The following list presents a summary of the results from the discussions.

- Cu and Al contacts: Contact between the different materials, copper and aluminum, which can occur in joints and could lead to a short circuit.
- Water trees: A tree-like phenomenon which allows water penetration through the insulation, and which occurred primarily in XLPE insulation cables produced in the mid-1970s.
- Corrosion of the screen or jacket.
- Lack of oil insulation due to poor refilling of oil in joints or terminations of cables laid up and down hillsides.
- Bending damage can occur because of digging holes too shallowly when laying.
- Scratches in the insulation, for example, in XLPE insulation because of the incorrect method or handling during laying.

The discussions showed that more useful information is known about the cable system failures than is actually being reported, which could help in identifying and preventing failures. Information revealed at the failure location, but not reported includes, for example: (i) repeated laying of the cable, (ii) signs of aging or earlier repair, (iii) dry insulation paper, (iv) stones in the filling material, (v) fractures in the jacket, (vi) new asphalt, (vii) joint dose to failure, (viii) taped cable, (ix) convolution and more. The recommendation is to make a checklist to report what is seen at the failure location. These lists could be evaluated after a test period to support further planning of cable maintenance. An important point is that several individuals are often involved in the actions following a failure. Observations by all these individuals could be useful inclusions in the reporting system.

6.3.4 Results

6.3.4.1 Summary of the Results from the Survey

Table 6.13 presents a summary of the results from the statistical analysis. The statistics used for the first set of analysis came from the FAR system (1982–1990) and for the second set of analysis from the FAR and DAR systems (for 1991–1999). The last results, that is for the type of insulation that contributed most to failures was only supported by the DAR data due to lack of reported information in the FAR reports.

Figure 6.4 presents a summary of the results from the survey and also the logic of the approach. The knowledge gained can be divided into three levels:

TABLE 6.13

Summary of the Results from the Analysis of Disturbance Statistics for the Stockholm City Network and Substation LH11, Statistics Based on Data from the Reporting Systems FAR and DAR

Factors	Results	Supporting Statistics
Critical area	11 kV network	1982–1990
Critical component	Underground cable	1982–1990
Causes of failures	Material and method	1982–1999
Part of cable	Cable	1982–1999
Type of insulation	About half of XLPE and impregnated paper	1991–1999

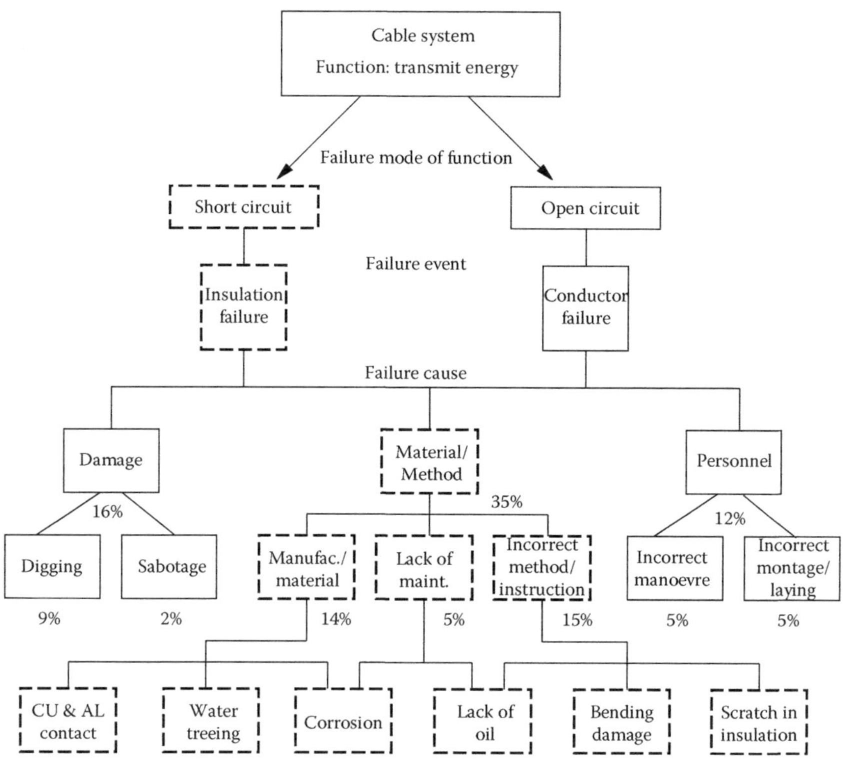

FIGURE 6.4
Resulting FMEA from the survey of data for RCAM analysis for the critical component of the electric distribution system, the Birka System. The boxes shown with dashed lines are those which could be considered to be affected by preventive maintenance. *Note*: the percentage for the lower failure re causes do not summate to that for the higher level because some items, specifically *unknown*, have been omitted. (Adapted from Bertling L. et al., Survey of Causes of Failures Based on Statistics and Practice for Improvements of Preventive Maintenance Plans, in *Proceedings of the 14th PSCC*, Seville, June 2002 [113].)

1. Level 1, the system and its components
2. Level 2, failure causes based on statistics
3. Level 3, details about failure causes based on experience from maintenance personnel

First, statistics that would support the RCM analysis were studied and deduced for both the whole system (the Stockholm City power delivery system) and for a critical voltage level (the 11 kV network system). Second, the focus was placed on the cable system component and a specific station LH11. Available data was extracted. The results showed that the data was not sufficient and specifically that causes were not sufficiently broken down into detail. To achieve a deeper level of understanding of the causes of failures, discussions with maintenance personnel were conducted. This resulted in a deeper level of knowledge about the causes of failure for the 11 kV cable component.

6.3.4.2 Recommendations for the Documentation System

The need for and benefits of reporting more as well as better information have been identified, and could lead to, for example, a more complete use of the reporting system.

The following list summarizes some of the resulting recommendations from discussions with maintenance personnel as well from the investigation of the current reporting system.

- Report information for components treated individually. This would help to easily identify trends and possible actions, for example, replacement of a type of joint that appeared to be prone to failure.
- Use a checklist and visual inspection during failure events. Include useful information about the environment to further support the reporting of failure incidents (see Section 6.3.3.2).
- Incorporate several information systems to create a comprehensive understanding of the system and its components, for example, include information about: material, maintenance times, remarks, and actions.
- Improve feedback and communication within the different active parts of the reporting procedures, for example, by sharing and discussing the statistics resulting from the reporting systems.
- Provide the necessary support for the organization to assist in long-term work with documentation led by their own expertise.

In summary, it was proposed that an extensive documentation system be created to provide decision-making support for maintenance planning, and furthermore that long-term responsibility for this system be prioritized in the organization. The motivation for this is and must be based on economics. This paper has shown the benefit of such a system in providing decision support for taking the right maintenance actions for components (RCM). This in turn would help make optimal use of the economic resources that would be an incentive in regard to requirements from authorities and owners.

6.3.4.3 Application of the Results in RCM

The data and knowledge obtained from the survey can be put into an RCM and RCAM framework with the aim of achieving an optimal maintenance strategy. Main results, in the format of an FMEA, from this cable application study are shown in Figure 6.4. It should be noted that the intention is not to suggest this as the only solution for the cable component, but to provide one possible solution for a real component based upon real data.

The cable application study can be summarized as follows. The cable system has one main function, being to deliver electrical power. This function can fail in one of two failure modes: (i) short circuit (e.g., one/two phase/s to earth) or (ii) open circuit (e.g., interruption in supply). Furthermore, a failure event leads to failure of function. A short circuit is caused by imperfections in the insulation, in other words by insulation failure. A failure event occurs due to a failure cause. An insulation failure is due to deterioration in the insulation and could, for example, be caused by water treeing in XLPE cables.

The results show the logic between detailed knowledge about failure causes and system function failure. In this application using an 11 kV cable system, the conclusion is that 59% of causes of failure are due to material and method and of these 34% have known causes. These failure causes were further identified in detail based on knowledge from maintenance personnel, which would suggest that these could be reduced by PM. Consequently, the application of these results could be used to show the benefit of preventing component failures with the right maintenance activity.

A reliability analysis of the 11 kV system defined (LH11) was made and presented in [117]. The results clearly demonstrate the system reliability benefits obtained by focusing maintenance on the critical components, and on the dominant causes of failures on which maintenance can have an impact. This forms a significant input to achieving cost-effective maintenance programs. Furthermore, this was supported by the use of data and information about the specific system as a result of the comprehensive survey of causes of failures as presented in this paper. Consequently, this survey has shown a contribution for improving PM plans.

6.3.4.4 An Analysis of the Maintenance Situation

As a result of the cooperation with the local DSO, an understanding of the current maintenance situation has been gained and is presented in the following analysis.

The following three major parties are involved in this provision of service.

1. The authority that regulates the market and supervises the utilities in fulfilling their certain requirements.
2. The customer who receives energy supplied via the electric distribution system.
3. The utility that owns, operates, and maintains the distribution system.

The following points summarize the changing situation they face.

- Deregulation $\Rightarrow$ market conditions $\Rightarrow$ requirements for effectiveness, for example, the benefit in reliability of a PM measure must balance the cost of applying the same.
- Market conditions $\Rightarrow$ new ways of handling a company effectively, for example $\Rightarrow$ larger utilities mean more assets resulting in the employees no longer know their systems, instead face new routines and systematic approaches $\Rightarrow$ subcontracting out maintenance; a need to specify in detail what the needs and benefits are.
- Customers and society require generally higher levels of availability than experienced in the past but also clearly visible incentives in receiving a certain availability at a certain cost, that is production of cost incentive goods.
- Authority supervises, and compensates or penalizes the utilities depending on how they meet their requirements and fulfill their obligations towards their customers.
- Power demand in most Western countries is on the increase but not at the same rate as before, which implies a decline in the rate of extension of power systems and the increasing age of the existing ones.

Conclusions about the maintenance situation:

- Requirements for PM measures will increase in the future
- There is a requirement for quantifying the benefit of expenses made, for example, PM in cost and reliability
- A systematic methodology is required to support PM strategy decisions

6.4 XLPE Cable Insulation Behavior

This section presents cable component behavior focusing on the degradation of the insulation due to the water-tree phenomenon, and also on the diagnosis and PM methods used to characterize, measure and prevent these. Consequently, the polymeric insulation material that they occur in is also focused on, especially, XLPE. The understanding of and the supporting data for component behavior presented in this chapter provides input data for the modeling and analysis of cable component behavior presented in Section 6.6.

6.4.1 Introduction

The main objective of the analysis in this book is to support an effective PM plan. At this point of the analysis, this issue relates to cable insulation and defining an optimal choice of PM method to prevent failures caused by water treeing. This chapter aims to provide a short introduction to the area of cable insulation in XLPE cables, though enough detail to support the following analysis that uses these results. For a comprehensive presentation of the different phenomena and relevant research, the interested reader is referred to more specific literature.

Electrical insulation as indicated in the introduction to the cable system component (see Section 6.2) is both a crucial and a complex area for the understanding of the cable behavior. Moreover, cable insulation failures are the effects of different aging mechanisms in the insulation material. This matter has been the subject of detailed studies and research, for example, CIGRE has published several reports on the subject of aging factors and diagnostics of cable systems [118–120].

Extensive research has been undertaken by the Electric Insulation Group at KTH. The original aim was to develop a method for defining the remaining lifetime of XLPE cables. The project resulted in the development and implementation of a tool, where permittivity and losses are measured as a function of voltage and frequency. This technique is suitable for measuring cables in the field as well as samples in the laboratory. Field-aged cables have been analyzed both in laboratory and field. In addition, a method for growing and analyzing water trees in small XLPE samples has also been developed. The aim is to continue the research and for the techniques developed to be put to practical use by the power utilities. The following book and publications summarize the results [122].

6.4.2 Aging Mechanisms

6.4.2.1 Aging Factors and Effects

An overview of different aging mechanisms and diagnostic methods for both XLPE and paper-insulated cables can be found in [89]. That study suggested that aging factors that affect cable insulating include the following (examples of each are provided in brackets): thermal (temperature cycling), electrical (voltage), environmental (humidity), and mechanical (bending). Moreover, an attempt has been made to define possible aging mechanisms for the cable system relating to these four different aging factors. One of these is the phenomenon of water treeing which is said to be caused by either electrical or environmental factors. Moreover, these aging factors are said to cause the aging mechanisms of electrical treeing, water treeing, and partial discharging (PD).

Important to note here is that some of the aging mechanism may take several years to develop while others may occur within days or weeks [89]. Another issue to note here is that this area is very complicated, which has led to a number of different theories, and no single correct model for aging exists.

6.4.2.2 The Water-Tree Phenomenon

Water treeing is considered to be the most important aging phenomenon in polymeric insulated cables [90].

6.4.2.2.1 The Water Tree Concept

The concept of a water tree is used to refer to structures in the polymeric insulation said to resemble the shape of a tree. Two types of water trees can be distinguished, referred to as bow-tie and vented trees, and are described later.

Bow-tie trees start to grow within the insulation and grow along the electrical field lines. The initiating point is usually a void or impurity in the insulation. The growth is rapidly from the initiating point but decreases dramatically and can almost stop at the end. This growth process implies there is a limited length of bow-tie trees and therefore these trees are seldom considered to be the cause of insulation failure.

Vented trees (also referred to as boundary or bush trees) start to grow from the interfaces between the insulation and the semiconducting screens. Consequently, these trees are in contact with the screens so water and impurities can easily diffuse into the water-tree structure. The initiating point is often difficult to find. Possible initiating causes of these include mechanical damage to the insulation surface, voids and irregularities or heterogeneities in the semiconductor screens. The growth of vented trees can be described in three stages: (i) the inception stage before growth begins, (ii) a rapid growth stage, and (iii) a slowing down in growth stage. This growth pattern can allow water trees to grow through the entire thickness of the insulation. Therefore, the vented trees are considered to be the more dangerous of the two types of water trees.

FIGURE 6.5
An example of a cable exhibiting vented water trees, where some of these have grown through the complete thickness of the cable insulation. (From Holmgren, B., Dielectric response, breakdown strength and water tree content of medium voltage XLPE cables. Licentiate Thesis KTH, Stockholm, TRITAEEA-9705, with permission [90].)

The following analysis and discussion sections of this book refer to vented trees, unless stated otherwise. Figure 6.5 shows an example of vented water trees.

6.4.2.2.2 Properties of Water Trees

It has been observed and is generally accepted that water trees reduce the 50 Hz, 0.1 Hz, DC, and impulse breakdown strength [108]. Though the water itself appears as an insulating material, cable samples with water trees bridging the whole insulation still exhibited a breakdown strength above service stress 2 kV/mm [90]. The breakdown voltage of water tree-deteriorated cables can he restored by drying the insulation. Breakdown voltages of up to approximately 50% of the original level have been reported. However, the trees do not disappear, and when water returns the breakdown voltage reduces again [90].

6.4.2.2.3 Water Tree Mechanisms

The following are the two most commonly only presented water tree growth mechanisms.

1. The electrochemical mechanism assumes that the water treeing is a result of chemical reactions that are strongly enhanced by electrical stress. The initiating process is caused by water entering the

regions of the polymer. This is followed by chemical reactions and electrolysis reactions at the border of the PE and water that lead to the degradation of the PE. This degradation causes oxidation and polar groups are created which enhance diffusion and further degradation take place.

2. The electromechanical mechanism assumes that water treeing is caused by Maxwells forces which are high enough to damage the PE.

6.4.2.2.4 Artificial Aging

Water treeing under service-voltage conditions is a very slow process. This means that accelerated or artificial aging is required in measurement studies. Application of such a procedure that accelerates the aging must also result in conditions that reflect the effects that would normally occur during service conditions. The parameters normally used to accelerate aging include application of electric stress, frequency, and temperature. Internationally proposed and generally accepted artificial aging factors include, for example, 30°C and $2.5U_0$ [90].

6.4.3 Diagnostic Measurements

The purpose of diagnosis is to evaluate and locate degradation phenomena causing cable failure [89]. Diagnostic measurements can be done using either destructive or nondestructive tests, depending on whether the measured object is destroyed by the test or not. Furthermore, these can be made either on-site or in the laboratory (where accelerated aging can be achieved). Diagnostic tests measure or monitor one or more properties of the insulation that are related to aging or failure, for example, a failure occurs when the property reaches a critical level after a certain age or aging period.

6.4.3.1 Diagnostic Methods

6.4.3.1.1 Nondestructive Method

Figure 6.6 provides a simple illustration of both (i) the problem and (ii) the method used to measure the status of water-tree affected cable insulation. Furthermore, this figure summarizes the method that has been developed and implemented by the former Insulation Research Group at KTH.

The upper part of the same figure (the process) can be summarized as follows. Due to the entry of water into the insulation, the phenomenon of water treeing is initiated. Water trees start to grow through the cable insulation, which in turn leads to a decrease in the breakdown voltage in the cable. The resulting effect is therefore a deterioration in the condition of the cable. To achieve effective maintenance and operation of the cable system, it is desirable to predict the remaining lifetime of the cable. For example, when a cable faults the decision has to made to either replace or repair the cable, and this

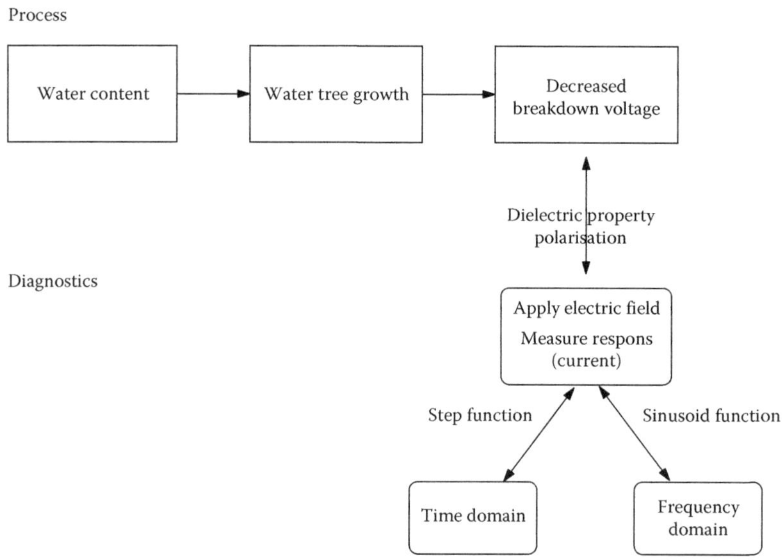

FIGURE 6.6
Illustration of the dielectric diagnostics of water-tree affected electric insulation, with two alternative functions for the electric field applied.

decision is influenced by the overall condition of the cable. Consequently, there is a need for a method able to diagnose cable condition.

The lower part of the same figure illustrates the diagnosis method that uses dielectric diagnostics of the insulation to gain an understanding of the condition of the insulation. Therefore, information has to be gathered to determine this, which is done by exciting the insulation and measuring the response. Importantly, a property that reflects the general status of the insulation should be used for these measurements. It has been proven that a change in the insulation has dielectric properties [121]. Moreover, these properties can be effected by dielectric polarization. If the polarization is changed in time it gives rise to an electric current. This current could be measured in a diagnosis technique to indicate insulation status.

The electric field that the insulation is subjected to can of course be any of a number of different types. Two of the most commonly used are voltage by a step function (e.g., Heaviside's step function [48] and voltage by a sinusoidal function [121]. Moreover, for these two different assumed functions, the two different methods for measuring the current are obtained in either the time domain or the frequency domain. Just as the name suggests, these methods are either performed as functions of time or frequency.

The frequency domain method has been discussed further here, and the results have been used in the analysis found in Section 6.6.

The resulting response current (which has phase and amplitude) can after some mathematical analysis [121] be measured through the use of the permittivity (ϵ), which is defined in complex form as follows:

$$\epsilon(\omega) = \epsilon_0(\epsilon_r' - i\epsilon_r''') = (\epsilon_r' - i\epsilon_r''') \tag{6.1}$$

where ϵ_r' is the real part of the relative permittivity, therefore, $\epsilon/\epsilon_0 \ - \ \epsilon_r''$ is the imaginary part of the relative permittivity (ϵ/ϵ_0). The imaginary part of the relative permittivity is also referred to as dielectric loss.

Furthermore, the ratio between the loss part and the capacitive part is given by

$$tan(\delta) = \frac{\epsilon''(\omega)}{\epsilon'(\omega)} \tag{6.2}$$

which is defined as the loss factor.

This method has been developed, implemented, and used for experiments utilizing the frequency domain method, in high-voltage applications ($U_0 = 12$ kV), with the frequency interval (0.0001–100) Hz, as presented in [121]. The resulting method is called dielectric spectroscopy.

Results from these experiments showed that non-water-treed cables had low losses and linear responses. However, the water-treed cables had increased losses and significantly nonlinear responses both in capacitance and losses [90]. This method can consequently be used for distinguishing between non-water-treed and water-treed cables, and how to relate this effect to insulation status. The method has been further developed and implemented as presented in [90] and [91]. The following relationship is presented for the analysis of the nonlinear response:

$$\begin{aligned} \Delta\epsilon'_{nonlin} &= \epsilon'(U_o) - \epsilon'(0.5U_o) = \Delta\epsilon!(U_o) - \Delta\epsilon'(0.5U_o) \\ \Delta\epsilon''_{nonlin} &= \epsilon''(U_o) - \epsilon''(0.5U_o) \end{aligned} \tag{6.3}$$

6.4.3.1.2 *Destructive Methods*

Two generally used "destructive" methods have been presented in following paragraphs.

Electrical breakdown testing can be done using AC, DC, or impulse voltages. This is a simple method and one of the most widely used. The most common tests are done using impulse and AC impulses at 50 Hz. A limitation with this method is that it requires many samples [89].

Water-tree testing is done on thin slices of cable, about 0.1–1 mm in thickness. Before doing the analysis, the trees must be made visible. In the method used in [90], the slices are stained and then investigated using a stereo microscope, or with the aid of image analysis using a video camera connected to a computer. To simplify the tedious measuring work, an evaluation method

TABLE 6.14

Diagnostic Criteria for Defining Cable Status

Category	Normalized Breakdown Voltage [V]			Comment about the Cable
Good		U_{bd}/U_0	>4	No measures necessary
Risk	2.5 ≤	U_{bd}/U_0	≤4	Risk zone keep for observation
Bad		U_{bd}/U_0	≤2.5	Poor condition replaced ASAP
Bad/leakage		U_{bd}/U_0	≤2	Very poor condition replaced ASAP

Source: Holmgren, B., *Dielectric response, breakdown strength and water tree content of medium voltage XLPE cables.* Licentiate Thesis KTH, Stockholm, TRITA-EEA-9705, ISSN 1100-1593, 1997. [90]; Werelius, P., *Development and Application of High Voltage Dielectric Spectroscopy for Diagnosis of Medium Voltage XLPE Cables.* Doctoral Thesis KTH, Stockholm, TRITA-EEK-2001-02, ISSN 1650-674x, 2001. [91]

based on the assumption that it is the length of the longest water tree that is of critical importance has been developed [124].

All investigations in [90] were made on field-aged water trees. One reason for this was that it had been noted when using the previously presented non-destructive method and the breakdown test method that these gave different results in laboratory-aged water trees.

6.4.4 Preventive Maintenance in Cables

6.4.4.1 Replacement Schemes Based on Diagnosis Methods

The nondestructive diagnosis method presented earlier and summarized in Figure 6.6 has been used together with a criterion for the condition of the cable to define replacement schemes for cables. This means that the result from the diagnostic measurements together with the resulting loss factor was used for the classification of the status of the cables. Results from these studies are presented in [122]. This method is now a commercial product. Table 6.14 summarizes these criteria.

A report by Elforsk [125] presents a comparison of four commercial methods based on the diagnosis of the condition of the insulation and water tree aging for different cable types. The report also indicated which type of information is essential to be collected in a system for a continuous diagnosis of component condition.

6.4.4.2 Rehabilitation Method for Water-Treed Cables

It has been shown that deterioration of cable insulation due to water treeing can be prevented by a rehabilitation method. This method for rehabilitation of XLPE cables with long water trees has been developed and patented in the USA [126]. A large amount of XLPE cables have been rehabilitated using this method in both Europe and the USA, totaling some 700–800 km of three-conductor cable. The utilities that have utilized the method have also to a large extent continued to use it in their installations and infrastructure [126].

This method has also been applied and evaluated in by the Foundation for scientific and industrial research at the Norwegian Institute of Technology (SINTEF) in Norway as presented in [127]. The same study includes application studies from the field and laboratory, and evaluation of the water-tree growth after rehabilitation (laboratory study). A full-scale failure test done at Sandnes Energi provides useful experience and results for the method. An aging process is studied 2 years after PM activities in laboratory studies. It has been verified that rehabilitation leads to a significant improvement in both AC and impulse voltages.

6.4.4.2.1 *Rehabilitation by Silicon Injection*

The main principle in this rehabilitation method involves injecting a silicon-based liquid between the wires in the conductor. It is, therefore, not possible to rehabilitate cables with massive conductors or cables with swallow powder in the conductor.

Rehabilitation is brought about when the component in the silicon-based liquid diffuses in and is absorbed into the semiconductor and the XLPE insulation. With heating during current loading, this rehabilitation measure works quicker since the diffusion speed increases with temperature.

When the rehabilitation liquid comes into contact with water from the water tree in the insulation, a chemical polymerization reaction takes place, and the liquid consumes the dampness and the diffusion of the molecules becomes slower [127].

6.4.4.2.2 *SINTEF Experience with This Rehabilitation Method*

Results from the rehabilitation of three major water-tree-aged XLPE cables (24 kV $3 \times 150\,mm^2$) performed at Sandnes Energi are presented in [127]. The cables were manufactured and installed in 1975. One of the lengths experienced a supply interruption. Using water-tree analyses on pieces of the cable near the failure, it could be ascertained that it contained vented water trees from the outer semiconductor of tape that equated to some 95%–99% of the insulation strength over a 5.5 mm distance. The cables that were rehabilitated were consequently strongly water-tree-aged. To obtain the best possible rehabilitation outcome, the cable was heated in DC current.

The results from the voltage strength testing of the cable at different points of time after the rehabilitation have been presented using, for example, a 5m length of cable containing water trees that was tested at a strength of $U_{g}j = 30\,kV$. After the rehabilitation, a cable with water trees could carry $U_{g}j = 156\,kV$, which implies that the electrical strength had been improved by 380%.

TABLE 6.15

Cost Values for the SINTEF Evaluation of the PM Method for Water-Treed XLPE Cables

Utility	Cable Length [m]	Cost for Rehabilitation [NKR]	Cost for Renewal [NKR]
Sandnes	1700	500,000	770,000
Fredrikstad	650	170,000	370,000

Source: Faremo, H., Rehabilitation of XLPE Cables with long Water-trees (Rehabilitering av PEX- kabler med lange ventilerte vanntrae fra ytre halvleder. Feltog laboratorieforsok.), *in Technical Report, Energiforsyningens Forskningsinstitutt (EFI)*, Trondheim, Norway, 1997. EFI TR A 4512 (in Norwegian), October 1990. [127]

Note: SEK 1.5≈ NKR 1 and SEK 10≈ Euro 1.

A common procedure used to evaluate the condition of an XLPE cable is to repair the first and often also the second failure without making any special arrangements. As a rule, failures in XLPE cables produced after 1982 are one-off occurrences. If the cable continues to fail, the utility evaluates other alternatives such as sampling for water-tree analysis, replacement, rehabilitation.

It is desirable to develop secure nondestructive methods for the diagnosis of cables for rehabilitation [127].

Laboratory aging of XLPE cables has shown that water-tree growth begins again after rehabilitation.

Rehabilitation liquid has a positive effect on the impulse strength, the initial voltage of partial discharges, and electrical trees.

The dielectrical losses increase at rehabilitation. This can have an impact on the performance criterion of nondestructive tests on rehabilitated cables. This is further discussed in Section 6.6.

6.4.4.2.3 Economical Benefits of Rehabilitation

The costs of rehabilitation compared to renewal of cable have been evaluated in two practical cases: (i) the Sandnes Energi case and (ii) a Fredrikstad Energiverk installation [127]. The utilities have provided costs for these relatively dissimilar installations, shown in Table 6.15. It can be seen that the rehabilitation cost is less than the replacement cost in both cases.

These calculations do not consider differences in quality, manufacturing, and installation methods between cables manufactured in the mid-1970s and those of today and in other words the loss in profit by having to install new cables.

The conclusion here is a recommendation to the utilities to adopt the rehabilitation method, because after evaluation, it has been shown to be an economically and technically beneficial method.

6.4.5 Survey about XLPE Cables Affected by Water Treeing

6.4.5.1 Introduction

In 1981, a working group was established with the objective to collect and analyze data on damaged XLPE cable stations. This working group delivered a report at the end of 1985, which included all known types of faults on XLPE cable in 12 and 24 kV systems, that had occurred during the period 1965 to the beginning of 1985. This report established that the failure rate was at an acceptable level, and that "the expected lifetime was essentially in excess of 15 years." Furthermore, it recommended the continued evaluation of the XLPE cables. In January 1988, another working group was appointed with the objective to determine lifetimes and the uses of XLPE cables, and to recommend any steps that needed to be taken. This section introduces the main features of this survey [94], from which data has been used for further analysis.

6.4.5.2 Collection of Basic Data

As discussed earlier in Section 6.2.3.3, it was established that the disturbance statistics available from Swedenergy (formerly the Swedish Power Utility Association) were too incomplete to be used as a source of data for the investigation. Therefore, the working group decided to perform a national data collection procedure for failures occurring in XLPE cable systems that would include failures at joints and terminations. Furthermore, it was considered that all failures should be included, even though they were caused by, for example, digging, and that it would include operation voltage levels of 12 kV and higher.

In March 1988, the request for information was sent out to 257 utilities. Of these, some 251 responded. The utilities that had reported failures received a questionnaire asking for a detailed description of these failures. The quality of the failure descriptions was of course varying, and in same instances failure descriptions were not received at all.

6.4.5.3 Analysis of the Statistics

A cable system is assumed to consist of a three-conductor cable or three one-conductor cables, joints, and terminations. A failure is defined as one or several failures that occur in the same cable system and implies voltage interruptions. Furthermore, the cable faults were grouped into two different causes of failures, described later.

1. Insulation material failure caused by treeing including:
    - Faults established by the manufacturer as treeing.
    - Repeated inexplicable faults of a cable system.

- Occasional inexplicable faults where the cable system had been in operation for more than 1 year and the working group considered the likelihood of tree faulting as high.

2. Other cable faults considered including:
    - Manufacturing faults that include faults established by the manufacturer such as production faults, as well as inexplicable faults occurring within less than a year after being put in operation.
    - Inexplicable faults that include faults occurring after a year in operation and that were considered not to be insulation material failure caused by treeing.
    - Failure caused by damage that includes digging and skewing and such like, and handling and laying not following the appropriate norms.

6.4.5.4 Results

The average failure rate of the XLPE cable systems in total has been shown to be somewhat more than half of that of the impregnated paper cables. The failure rate for modern XLPE cables developed after 1975 is only one-seventh of the failure rate of the total XLPE cables in the study. With this background, it is concluded that the XLPE cable is expected to reach the same lifetime expectancy as impregnated paper cables.

This also shows that the inclusion of the extruded outer conducting layer that protects the insulation from water penetration has been able to significantly reduce the number of failures. Consequently, the results indicate that the problem with water trees was significantly reduced with the introduction of extruded layers.

An interesting question is, "What was the cost of this benefit?" However, the issue has not been treated in the study.

These statistics have been analyzed further and used in the modeling of failure rate presented in Section 6.6.

6.4.5.5 Conclusions and Further Analysis

The conclusions from the Svel study are that treeing occurs primarily in failures in cables that were manufactured before 1974 and that have an outer semiconductor of tape (this is before the introduction of the extruded layer), are low loaded and have dampness or water in the conductor, and that these often occur in connection with overvoltages in the network, for example, due to lightning.

Another report was presented analyzing the same issue in 1995 [128], where statistics for the improved XLPE cables were evaluated. The result showed during a 5-year period no water treeing failures for these new type of cables, with the extruded outer layer.

6.5 RCM Application Studies Using Approach I

This section presents the results of a reliability analysis on two different power distribution systems: (i) a rural system referred to as Flymen and dominated by overhead lines and (ii) an urban system referred to as Birka system with mainly underground cables. This analysis focuses on relating system reliability to the PM of the included components such as lines or cables. In the approach presented, referred to as Approach I, it has been assumed that the impact of maintenance results in a percentage reduction in the causes of component failure.

6.5.1 Approach

The approach used in the RCM analysis of the two case distribution systems has been presented below. This approach involves running through all the steps defined in the RCM methodology presented in Table 4.1, except for the last step, which is the cost and benefit analysis analyzed in detail in Section 4.2. The overall approach has been outlined below.

1. Choose and define a system to analyze, consisting of
   - Boundary
   - Included components
2. Collect, select, and define input data:
   - System network data
   - Customer and power data
   - Component reliability data
3. Evaluate system and load point reliability with RADPOW:
   - Identify critical components
   - Identify critical load points, voltage levels, and so on
4. Evaluate component behavior:
   - Conduct a failure mode and effect analysis
   - Collect statistics and experience about the causes of failures
5. Model the effect of PM on reliability.

   Approach I has been used in accordance with Chapter 4 and can be described as follows. It is assumed that a PM activity results in a percentage reduction in the causes of failures for affected components. Furthermore, it is also assumed that the failure rate is reduced by an equivalent percentage. The resulting model of the relationship between failure rate and PM is referred to as $\lambda(PM)$.

6. Evaluate the effect of PM on system reliability.

6.5.2 Flymen: A Rural Overhead Line System

This section describes the analysis and presents the results of a reliability and maintenance study done on a rural Swedish distribution system called Flymen. Results from this study have been published in [132].

6.5.2.1 Definition of the Flymen System

6.5.2.1.1 The Flymen Network

The Flymen network is a 10-kV distribution system located in Blekinge in southern Sweden and belongs to the E.ON Elnat distribution area network. The Flymen network supplies just over 900 customers at 112 load points, with a total energy demand per year of 8890 MWh. The Flymen distribution system is supplied via two 4 MVA 20/10 kV transformers, which constitute a reciprocal reserve for each other. The system consists of four outgoing feeders, normally fed radially from the transformer station in the Flymen network. This network has coupling possibilities via disconnectors (see Figure 6.7). The Flymen network includes manual and automatic disconnectors. Moreover, it is a simple country distribution system compared with the more complicated urban networks.

The maintenance policy covering the Flymen network follows the investment and maintenance policy for local networks in E.ON [133]. The maintenance of the overhead lines in streets is included in a separate document [134]. The aim of the maintenance efforts is to keep the equipment in a state that fulfills the requirements for operation and personnel safety at the lowest possible cost [133]. According to E.ON, there are no special maintenance problem areas or where failure occurrences are especially high in the Flymen system. However, there are major changes underway in maintenance planning at E.ON and at other utilities. This could indicate that there are weaknesses in the existing procedures for maintenance planning. A new process-oriented approach to this work is being developed at E.ON, and a new policy for maintenance is being drafted. Life-cycle cost (LCC) is to be used as a tool for analyzing the inspection results according to the policy [28].

6.5.2.1.2 The Flymen Model

As can be seen in Figure 6.7, the Flymen network can be separated into northern and southern parts, which are independent of one another in respect to

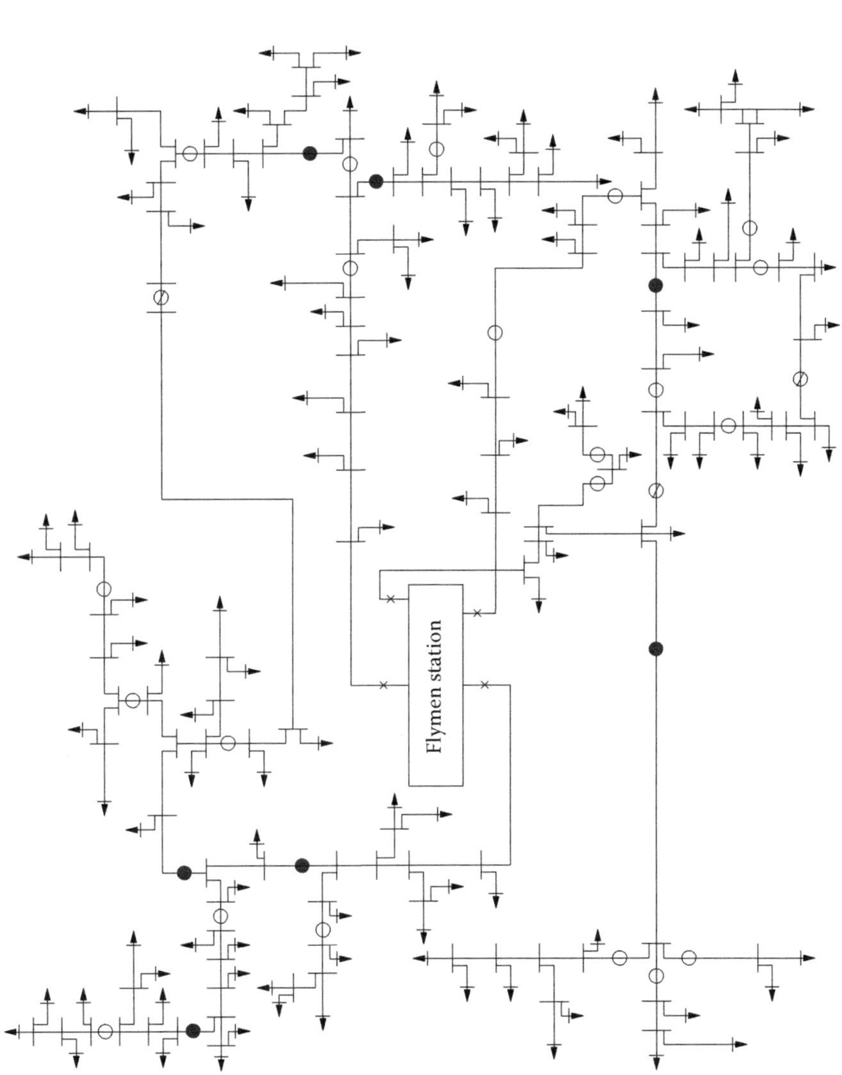

FIGURE 6.7
The Flymen system.

reliability. Therefore, distribution in these two parts can be analyzed separately. Since it is desirable to make the analysis on a limited system, one of these parts has been chosen for this analysis, being the northern part referred to as FlymenN (see Figure 6.8). The system is modeled from the supply point to the load points, and the Flymen station is not included in the analysis. Figure 6.9 illustrates the components included in the analysis. Note that the component type referred to earlier as a bus is called a junction here. The

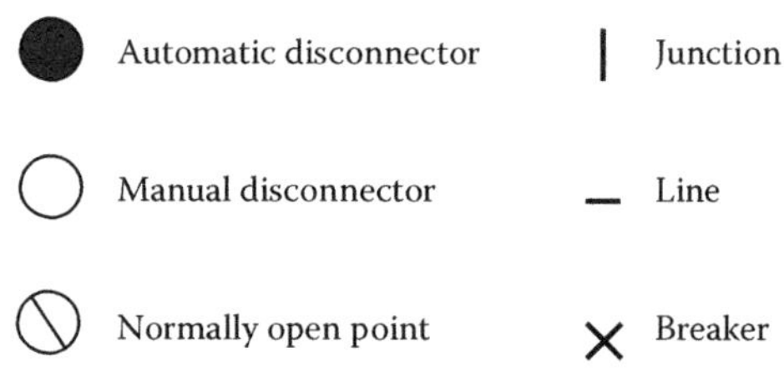

FIGURE 6.8
Component types in the Flymen system.

system is defined by its load points and included components (as illustrated in Figures 6.10 and 6.11). All these points have been identified by a number.

As can be seen in Figure 6.8, two load points (denoted 295 and 296) have been introduced that are not included in the Flymen network. These load points are the dummy load points referred to previously for ensuring correct deduction of failure events for other load points, but are not themselves included in the reliability analyses. By introducing these two load points, the failure of any component in the normally open path between the subsystems FlymenNW and FlymenNE (via branches 80 and 137) has been included in the analysis.

6.5.2.2 Network, Power, and Customer Input Data

The network, power, and customer input data for the reliability analysis of the Flymen system have been provided by E.ON, and are presented below.

1. Connection diagram of the network for the Flymen station including:
   - Load points with individual notations
   - Section (switching) points with disconnectors
2. Operation diagram showing the topology of the network including:
   - Load points with individual notations
   - Section (switching) points with disconnectors
3. Data for the overhead lines including:
   - Connection points
   - Length
   - Type
   - Impedance data
4. Table of the included load points including:
   - Name and number
   - Number of customers

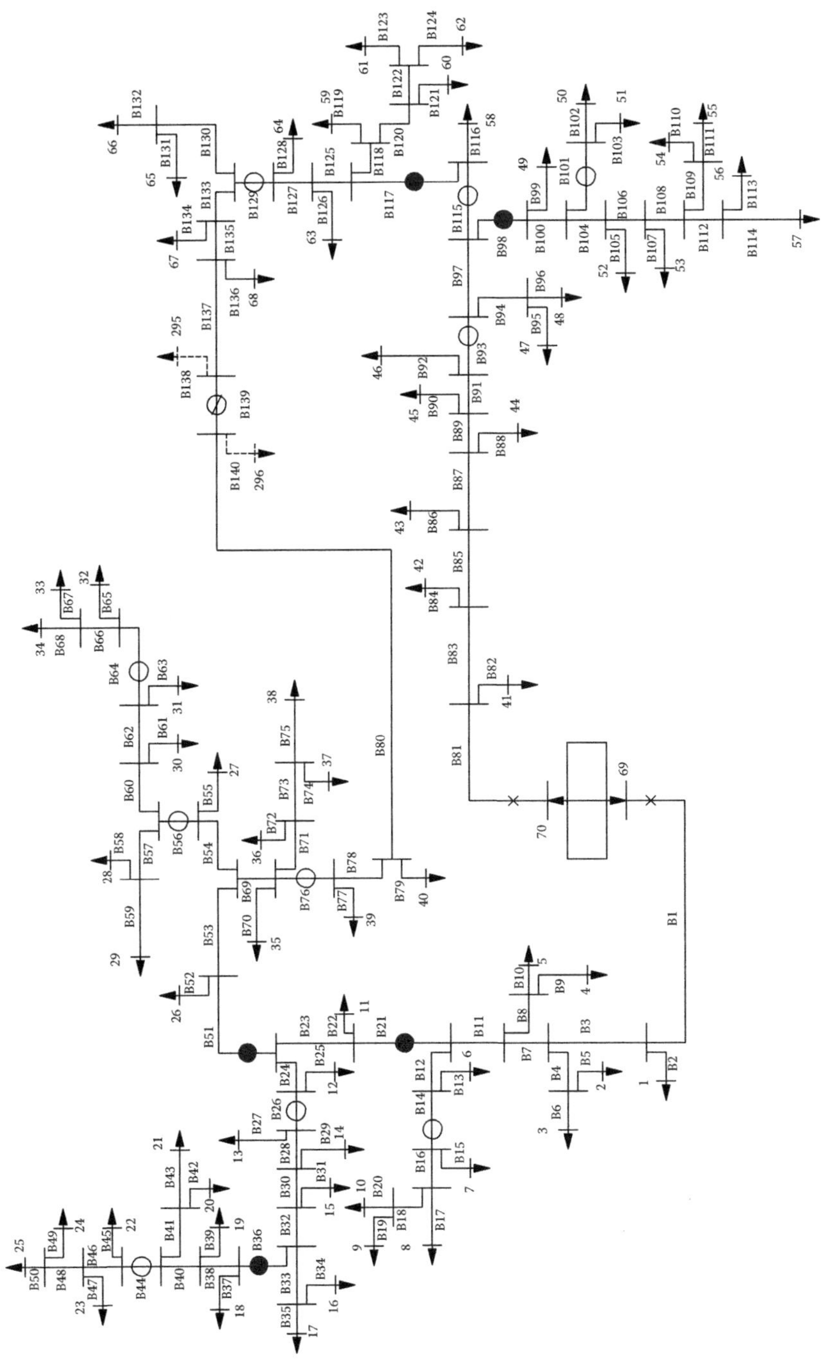

FIGURE 6.9
The FlymenN system.

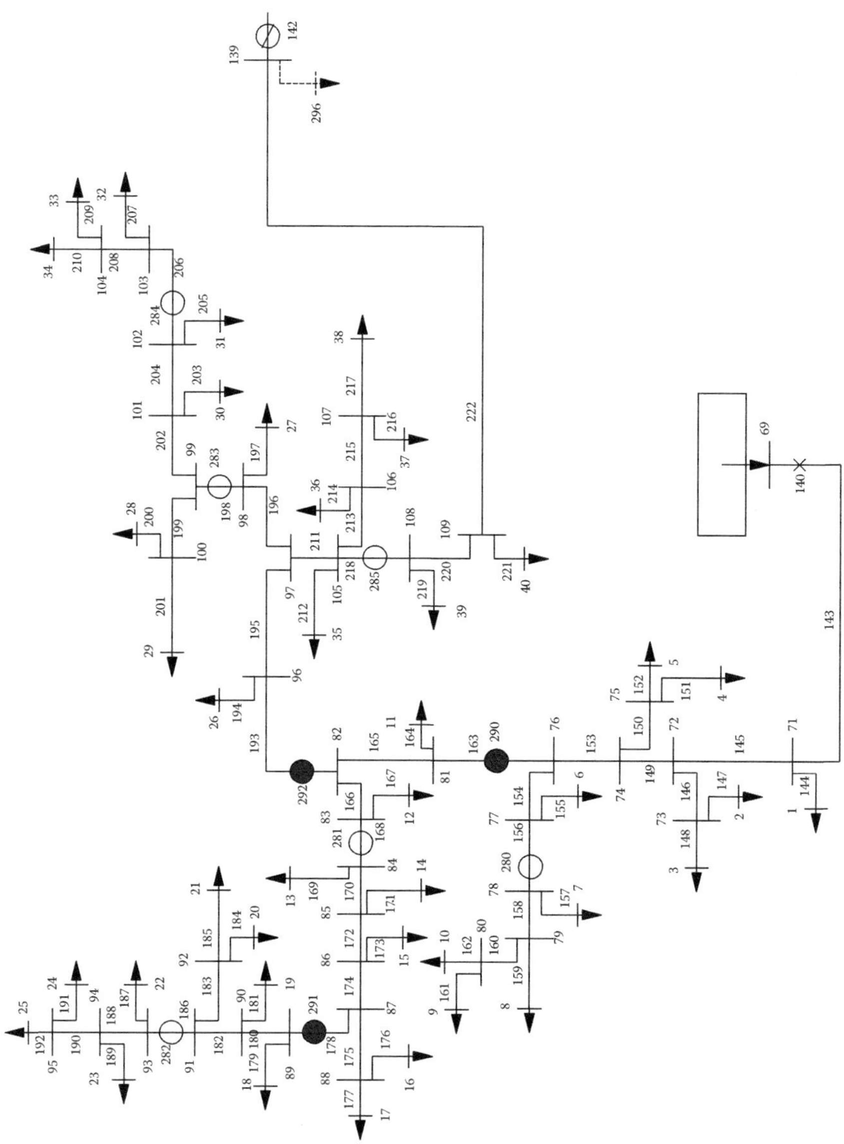

FIGURE 6.10
The FlymenNW system.

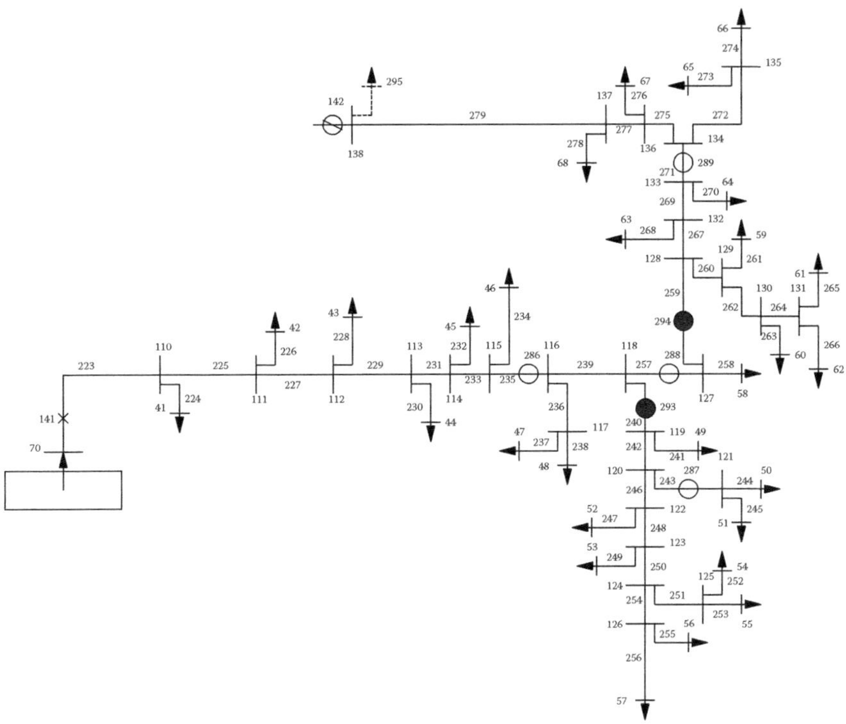

FIGURE 6.11
The FlymenNE system.

- Yearly energy consumption [kWh]
- Type of customers connected to each load point by:
  - NE: bungalows, summer cottages, small farms, poultry farm
  - NW: bungalows, summer cottages, small farms, country shop

The network topology input data for RADPOW can be summarized as follows:

- 140 branches and 296 components
- Two supply points (components 69 and 70)
- One normally open point (component 142)
- Component types (according to Figure 9.2)

The customer and power data-topology-input-data for RADPOW can be described as follows. The customers are mainly residential, but as can be seen in the input data from E.ON presented earlier, there are also other categories of customers. From the Flymen input data (see Appendix B), it can be seen that four load points have a distinctly higher power demand per customer than the other load points. Therefore, it has been assumed that these load points (6, 7, 24, and 68) are industrial customers. These data have been divided into different categories of customers (see Appendix B for a summary of these).

Further information of interest from E.ON is that there are two proposed switching points in the northern part of the network:

- NW: cno 280 (F14-2870) and cno 142 (F142875)
- NE: cno 287 (F14-2880)

These proposed points could possibly effect the future reliability of the Flymen system.

6.5.2.3 Reliability Input Data

E.ON has provided reliability data for the Flymen system which has been summarized below.

1. Switching time:
    - Manual switching—1 h
    - Remote control—10 s ≈ 0.17 h
    - Automatic switching—5 s ≈ 0.083 h
2. Failure rate:
    - Line—1.8 f/yr per feeder
    - Breaker—0.005 f/yr
    - Manual disconnectors—0.05 f/yr
    - Equipment for remote control—0.0005 f/yr
    - Equipment for automatic switching—0.001 f/yr
3. Repair times:
    - Line—3 h
    - Breaker—12 h
    - Remote or automatic disconnectors—6 h
4. Probability of nonfunction via:
    - Remote—0.0005
    - Automatic—0.0014

Further input data required by RADPOW have been defined based on the following assumptions.

- Based on the failure rate for the disconnectors, the following assumptions have been made for the failure rates: (i) remote disconnectors $0.05 + 0.0005 = 0.0505$ f/yr and (ii) automatic disconnectors $0.05 + 0.001 = 0.051$ f/yr.
- Failure statistics for the load points have been estimated as follows. During a period of 3 years (1996–1998), 900 failures for E.ON's total number of load points (approximately 17,000) have been registered. This gives $900/(17{,}000 \cdot 3) \approx 0.018$ f/yr per load point (bus bar type).

The statistics E.ON have today are not broken down to component level. This will be changed in their future maintenance planning, since it is necessary for providing correct analyses.

6.5.2.4 Definition of Failure Rates for Overhead Lines

This section analyzes and defines some additional reliability input data for the Flymen system, including different failure rates.

6.5.2.4.1 Evaluation of Permanent Failure Rates for Overhead Lines

The failure rates for overhead lines have been provided by E.ON as the average failure frequency per feeder. However, input data of RADPOW for failure rates are for each component, that is for each line. Therefore, this average value has had to be converted into failure rates per km for each feeder. Since the feeders are of different total lengths, the result is a different failure rate per km and feeder. A more correct way statistically would have been to use the same failure rate per km for both feeders, since the statistics are known to be very limited. It should, however, be remembered that these failure rates are an overall average, and there will be differences occurring in the actual rates of different feeders due to differences in terrain, tree coverage, weather patterns, and so on. Therefore, it is reasonable to incorporate this effect by having different values for different feeders, which was the approach used in this analysis of the Flymen system.

Assuming that the failure rate for the feeders is proportional to its length, then the following E.ON data are known:

- The average failure rate per feeder is 1.8 failures per year
- The lengths of individual overhead lines

With this information, calculations have then been made to obtain an average failure rate for different lengths of overhead lines. FlymenN consists of two main feeders: (i) the north eastern (NE) and (ii) the north western (NW). The total length for these feeders was deduced and the resulting average failure rates per year and kilometer were evaluated as follows.

TABLE 6.16

Failure Rates for Overhead Lines and Feeder in the NW Subsystem with 0.046 f/yr and km

Line Type	Average Length [km]	Average Failure Rate [f/yr]	Number of Lines	Resulting Failure Rate [f/yr]
Line 1W	0.125	≈0.006	33	0.198
Line 2W	0.375	0.017	15	0.255
Line 3W	0.625	0.029	13	0.377
Line 4W	0.875	0.04	10	0.4
Line 5W	1.125	0.05	3	0.15
Line 6W	1.375	0.06	7	0.42
Total			81	1.8

TABLE 6.17

Failure Rates for Overhead Lines and Feeder in the NE Subsection with 0.064 f/yr and km

Line Type	Average Length [km]	Average Failure Rate [f/yr]	Number of Lines	Resulting Failure Rate [f/yr]
Line 1E	0.125	0.008	19	0.152
Line 2E	0.375	≈0.026	18	0.468
Line 3E	0.625	0.04	7	0.28
Line 4E	0.875	0.06	7	0.42
Line 5E	1.125	0.07	3	0.21
Line 6E	1.375	0.09	3	0.27
Total			57	1.8

1. NW: Total length 39 km $\Rightarrow$ $1.8/39 \approx 0.046$ f/yr and km.
2. NE: Total length 28 km $\Rightarrow$ $1.8/28 \approx 0.064$ f/yr and km.

Furthermore, assuming that all the overhead lines in the NW and NE subsystems have the appropriate average failure rate, then if one average value would have been used for all the overhead lines, the resulting rate would have been 0.054 f/yr and km.

To obtain failure rates for individual overhead lines, all overhead lines were grouped according to length. The overhead line lengths vary between 10 and 1700 m. Six groups have been chosen in increments of 250 m. For each feeder, six groups have been defined and average failure rates calculated. The results are summarized in Tables 6.16 and 6.17. Consequently, the groups with length l have been grouped as follows:

1. Line 1, $0 < l \leq 0.25$ km
2. Line 2, $0.25 < l \leq 0.50$ km
3. Line 3, $0.50 < l \leq 0.75$ km
4. Line 4, $0.75 < l \leq 1.00$ km
5. Line 5, $1.00 < l \leq 1.25$ km
6. Line 6, $1.25 < l$ km

The average length for each group has been used when calculating the average failure rate.

The failure rates were then rounded off to obtain the total average value of 1.8 f/yr per feeder.

6.5.2.4.2 Permanent Failure Rates for Junctions

There was one component type included in the system model for which no data was provided i.e. the junctions. A junction is a connecting link in the system. This component could be a simple bus bar construction or be more complicated including several components.

Input data for the failure rate have therefore been based on other sources of statistics. The selected data were obtained from statistics from Svl (Swedish electrical power utilities) and the Swedish National Outage Data Bank (DAR). The failure rate used was 0.014 f/yr and refers to the component named kopplingsskåp (switchbox) at 11 kV [95].

6.5.2.4.3 Passive and Active Failure Rates

The input data for failure rates discussed earlier in this section relates to permanent failure rates. The permanent failure rate (A_t) is a total failure rate for passive and active failure rates (see the definitions in 6)

$$\lambda_t = \lambda_p + \lambda_a$$

However, input data for RADPOW also include active and temporary failure rates, which have been defined below.

- For all static components (overhead lines, transformers, and such like), the number of passive failures is generally negligible, therefore, $\lambda_p \approx 0 \Rightarrow \lambda_t \approx \lambda_a$.
- Switching devices (particularly breakers and automatic or remote devices) can be prone to operate when not required, hence $\lambda_p > 0 \Rightarrow \lambda_t > \lambda_a$. For manual switches, however, $\lambda_p \approx 0$ [48].

  Conclusions from the above discussion are hence: (i) that for all components, except the breakers and disconnectors, the active failure rate approximately equals the permanent failure rate, and (ii) that for the breakers and disconnectors, the active failure rate is lower than the passive. Appendix B contains the assumed values for active and temporary failure rates.

6.5.2.4.4 Maintenance Frequency Data

RADPOW includes one set of reliability input data for maintenance (λ_m), as presented earlier in Chapter 5. However, this input data set relates to the outage frequency and average time for scheduled maintenance. These activities are planned and therefore the outage will be known in advance and the customers concerned can be informed. Consequently, a scheduled outage is not usually included in the evaluation of a load point's reliability if this activity causes disconnection [79] (see also Chapter 4). However, where events overlap several systems, these cause interruption of supply. Therefore, failures caused by a "maintenance outage followed by a component failure" must be analyzed. This aspect was analyzed by RADPOW. However, the Flymen system does not include overlapping failure modes, and therefore these values will not have an impact on the results for this system. Therefore, a detailed analysis of the maintenance procedures has been omitted at this stage, and λ_m has been set to zero for all components in the Flymen system model.

6.5.2.5 Cost Input Data

The following cost data presented in thousands of Swedish kronor (KSEK) were obtained from E.ON:

1. Installation costs for breakers:
   - Material, KSEK 120
   - Labour, KSEK 20
2. Installation costs for manual disconnectors:
   - Material, KSEK 20
   - Labour, KSEK 10
3. Installation costs for remote or automatic disconnectors:
   - Material, KSEK 20 + 60 = 80 (namely the cost of the component plus surrounding equipment)
   - Labour, KSEK 10

6.5.2.6 Reliability Analysis

Appendix A presents the load point indices resulting from the RADPOW analysis of the Flymen system.

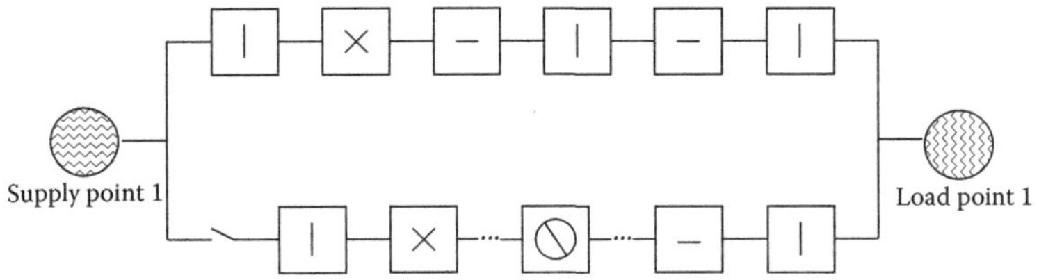

FIGURE 6.12
Minimal cut set scheme for Load Point 1.

6.5.2.6.1 Failure Modes

Figure 6.12 illustrates the deduced minimal cut sets for Load Point 1. Like all load points in the Flymen system, this load point has one radial path and one normally open path. Therefore, the system only includes first-order failure events. The minimal cut sets of the first order for Load Point 1 include components 69, 140, 143, 71, 144 and 1. The normally open path is connected by the closure of the normally open disconnector that is component 142. The second group of failure modes for Load Point 1 that is the primary system failure modes according to Section 4.5 includes all components that are connected with the load point but not included in its path.

6.5.2.7 Results from RADPOW

Figure 6.13 shows the output from analyzing the FlymenN subsystem with RADPOW. Furthermore, Appendix A presents the results for each individual load point. The following paragraph contains some remarks about these results.

The failure rate seems to be a straight line at two levels: one for the NW subsystem (Load Points 1–40) and the other for the NE subsystem (Load Points 41–68). The diversity between the subsystems reflects the assumption of using different failure rates for the feeders and the overhead lines in accordance with Section 9.2.4. The failure rate is, however, not constant for the load points, and this can be seen in a more detailed diagram (Figure 6.14) from the NW subsystem. In the FlymenNW system, the failure rate varied between 4.360 and 4.362 f/yr, and with an increasing failure rate with the number of components from the supply point. In the FlymenNE system, the failure rate varied from 3.605 to 3.606 f/yr. It should be stressed that these differences were less than the accuracy in the input data, however, they provide information about the different behaviors in the system, which is the aim of the analysis. Furthermore, the almost constant failure rate occurred because all the component failures would result in an outage at all load points. This is in turn the result of the characteristics of the system, for example, it is a serial system, has one normally open path that can be used when failure occurs, and two breakers situated at the

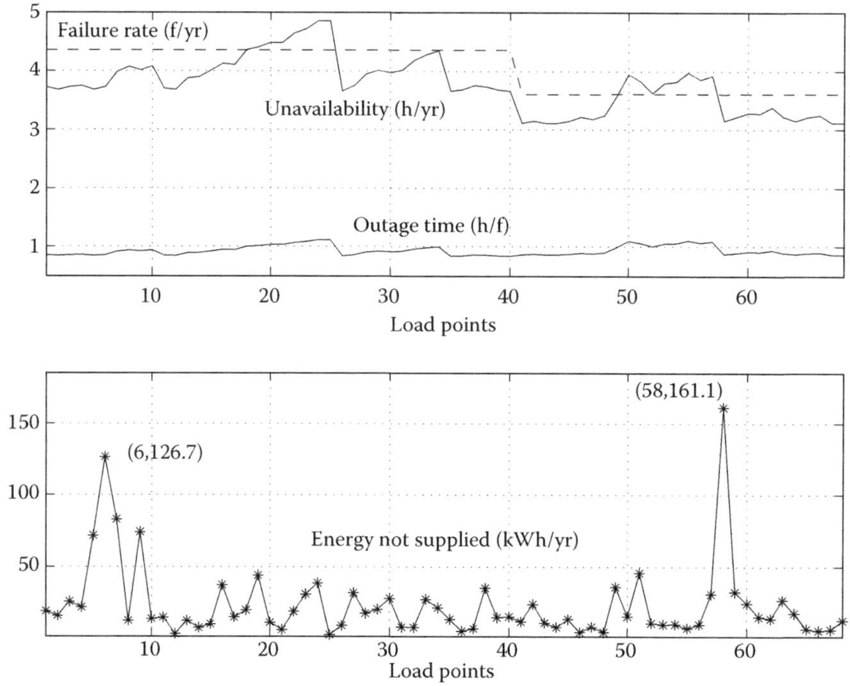

FIGURE 6.13
Load point indices for the FlymenN subsystem.

supply points. Figure 6.15 illustrates the different categories of the failure rate, which in total provide the nearly constant rate. This further illustrates that when the number of permanent failures increases, the active failures decrease, and that the number of temporary failures are constant. However, the outage time caused by these failures varies for different restoration activities.

The outage time varies between 0.8380 and 1.1132 h, where Load Points 24 and 25 exhibit the maximum values, and 26, 35, and 40 exhibit the minimum values. Unavailability varies between 3.1121and 4.8559 h/yr, where Load Points 41–44 and 67–68 have the lowest values, and 24 and 25 the highest.

The results for the energy not-supplied indicate two exceptionally high values for the Load Points 6 and 58. As can be seen from the input data in Appendix A, this reflects divergent customer and power data for these two load points compared with the other load points; for Load Point 58 because of a large number of customers, and for Load Point 6 because of a combination of high total active power and large numbers of customers.

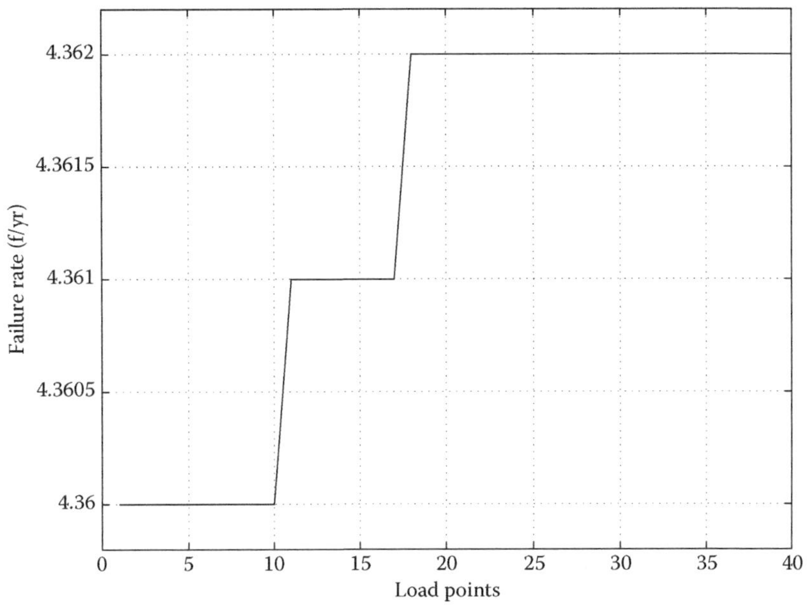

FIGURE 6.14
Close-up view of the failure rate in the FlymenN subsystem.

6.5.2.7.1 Validation of the Results

The correctness of the results can be checked using a number of estimations. Consider for example Load Point 1, which has the minimal cut sets of components 69, 140, 143, 71, 144, and 1. All other components in the system constitute additional active failure modes. The same reasoning can be applied to all of the load points. The temporary failure events occur as both minimal cut sets and additional active failures. According to the input data (see Appendix B), the permanent failure rate equals the active failure rate for all components except the two breakers and the disconnectors that are automatically operated ($\lambda_t = \lambda_a$). This system only contains first-order failure modes and the breakers have no "stuck probability." According to the above discussion, by letting ntype be the total number of components in the system of a certain type, then an estimate of the failure rate for a load point can then be evaluated by

$$\lambda_{lp} \approx \sum_{type} n_{type} \cdot (\lambda_t + \lambda_{te}) \tag{6.4}$$

where type = load point, junction, breaker, manual disconnectors, automatic disconnectors, and overhead lines.

Table 6.18 summarizes data used for the validation. Moreover, with data from Appendix A, an estimate of the failure rate in the Flymen system

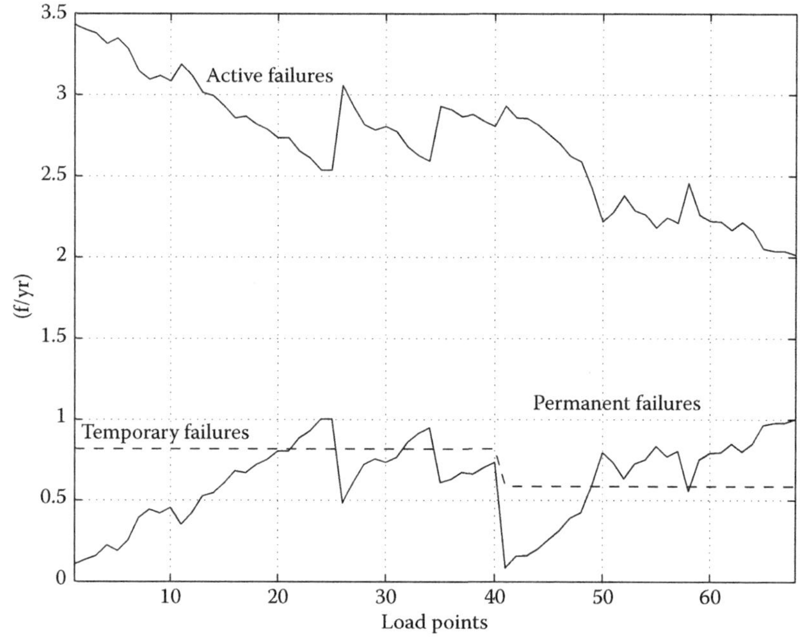

FIGURE 6.15
Close-up view of the failure rate in the FlymenN subsystem.

NW can be evaluated according to the following: $\lambda_{NE} \approx 40 \cdot (0.018 + 0.01) + 40 \cdot (0.014 + 0.01) + 1 \cdot (0.05 + 0.02) + 6 \cdot 0.05 + 6 \cdot 0.05 + 3 \cdot 0.051 + 33 \cdot 0.006 + 15 \cdot 0.017 + 13 \cdot 0.029 + 10 \cdot 0.04 + 3 \cdot 0.05 + 7 \cdot 0.06 = 4.358$ f/yr.

Equivalent evaluations for the NE subsystem result in 3.607 f/yr. Comparing these results with the output from RADPOW, which are 4.36–4.362 f/yr and 3.605–3.606 f/yr (see Appendix B), leads to the conclusion that the evaluated failure rates from RADPOW seem to be correct.

The unavailability in this system model is caused by three types of failures: permanent, temporary, and active. The outages for the permanent failures consist of repair time for the component if the normally open path cannot be used or the switching time of 1 h. The outage time for an active failure is also the switching time. The unavailability for the load point is the sum of the unavailabilities for each failure mode. The contributions from the temporary failures to the uncertainties are small (no more than 8/100 h/yr), and these can be ignored when making an estimate.

Continuing on with Load Point 1, the major contribution to the unavailability is caused by the permanent and the active failures. The restoration activity for a minimal cut set failure mode (69, 140, 143, 71, 144, and 1) is either repair or reclosure of the normally open disconnectors (with switching time of 1 h). The additional active failures are restored by switching. All components that

TABLE 6.18

Data Used in the Validation of the RADPOW Result for the Flymen System

Component Types	Number of Components	λ_t [f/yr]	λ_{te} [f/yr]
BUSL	40	0.018	0.01
BUST	40	0.014	0.01
BREAK	1	0.005	0.02
DISCM	6	0.05	0
DISCA	3	0.051	0
Line 1W	33	0.006	0
Line 2W	15	0.017	0
Line 3W	13	0.029	0
Line 4W	10	0.04	0
Line 5W	3	0.05	0
Line 6W	7	0.06	0

are not a minimal cut set constitute an additional failure. Consequently, Load Point 1 has 171 − 6 = 165 additional failures (see also Section 4.5). The failure rate for the total components in the subsystem is calculated to be 4.358 f/yr with the temporary failures. An estimate of the unavailability at Load Point 1 can then be estimated as follows:

$$U_{lp} \approx U_{active} + U_{permanent} \tag{6.5}$$

and where

$$U_{active} \approx (\lambda_{lp} - \lambda_{te} - \lambda_t) \cdot r_s. \tag{6.6}$$

Note that the permanent failure rate (λ_t) refers to the actual minimal cut sets for the load point that is analyzed.

Active failures:

$$(4.358 - (40 \cdot (0.01 + 0.01) + 0.02) - (0.018 + 2 \cdot 0.014 + 0.017 + 0.04 + 0.004)\,\frac{\text{f}}{\text{yr}} \cdot 1 \cdot \frac{\text{h}}{\text{f}} = 3.431\,\text{h/yr}.$$

Permanent failures:

$$0.014 \cdot 2 + 0.04 \cdot 3 + 0.018 \cdot 2 + 0.017 \cdot 1 + 0.005 \cdot 1 + 0.014 = 0.22\,\text{h/yr}.$$

Total failures:

$$U_{lp} \approx 3.41 + 0.22 = 3.651\,\text{h/yr}.$$

This value for the total failures compared with the result from RADPOW (3.72016 h/yr) confirms that the results seem correct. The results for unavailability vary from 3.65 h/yr up to 4.85 h/yr for load points in the NW subsystem. The divergence reflects the difference in the number of components that are needed to be repaired when the load point fails, which in turn reflects the number of components in a minimal path to the load point. For example, Load Point 25 has 27 components that when they fail, need to be repaired to achieve restoration of supply. Load Point 25 has an unavailability of 4.85 h/yr. The other load point indices resulting from RADPOW (the average outage duration, average energy not-supplied, and load not-supplied) are evaluated based on earlier calculations together with customer input data. The conclusion is that the RADPOW results for the FlymenN subsystem seem to be correct.

6.5.2.8 Identifying the Critical Components

The first step in deducing an effective maintenance plan is to decide which parts of the system to maintain and how to perform the maintenance activities. To achieve this, a sensitivity analysis is performed for the system with the aim of identifying the critical components that are the components that have a large influence on system reliability.

The approach that has been used is described in the following. The components were categorized according to their type, then the failure rates were varied for one type at a time, and the resulting indices then evaluated. The different cases have been listed below:

1. cl base case
2. c2 junctions
3. c3 breakers
4. c4 disconnectors
5. c5 overhead lines

If in cases c2–c5 the associated component category was assumed to be 100% reliable, then the resulting load point indices presented in Figures 6.16 through 6.18 were obtained.

Results have been provided for Load Points 25, 6 and 58, which strongly indicate that the overhead lines have the primary impact on system reliability. The overhead lines show a significant contribution to the overall failure rate, unavailability, and energy not-supplied. The overhead lines have a minor impact on the outage time since all components have a similar repair time. This result is, however, a logical conclusion since the system consists mainly of overhead lines.

The apparent impact of the junctions that can be seen from the figures is somewhat more surprising. For a distribution system, these components are

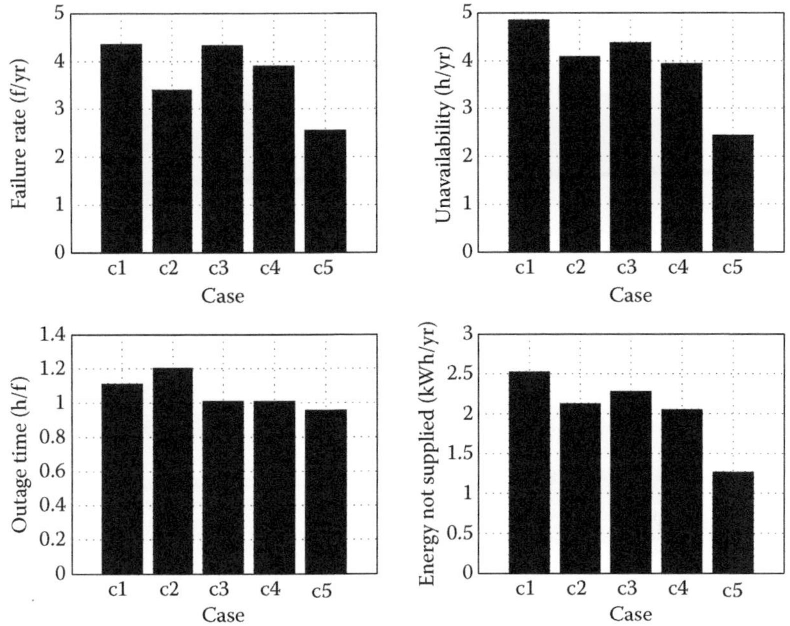

FIGURE 6.16
Identifying critical components showing results for Load Point 25.

generally very simple and therefore do not have a large impact on system reliability. However, the result depends on the assumed failure rates. The failure rate that is assumed for the junctions is an average value from the DAR statistics, being 0.014 f/yr. This value has been comparable with many other overhead lines. A reasonable conclusion would therefore be that the assumed failure rate for the junctions is too high.

Note that the average outage time increases when the junctions are assumed to be totally reliable that is in Case c2. This reflects the fact that these events have repair times that are shorter than the majority of the other components, particularly the overhead lines which have a dominant effect.

6.5.2.9 Analysis of Critical Load Points

The Load Points 6 and 58 were identified as having a much higher "not supplied energy" value than the other load points. It could also be seen from the results that the differences were not due to lower unavailability at these points. However, the energy not-supplied could of course be decreased at these points if the unavailability was decreased. According to the statistics from Svel and Table 6.3, it can be seen that failures caused by trees falling are dominant, representing ($17+8 \approx 25\%$) of the causes of failures. Therefore,

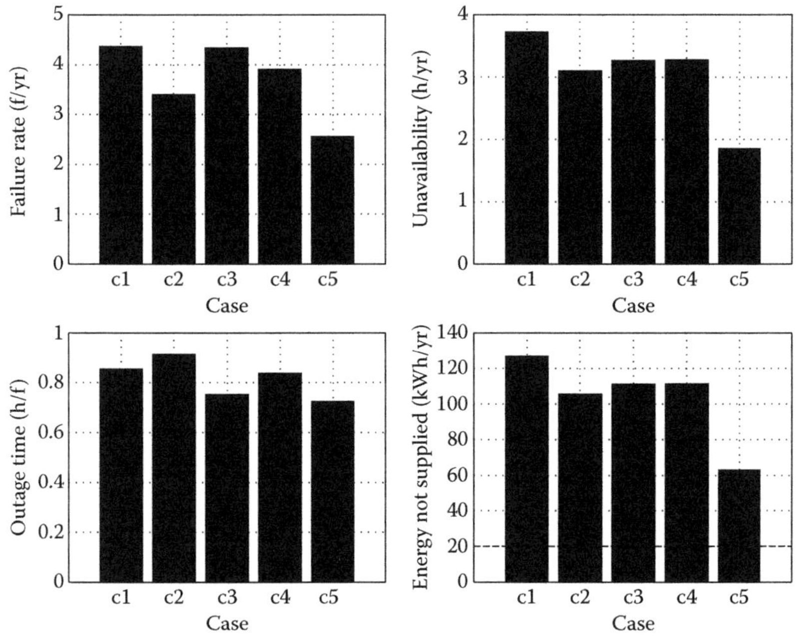

FIGURE 6.17
Identifying critical components showing results for Load Point 6.

one effective way to improve the unavailability would be to eliminate failures caused by falling trees.

Assuming that maintenance activities are performed to prevent trees from falling, and assuming that this leads to "no failures caused by falling trees," this would then imply that the failure rate would decrease by 25%. Now assume first that in Case c6, only overhead lines connected to the Load Points 6 and 58 from the supply point through the minimal path are maintained, then this would imply a prioritizing of the maintenance for critical load points. Second, assuming in Case c7 that all overhead lines in the system were maintained with the same effect, and with Case cl being the base case, then the different cases can be summarized as follows:

1. cl denotes the base case
2. c6 denotes the maintained overhead lines in minimal paths (Load Points 6 and 58)
3. c7 denotes all the maintained overhead lines

The results have been presented in Figures 6.19 and 6.20. Because all overhead lines contribute to the failure rate, and most are not affected by line maintenance in Case c6, the benefit in this case is not significant. Case c7, however, shows considerable improvement.

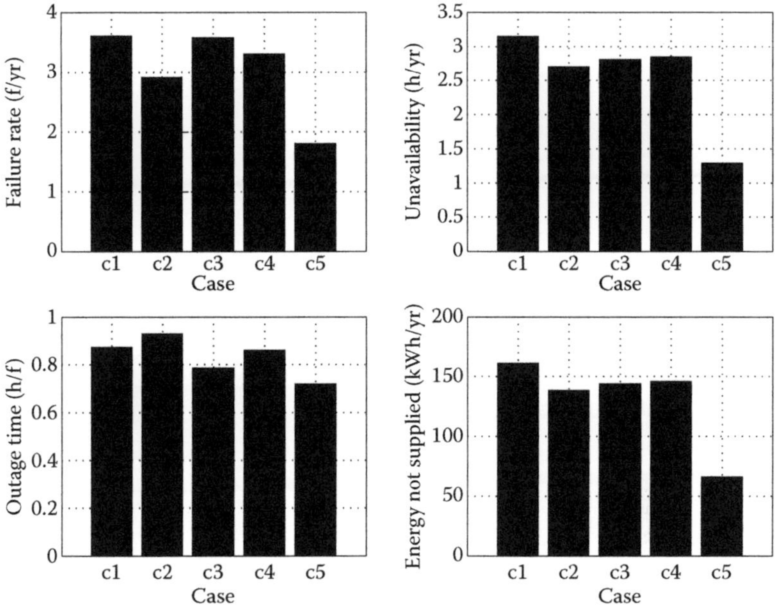

FIGURE 6.18
Identifying critical components showing results for Load Point 58.

6.5.2.10 Benefits of Maintaining the Overhead Lines

As discussed previously, one way to reduce the unavailability of supply is to reduce the frequency of interruptions, and this in turn can be achieved by focusing on the causes of failures. One way to gain an understanding of the relationship between maintenance and reliability would then be to study the causes of failures. This, however, requires knowledge about the behavior of components, and that can be gained from statistics, as discussed earlier in Sections 6.2 and 6.3.

Statistics based on the NAFIRS from the UK was provided in Table 6.5. This table shows the causes of failures for 11 kV distribution systems and overhead lines.

The causes that may be affected by maintenance have been identified and studied further in the following.

6.5.2.10.1 Reducing Causes of Failures in Overhead Lines

Assume here that maintenance is performed on overhead lines to reduce certain causes of failures categorized into the following cases:

a. Base case
b. Wind-borne materials

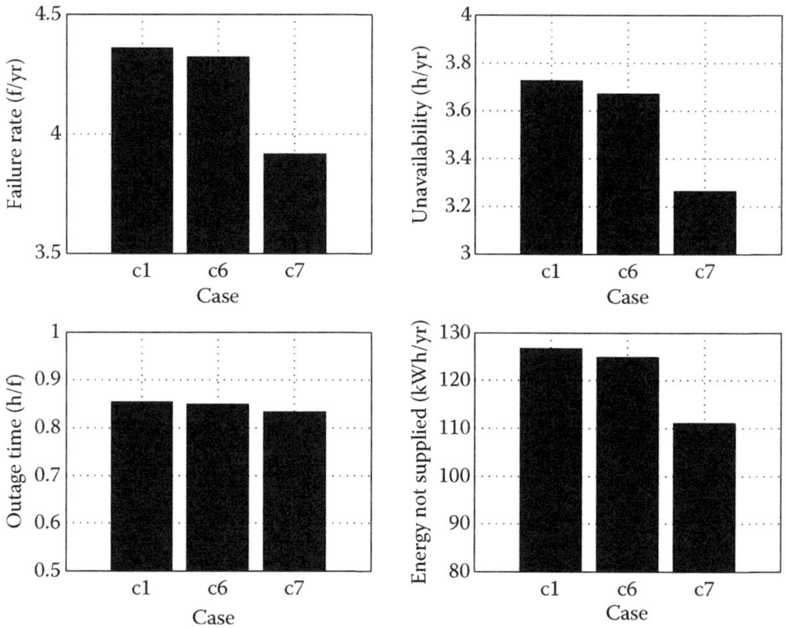

FIGURE 6.19
Failure rate decreases for overhead lines for Load Point 6.

c. Corrosion
d. Vibration
e. Trees (e.g., falling trees)
f. Vermin
g. Farm animals
h. Willful damage
i. Faulty construction and installation
j. Aging and wear
k. Faulty manufacture and design

Furthermore, assume that these causes of failures are eliminated by the maintenance activities. The corresponding failures will then be eliminated and the reliability indices influenced accordingly.

Figure 6.21 reveals the benefit to system indices attributable to the different cases, and Table 6.19 provides a ranking of these results. It can be seen that two causes of failures have a major impact on system reliability, namely trees (Case e) and aging and wear (Case j). Importantly, the different causes of failures have been studied separately here, though in reality more than one

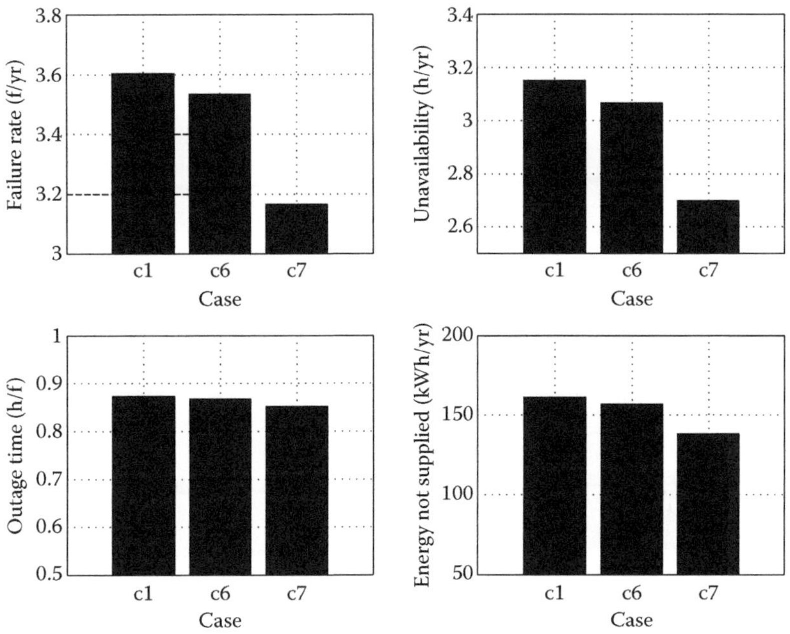

FIGURE 6.20
Failure rate decrease for overhead lines for Load Point 58.

set of causes of failures could be affected by maintenance. Therefore, combinations of causes have been studied to observe the cumulative effect of maintenance.

6.5.2.10.2 *Reducing Causes of Failures by Trees and Aging and Wear in Overhead Lines*

From the analysis, it can be seen that two causes of failures have a major impact on the reliability of the system: trees (Case e) and aging and wear (Case j).

Assuming that maintenance is performed and has the effect that failures caused by trees and aging and wear are eliminated, then according to Table 6.5, it can be seen that the failures can be reduced by $10.6 + 13.1\% = 23.7\%$. Therefore, the corresponding failure rate would be reduced by the same amount. If the failure rates for the overhead lines (permanent and active failure rates) were cut by 23.7%, then the failure rates would change according to Table 6.20.

Figure 6.22 presents the RADPOW results for the NW subsystem. The results are shown for the base case and for when the failure rates for overhead lines are reduced. As can be seen in the figure, there is a significant decrease in unavailability. It can, however, be observed that the value of

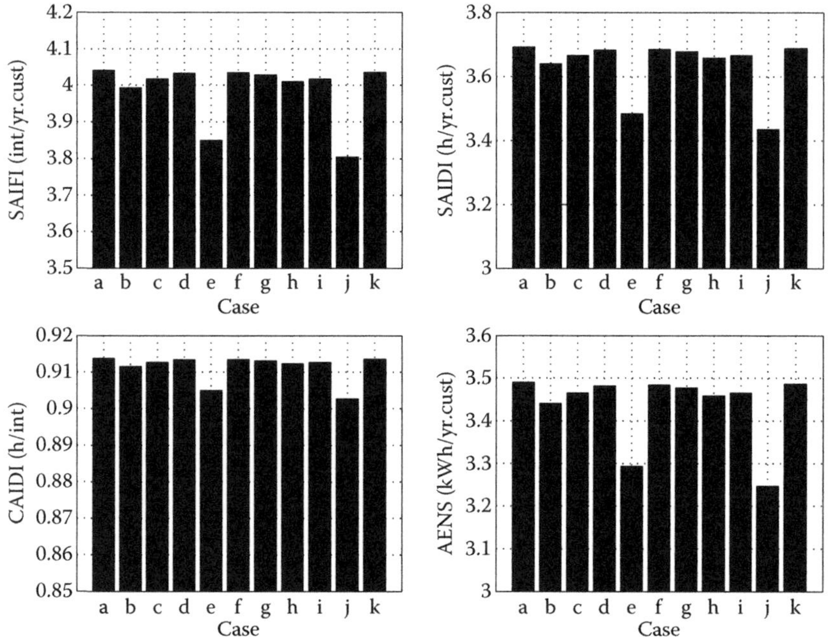

FIGURE 6.21
Sensitivity analyses for determining the benefit of maintaining overhead lines.

energy not-supplied decreases slightly due to the outage times that remain nearly unchanged.

The system indices for this system will also change with the decreased failure rate. Table 6.21 presents these changes.

6.5.2.11 Summarizing Analysis

It must be emphasized that the information presented in Table 6.5 includes average values for all overhead lines including those for which falling trees may have no impact at all, and those where trees may have a major impact. It should also be noted that the statistics come from the UK, and in Sweden tree coverage is generally higher, and therefore the impact of trees is probably even more significant. Corresponding statistics from Sweden in Table 6.3, which are not separated for different component types, showed an average value of about 25% for causes of failures due to trees. The conclusion from both observations is that for those overhead lines that can be affected by trees, the percentage of faults due to this cause may be much greater than the average figure quoted in Table 6.5. To estimate the influence of this aspect, a number of calculations have been made using percentages of failures caused

TABLE 6.19

Ranking of the Case Study for Overhead Lines

Case	SAIFI [int/yr cust]	SAIDI [h/yr cust]	CAIDI [h/int]	AENS [kWh/yr cust]
j	3.80506	3.43481	0.902695	3.24763
e	3.85027	3.4841	0.904898	3.29412
b	3.99312	3.6398	0.911517	3.4409
h	4.01046	3.65871	0.912292	3.45875
c	4.01732	3.6662	0.912598	3.46581
i	4.01732	3.6662	0.912598	3.46581
g	4.02845	3.67835	0.913093	3.47733
d	4.03307	3.68337	0.913293	3.48204
f	4.0352	3.6857	0.913387	3.48421
k	4.03707	3.68771	0.91346	3.48612
a	4.04071	3.6917	0.913625	3.48988

TABLE 6.20

Reduction of Failure Rates in Overhead Lines

Line	Failure Rate [f/yr]	
Type	Old	New
Line 1W	0.006	0.005
Line 2W	0.017	0.013
Line 3W	0.029	0.022
Line 4W	0.04	0.031
Line 5W	0.05	0.038
Line 6W	0.06	0.046
Line 1E	0.008	0.006
Line 2E	0.026	0.020
Line 3E	0.04	0.031
Line 4E	0.06	0.046
Line 5E	0.07	0.053
Line 6E	0.09	0.069

by trees varying between 5% and 95%. The results of this analysis are presented in Figure 6.23, which shows that the span between the system indices is significant. For SAIFI, the divergence is about 1.5 interruptions per year and customer.

Figure 6.24 provides a summary of results from the analyses for the NW feeder, where the base case is compared with two cases. It shows the cumulative effect of maintenance when reducing faults for overhead lines caused

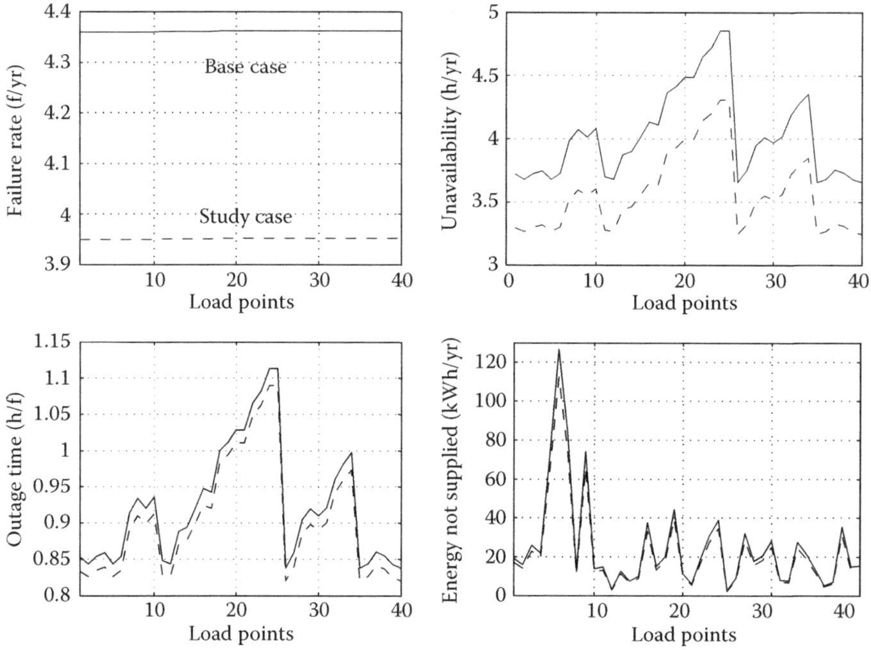

FIGURE 6.22
Reduction in the causes of failures due to trees and aging and wear in the FlymenNW system.

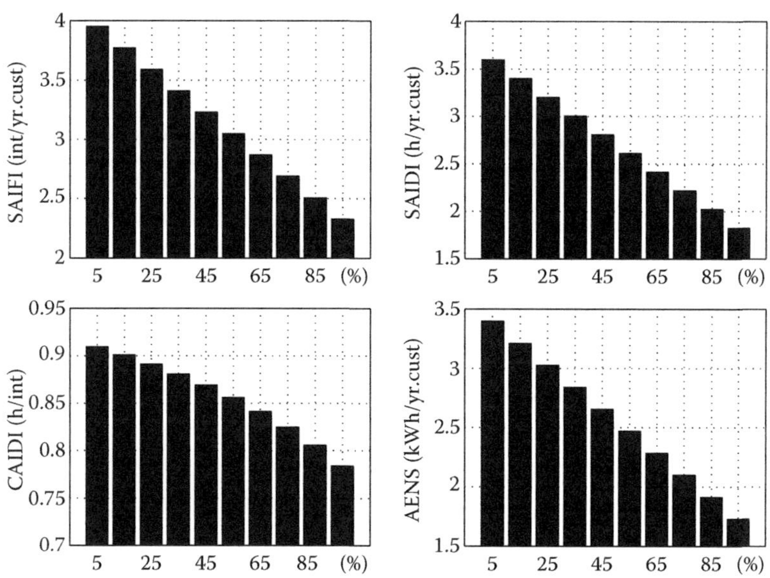

FIGURE 6.23
Benefits of maintenance with the causes of failures attributed to trees ranging from 5% to 95%.

TABLE 6.21
System Indices before and after a Reduction in the Causes of Failure for Overhead Lines

Index	Unit	Old	New
SAIFI	[int/yr and cust.]	4.04	3.63
SAIDI	[h/yr and cust.]	3.69	3.24
CAIDI	[h/int.]	0.91	0.89
AENS	[kWh/yr and cust.]	3.49	3.06

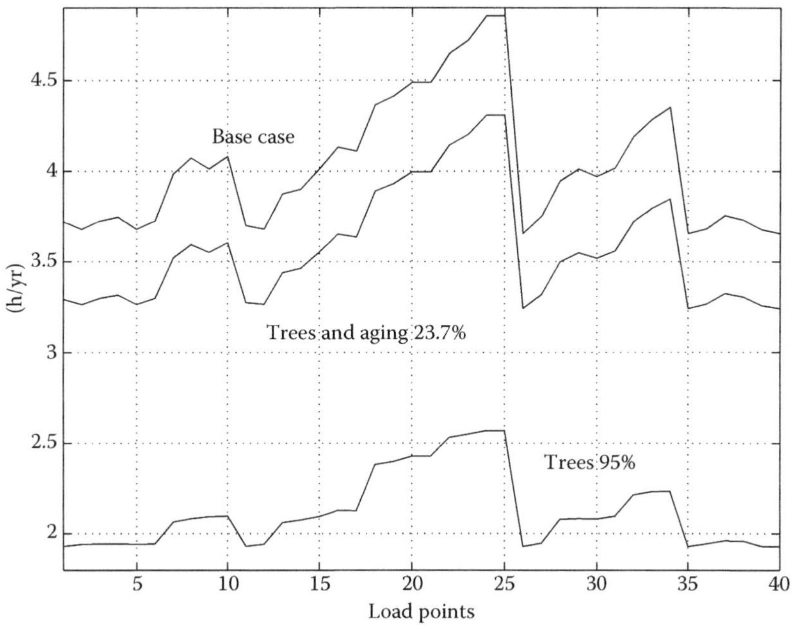

FIGURE 6.24
Decrease in unavailability by preventing failures.

both by trees and aging. It also shows the effect of maintenance when the overhead lines have a cause of failure of 95% for trees. The latter shows that for those overhead lines that can be affected by trees, the percentage of faults by this cause may be much greater than the average figure quoted in the statistics. The conclusion is that the individual behavior needs to be studied in order to understand the relationship between maintenance and failure rate.

TABLE 6.22

Failure Rates for Underground Cables in the FlymenN Subsystem

Underground Cable	Number of Cables	Failure Rate [f/yr]	
		Average	Total
Cable 1	52	0.004	0.208
Cable 2	33	0.011	0.363
Cable 3	20	0.018	0.360
Cable 4	17	0.026	0.442
Cable 5	6	0.030	0.180
Cable 6	10	0.041	0.410
Total	138		1.963

6.5.2.12 Overhead Lines Replaced by Underground Cables

One way to improve the reliability of a distribution system is by using underground cables instead of overhead lines. This is, however, a very expensive solution. Therefore, it is interesting to study quantitatively what the values of this benefit in system reliability would be. Assume here then that the network consists of underground cables instead of overhead lines. According to Svel's statistics for 1995–1997, the average value for failures in underground 11 kV cables is 0.0293 f/km and yr. This value should be compared to the corresponding value for overhead lines, which is 0.125 f/km and yr. However, in the FlymenN subsystem, the average failure rates for overhead lines are 0.046 f/km and yr and 0.064 f/km and yr (in 6.5.2.4). Therefore, it can be seen that the average failure rate for underground cables is significantly lower than for overhead lines.

If the same procedure used for overhead lines (see 6.5.2.4) is used, then the resulting failure rates for the different underground cables can be evaluated for example as follows: 0.0293 f/yr and km $\cdot$ 0.125 km $\approx$ 0.004 f/yr. The total length of the overhead lines according to Section 6.5.2.4 is $39+28=67$ km. The resulting failure rate is then $0.0293 \cdot 67 \approx 1.963$ f/yr. This value should be compared with 1.8 f/yr and feeder, which is $2 \cdot 1.8 = 3.6$ f/yr. Consequently, the underground cables have approximately half the failure rate for the system. Table 6.22 presents the resulting failure rates per underground cable.

Figure 6.25 summarizes the resulting load point indices for the FlymenN subsystem with underground cables instead of overhead lines. The large reduction in the failure rate when underground cables are used has a great impact on the load point indices. For example, the energy not-supplied at Load Point 58 is now 108.2 kWh/yr instead of 161.2 kWh/yr (see Figure 6.13). Figure 6.25 presents the resulting system indices for the same cases as used in the study of the overhead lines. As can be seen from the results and according

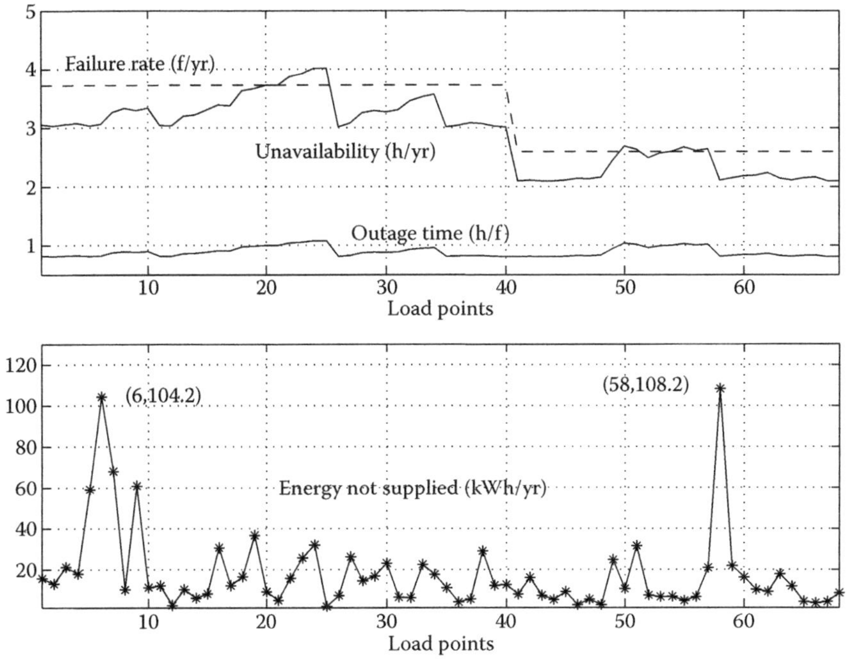

FIGURE 6.25
Results for the FlymenN subsystem using underground cables.

to the statistics (see Table 6.5), aging and wear (Case j) are the primary causes of failure that can be reduced by maintenance (Figure 6.26).

6.5.2.13 Conclusion

This section has presented the results from reliability analysis and the effects of maintenance on the Flymen rural distribution system. From sensitivity analyses, the effect of critical components has been determined. By the change in system indices, the lines were identified to be the critical component reflecting the behavior of a rural distribution system. The relationship between maintenance and reliability was studied through sensitivity analyses of the benefit when reducing causes of failures. Failures caused by trees and aging were identified to have the most significant impact on reliability indices.

The conclusion is that PM can have a great impact on system reliability and the relationship between failure rate and maintenance should be investigated in more detail. Improvement in reliability indices must be compared with the investment costs required and the benefit in outage costs. Therefore, the relationship between reliability levels and economics should also

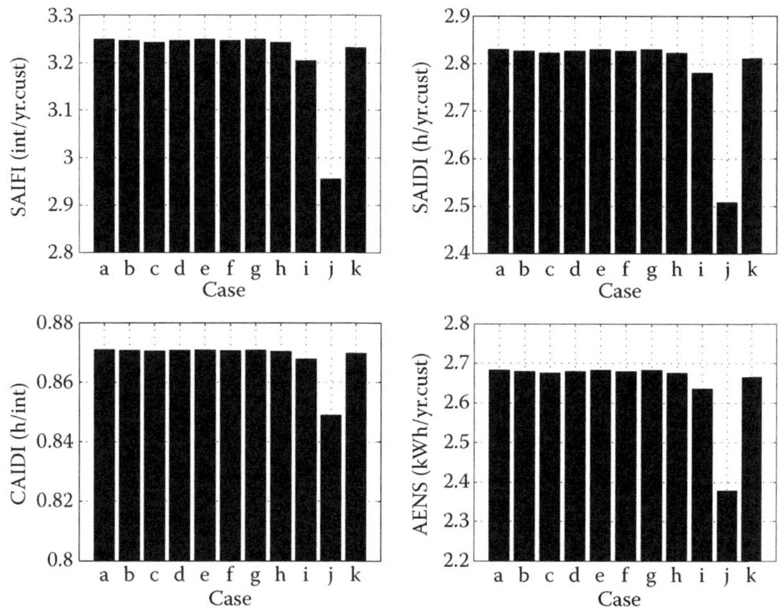

FIGURE 6.26
System indices for the FlymenN subsystem using underground cables.

be investigated. The crucial need for appropriate data is another matter that needs to be recognized.

6.5.3 Birka: An Urban Underground Cable System

This section describes the analysis of and presents the results from a reliability and maintenance study of a Swedish distribution system belonging to the Birka Nat distribution system in the Stockholm City area. Some of the results from this study have also been published in [117].

6.5.3.1 Definition of the Birka System

The Stockholm City distribution system supplies power to about 450,000 customers and belongs to the utility Birka Nat, which has the greatest number of customers of any power utility in Sweden [135]. A survey of disturbance statistics and maintenance practices focusing on the causes of failure in underground 11 kV cables has been undertaken in cooperation with Birka Nat, in accordance with the approach presented earlier in Section 6.3. In this study, a special part of the network has been defined and studied in greater detail as a case system. This special part of the network has been modeled and is referred to as the Birka system.

This case system includes the Birka Nat 220/110 kV Bredang station and the 33/11 kV Liljeholmen station, which are connected to each other via two parallel 110 kV cables. From the Liljeholmen station (LH11), there are 32 outgoing 11 kV feeders that supply the southern part of central Stockholm City. The system is an urban network whose distribution feeders consist mainly of underground cables and is shown in Figure 6.3.

Figure 6.27 presents the model of the case system called the Birka system. The simplified network consists essentially of bus bars where double bar arrangements are represented by single bus bars (note that this simplification is taken into account when defining the reliability input data as discussed in Section 6.5.3.5). Moreover, the customers are represented by two 33-kV load

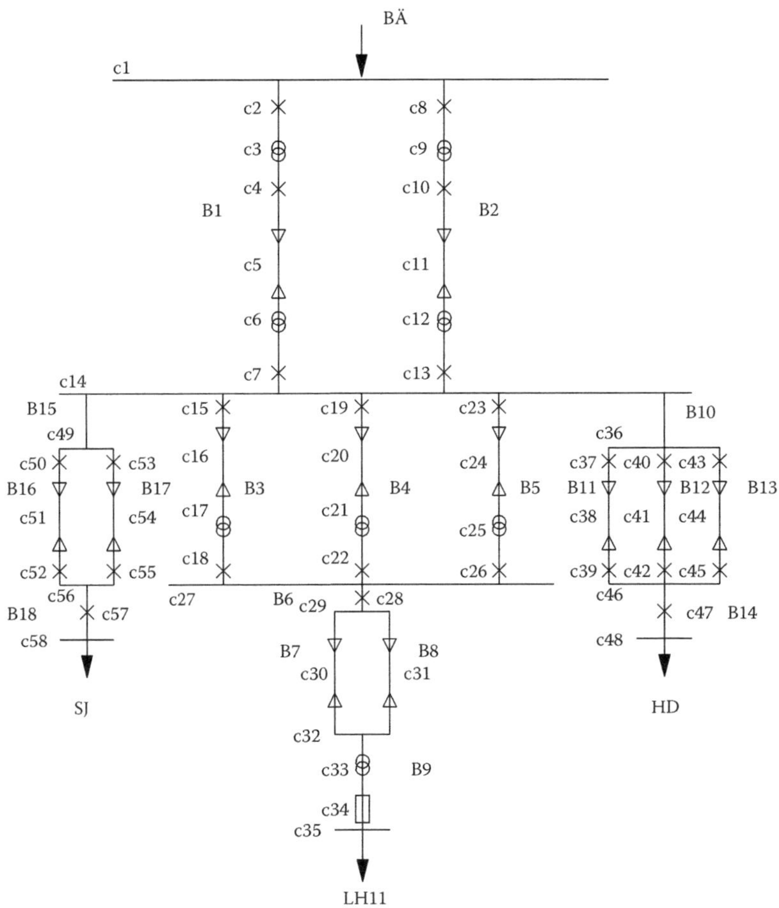

FIGURE 6.27
The model of the Birka system used in the RADPOW reliability analysis of the Birka Nät case system.

points, referred to as Hogalid station (HD), and Statens Jarnvagar railway line (SJ), and one 11-kV load point, referred to as Liljeholmen station (LH11). LH11 includes customers connected to a total of 32 outgoing 11-kV feeders, in parallel from the Liljeholmen station with a total of 14,300 customers. These distribution feeders are represented by the one "averaged" component. HD supplies a total of 23,400 customers. The cable lengths and components in HD are of a mixture of mass-impregnated paper and XLPE cables. The 11 kV cable within the Liljeholmen station has been ignored since it is of minimal length compared with the 11 kV network. The resulting model of the Birka system has in total 58 components and 18 branches. The following components are included in the system model:

- 220, 33, and 11 kV bus bars
- 220, 110, 33, and 11 kV breakers
- 110, 33, and 11 kV underground cables
- 220/110, 110/33, and 33/11 kV transformers
- 11/0.4 kV fuse

The data defined above constitute the network input data for the RADPOW reliability analysis. These input data are discussed further in the following sections.

6.5.3.2 Sources of Input Data for the Birka System

The model of the Birka system has been provided with input data from the following data sources.

- Birka Nat:
  - Results from the survey at Birka Nat are presented in [112] and summarized in Section 6.3.
  - Statistics collected in connection with a previous study are presented in [136], in which the author participated in connection with some reliability evaluations.
  - Estimated input data from an earlier reliability analysis are presented in [101].
  - Verbal discussions with Kjell Gustafsson [137] (the customer and power data), Lars-Åke Gustafsson [114], Daniel Terranova [138] and Mats Åhlén [116].
- Other sources:
  - Nordel
  - DAR and FAR
  - T-book, which were introduced in Section 6.2

6.5.3.3 Customer and Power Input Data

The Birka system introduced in Section 6.5.3.1 represents the distribution network from the Bredang (BA) 220/110 station to the Liljeholmen (LH) 33/11 station. Furthermore, the HD and SJ 33 kV stations are connected with the network as shown in Figure 6.27. The customers are represented by two 33 kV customers (HD and SJ), and one 11 kV customer (LH11). LH11 includes the customers connected to the roughly 30 outgoing 11 kV feeders leading from Liljeholmen station. HD supplies a number of customers.

The Stockholm City network was introduced in Section 6.5.2.1. This system includes stations and feeders from the 220 kV level down to the 400 V level that is from the transmission voltage to the household customer. Table 6.23 shows the number of stations of each type included in the Stockholm City system. There are, for example, 24 stations of the type denoted *Fs* (distribution stations or fördelningsstation in Swedish) at 33 or 11 kV including the HD and LH11 stations. The table also includes the total number of customers for Stockholm City (450,000), LH11 (14,300), and so on. An average station at the 33 or 11 kV level has consequently about $315{,}500/24 \approx 13{,}000$ customers, which indicates the general size of the HD and LH11 stations, at Load Point LH 11.

Note that for the reliability analysis, one average 11 kV feeder connected to LH11 has been analyzed. This means that the average number of customers is about $14{,}300/32 \approx 446.875 \approx 447$. The power data for the load points have

TABLE 6.23

Different Types of Stations and the Number of Customers in Stockholm City

Station Type	Voltage Levels [kV]	Total Number of Station Type	Station Name	Number of Customers
Fs	33 or 11	24	HD	23,400
			LH	14,300
			...	...
			Total	315,500
Tp	110	4	LH	37,700
			...	...
			Total	155,500
Sn	220	5	BÄ	73,600
			...	...
			Total	325,800
220*Fs*	220	4	...	...
			Total	114,100
			Total	450,000

Source: Birka Nät 2000.

TABLE 6.24

Customer Data for the Birka Case System

Station	Number of Customers	Power [MW]		Power [kW/cust]	
		Average	Max.	Average	Max.
LH11	14,300	24.6	39.4	1.7203	2.7552
HD	23,400	23	37.8	0.9829	1.6154
SJ	1	0.8	4	0.8	4
Total	37,701				

TABLE 6.25

Estimates of Customer Categories for the Birka Case System

	Customer			Active Power		
Station	ind. [%]	res. [%]	com. [%]	ind. [%]	res. [%]	com. [%]
LH11	25	50	25	40	40	20
HD	10	80	10	20	70	10
SJ	0	0	100	0	0	100

also been analyzed. Table 6.24 presents the resulting values. These data are partly based on estimations [137], since measurements are difficult to make at the actual point. For example, power up to 2 MW may be backfed to the network at SJ [137]. Furthermore, it is obvious that the number of customers changes (the values used are from 2000), but this difference is considered to have little in this context and is therefore ignored.

For a reliability analysis, it is also of interest to divide the customers into different categories. This is for example useful when valuing the benefits from changes in not-supplied energy. The customers should be categorized into the following three different groups for RADPOW input data: industrial (ind.), commercial (com.), and residential (res.). Estimations about the percentage distribution between these groups have been made with knowledge about the system. Table 6.25 presents these results. It can be seen that LH11 shows a fair spread for different customers, however, HD is dominated by residential customers and SJ is the sole commercial customer (the railway).

Table 6.26 presents the resulting RADPOW customer and power input data for the Birka system. The input data for the reactive power is not considered in the analysis (trppc = irppc = rrpp = crpp = 0), and therefore not included in the table.

6.5.3.4 More about the Input Data

For some of the components in the Birka case system, the voltage levels belong to some extent to the transmission system instead of the distribution

TABLE 6.26

Customer and Power Input Data for RADPOW for the Birka System

		Number of Customers				Active Power [kW/cust]			
Load Point		**Total**	**ind. [%]**	**res. [%]**	**com. [%]**	**Total**	**ind. [%]**	**res. [%]**	**com. [%]**
LH11	c35[a]	447	25	50	25	1.7203	40	40	20
HD	c48[a]	23,400	10	80	10	0.9829	20	70	10
SJ	c58[a]	1	0	0	100	0.8000	0	0	100

[a]Component number according to Figure 6.27.

TABLE 6.27

Selected Component Statistics from Nordel (Sweden 1990–1999)

	Cable [kV]	Breaker [kV]	Transformer [kV]	
Component Char.	**110–150**	**110–150**	**220–300**	**110–150**
λ [f/yr and km]	0.0098	0.0087	0.0261	0.0205
Percentage distribution of causes of failures				
Lightning	0%	15%	8%	8%
Weather	0%	1%	0%	7%
Damage	33%	0%	0%	0%
Personnel	0%	13%	16%	16%
Equipment	67%	64%	47%	41%
Other	0%	7%	29%	28%

Source: Nordell, *Statistics for Interruption of Voltage (Driftstörningsstatistik)*. Sorgenfri Tryk, Norway, 1999 (in Swedish). [103]

system. Therefore, additional statistics were obtained from Nordel and the Scandinavian transmission systems [103]. Table 6.27 presents selected statistics for the actual components in the model system. In addition to the failure rates that would be used for RADPOW input data, causes of failures were also summarized. The results show that weather (excluding lightning) makes a small contribution to failures.

Moreover, it can be seen that both the cables and breakers are dominated by failures due to equipment, being 67% and 64%, respectively. However, failures due to damage make a dominant contribution to cable failures. It is interesting to relate these results to the results for the 11 kV cable presented in Section 6.3. For both of the cables, weather has little or no impact, but damage and equipment (material/method), respectively, have a significant impact. The only significant difference is for personnel, where the results for the 11 kV cable show a contribution of about 12% for personnel, but Table 9.12 shows a 0% contribution, which is difficult to draw any conclusions from.

6.5.3.5 Component Reliability Input Data

Table 6.28 summarizes the component input data for the Birka system. The collection and selection of input data are presented in more detail below.

TABLE 6.28

Component Reliability Input Data for the Birka Case System Used in RADPOW

Component	$\lambda_{permanent}$ [f/yr][a]	Source	λ_{active} [f/yr][a]	r_{repair} [h]	$r_{replace}$ [h]	P_{stuck}
Bus						
220 kV	0.00964	T-book	0.00964	1	0	0
33 kV	0.00964	T-book	0.00964	1	0	0
11 kV	0.00867	T-book	0.00867	1	0	0
Breaker						
220 kV	0.00870	Assume = 110	0.00861	168	24	0.001
110 kV	0.00870	Nordel99	0.00861	168	24	0.001
33 kV	0.00089	FAR81	0.00088	72	24	0.001
11 kV	0.00243	FAR81	0.00241	48	24	0.001
Cable						
110 kV	0.00980 km	Nordel99	0.00980 km	168	0	0
33 kV	0.00937 km	FAR81	0.00937 km	48	0	0
11 kV	0.01242 km	LH11	0.01242 km	6	0	0
Transformer						
220/110 kV	0.02610	Nordel99	0.02610	504	24	0
110/33 kV	0.02050	Nordel99	0.02050	504	24	0
33/11 kV	0.01989	FAR81	0.01989	504	24	0
11/0.4 kV	0.00331	FAR81	0.00331	48	24	0
Fuse						
11 kV	0.01340	T-book	0.01340	4	0	0.001

Source: *T-book (3rd edition) Reliability Data of Components in Nordic Nuclear Power Plants.* The ATV Office, Sweden, 1992. 235pp. https://inis.iaea.org/search/search singlerecord.aspx?recordsFor=SingleRecord&RN=24057248 [111]; LH11, Bertling, L., RCM-study Birka Nt and KTH-a study of causes of failures based on statistics and practice (Original title: *RCM-studie Birka Nät och KTH-en studie av felorsaker baserad påstatistik och praktik,* in Swedish). Technical Report A-EES-0104, Department of Electrical Engineering, KTH, 2001, p. 64. [112]; Nordell, Statistics for Interruption of Voltage (Driftstörningsstatistik). Sorgenfri Tryk, Norway, 1999. (in Swedish). [103]; FAR81, Nitell, P., *Availability analysis for the Stockholm Distribution system (Tillgängligh etsanalys för Stockholms Elnät).* Master Thesis KTH, Stockholm, MAT-STAT-1998-03, 1998. (in Swedish). [101]

[a] For cables the failure rate has unit [f/yr and km].

- Input data for the permanent failure rate ($\lambda_{permanent}$) originate from the following sources: T-book [111], LH11 [112], Nordel99 [103], or FAR81 [101].
- The active failure rate is defined by $\lambda_{permanent} = \lambda_{active} + \lambda_{passive}$.

  For all switching devices, that is, breakers and the fuse, it is assumed that

$$\lambda_{passive} = 0.01 \cdot \lambda_{permanent} \Rightarrow \lambda_{active} = 0.99 \cdot \lambda_{permanent}$$

  For all other components, it is assume that $\lambda_{passive} = 0 \Rightarrow \lambda_{active} = \lambda_{permanent}$.
- Input data for the average time of repair (r_{repair}) equals that of a previous reliability analysis at Birka Nat, presented in [101], with the exception that 1 h has been used for the bus bar repair time instead of 72 h. The reason for this is the simplification in the RADPOW model system, in which the double bus bar arrangements have been represented by one single bus bar. To compensate for this, the repair times have been reduced to the estimated effective switching time to transfer load from one bus to the other following a bus failure.
- The effect of the cable component failure rate on the system unavailability was not as significant as had been expected. Furthermore, this was caused by the input data for the restoration times. To better model the system, input data was changed by including replacement times for some components for high-voltage breakers and transformers. These components would most probably be replaced by spares, which means that the repair procedures would involve replacement rather than repair. Replacement time of 1 day was considered to be a reasonable estimate. Consequently, replacement times of 24 h have been used for the breaker and transformer input in RADPOW.
- A "stuck probability" of $P_{stuck} = 0.001$ has been assumed for the breakers.

  The component reliability input data required by RADPOW that is not presented in Table 9.13 is either assumed to be zero or equal for all the components. The following contains some detailed comments about these.
- Short interruptions have not been included in the source statistics studied. Therefore, the transient failure rates are assumed to equal zero.
- It has not been possible to identify temporary failure rates. These would either equal zero, since the component does not cause temporary forced outages, or these failures would be included in the assumed permanent forced outages. Therefore, the input data for the temporary failure rates are also assumed to be zero.

- The maintenance frequency input data has been set to zero. This might appear strange, but is just a reflection of an effect of using PM based on component reliability rather than predefined intervals. First, it should be realized that this input data only has an effect on the reliability indices (with the logic implemented in RADPOW) for overlapping faults. This implies that the PM intervals with predefined times for PM only lead to system unavailability if first the PM activity is applied and then a fault randomly occurs but not the contrary (see also the discussion in Section 6.5.2.4).
- The reclosure time is assumed to be 5 seconds and the switching time 1 h for all components.

  The failure rates for the cables depend on both the length and the type of cable. Detailed information for the specific system (LH11) has been studied to define the input data for the actual system. The lengths of the different cables in the Birka system have been provided by Birka Nat. Table 9.14 illustrates the evaluation of the resulting failure rates.

6.5.3.6 Impact of the Length of the 11 kV Cable

As discussed earlier, the Birka case network has been modeled assuming that one 11 kV cable represents all the 32 outgoing 11 kV feeders leading from LH11. However in reality, these cables have different lengths and are a mixture of different types with XLPE or oil-impregnated paper insulation. The effect of these different lengths has been analyzed and some of the results presented in this section. Detailed data for the cable lengths have been provided by Birka Nat and relevant information for this study are presented in the previous section. For the 11 kV cables, these data can be used in an analysis as follows (Table 6.29).

1. Cable feeders:
   - Total number = 32
   - Total length = 259.387 km
   - Average length = 259.387/32 ≈ 8.106 km
   - Minimum length = 0.838 km (Liljeholmen Ls)
   - Maximum length = 15.654 km (Tellusborgsvagen 34)
2. *Cable insulation type and length:* An estimation of cable type is made using an heuristic approach based on the evaluation of the quotient ampere/kilometer [A/km]. If the cable length contains mainly mass-impregnated cable insulation, then the quotient is about 1 A/km, and with XLPE insulation is about 2–4 A/km (for 150–240 mm^2) [137]. Based on this approach, the cable in the Birka case system has been evaluated as follows.

TABLE 6.29

Cable Reliability Input Data from the Birka System Model Used in RADPOW

Cable id	Voltage [kV]	Failure Rate [f/yr and km]	Length [km]	Failure Rate [f/yr]
110 kV a)	110	0.00980	7.1555	0.07012
110 kV b)	110	0.00980	7.1745	0.07031
LH33 kV	33	0.00937	0.0300	0.00028
HDa)	33	0.00937	2.4445	0.02291
HDb)	33	0.00937	2.4375	0.02285
HDc)	33	0.00937	2.4170	0.02265
SJa)	33	0.00937	0.9210	0.00863
SJb)	33	0.00937	0.8930	0.00837
LH11	11	0.01242	8.1058	0.10069

- Liljeholmen Ls: $2.346/0.838 \approx 2.80$ A/km * XLPE cable.
- Hornstullstrand 4: $8.091/15.654 \approx 0.5$ A/km * paper cable.
- 13 feeders are paper cables with a total length of 160.596 km, which gives an average length of 12.35 km.
- 19 feeders are XLPE cables with total length of 98.79 km, which gives an average length of 5.20 km.

3. Failure rates ($\lambda_{permanent}$ f/yr) can be estimated based on knowledge about failure occurrences and cable lengths as follows:
   - $58\,failures/18$ yr $\cdot$ 259.387 km $\approx 0.01242\,failures/$yr and km
     - $\lambda_{min\ length} = 0.01041$ f/yr
     - $\lambda_{average\ length} = 0.10069$ f/yr
     - $\lambda_{max\ length} = 0.19446$ f/yr
     - $\lambda_{average\ paper\ length} = 0.15342$ f/yr
     - $\lambda_{average\ pex\ length} = 0.06460$ f/yr

Consequently, the 11 kV cables leading out from LH11 have a length of $l_{LH11} \in [0.838, 15.654]$ km and a failure rate of $\lambda_{LH11} \in [0.01041, 0.19446]$ f/yr.

For the purpose of these studies, a single LH11 load point (Load Point LH11) has been assumed that represents an average cable and consequently is connected to the 11 kV station with a pair of cables, each about 8.1 km long. The real system has 32 such load points connected by cables varying in length from 0.84 to 15.7 km. Figure 6.28 shows the effect on system reliability resulting from varying this cable length. This figure shows the results from RADPOW analysis of Load Point LH11. It can be seen that the reliability does differ depending on which cable is modeled in regard to both length and type of insulation, where for example the XLPE insulation was shown

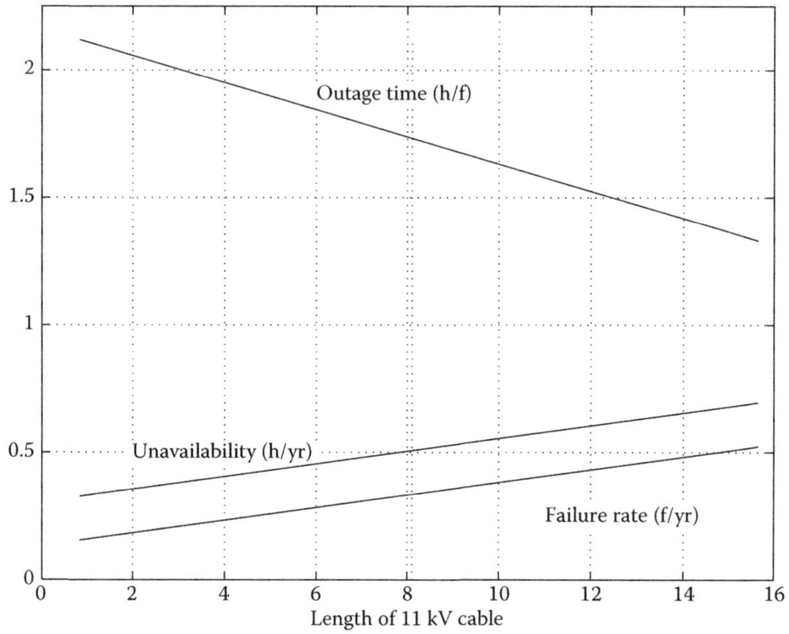

FIGURE 6.28
The effect of the individual cable failure rates on the system reliability, showing results from RADPOW for the Birka case system and Load Point *LH*11, where $\lambda_{LH11} \in [\lambda_{minlength}, \lambda_{maxlength}]$.

to result in fewer faults than the mass-impregnated paper cable. The average cable representation has been used in the following analyses.

Different sources of supply interruption statistics were introduced in Section 6.3. At this point, it is important to assess whether or not the failure rate value found for the 11 kV cable in this study (about 0.01242 f/yr and km) is reasonably compared with general statistics. Table 6.4 presents national statistics from the DAR and FAR systems, as well as internal statistics from Birka Nat. Comparison with the national values shows that the resulting value for this 11 kV cable has a higher failure rate than the corresponding results from DAR. However for higher voltage levels, the national statistics reach the same values, and for the 33 kV level, the national DAR statistics correspond with a value of about 0.01279 f/yr and km. For the internal Birka statistics that cover a larger area than the statistics analyzed in Section 6.3 (the input data), there is good correspondence between the results. These provide a value of 0.01201 f/yr and km for the 11 kV cable. However using the corresponding data from FAR, the value is as high as 0.03057, which could indicate the use of inferior cable types as in earlier times, but may also indicate differences in the reporting systems. Consequently, it has been shown that the

TABLE 6.30
Reliability Indices Evaluated for the Birka System Base Case

Load Point	λ [f/yr]	*U* [h/yr]	*r* [h/f]	*L* [kW]	*LOE* [kWh/yr]
LH11	0.334494	0.506714	1.51487	768.974	389.65
HD	0.111307	0.133388	1.19837	22,999.9	3067.9
SJ	0.110706	0.150543	1.35984	0.8	0.120434

SAIFI = 0.115491 [int/yr, cust.] SAIDI = 0.140386 [h/yr, cust.]
CAIDI = 1.21556 [h/int.] AENS = 0.144988 [kWh/yr, cust.]

resulting average failure rate for the 11 kV cable (LH11) is a reasonable failure rate compared with the general statistics (Table 6.42).

6.5.3.7 Reliability Analysis

Table 6.42 presents the base case results from the reliability analysis.

The failure rate is significant at the 11 kV level compared with the two 33 kV load points, with the average failure rate for an 11 kV customer being about 0.33 f/yr. For the network, this would imply about 11 f/yr, and for the system with 24 similar stations, about 5 f/wk. Real statistics (see Section 6.3 and [113]) indicate about 2 f/wk for this area. The reason for this is that the average length of the 11 kV cables in the case system (8.1 km) is considerably longer than the average for the total 11 kV network (1.6 km). The loads at the different load points vary significantly, which is partly caused by the different number of customers for each one (SJ is used by one and HD by 234,000), and partly by the difference in load per customer (where the average customer at LH11 has a power consumption of about twice the average customer at HD).

6.5.3.8 Identifying the Critical Components

Critical components for the reliability of the system have been identified in similar way as in the analysis of the Flymen system (see Section 6.5.2). The components are first categorized according to their type, then the failure rates are set to zero for one type at a time, and then the resulting indices are evaluated. The different component cases are:

1. Base case
2. Bus bar
3. Breaker
4. Cable
5. Transformer

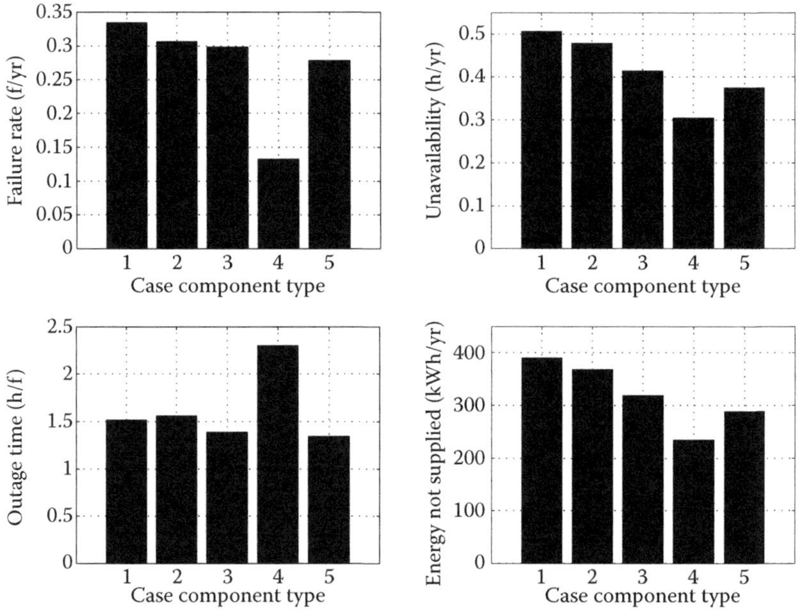

FIGURE 6.29
Impact of critical components showing RADPOW output results for Load Point LH11.

The results for Load Points LHll, HD and SJ are presented in Figures 6.29 through 6.31. These results strongly indicate that the underground cables have the primary impact on system unreliability at the Load Point LH11. The significant rise in the outage time is because the repair time for the underground cables is much lower than that for the other components. Therefore, the average restoration time increases when the number of short interruption times is reduced. However, at the 33 kV load points (see Figures 6.30 and 6.31), other components are deemed to be more critical for this result, namely the breakers. This is an extremely important point because the impact of PM on component behavior should not be determined by considering the components in isolation (most PM programs do this) but should consider the impact that the component has on the particular system in which it is being used. In this case, the conclusion is that cables are the critical components at the 11 kV level.

6.5.3.9 Impact of Different Cable Voltage Levels

The analysis above indicates that the 11 kV cables have a significant impact. Therefore, it is of interest to study the cable components in more detail. A similar sensitivity study has been done to analyze the effect of cables at different voltage levels. The following cases have been considered.

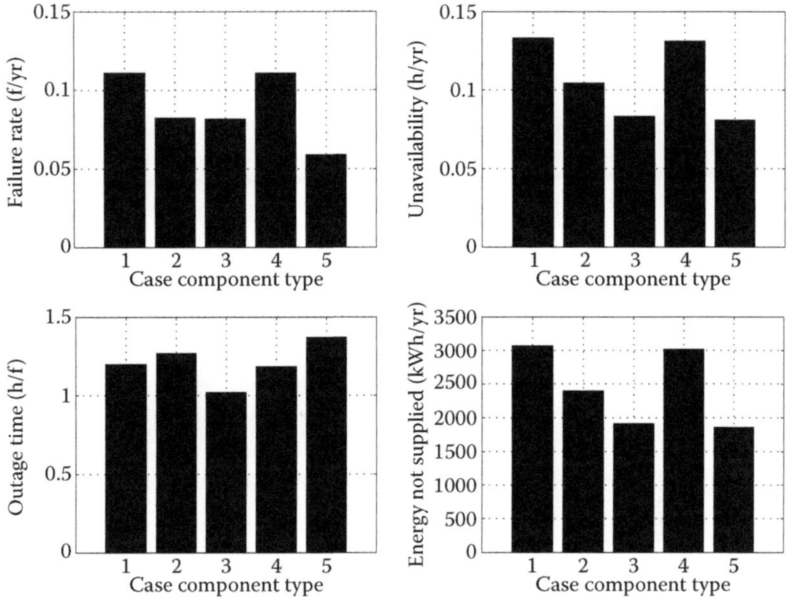

FIGURE 6.30
Impact of critical components showing RADPOW output results for Load Point HD.

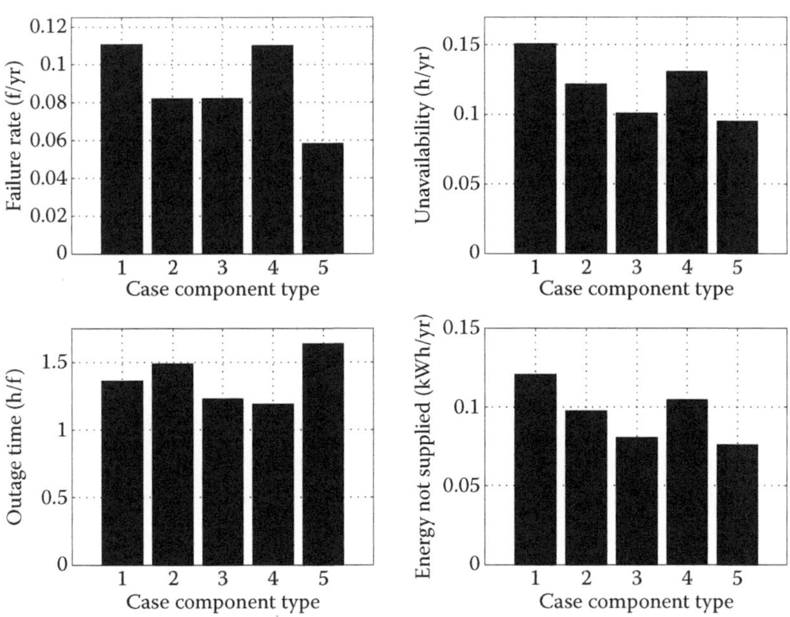

FIGURE 6.31
Impact of critical components showing RADPOW output results for Load Point SJ.

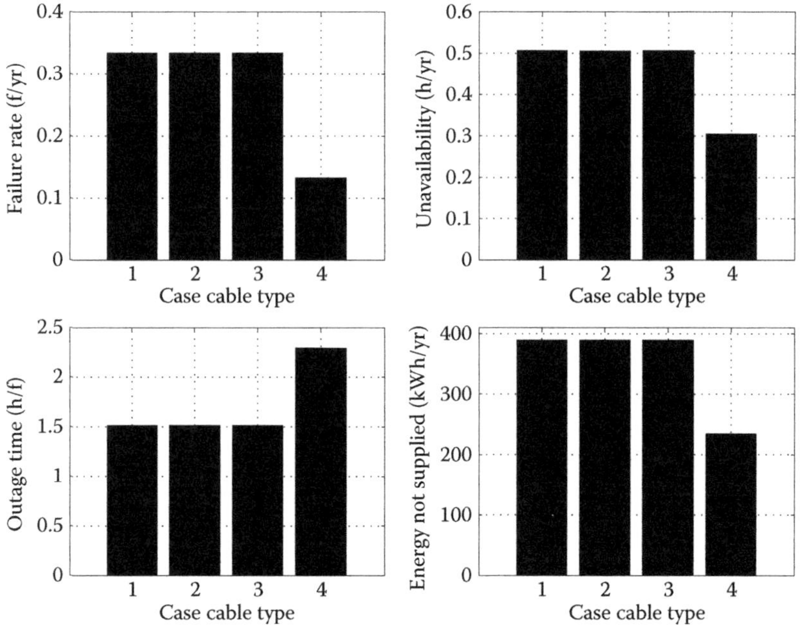

FIGURE 6.32
Impact of cable operation voltage [kV] showing RADPOW output results for Load Point LH11.

1. Base case
2. 110 kV cables
3. 33 kV cables
4. 11 kV cables

Figure 6.32 illustrates the results for the three load points in the Birka system. It further confirms that the 11 kV level has a significant impact, and that the 11 kV cables are the critical components in this system at Load Point LH11.

6.5.3.10 The Benefit of Maintaining the 11 kV Cable

The previous sensitivity studies indicated that the cables have a significant impact on the reliability of the system and are therefore the critical components.

This concludes Step 2 of the procedures defined in Table 4.1. The next step is to analyze the failure modes and causes, in order to identify the improvement that PM could make. Figure 6.4 presented in Section 6.3 summarizes results from a study on the causes of failures for these cables [112]. The cable system in the study included the cable itself, the joints and the terminations.

Figure 6.4 puts the results into the framework of an RCM study deducing the following: function, failure modes, failure events, and failure causes. Finally, the numerical information gained from the statistical survey has been included. It is at this level that the maintenance activities can be considered and applied. The results show that the causes of failures that have significant impact on the 11 kV cables are: Damage (16%), Personnel (12%), and Material/Method (59%). The causes denoted Damage and Personnel are not likely to be affected by PM but Material/Method could be greatly affected by PM. As this category constitutes 59% of the total, this should be considered in greater depth, as also shown in Figure 7.3. This indicates that the main contributions to this category of causes of failures are: fabric and material (14%), lack of maintenance (5%), and incorrect method or instruction (15%).

All of these would benefit by proper maintenance procedures.

Further sensitivity studies have been made to see the effects at the system level if each of these causes of failures could be decreased individually or in combination (in a similar way to what was done in the Flymen system).

The different cases have been listed below:

1. The base case
2. Manufacturing or material faults = 14%
3. Lack of maintenance = 5%
4. Incorrect method or instruction = 15%
5. The total for 2–4 = 34%
6. The total for material and method = 59%

Figure 6.33 shows the benefit of these different cases on system indices. It has been assumed for each case that the causes of failures can be eliminated by the maintenance activities. Thus the corresponding failures would be eliminated and the reliability indices influenced.

The difference between the two last cases (34% and 59%) lies in the reported failure causes material and method but has no further level of classification. This indicates the need for greater focus on completing fault reports with as much detail as possible in order to facilitate the establishment of effective PM programs.

The results establish that there is a benefit in reliability from well-chosen PM actions, however, these improvements must be compared with the cost required to achieving it.

6.5.3.11 Concluding Remarks on the Birka System

The relationship between PM and reliability in an urban distribution system has been studied in order to demonstrate the need for RCM. The results clearly demonstrate the benefits in system reliability obtained by focusing

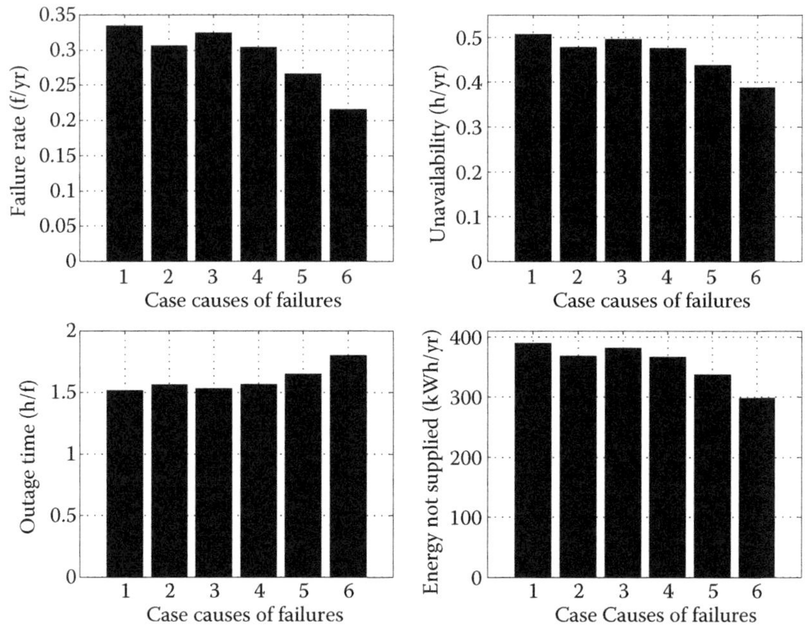

FIGURE 6.33
Effect of reducing the causes of failures in 11 kV cable showing RADPOW output results for Load Point LH11.

maintenance on the critical components, and on the dominant causes of failures for which maintenance can have an impact. These studies involved 11 kV cables and causes of failures related to "material and methods."

The conclusion is that PM can have a great impact on system reliability and the relationship between failure rate and maintenance should be investigated in detail. The results clearly demonstrate the need to identify the critical components, to comprehensively understand the causes of failures, and not to create PM programs by considering components in isolation but always as part of the system in which they are used. This forms a significant input to achieving cost-effective maintenance programs.

6.5.4 Conclusion

This section has presented results from reliability analysis of two different power distribution systems. Approach I was used for relating the PM of components to system reliability, based on statistics on the causes of failures and the assumption that maintenance has a percentage impact on reducing the cause of component failures.

In both of these studies, the critical components for system reliability have been evaluated as well as the failure causes dominating the failure of those component types. The results of the Flymen System (a rural system) show overhead lines to be a critical component in that system, and that failure caused by trees and aging contributes significantly to overhead line failure. The results from the Birka case system (an urban system) show that underground cables of 11 kV are a critical component in that system, and that failure caused by material and methods contributes significantly to failure in underground cables. Both these RCM application studies indicate that the effect of PM on system reliability is significant when maintenance measures are focused on the correct method, allowing them to prevent the correct causes of failures and to target the correct and critical components.

The RCM application studies also provided the following general conclusions.

- The crucial need for the appropriate data should be recognized in relation to both system analysis and the lack of detailed information about the components (e.g., capturing the individual characteristics of a certain type of component).
- For the reliability sensitivities, it could be argued that the differences are less than the accuracy of the input data. However, it is important to recognize that these provide information about the different behaviors in the system, which is the aim of the analysis rather than to give a true picture of the system state.
- For the analysis of the Birka case system, it was shown that critical components are different for different load points. This is an important result because the impact of PM on component behavior should not be determined by considering the components in isolation, but should consider the impact that the component has on the particular system in which it is being used.

The conclusion is that PM can have a major impact on system reliability and that the relationship between failure rate and maintenance should be investigated in more detail. Improvements in reliability indices must be compared with the investment cost required and the benefit in outage cost. Therefore, the relationship between reliability levels and economics should also be investigated. This leads on to the next phase of the analysis that includes a comprehensive analysis of both component behavior in Section 6.6, and system and benefit analysis in Section 6.7.

6.5.5 Concluding Remarks about the Approach for Relating Reliability and PM

The application studies presented in this chapter were based on Approach I, which constitutes the first attempt to solve the question of relating reliability and PM. A fundamental limitation with this approach is that the time aspect is disregarded. The growth of trees over time for example impacts on the causes of failures, which has been disregarded in this approach. This raises the following questions: "How can this approach be improved?" "Could a theoretical relation between failure rate and maintenance be proposed?" and "Could a generic theoretical approach be formulated?"

The time aspect of maintenance has been included in diagnostic methods, therefore, KTH's research activities in cable diagnosis mentioned in Section 6.4 should be helpful for including this aspect. This leads onto Approach II, which is analyzed further in the following two chapters where the underground cable component and the Birka system have been analyzed further in the Cable Application Study.

6.6 Modeling the Relationship between Failure Rate and Maintenance: Approach II

This section defines a quantitative relationship between reliability and PM for an underground cable component susceptible to faulting due to water treeing. This chapter is based on the results from experience, experiments, and physical expressions. The resulting model includes the time parameter and uses the approach referred to as Approach II. The modeling of the cable component is done for the specific purpose of demonstrating how to develop theory for generating generic principles.

6.6.1 Introduction

6.6.1.1 Definitions of Parameters for Component Behavior

Section 6.4 summarized results from different research studies and analyses of the cable component behavior that have specifically looked at both the phenomenon and prevention of water trees. Together with this knowledge about component behavior and previous studies of the failure rate, the following parameters for the component behavior can be defined.

- Water-tree length denoted by l [mm or %] is a measurement of the longest water tree in a cable sample; the percentage unit refers to the proportion of maximum cable insulation thickness.
- Normalized breakdown voltage denoted by U_{bd}/U_0 is a measurement of the cable condition that is its ability to withstand overvoltages, where U_{bd} is the breakdown voltage and U_0 is the phase-to-ground voltage and the quotient is consequently a normalized value.

- Nonlinear loss parameter denoted by $\delta \in''$ is the increase in E11 (typically from $0.5U_0$ to U_0) which is the dielectric loss (the imaginary part of the relative permittivity).

6.6.1.2 Objective

The objective here is to formulate a relationship between system reliability and the PM of components. It has been found that these factors are connected through causes of failures of components. One of the primary aims of the Cable Application Study and the development of a model for a quantitative definition of this relationship is to show that it is actually possible to make the model. Furthermore that rather than being based on assessments, this model can be based on knowledge of the following: actual experience, experiments, and physical models. Consequently, the modeling of $\lambda(t, PM)$ that is presented in this chapter is based on real data and theories that connect preventing failures and improving failure rate. The component focused on in the application study is the underground cable. It has been found that insulation failure is one of the major causes of system failure for cable components. The specific aging mechanism in insulation material is known as water treeing, which was investigated in Section 6.4.

6.6.1.3 Process and Logic

The physical process studied can be summarized as follows: water trees grow through the cable insulation which results in a decrease in the breakdown voltage of the cable. The breakdown voltage of a cable gives a measure of the condition of the cable in the form of resisting overvoltages that is the cable's ability to fulfill component function. The decrease in cable condition leads to a higher failure rate for the component. The parameters included in this process, expressed as functions of time, are (Figure 6.34):

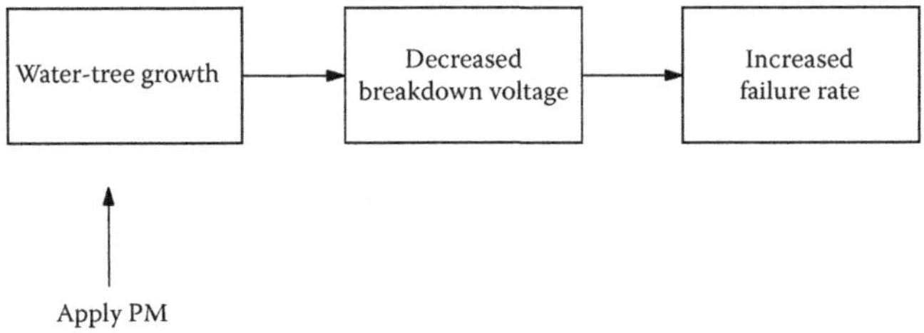

FIGURE 6.34
Illustration of the process relating water-treed cable insulation as the cause of failure to application of a PM activity.

1. The water-tree length denoted by $l(t)$ [mm or %]
2. The normalized breakdown voltage denoted by $U_{bd}/U_0(t)$
3. The failure rate denoted by $\lambda(t)$ [f/yr and km]

The connection between resulting failure rate and cause of failure through water-tree growth is the breakdown voltage:

$$l(t) \Rightarrow U_{bd}/U_0(t) \Rightarrow \lambda(t).$$

In this application study, PM efforts should be applied to prevent the growth of water trees. The logic for preventing the process described earlier can be summarized as follows.

1. PM by rehabilitation methods is applied by injecting silicon to fill the water trees, which results in:
   - Stopping of the existing water-tree growth
   - Improving the condition of the cable
2. The effect of the PM efforts is an improvement in the cable condition
3. The improvement in the cable condition is assumed to stabilize the failure rate

The functional relationships can therefore be expressed as follows:

$$PM \Rightarrow l(t) \Rightarrow U_{bd}/U_0(t) \Rightarrow \lambda(t).$$

Another parameter, the nonlinear loss parameter ($\delta\epsilon''$), has been identified to be of great interest. This parameter provides an alternative indicator (measure) of cable condition without itself being destructive as the breakdown voltage tests (U_{bd}/U_0) [91]. Therefore, this parameter would be of great use for defining when to perform the PM task.

6.6.2 Procedures for Modeling $\lambda(t, PM)$

It should be emphasized that the main difficulties with modeling the relationship between failure rate and PM lie in finding supporting real data and understanding the related physical process. The modeling includes using the available knowledge in a suitable way. The result would be both the resultant model itself and proof that it is possible to achieve.

6.6.2.1 Supporting Data Sources

To relate failure rate with PM, comprehensive knowledge about the component is required. To understand the cable component behavior, a literature study has been undertaken to identify available and possibly useful knowledge. The list below summarizes the useful sources obtained to support the model development.

1. A survey and analysis of XLPE cable faults exhibiting a high number of water trees, reported by Svel (Association of Swedish Electrical Power Utilities) [94]. This supports the modeling of failure rate ($\lambda(t)$).
2. Research results analyzing XLPE cable component behavior in terms of, for example, breakdown voltage (U_{bd}/U_0), loss parameter, and the relationship between these. This is supported by studies made at KTH [122].
3. The relationship between cable condition and loss parameter ($\delta \in ''$) as presented in an ELFORSK report [125], and also in [91].
4. The effect on breakdown voltage (U_{bd}/U_0) of a PM method supported by SINTEF, presented in [127].

   These supporting sources of data have been discussed further in the following sections, while presenting the modeling of the different parameters required in greater detail.

6.6.2.2 Modeling Steps

The aim here is to develop a model of failure rate as a function of time and a PM method ($\lambda(t, PM)$), is based as far as possible on existing data and theories. For the XLPE cable component, this means that the model should be based on the data sources presented in Section 6.6.2.1. Furthermore, this model attempts to relate the following parameters (all expressed as variables of time) to PM:

- Failure rate denoted as $\lambda(t)$ [f/yr and km]
- Water-tree length denoted as lt [% or mm]
- Cable condition defined by the normalized breakdown voltage, denoted as $U_{bd}/U_0(t)$, which could be measured by either,
  - Breakdown voltage tests denoted as U_{bd}/U_0
  - Nonlinear loss parameter denoted by $\Delta\epsilon''$

$$\lambda(t, PM) = \cdots = f(\ldots, \Delta\epsilon'') \tag{6.7}$$

where $\lambda(t, PM)$ is used for prediction, and for example, $\Delta\epsilon''$ is measured. In order to obtain this functional relationship, several relationships between the various parameters have had to be deduced. Which parameter to start with is a matter of choice and the choice here has been to start with the failure rate as a function of time, which would be obtained from statistics.

TABLE 6.31

Summary of Steps in the Deduction of the Model Describing Failure Rate as a Function of PM ($\lambda(t, PM)$)

Step	Activity	Resulting Function	Source
Stage A: Water-tree length			
1	Modeling failure rate	$\lambda(t)$	Svel [94]
2	Modeling water-tree length	$l(t)$	KTH [122]
3	Superposition of results	$\lambda(t, l)$	Steps 1 and 2
4	Modeling breakdown voltage	$U_{bd}/U_0(l)$	KTH [122]
5	Superposition of results	$U_{bd}/U_0(t)$	Steps 2 and 4
6	Modeling loss parameter	$\Delta\epsilon''(U_{bd}/U_0)$	KTH [122]
7	Superposition of results	$\Delta\epsilon''(t)$	Steps 5 and 6
8	Summarize results	$\lambda(t,\ \Delta\epsilon'',\ l,\ U_{bd}/U_0)$	Steps 1, 2, 5, and 7
Stage B: Effect of maintenance			
9	Modeling effect of PM	$U_{bd}/U_0(PM)$	SINTEF [127]
10	Superposition of results	$U_{bd}/U_0(t,\ PM)$	Steps 9 and 5
11	Modeling effect of PM	$\lambda(t, PM)$	Steps 8 and 9
Stage C: Water-tree length			
12	Repetition of results	$\lambda(t, PM)$	Stage A

Table 6.31 presents a summary of the steps required to deduce the failure rate model. The process can be defined in three stages where, in the first and third steps, the length of water trees describes the process, and in the second step, the effect of the PM activity (rehabilitation) is described. Figure 6.35 presents a summary of these three steps as well supporting data and the assumptions made.

Each one of these stages and steps are described in more detail in the following sections.

6.6.3 Modeling Failure Rate

6.6.3.1 Introduction

The objective here is to model the failure rate as a function of time and to base this model on real data. A detailed study of XLPE cable failures has been

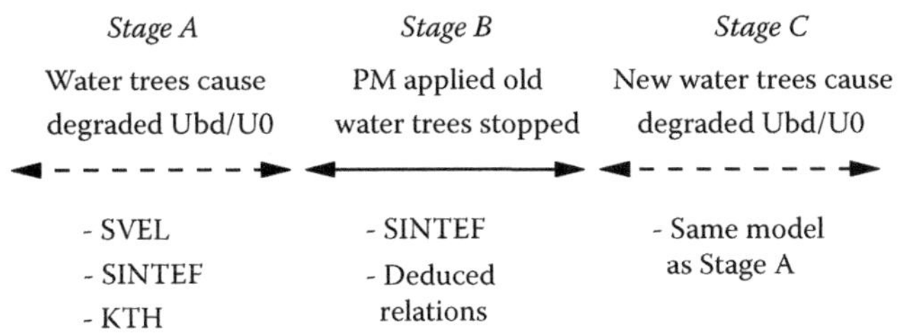

FIGURE 6.35
A three-stage process for the modeling of failure rate as a function of PM ($\lambda(t, PM)$), where the information in the lower part shows sources supporting the model.

done by Svel and presented in [94]. This report has been identified as giving support to a model of failure rate related to failures due to the phenomenon referred to as water treeing.

6.6.3.2 *The Svel Study of XLPE Cables*

Data for XLPE cables exposed to water treeing has been collected for the period 1965–1988 [94]. Two changes in cable production in 1974–1975 have greatly impacted on the water-treeing phenomenon. The first was a change in the outer semiconductor from being painted and taped to being extruded. This decreased water penetration. The second was a change in the binding process from steam vulcanizing to a dry method. This involved a lowering of the water content in the insulation. Consequently, both these changes resulted in a reduction in the initiation of water-tree growth. Based on these changes in the cable component, the statistics can be treated as two different study objects: (i) cables produced up to 1975 and (ii) cables produced after 1975. The causes of cable failures were classified as presented previously, and it has been shown that the significant cause of cable failure is water treeing, averaging 38% of all the causes in the total population of analyzed cables.

In addition to this, cables can be of two different categories, those with and those without the extruded layer. From the statistics presented, it was identified that the inclusion of the extruded layer led to a significant change in failure rate. Of the total 228 cable failures, there were 211 in the 2559 km of cable without an extruded outer semiconduction layer, and 17 in the 10,500 km of cable with the extruded layer.

The total length of cables was estimated and the results have been presented as an average kilometer of cable in service per year. The results show that the first group of cables (produced before 1975) represents a comparatively small part of the population in kilometers, but a comparatively large number of failures.

6.6.3.3 Input Data for the Modeling

The failure rate function ($\lambda(t)$) for an XLPE cable component can be estimated based on the data presented in the previous section. The supporting data selected for the analysis covered:

- The period 1970–1988 (19 years)
- Failures caused by water treeing
- Cables without an extruded layer

Moreover, the cables that were selected to be studied were those that were put into service 1970–1975 (as listed above) and had data collected on them during the period 1970–1988. Cables put into service earlier than this are so small in number and length that the information that would be gained was considered to be not significant. For the cables included in the study, the data collection period is 19 years, which includes the failure process. However, this is not the case for the large population of cables with the extruded layer, which have not yet started to fail given the limits set by the study period and are therefore not included.

Based on the Svel report [94], the age of the cable when the failure occur (in years after being put into service) was identified. These results show that most failures occur after 9–15 years of service. Estimated average length of cable (km) as presented in the report was based on the average length that was extracted for cables put into service during 1970–1975. These data constitute the input source for providing failure rate changes during this 19-year period. Table 6.32 presents the resulting failure data and calculated failure rates. The cables put into service in 1970 have a total aging of 18 years. However, those put into service in 1975 have only a total aging of 13 years. Consequently, the full samples of data cover the years until year 13. This time is defined as follows:

$$t_{svel} = 13 \quad [\text{yr}]. \tag{6.8}$$

6.6.3.4 Mathematical Model and Analysis

The data chosen for the analysis are cables put into operation 1970–1975 and data collected for these during the period 1970–1988. The cable population (total length) is not constant during the time period, however, for a period of 13 years, the data provide results for the complete population. The time parameter in this context could therefore be presented in two different ways, as shown in Figure 6.36; either the real time for the system or the aging time for the different cables. When using this data in the context of modeling the behavior of a cable component, the aging time has been used. Using numerical methods, the experience data presented in Table 6.32 can be adjusted to a

TABLE 6.32

Failure Occurrences (in Years after Being Put into Service) Caused by Water Trees, in XLPE Cables Put into Service during the Early to Mid-1970s

	Number of Failure Occurrences						
Aging [yr]	**1970**	**1971**	**1972**	**1973**	**1974**	**1975**	**λ [f/yr and 100 km]**
0	0	0	0	0	0	0	0.0000
1	0	0	0	0	0	0	0.0000
2	0	0	0	0	0	0	0.0000
3	0	1	0	0	0	0	0.0556
4	0	0	0	0	0	0	0.0000
5	0	1	0	0	4	0	0.2778
6	0	0	2	2	1	1	0.3333
7	0	0	0	1	1	2	0.2222
8	0	0	0	0	2	1	0.1667
9	1	4	1	0	5	1	0.6667
10	0	4	0	1	10	3	1.0000
11	2	5	2	4	3	1	0.9444
12	2	0	5	7	5	0	1.0556
13	6	1	6	9	7	0	1.6111
14	6	3	6	2	5	–	1.3333
15	1	3	4	8	–	–	1.0667
16	4	6	2	–	–	–	1.1429
17	2	2	–	–	–	–	0.6154
18	8	–	–	–	–	–	2.6667
Length	3	3.5	4	4.5	1.5	1.5	(100 km)

curve. In this situation, the curve would present the failure rate as a function of time ($\lambda(t)$).

As mentioned in the previous section, the data set is complete up until year 13 of aging. In the following modeling, results up until 13 years of aging have been used. However, graphs are first plotted for the complete data set.

Figure 6.37 presents results for cables put into service over different years. The plots are results from simple linear piecewise interpolations. The graphs show an initial trend of an almost constant failure rate for about the first 4–6 years, then it increases to a maximum level at about 10–13 years. Figure 6.38 demonstrates the total failure rate over time for the whole cable population. Note that it is only the population of cables put into service in 1970 that are included in the system at the 18th year of aging, and consequently the population decreases seen in that year.

Figure 6.38 shows the behavior of an average cable component from the Svel study. In this analysis here, the objective is to formulate a general

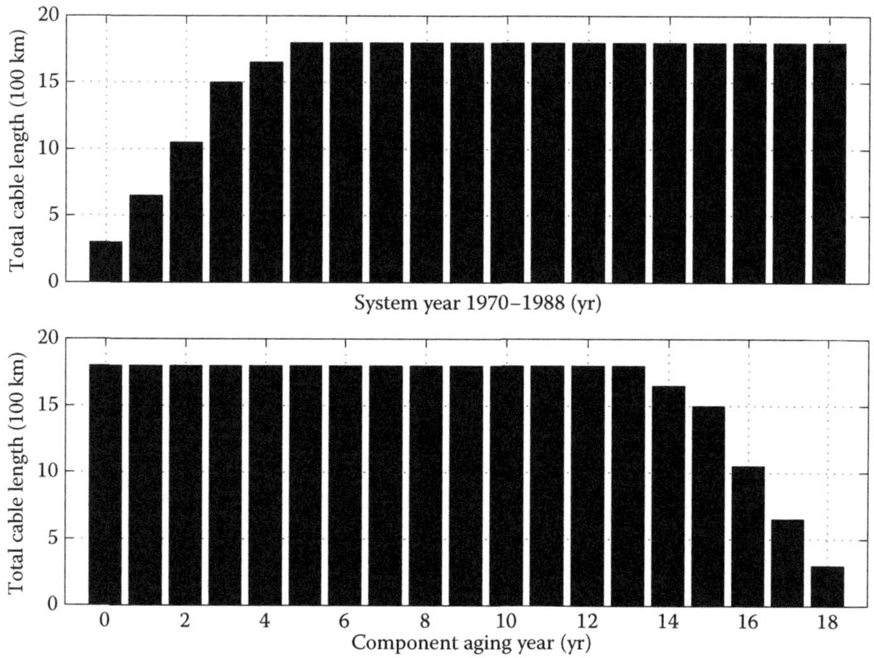

FIGURE 6.36
Total cable length for different time aspects.

function for the failure rate. Therefore, the next step in the analysis is to approximate the available data with a function.

6.6.3.5 Approximation of the Data to a Function

This section introduces two different approaches to solving this task. First, a least square method, and second, a polynomial curve fitting method.

The least square method implies that a known function can be approximated with observed data in such a way that the sum of the square of approximation error can be minimized. Figure 6.38 illustrates the behavior for the complete data set and shows a trend similar to that of an exponentially distributed function. Based on this observation, the following choice of function for the data has been assumed, where the function $F(t) \approx y$, and y is the set of observed values constituting the collected statistics:

$$F(t) = \left\{ \; x_1 \cdot e^{x_2 \cdot t} \quad : \quad 0 \leq t \leq t_{svel}. \right. \tag{6.9}$$

The unknown parameters for the function (x_1 and x_2) can be defined by curve fitting by the least square method. This implies that an abstract model (Ax) is made for the observed data (b). The resulting solutions, that

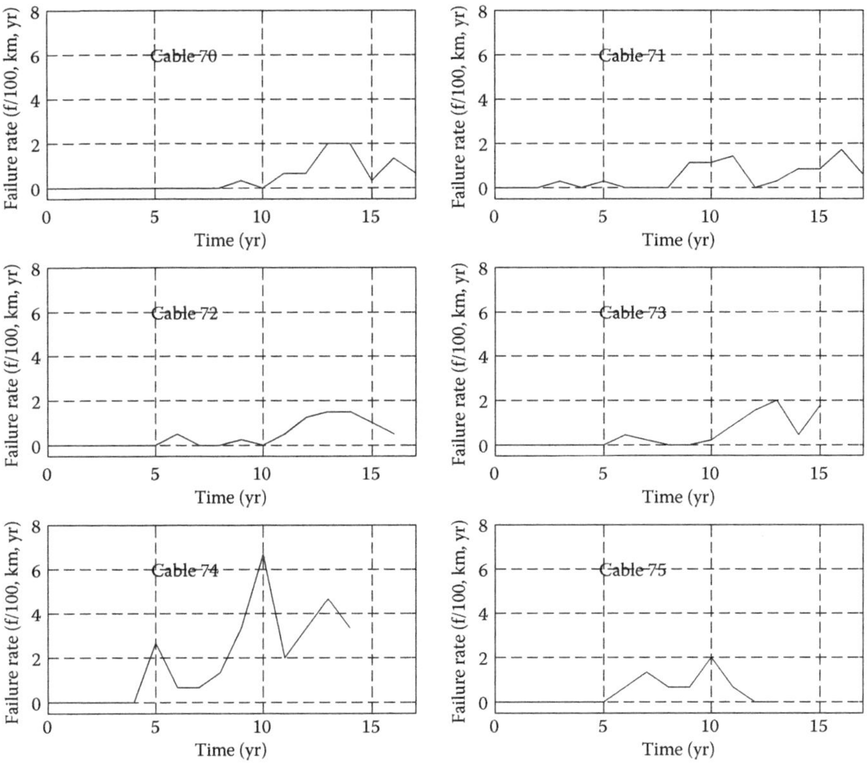

FIGURE 6.37
Failure rate as a function of time for XLPE cables put into service (1970–1975). The cables are categorized into different years being put in operation.

are the normal equations, are written as $A^{T}Ax = A^{T}b$. The vector ($x = (A^{T}A)^{-1}A^{T}b$) is the least square solution to $Ax = b$. In this example, the model function is not a linear function and it is therefore necessary to rewrite the function, for example, by use of the *ln* function. However, the resulting function did not give a good approximation of this data, therefore, an alternative method has been analyzed as follows.

The data has been fitted to different degrees of polynomials, using the MATLAB® standard functions for polynomial curve fitting. Figure 6.39 demonstrates six of these with functions approximated with the data for the full cable population data sets (for 13 years in operation). The accuracy of the model would increase with a higher degree of polynomial, however, at six degrees, too much data are included for capturing the smooth increase in the failure rate. Consequently, the five degree polynomial has been chosen for the following analyses. The error gained using these approximations has been analyzed and no trends have been identified. Table 6.33 summarizes the errors using time steps of 1 h (note that for the following analysis

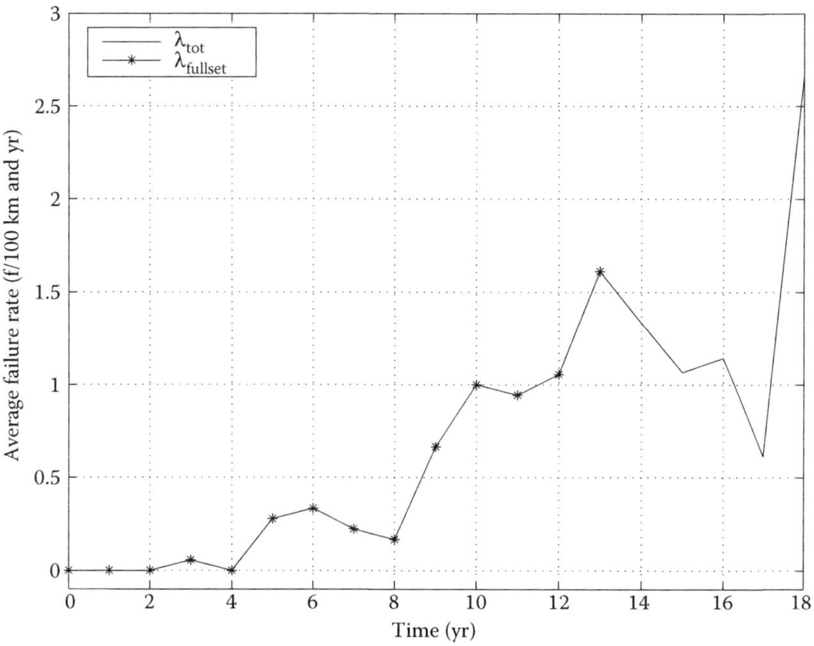

FIGURE 6.38
Failure rate as a function of time for XLPE cables put into service (1970–1975).

TABLE 6.33
The Accuracy of Function Fitting for $\lambda(t)$ and Time Step 1

t [year]	$\lambda_{observed}$	$\lambda_t = f$(poly5)	$\lambda_{observed} - \lambda_t$
0	0	− 0.0080	0.0080
1	0	0.0030	− 0.0030
2	0	0.0189	− 0.0189
3	0.0556	0.0453	0.0103
4	0	0.0860	− 0.0860
5	0.2778	0.1437	0.1341
6	0.3333	0.2200	0.1133
7	0.2222	0.3168	− 0.0946
8	0.1667	0.4361	− 0.2694
9	0.6667	0.5809	0.0858
10	1.0000	0.7559	0.2441
11	0.9444	0.9676	− 0.0232
12	1.0556	1.2254	− 0.1699
13	1.6111	1.5418	0.0693

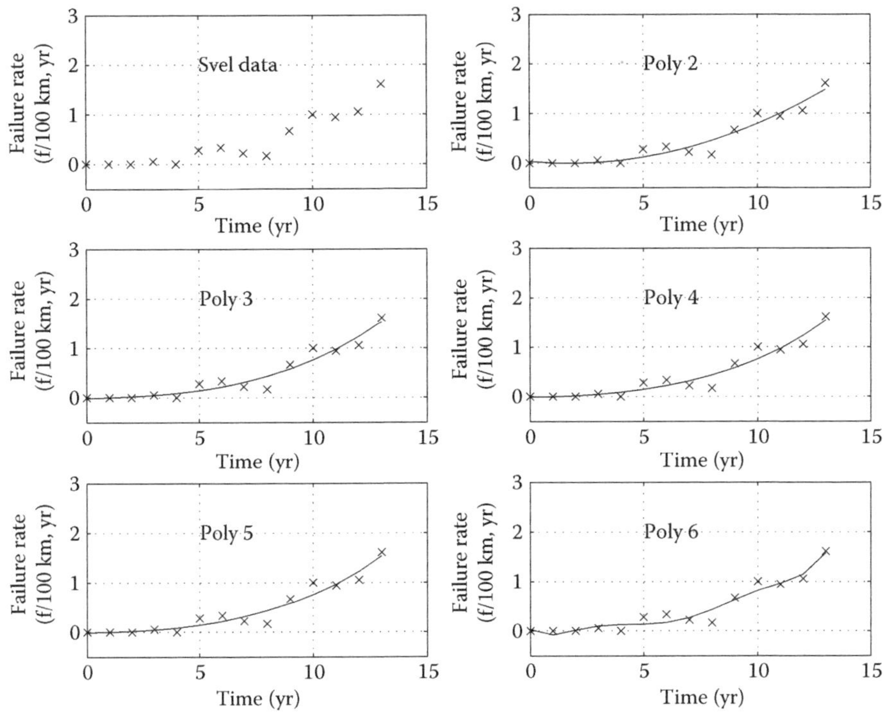

FIGURE 6.39
Polynomial curve fitting functions.

the much smaller time step of 0.01 h has been used). This accuracy table also shows another factor to be considered in the modeling, where the first element shows negative values. This characteristic is obviously not possible for the failure rate function that is being modeled.

To correct this in the model, the resulting approximated function has been adjusted by replacing all the negative elements by zero.

Figures 6.39 and 6.40 show this polynomial fit in detail. The data set including the total length of cables covers the period of time until $t_{svel} = 13$ years. For the modeling of the failure rate, the data set of the total population has been used. The major reason for not including the later time period is because the cable population is smaller and the last point actually only includes one set of cables that were put into operation in 1970.

6.6.3.6 Assumptions about the Failure Rate Characteristics

The approximation of a function of the cable data as presented in the previous section included the period until $t_{svel} = 13$ years of operation or aging. How the failure rate behaves after this period of time is fairly unknown. The

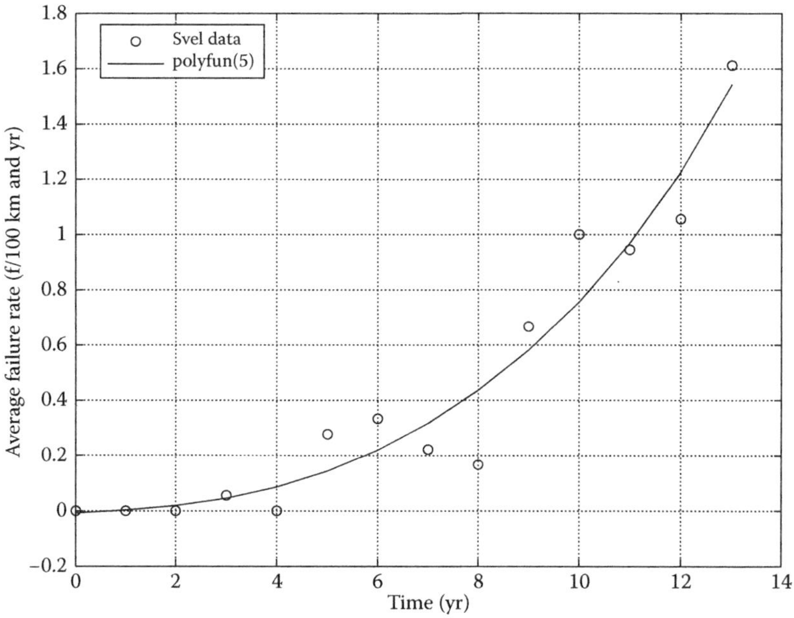

FIGURE 6.40
Polynomial fitting with 5° for the failure rate in the full population data.

data available for the years up until 18 years of operation are poor as these cover such a small part of the cable population. However, it is possible to argue that one characteristic could be better than another. This failure rate function corresponds to the failure due to water-tree length. Svel data support a model for an average behavior between aging years 0–13. However, data do exist for some cables up to the aging time of 18 years. In the model proposed, a polynomial function has been approximated with the data up to year $t_{svel} = 13$, which correspond to a complete population of cables (total cable length). An assumption has to be made about the characteristic beyond this year.

Figure 6.41 shows two different alternatives.

1. Increasing failure rate function, where the assumed function implies a continuation of the polynomial function.
2. Constant function for failure rate, where the assumed constant value is based on the polynomial function value at the time t_{svel} (the end of the full data set).

In the Cable Application Study, the latter of these two alternatives has been used, however, the true characteristic is likely to be somewhere between these two. This is based on the assumption that the water trees have penetrated the

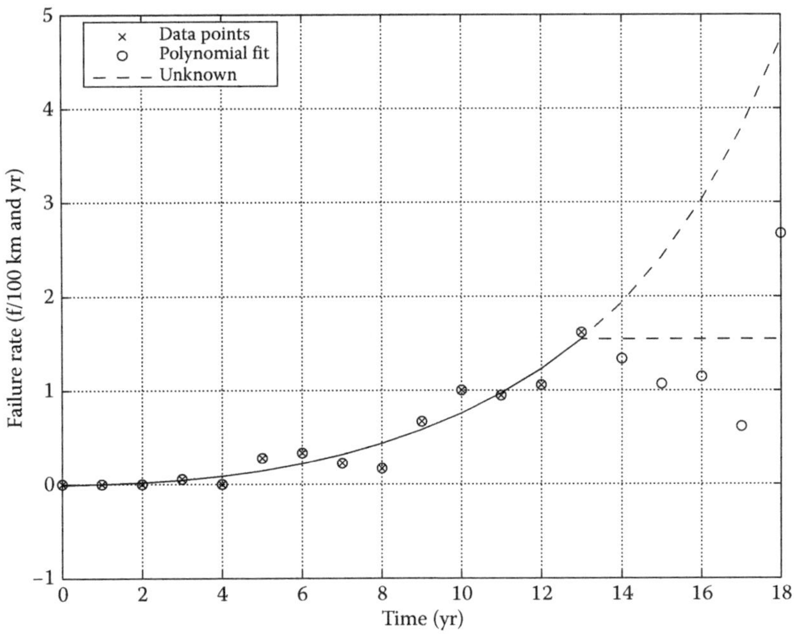

FIGURE 6.41
Assumed characteristics for the failure rate following from the polynomial function.

whole insulation width and therefore, no further decrease in cable condition due to water trees is expected.

For the first alternative, it is interesting to note that the need and the cost of corrective maintenance would escalate with time. Consequently with this model, a greater benefit would be achieved by using PM. However, one argument against the first alternative is that the water trees have a maximum length. Physically then, it is therefore reasonable to assume that the failure rate would reach a maximum value as well. On the other hand, if it is considered that the number of water trees has an impact, it could be argued that the failure rate would continue to increase. Moreover, a second argument against using the first alternative is that the available data based on experience show a decreased rather than an increased behavior in the failure rate. This could, however, depend on that the cables have been replaced.

The following remarks can be made for the second alternative. First that the resulting time period is supported by the SINTEF material [127] which says that it takes water trees about 12–15 years to grow through the insulation. Consequently, the assumption seems reasonable. Second, with respect to the transition between the upward curve given by the polynomial assessment and the constant value, there is a discontinuity that is improbably in a practical sense. Instead it is more likely that there is a smooth transition from the upward gradient curve to the constant. Therefore, the gradient will gradually decrease from the final polynomial value to zero, although over a small time period. However, this problem with the model is compensated by the fact that the assumed function is based on a very limited use of assumptions. Actually the only assumption made is that the end value of the data defines the continued function. Consequently, the smooth function would be achieved at the expense of introducing unknown data, which is not considered sufficient enough reason for doing so.

6.6.3.7 Resulting Model of the Failure Rate in Time $\lambda(t)$

A failure rate function has been defined based on the data from the Svel study and approximations presented in the previous section. Consequently, the first step in the process of modeling $\lambda(t, PM)$ has been completed. The resulting failure rate function with units of failure occurrences per 100 km and yr is defined as follows in general terms:

$$\lambda(t) = \begin{cases} 0 & : \quad 0 \le t \le t_{p0} \\ \text{poly}(f(t), 5) & : \quad t_{p0} \le t \le t_{svel} \\ f(13) & : \quad t_{svel} \le t \le T, \end{cases} \tag{6.10}$$

where $f(t)$ refers to a polynomial that approximates to the Svel data of five degrees. One assumption made for the data was that if the initial function values become negative, they are replaced with zero values. The time t_{svel} refers to the end period of the data from Svel that was identified to provide significant input data.

The resulting failure rate function with the input of data and the use of time steps of 0.01 can therefore be presented as follows:

$$\lambda(t) = \begin{cases} 0.0000 & : \quad 0.0000 \le t \le 0.7400 \\ -0.0080 + 0.0109t - 0.0014t^2 + & : \quad 0.7400 \le t \le 13.0000 \\ +0.0016t^3 - 0.0001t^4 & : \\ 1.5418 & : \quad 13.0000 \le t \le T. \end{cases} \tag{6.11}$$

Figure 6.42 illustrates this function together with water-tree growth over time. This figure shows the resulting failure rate function for the Cable Application Study, that is, where the maximum failure rate level is reached after 13 years of operation. Numerical values for the function are presented in the first column noted $\lambda_{old}(t)$ in Tables 6.39 and 6.41. Note that the tabled values represent function values for integer numbers for time.

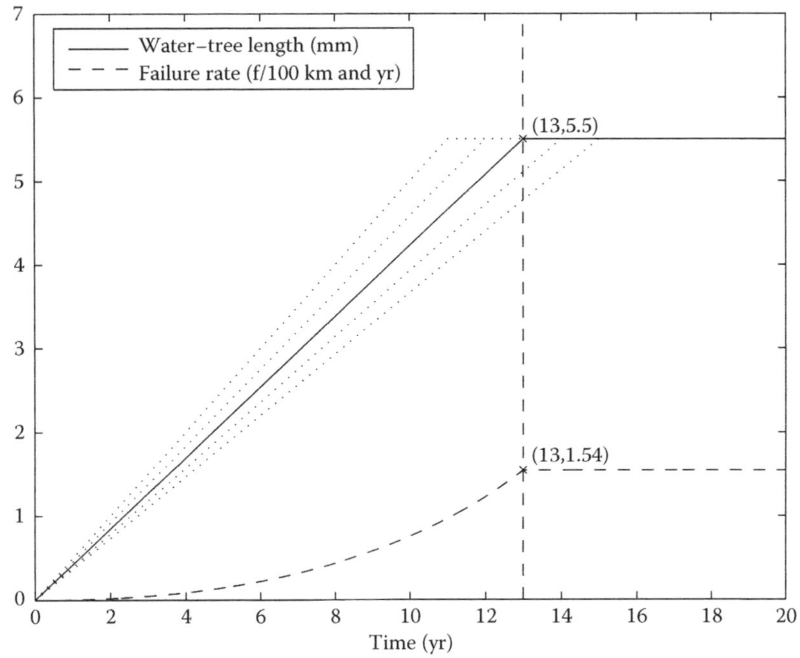

FIGURE 6.42
Relationship between models of water-tree length and failure rate for the XLPE cable component. (The dotted lines show different possible functions for reaching the maximum level of water-tree length.)

6.6.4 Modeling Water-Tree Length

6.6.4.1 Characteristics of Water-Tree Length

The concept of water treeing was introduced in Section 6.4. It was also identified that the type of water trees that are most likely to cause cable faults are the so-called vented trees. Furthermore, it was recognized that these have a characteristic growth involving an initial inception stage, followed by a rapid growth stage, that in a final stages slows down but still continues. However, data supporting such a function in real time is not available. Results have been presented in [122] that show both the length and rate of water-tree growth linear with laboratory aging time. This statement confirms the previous concern about the transitions in the failure rate characteristics.

The evaluation of a method for the rehabilitation of water-treed XLPE cables has been made by the SINTEF in Norway as introduced in Section 6.4. In that study, attempts were made to model water-tree length after fault rehabilitation. The time parameter used was, however, the aging time in the laboratory. The water-tree length was modeled based on the assumption of constant linear growth of water trees.

The conclusion is that it is a very difficult task to model the water-tree length in real time. Even though much research has been conducted within this specific area, the problem has still not been solved. However, that does not necessarily present a reason for not making an initial assumption to see where it would lead. A reasonable assumption seems to be that once the trees start to grow they grow uniformly and with a linear increase until the maximum length is reached. The upper plot in Figure 6.42 presents a set of possible functions based around this assumption.

Data found to support the slope of this function can be found in the SINTEF report [127], which states that maximal water-tree length is reached after about 12–15 years. The report shows results from an application where maximal water-tree length is 5.5 mm (the entire width of the insulation). Based on this result as well as on the previously developed model of the failure rate (see Section 6.6.3.7), the time limit of 13 years ($t_{lmax} = 13$) has been chosen for the Cable Application Study. It is, however, important to recognize that the model being developed does not require such an assumption, and more complex functional relations could also be incorporated. Consequently, the second step in the process of modeling $\lambda(tPM)$ has been completed. The following is the resulting function for water-tree length over time:

$$l(t) = \begin{cases} (l_{max}/t_{l_{max}}) \cdot t & : \quad 0 \le t \le t_{l_{max}} \\ l_{max} & : \quad t_{l_{max}} \le t \le T \end{cases} \tag{6.12}$$

Figure 6.42 includes an illustration of this function. The figure shows both the different possible functions for reaching the maximum level of water-tree length, as well as the specific function for the application study that is with $l_{max} = 5.5$ mm and $t_{lmax} = 13$ years.

6.6.4.2 Relationship between Failure Rate and Water-Tree Length

It has been established in [127] that water-tree length relates to increasing failure rate. Furthermore, failure rate reaches its maximum when water trees cross the whole insulation layer. However, a relationship between these has not yet been defined.

Section 6.6.3.7 defined the function for > ..$(t)f$ or a cable exposed to water treeing, and Section 6.6.4.1 modeled the length of water trees ($l(t)$). Assuming that there is a proportionality between these, and by superposition of these two results would therefore give a relation for the failure rate as a function of water-tree length. Figure 10.9 illustrates the resulting relationship between failure rate and water-tree length. Of critical importance in this context is defining the time period over which the water-tree length has a significant impact on the increase in failure rate. For this example, it seems to be around 6–13 years.

6.6.5 Modeling Breakdown Voltage Function

The condition of a cable can be defined by measuring its breakdown voltage (its capacity to resist overvoltages). These could be due to for example faults in the AC system, switching operations in the network, or lightning.

Results from studies of breakdown strength testing of water-treed XLPE cables have been presented in [122], and report the results of two of breakdown test investigations: the first with lightning impulses and the second with AC impulses. The dielectric loss (ϵ'') at U_0 and 0.1 Hz, and normalized breakdown voltage were plotted for these two investigations. From the first investigation, results show a normalized breakdown voltage varying between 5 and 35 times the service voltage. The second investigation showed corresponding results of about 2–10 times. Consequently, it is of great importance to know what kind of faults are being considered in the analysis of the component behavior. In this present study, the faults of the AC loads are considered. Therefore, the normalized breakdown voltage is expected to lie within the interval of 2–10 times the service voltage.

6.6.5.1 *Breakdown Voltage as a Function of Water Trees*

A correlation between electrical breakdown voltage and water-tree length has been shown and discussed in Section 6.4.3.1. Breakdown voltage and the water-tree content have been investigated in work done to develop the dielectric spectroscopy measurement system for the diagnosis of XLPE cables. Based on these results [90], assumptions have been made about the relationship between breakdown voltage and water-tree length. Further refinements would be possible with access to source data.

Figure 6.43 (after [90]) defines a relationship between breakdown voltage (U_{bd}/U_0) and the longest observed water tree (l_{max}). This figure shows that the breakdown voltage decreases from a maximum value of around 10 to a minimum value of about 2. The water-tree length on the other hand increases from 0 to maximum of 100%.

A simple assumption in the modeling is therefore to assume a linear relationship between breakdown voltage and water-tree length within the boundaries: $U_{bd}/U_0 : 10 \text{ –> } 2$ and $l : 0 \text{ –> } 100$. The resulting model shown in Figure 6.44 can consequently be formulated as follows:

$$U_{bd}/U_0(l) = U_{bd}/U_0(l=0) - \frac{(U_{bd}/U_0(l=0) - U_{bd}/U_0(l=100))}{l_{max}} \cdot l$$

$$0 \leq l \leq l_{max} \tag{6.13}$$

where $l : 0 \rightarrow 100\%$.

In the Cable Application Study, the input data used for this equation are:

$$\begin{aligned} &l_{max} = 5.5\,\text{mm} \\ &U_{bd}/U_0(l=0) = 10, \quad U_{bd}/U_0(l = 100\% \equiv 5.5\text{mm}) = 2, \end{aligned} \tag{6.14}$$

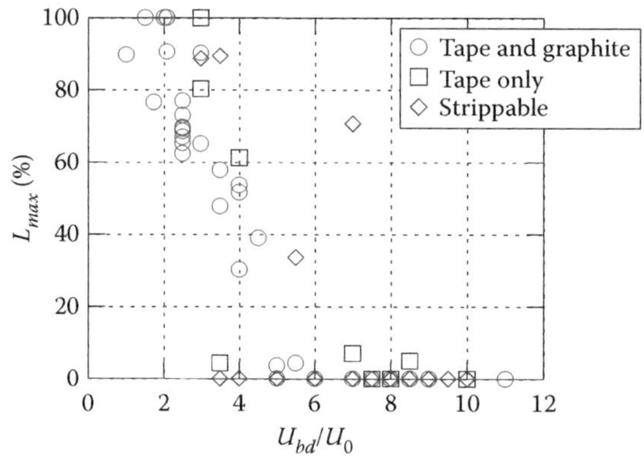

FIGURE 6.43
Relationship between normalized breakdown voltage and the longest water-tree observed in investigated cables. (Adapted from B. Holmgren, *Dielectric response, breakdown strength and water tree content of medium voltage XLPE cables*. Licentiate Thesis, KTH, Stockholm, TRITA-EEA-9705, ISSN 1100-1593, 1997 [90].)

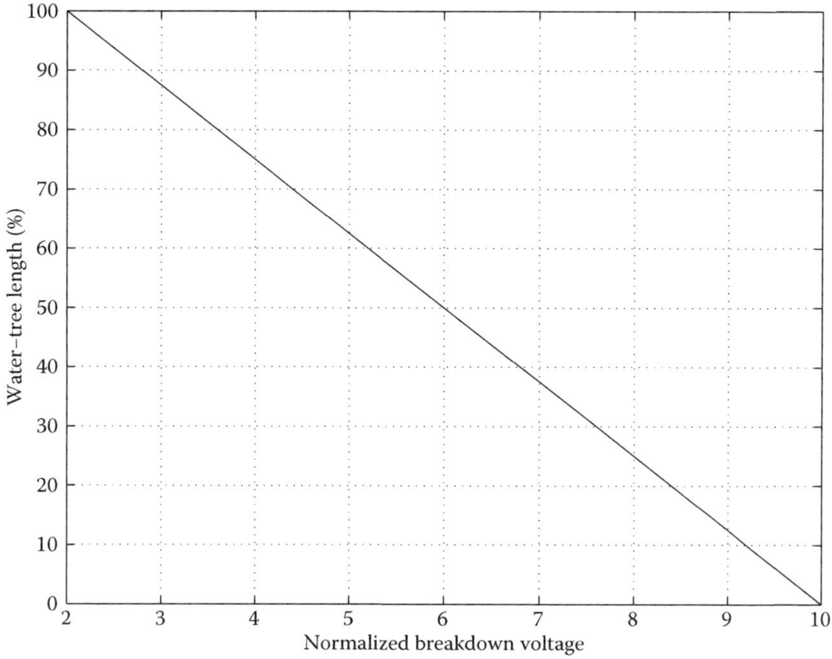

FIGURE 6.44
Functional relationship between breakdown voltage and water-tree length in XLPE cable components, based on the results from KTH and SINTEF.

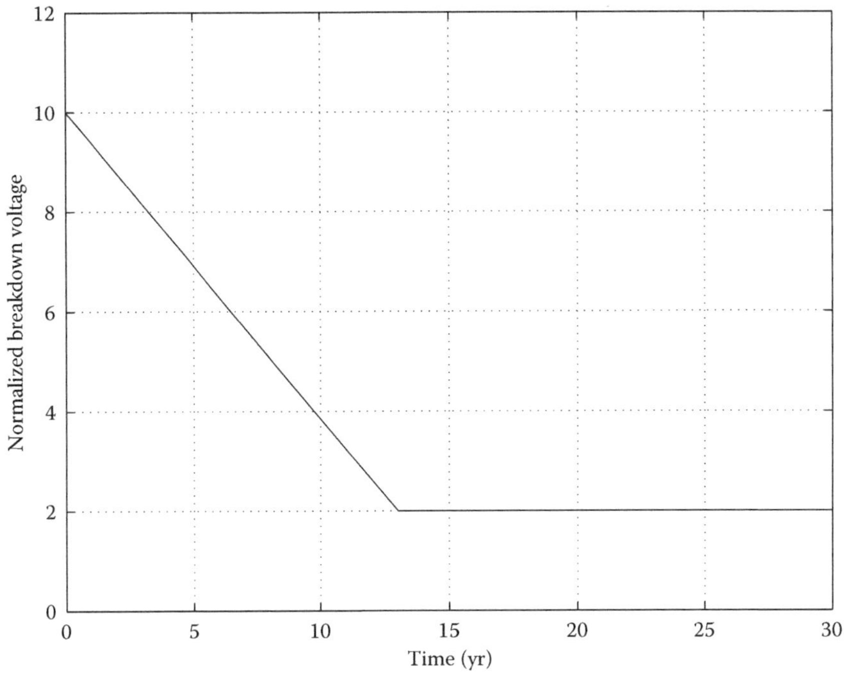

FIGURE 6.45
Model for normalized breakdown voltage as a function of time.

which are based on the results from research results at KTH and SINTEF (as introduced in Section 6.6.2.1). Figure 6.45 shows the results for the breakdown voltage as a function of water-tree length.

6.6.5.2 Breakdown Voltage as a Function of Time

The function for the water-tree length ($l(t)$) was defined in Section 6.6.4.1, and Section 6.6.5.1 presented a definition of a relationship between breakdown voltage and water-tree length. Superposition of these two results would therefore give a relation for the breakdown voltage as a function of time.

The resulting function for the normalized breakdown voltage varying in time is defined by where:

The resulting function for the normalized breakdown voltage varying in time is defined by

$$U_{bd}/U_0(t) = \begin{cases} bd_{max} - k_{bd} \cdot t & : \quad t_0 \leq t \leq t_{l_{max}} \\ bd_{min} & : \quad t_{l_{max}} \leq t \leq T \end{cases} \tag{6.15}$$

where

- bd_{max} is the maximum level for the normalized breakdown voltage
- bd_{min} is the minimum level for the normalized breakdown voltage
- $t_{l_{max}}$ is the year when the water trees reach their maximum length and the breakdown voltage is assumed to reach a constant level
- k_{bd} defines the slope for the decrease as

$$k_{bd} = (bd_{max} - bd_{min})/t_{l_{max}} \tag{6.16}$$

- t_0 is the starting point for time in the model
- T is the end point for time in the model

In the Cable Application Study, the input data for the model of the breakdown voltage consist of the following two values:

$$bd_{max} = 10 \quad \text{and} \quad bd_{min} = 2 \tag{6.17}$$

which are based on the results from KTH [122] discussed in Section 6.4.

The resulting graph of the breakdown voltage as a function of time for the cable is presented in Figure 6.45.

6.6.6 Modeling Loss Parameter Function

This section includes the nonlinear loss parameter ($\Delta\epsilon''$) in the modeling procedure. This parameter can be measured by a nondestructive method and is therefore very valuable to use and preferable to the breakdown voltage test method which is a destructive method (destroying the measured object).

Diagnosis methods for measurements of cable behavior were introduced in Section 6.4.4.1.

It is important to note that since this parameter can be measured without destroying the cable, it could for example be used as an indicator of when PM should be undertaken. Consequently, it is of great interest to include the loss parameter in this procedure for modeling the relationship between failure rate and PM in real time.

It is also relevant to comment that no previous results have been found that relate these parameters and specifically for this purpose. This is not only because it is difficult to accomplish but also because the benefits of doing it are unproven, which the work done here aims to rectify. Consequently, it has been shown that data from earlier investigations of the loss parameter and the breakdown voltage (as presented in [122]) can be used to obtain the relations necessary to establish the required functional relationship for the model being developed.

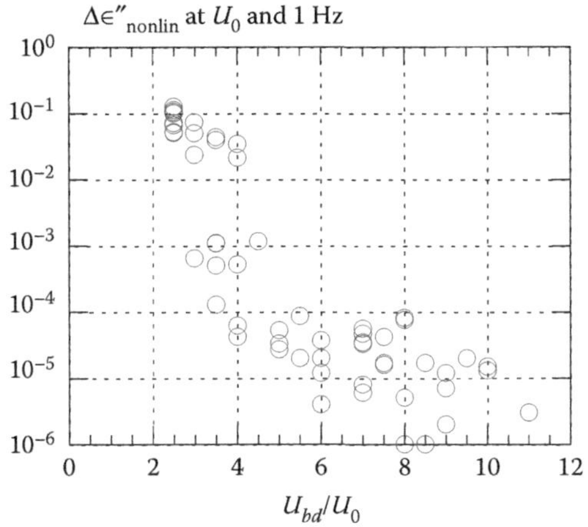

FIGURE 6.46
The nonlinear loss parameter as a function of normalized breakdown voltage. (Adapted from P. T. P. Werelius et al., *IEEE Transactions on Dielectrics and Electrical Insulation*, 8(1), February 2001 [123].)

6.6.6.1 Relationship between the Loss Parameter and Breakdown Voltage

Results from XLPE cable studies of more than 200 field measurements combined with laboratory investigations have been presented in [122]. These include an analysis of the relationship between the nonlinear loss parameter ($\lambda \in''$) and the breakdown voltage (U_{bd}/U_0). Figure 6.46 from [122] presents the resulting measurements.

The data in the above show a trend where the breakdown voltage decreases from about 10 asymptotically towards 2.5. $\Delta \in''$ varies from about 10^{-5} to 0.1 (note that the values for the nonlinear loss parameter are presented in log scale). An exponential trend can be seen for the relation. Based on such an assumption, an exponential function has been estimated to fit the measurement data. Figure 6.47 shows the resulting function, which is based on the data presented in Table 6.34. Similarities between the two figures can be seen, with a significant increase in the loss parameter function when the breakdown voltage decreases towards 2.

6.6.6.2 Loss Parameter as a Function of Time

The model for breakdown voltage over time presented in Section 6.6.6.1 shows an initial decrease from 10 and then a constant value at 2. The loss parameter on the other hand (seen in Figure 6.47) shows a relationship with breakdown voltage decreasing from the value of 10–2.5. Consequently, these two results together show the relations for these parameters as functions of time.

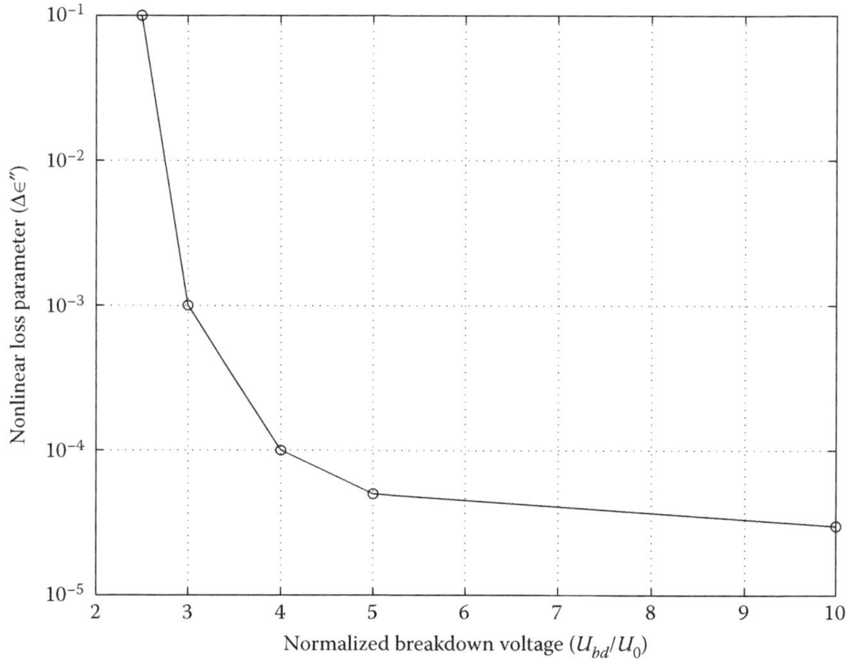

FIGURE 6.47
The nonlinear loss parameter as a function of normalized breakdown voltage.

TABLE 6.34
Nonlinear Loss Parameter and Normalized Breakdown Voltage Input Data

Loss Parameter $\Delta\epsilon''$	Breakdown Voltage U_{bd}/U_0	Equivalent Time [yr]
0.1	2.5	12.1875
0.001	3	11.3750
0.0001	4	9.7500
0.00005	5	8.1250
0.00003	10	0

Figure 6.48 shows the results when superposing the relationships defined earlier on. Compared with the previously modeled parameter functions, the loss parameter exhibits behavior with high sensitivity.

6.6.7 Resulting Model of Cable Component Behavior over Time

The preceding sections have treated the modeling of cable component behavior. For the Cable Application Study, this has meant modeling the

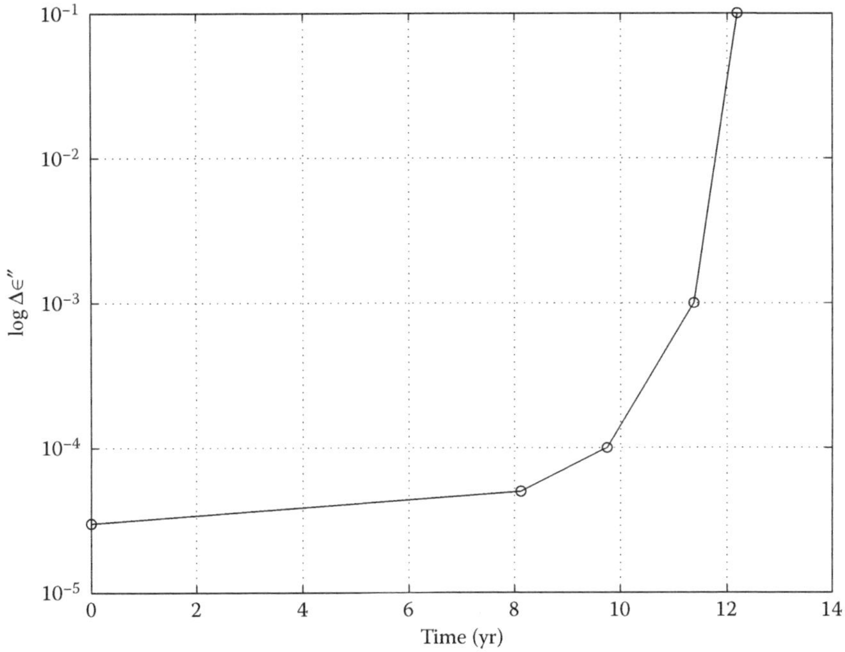

FIGURE 6.48
Model of the nonlinear loss parameter function over time.

characteristics of an XLPE cable that is effected by water-tree growth through its cable insulation. Furthermore, the cable component behavior has been defined by the following three parameter functions:

1. Failure rate ($\lambda(t)$)
2. Water-tree length ($l(t)$)
3. Normalized breakdown voltage ($U_{bd}/U_0(t)$)

where the last parameter can be measured by either breakdown voltage tests or measurements of the nonlinear loss parameter ($\delta \in''$).

Figures 6.49 and 6.50 illustrate these functions for the Cable Application Study. The results demonstrate a model for the effect of water treeing in an XLPE-insulated cable component.

Table 6.35 summarizes the resulting functions, the approximations used to obtain these, the assumptions made, and the supporting data sources. Some clarifications are required here in relation to the table. On the one hand, the approximation implies assumptions about the function characteristics. For example, in regard to the failure rate, the function is approximated to the data from Svel by a polynomial function. On the other hand, the assumptions refer to the specific adjustments for a realistic model of the parameter

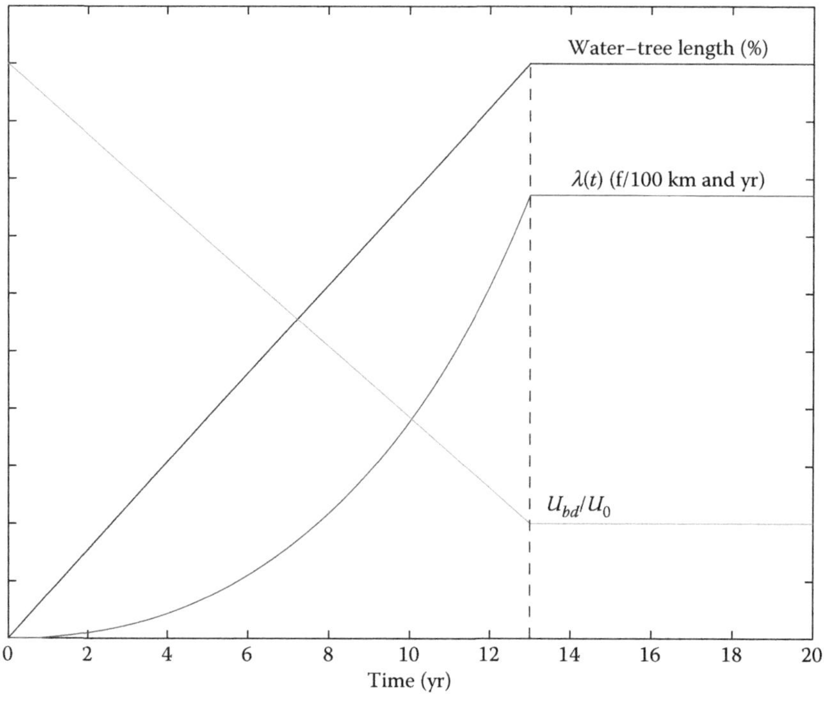

FIGURE 6.49
Resulting model for XLPE cable component behavior.

where: (i) the failure rate cannot drop below zero or (ii) the specific assumptions for providing a single result for the cable, being the maximum length of water trees after 13 years, which was shown to lie within the real time period. Finally, it should be recognized from the results presented in the table that this in some regards heuristic approach for defining a quantitative model for the component behavior is first supported by real data, and second is based on a limited set of assumptions that could easily be adjusted. This completes the first task in the modeling of component behavior, being to deduce a functional relationship for the component over time, that is, with aging.

6.6.8 Modeling the Impact of Rehabilitation Preventive Maintenance

6.6.8.1 Introduction

Preceding sections of this chapter have defined component behavior by modeling the two functions failure rate ($\lambda(t)$) and water-tree length ($l(t)$), and indicated the condition of the cable using the functions normalized breakdown voltage ($U_{bd}/U_0(t)$) and the linear loss parameter ($\delta \in ''(t)$). As the main

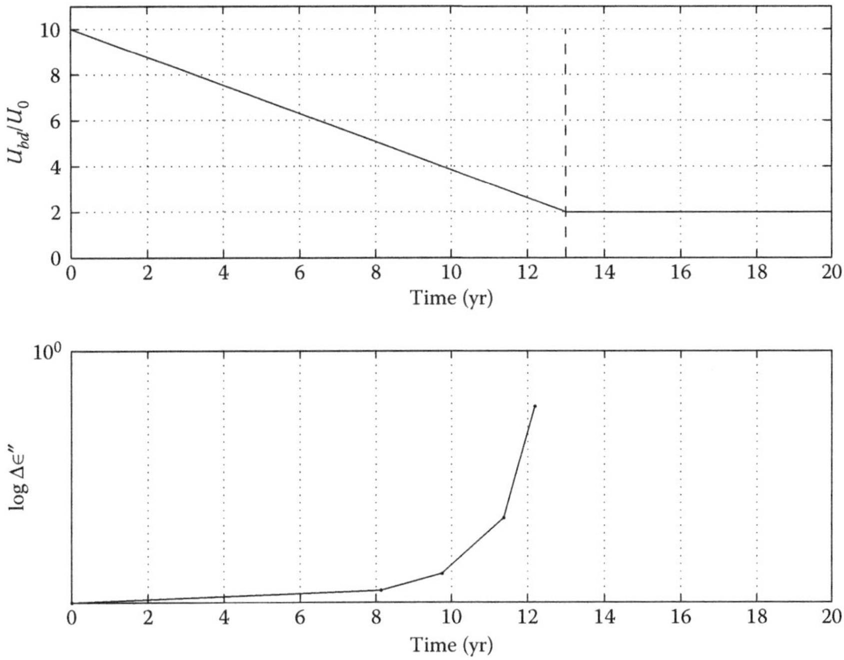

FIGURE 6.50
Resulting model for XLPE cable component condition.

objective of the work presented in this book is to achieve a model that provides a quantitative relationship between PM and reliability, the next step in the modeling procedure is to resolve how these defined functions can be extended to include the effect of PM.

It has proven quite difficult to find the literature or research results to support such a model, and no similar attempts to relate the information in this way have been found.

However, for this application study of an XLPE cable affected by water treeing, results have been found that can support such a model. SINTEF in Norway have analyzed the effect of applying a rehabilitation method for XLPE cables with the aim of preventing failures due to water-tree growth [127]. Experiences and results from this study have been identified to support the objective here of modeling the cable component behavior, which can be used in this study to translate the benefit of PM on breakdown voltage (U_{bd}/U_0) into a benefit in the failure rate ($\lambda(t)$).

6.6.8.2 SINTEF Study of XLPE Cables

The effect of PM on an XLPE cable exposed to water treeing has been studied at SINTEF (see [127] and in Section 6.4.4.2). The condition of the cable

TABLE 6.35

Summary of the Model for Cable Component Behavior

Failure Rate $\lambda(t)$	**Water-Tree Length $l(t)$**	**Cable Condition $U_{bd}/U_0(t)$, $\Delta''\epsilon(t)$**
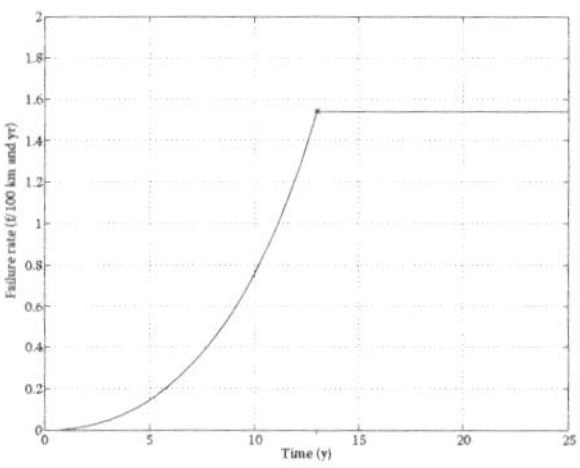	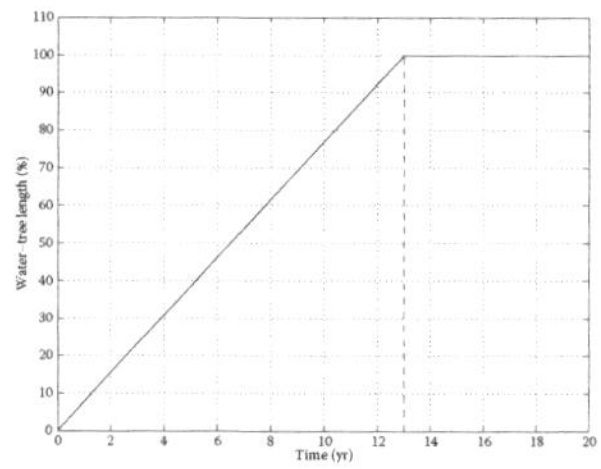	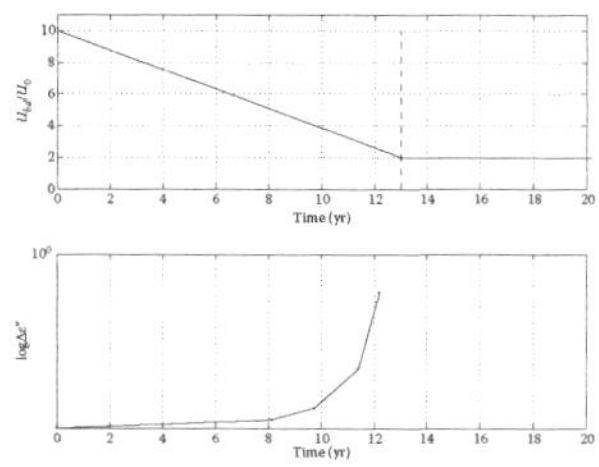
Approximation		
poly(f, 5)t:0→$t_{svel} = 13$	Linear to max. value	Linear from max. to min.
	Constant at max.	Constant at min.
Assumptions		
$f(t) < 0 \rightarrow f(t) \equiv 0$	$t_{l_{max}} = 13$	$bd_{max} = 10$, $bd_{min} = 2$
$t > t_{svel} \rightarrow f(t) \equiv f(t_{svel})$		$\Delta''\epsilon(U_{bd}/U_0)$
Data source		
Svel [94]	SINTEF [127]	KTH [122]

TABLE 6.36

Results from Breakdown AC Testing for Water-Tree Affected Cables (after SINTEF)

$U_{0.63}$ [kV]	95% Conf. Interval [kV]	t [mth]
52.2	46.2–57.3	0
64.3	58.2–69.4	4
78.5	71.8–84.2	16
73.7	64.7–81.3	28

was defined by measuring the breakdown voltage in different cable samples. Furthermore, a statistical analysis was made of the measurement test results. This analysis resulted in a probability of breakdown voltage occurring of 0.63 based on Weibull distribution with a 95% confidence interval. These measurements were made at four different times: (i) just before rehabilitation, (ii) 4 months after, (iii) 1 year after, and (iv) 2 years after rehabilitation treatment. The total measuring period was consequently 2 years and 4 months. Table 6.36 and Figure 6.51 present the results from the study. The change in breakdown voltage as a function of time has been shown and it can be seen that the last drop in breakdown voltage is not numerically significant. Therefore, only the first three points have been used in the modeling.

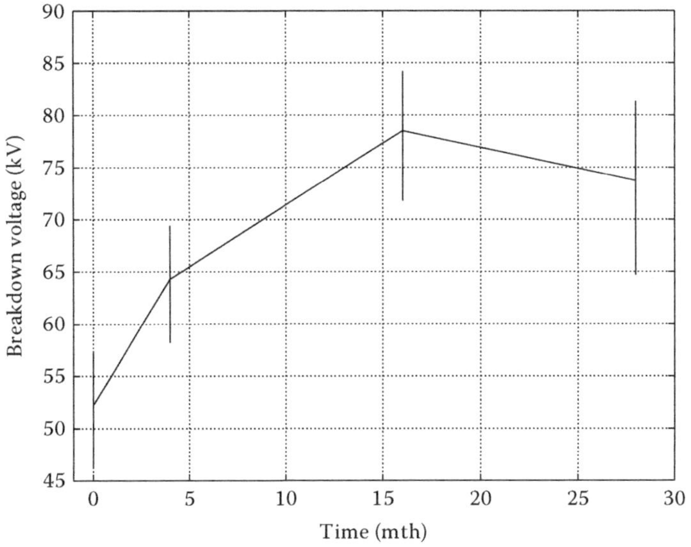

FIGURE 6.51
Change in the breakdown voltage of XLPE cables due to PM through rehabilitation (results from the SINTEF).

6.6.8.3 Input Data for Modeling the Effect of Rehabilitation

Results from investigations of XLPE cable conditions done at KTH [122] are presented in terms of normalized breakdown voltage. Normalization implies that the breakdown strength is divided by the rms value of the phase-to-ground voltage (U_0). In the SINTEF study, the rehabilitation was applied to a 24-kV cable that is with phase-to-phase voltage of 24 kV. For a symmetric three-phase voltage system (i.e., three sinusoidal voltages with a phase angle of 120 between the voltages and with the same peak value compared with ground), the phase-to-phase voltage equals the root square of three times the phase-to-ground voltage. Consequently, the resulting normalized breakdown voltage for this study is defined as

$$U_{bd}/U_0 = \{U_0 = 24/\sqrt{3}\,kV\} = U_{bd} \cdot \frac{\sqrt{3}}{24} \approx U_{bd}/14. \tag{6.18}$$

The results of the benefit in breakdown voltage shown in the SINTEF study were presented as measurements of the breakdown voltage. Therefore, these results have been evaluated as normalized values for further use in this study. Consequently, data for the defined significant data points (presented in Table 6.36) have been normalized for this application study (see Table 6.37). These data have been used as input data for the benefit of PM in this application study.

TABLE 6.37

Breakdown Voltage Input Data for the Application Study

Time t [yr]	$U_{bd}(t)$ [kV]	$U_{bd}/U_0(t)$
0	52.2	3.7672
1/3	64.3	4.6405
$1\frac{1}{3}$	78.5	5.6652

6.6.8.3.1 The Benefit in Breakdown Voltage

The benefit in breakdown voltage can be defined by the percentage increase in breakdown voltage, and evaluated as follows:

$$\Delta U_1 = \left(\frac{100}{U_{bd}/U_0(t=0)}\right) U_{bd}/U_0(t=1/3) - 100 \tag{6.19}$$

and in similar way for the next level of increase (ΔU_2). The resulting numerical values for the Cable Application Study are

$$\Delta U_1 \approx 23\% \quad \text{and} \quad \Delta U_2 \approx 22\%. \tag{6.20}$$

Figure 6.52 shows the resulting benefit in normalized breakdown voltage due to the rehabilitation of XLPE cable.

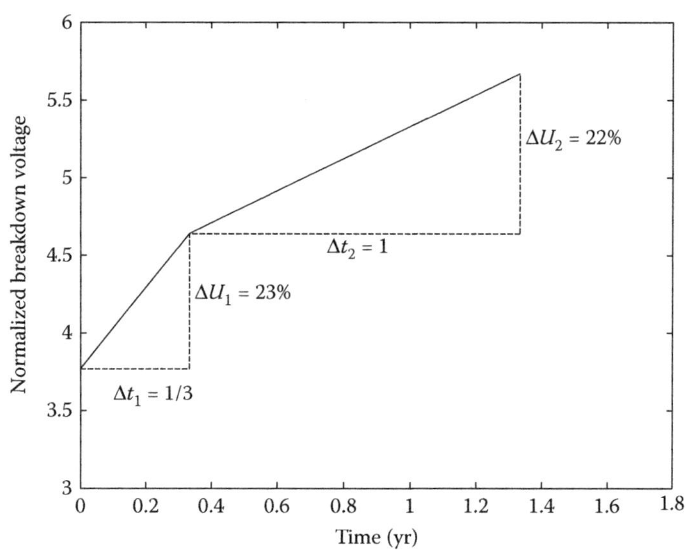

FIGURE 6.52
Input data for the analysis of the benefit in breakdown voltage due to PM through rehabilitation using input data based on the results from SINTEF.

6.6.8.3.2 Discussion and Validation of Results

Section 6.6.8.2 presented a definition of a model for the effect of PM on the breakdown voltage, based on the results from a SINTEF study. However, this model included only the time period after about 1 year following rehabilitation. These results show that this rehabilitation led to an initial rapid increase in breakdown voltage, which after about a year degraded somewhat.

This information should be related to the previously identified relationships for the cable component over time. Therefore, it is of interest to investigate whether or not any knowledge is available about at the time the rehabilitation activity was applied or the condition of the cable at that time. The breakdown voltage at the first measuring point could be used for this purpose. This is the point where $U_{63\%} = 52.2$ kV, which equals the normalized breakdown voltage of 3.767. Equation 10.7 gives a point that would agree with a degraded breakdown voltage after about 10 years of operation ($t = (10 - 3.767) \cdot 13/8 \approx 10$ yr). The corresponding water-tree length could be calculated from previous results as follows: $l = (100/13) \cdot 10 \approx 77\%$ or 4.2 mm. This value could be compared with the statement in the SINTEF report that the cables put through a test would have water trees of about 95%–99% of the maximum water-tree length, which would imply about 12 years of operation in the model developed here. Consequently, there is a difference of about 20% or 2 years. However, given the quite poor availability of data suitable for comparing these results with other sources, and the need to use several heuristic assumptions, differences are not so unexpected. However, the results show that the input data sources are comparative, which indicates that the resulting model could be fairly realistic.

6.6.8.4 Effect of Rehabilitation on Water-Tree Length

In the rehabilitation method, the injected silicon penetrates gradually through the insulation with the effect of stopping water-tree growth. However, new trees are not prevented from growing. The new water trees are assumed to start to grow with the same initial behavior as the previous trees as presented in Section 6.6.4.1.

Figure 6.53 shows the resulting behavior of water-tree length before and after cable rehabilitation. The parameter of water-tree length ($l(t)$) is consequently included in the failure rate model in the first and last stages, and with similar characteristics in both of these.

6.6.8.5 Modeling the Effect on the Breakdown Voltage

The rehabilitation process leads to both a stop in water-tree growth as the silicon penetrates through the insulation, as well as an improvement in breakdown voltage.

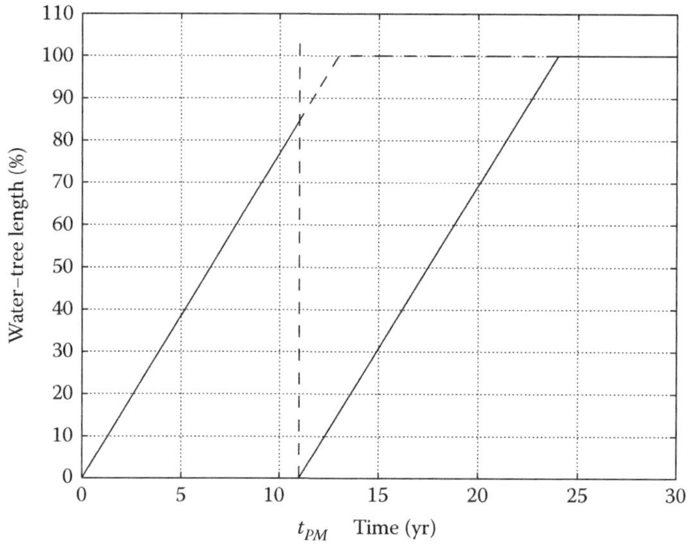

FIGURE 6.53
Model of water-tree length before and after the rehabilitation at time t_{PM}.

It could be asked why the breakdown voltage does not reach the values of new cables. Is the rehabilitation process inadequate for this? There seems to be no firm answer to this, and the SINTEF report supports this modeling, showing a certain but limited benefit. Although finding valid or certain answers may not be possible, some indications would still be useful anyway; for example, water trees still exist but are now filled with silicon and so may behave better than water trees otherwise, but not as well as pristine XPLE. However, the *silicon trees* are then prevented from growing further, which causes the failure rate to remain constant until more *water trees* grow to a substantial size.

After the rehabilitation, new water trees will start to grow and the breakdown voltage will degrade again because of these. By using some previously deduced relations between water-tree length and breakdown voltage, a point can be identified where the water trees would contribute to a further reduction in the breakdown voltage.

Based on the previous relationships between breakdown voltage and the effect of the PM through rehabilitation, a resulting function can be defined, built up of six different intervals. The characteristic for the first interval depends on the time that the PM activity is applied. For a useful effect, the PM activity should be applied before the cable reaches the constant low level of breakdown voltage that is during the period with a constant decrease of the function that identifies the first interval. The following two intervals

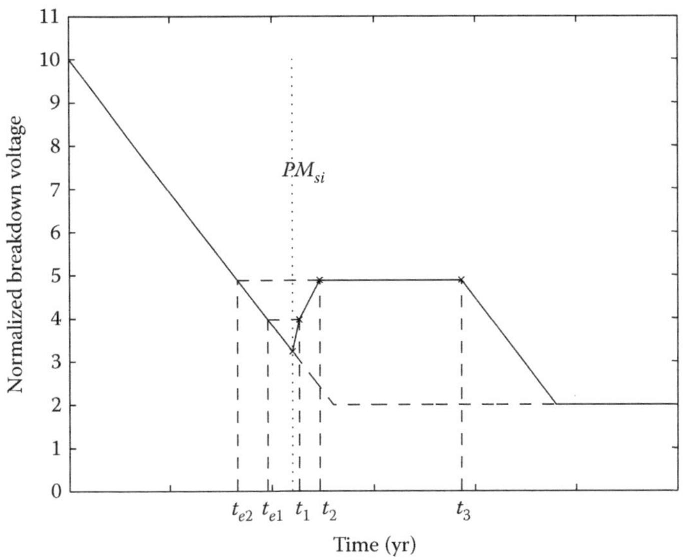

FIGURE 6.54
Time parameters for deducing the effect on breakdown characteristics due to PM through rehabilitation.

relate to the effect of increased breakdown voltage due to the PM activity within intervals Δt_1 and Δt_2 based on the SINTEF data. The third interval represents the new level of improved breakdown voltage that becomes constant when the water trees have been filled with silicon. However, due to the growth of new water trees, the breakdown voltage for the cable will continue to decrease when these new water trees have reached a significant level. Finally, the water tree reaches a maximum level which would result in maximal impact on the breakdown voltage which has then reached its low constant value.

The breakdown voltage is consequently affected by both a benefit from the rehabilitation as well as a decrease due to water trees before and after the maintenance activity. Figure 6.54 shows in detail the different time points and functional relationships required to obtain the resulting function for the breakdown voltage. This includes the time when PM is applied (t_{PM}), as well as the measured effect of the PM (t_1 and t_2). It also includes the equivalent times for obtaining the corresponding values of the breakdown voltage (t_{e1} and t_{e2}) that are based on the assumption that the cable behaves the same as a new cable with respect to the growth of new water trees after the PM activity. Finally, Figure 6.55 presents the resulting function for the normalized breakdown voltage in relation to the length of water trees.

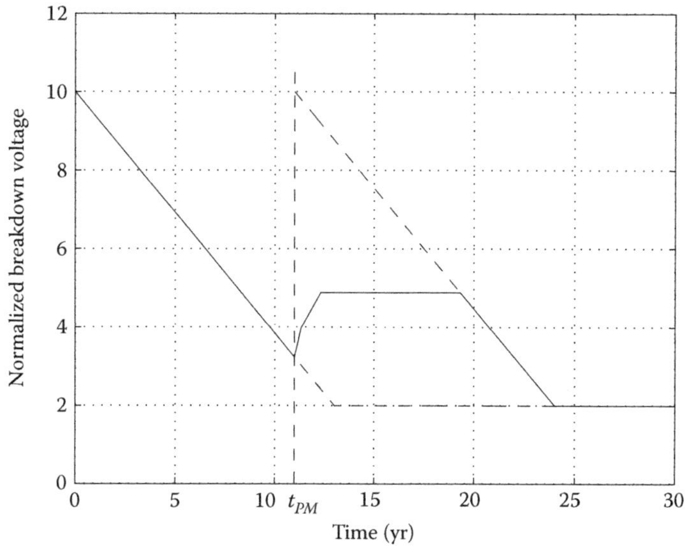

FIGURE 6.55
Model for the normalized breakdown voltage affected by PM through rehabilitation showing results for $t_{PM} = 11$.

Equation 6.21 summarizes the resulting function for the breakdown voltage as follows:

$$F_{U_{bd/U0}}(t, PM) = \begin{cases} bd_{max} - k_{bd} \cdot t & : t_0 \le t \le t_{l_{max}} \\ (bd_{max} - k_{bd} \cdot t_{PM})(1 + \frac{\Delta U_1}{\Delta t1} \cdot (t - t_{PM})) & : t_{PM} \le t \le t_1 \\ (bd_{max} - k_{bd} \cdot t_1)(1 + \frac{\Delta U_2}{\Delta t2} \cdot (t - t_1)) & : t_1 \le t \le t_2 \\ bd_{max} - k_{bd} \cdot t_{e2} & : t_2 \le t \le t_3 \\ bd_{max} - k_{bd} \cdot (t - t_{PM}) & : t_3 \le t \le t_{l_{max}} + t_{PM} \\ bd_{min} & : t_{l_{max}} + t_{PM} \le t \le T. \end{cases} \tag{6.21}$$

Additional parameters included in the modeling of the impact of the PM have been presented in the following list:

- t_{PM} is the time when PM is applied
- ΔU_1 and ΔU_2 are the benefits gained in normalized breakdown voltage due to the rehabilitation, where 0.23 implies an increase of 23%
- Δt_1 and Δt_2 are the measuring points for the change in breakdown voltage
- $t_1 = t_{PM} + \Delta t_1$

TABLE 6.38

Summary of the Input Data for the Breakdown Function for the Cable Application Study

Parameter	Value	Source
b_{max}	10	KTH
b_{min}	2	KTH
$t_{l_{max}}$	13	KTH and SINTEF
ΔU_1	23%	SINTEF
ΔU_2	22%	SINTEF
Δt_1	0.33	SINTEF
Δt_2	1	SINTEF

- $t_2 = t_1 + \Delta t_2$
- $t_3 = b_{max} - k_{bd} \cdot t_{PM}$, which is the intersection point when the decrease in the breakdown voltage caused by new water trees reaches the same level as the constant level gained after the rehabilitation

The input data for this specific study are summarized in Table 6.38, which also defines the values for k_{bd}, t_1, t_2, and t_3.

6.6.8.6 Modeling the Effect on the Failure Rate

New water trees start to grow just after rehabilitation. However, these do not initially lead to cable faults, since the normalized breakdown voltage (in this example about 10) gives the cable much better characteristics than required to withstand overvoltages (in this example about 5). Therefore, the cable faults start to occur first after a time when the water trees have grown a certain length, and the breakdown consequently decreases. After rehabilitation, the failure rate function over time is assumed to exhibit an equivalent effect as does the breakdown voltage. This is based on the relationships between the parameters illustrated in Figures 6.56 and 6.57.

This means that before the PM activity is applied, the failure function follows the original function, referred to further on as $\lambda_{org}(t)$. This function was presented earlier in Section 6.6.3.7, and Equations 6.10 and 6.11. A characteristic for this function is an increased failure rate function due to water-tree length that reaches a maximum value which defines a continued constant rate. Further on when the PM activity is applied, the effect from the rehabilitation is treated in similar way as for the breakdown voltage, which implies that the failure rate is decreasing within the period of t_1 and t_2, due to the benefits $\delta U1$ and $\delta U2$.

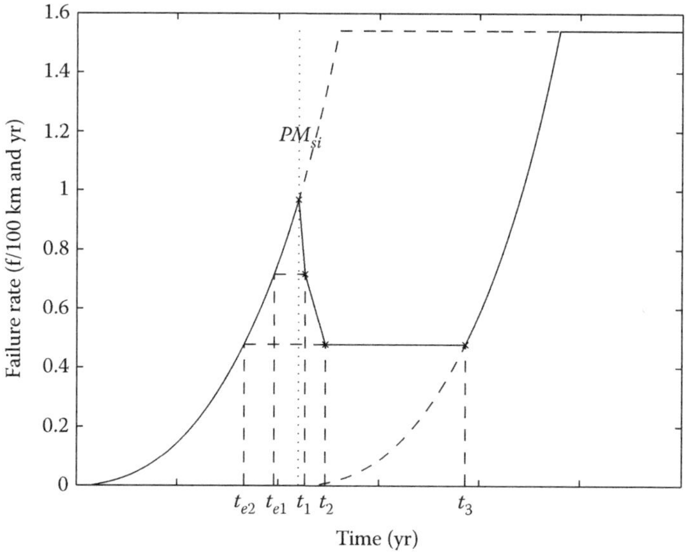

FIGURE 6.56
Time parameters for deducing the effect on the failure rate function of PM through rehabilitation, using the results for $t_{PM} = 11$.

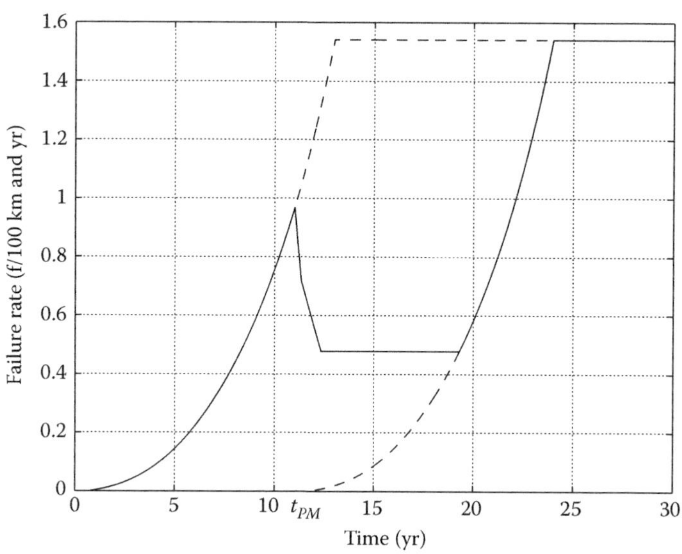

FIGURE 6.57
Model of the failure rate function affected by PM through rehabilitation, using the results for $t_{PM} = 11$.

The effect of failure rate due to the PM activity is defined based on relationships between breakdown voltage and failure rate over time deduced earlier. The corresponding time for the function to reach the same value before PM was decided for the new breakdown voltage values. This time is referred to as an equivalent time. Based on the SINTEF data, there are two such times: t_{e1} and t_{e2}. The corresponding failure rates at these times have been defined. A linear relationship has been assumed between the time period δ_{t1} and δ_{t2} as for the breakdown voltage. However, it should be noted that as previously described, the failure rate function defined follows a polynomial function rather than a linear one. Therefore, the function defined includes the assumption that the change in the failure rate within these points can be assumed to be linear. The difference in numerical values by using either of these functions interpolated between the equivalent times would be very small. Therefore, the conclusion is that the simplified function can be justifiably used for convenience.

Due to the growth of new water trees, the failure rate will increase again. This effect begins at the point t_3 when the effect of the new water trees on the failure rate becomes predominant on the overall failure rate for the cable. Consequently, this function ($\lambda(t, PM)$) can be formulated for five different periods as presented below.

The resulting failure rate function in response to rehabilitation is consequently affected by both (i) an increase in breakdown voltage (due to the PM which stop the growth of the existing water trees) and (ii) an increase in water trees before and after the maintenance activity. The following formula summarizes the function:

$$\lambda(t, PM) = \begin{cases} \lambda(t) & : \quad t_0 \le t \le t_{PM} \\ \lambda(t_{PM}) - \frac{\lambda(t_{PM}) - \lambda(t_1)}{\Delta t_1}(t - t_{PM}) & : \quad t_{PM} \le t \le t_1 \\ \lambda(t_{e1}) - \frac{\lambda(t_{e1}) - \lambda(t_2)}{\Delta t_2}(t - t_{e1}) & : \quad t_1 \le t \le t_2 \\ \lambda(t_{e2}) & : \quad t_2 \le t \le t_3 \\ \lambda(t_{e2} + t) & : \quad t_3 \le t \le T. \end{cases} \tag{6.22}$$

Figure 6.56 shows in detail the approach used for defining the failure rate with the benefit due to the rehabilitation. It specifically illustrates the different time points and corresponding function values.

Finally, Figure 6.57 presents the resulting failure rate function in an XLPE cable affected by and rehabilitated from water-tree growth.

6.6.8.6.1 The Effect of Rehabilitation on the Loss Parameter

The last parameter to define here is the nonlinear loss parameter. Section 6.6.6.2 defined a relationship for the nonlinear loss parameter as a function of time ($\Delta\epsilon''(t)$). Furthermore, this function was defined based on the results relating the loss parameter to the breakdown voltage. If the same relationship between the nonlinear loss parameter and the breakdown voltage is assumed to hold before as well as after the PM rehabilitation, then the resulting relationship for the nonlinear loss parameter including the effect of rehabilitation could be obtained. However, there are no available data to support this assumption, and it does not seem reasonable to make this assumption either, since the cable condition changes as a result of the rehabilitation, and a new characteristic would be expected for the loss parameter.

Nevertheless, it is appropriate to note the following. First, the loss parameter could be used primarily to define the point in time to apply the PM action, and second it should be possible to measure the effect of PM on the loss parameter to define a relationship if needed. In this model, it is suggested that the nonlinear loss parameter can be used for defining when to apply the activity. It would therefore support in defining maintenance strategies.

6.6.8.7 Resulting Model of the Rehabilitated Cable Component Affected by Water Treeing

Figure 6.58 presents the resulting functions for the three variables, $\lambda(t)$, $l(t)$, and U_{bd}/U_0, that characterize the behavior of the XLPE cable component. The resulting model relates the failure rate to the effect of the PM activity applied to prevent water-tree growth in an average XLPE cable.

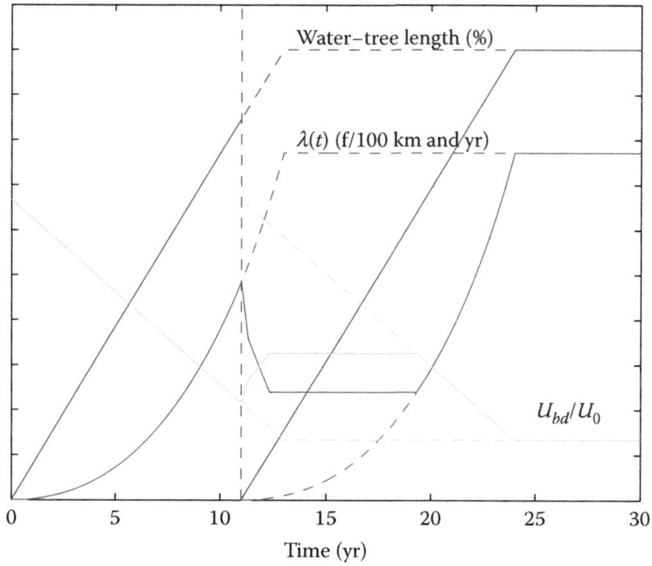

FIGURE 6.58
The resulting relationships for an XLPE cable component defining a model of $\lambda(t, PM)$.

6.6.9 Modeling the Impact of PM by Replacement

6.6.9.1 Introduction

In this section, an alternative PM method to rehabilitation has been studied that involves replacement of the cable component. This situation is used, for example, if a population of cables is to be replaced, but due to limits in expenses and the impact on system function, they must be replaced at various times during an investment period. Consequently, this step in the modeling procedure is an extension of the previously defined functions for component behavior including the effect of PM by replacement.

6.6.9.2 Assumption and Validation

It is assumed that the PM activity of replacement implies that the cable component behaves the same way as a new component after the replacement. This assumption implies that the component behavior including the effect of PM by replacement results in function characteristics like those for cable aging summarized earlier. Consequently, this assumption has produced a simple and straightforward solution for this step of the modeling. However in model validation, it should be asked whether or not this assumption is reasonable.

The practical implication of the assumption is that the replaced cable (the new cable) has the same behavior as the old cable originally did. However, this would not be the case in reality for an XLPE cable, as the problem with water treeing occurred primarily in the cables produced in the mid-1970s. If those cables are replaced with a cable produced today, the same water-tree development phenomenon would not appear. On the other hand, for other types of components such as overhead lines or transformers, the new component would be expected to closely resemble the old component. Consequently, the assumption is justified keeping in mind that the application study is being done with the primary objective of showing how the modeling can be realized rather than defining the true model for this specific case.

Another aspect that justifies the assumption is that it allows for the so-called worst case, since the new XPLE cables are less prone to failure caused by the water-tree phenomenon than the old cables. Therefore with this assumption, the failure rate function is expected to have worse component characteristics and consequently, the benefit of replacement would represent a worst-case situation.

The conclusion is consequently that the assumption made for the effect on component behavior by replacement is valid and should produce results with a pessimistic view of the benefit from the PM activity. Figures 6.59 and 6.60 show the results for the XLPE cable component using this model in the cable application example.

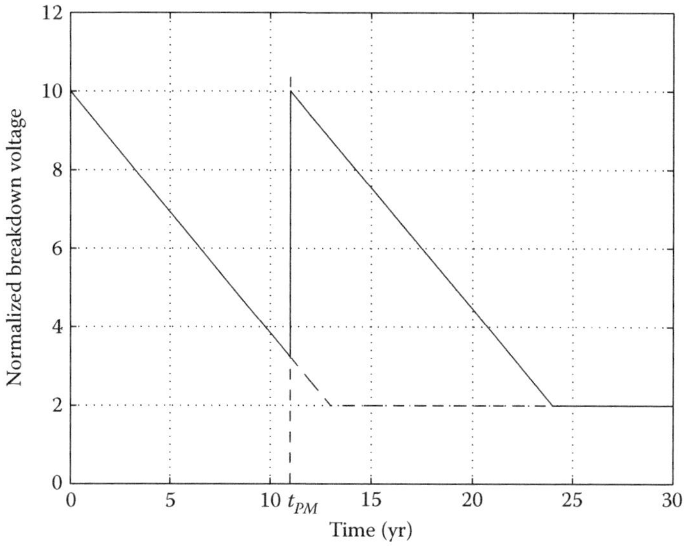

FIGURE 6.59
Effect of the replacement of a cable system component on the breakdown voltage.

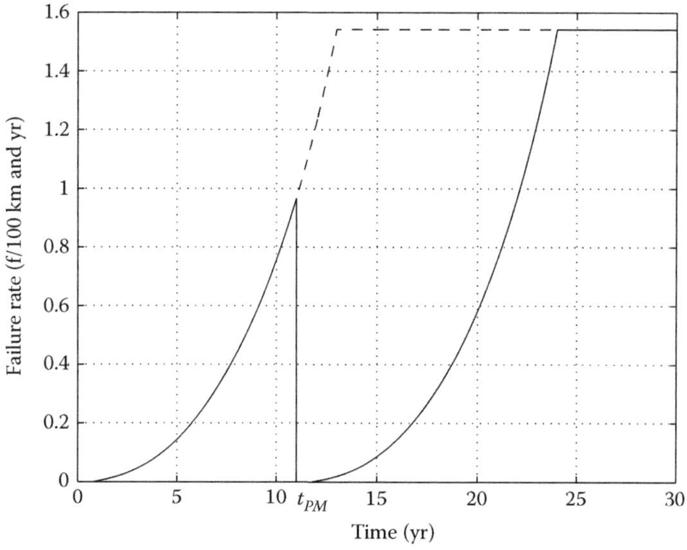

FIGURE 6.60
Effect of the replacement of a cable system component on the failure rate.

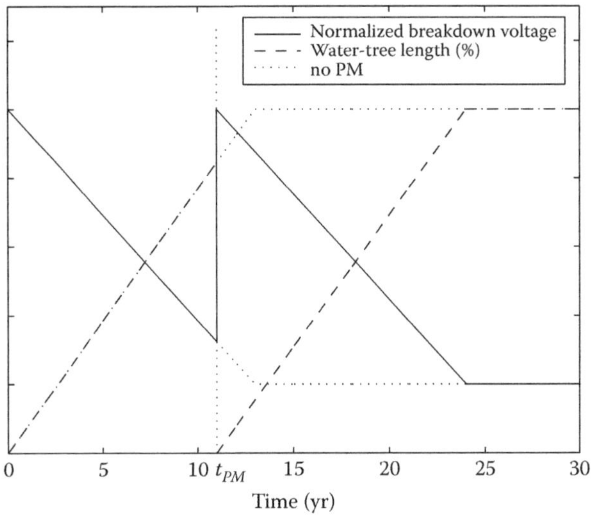

FIGURE 6.61
The effect of the replacement of a cable system component on breakdown voltage and water-tree length.

6.6.9.3 *Modeling the Effect of PM*

The effect of PM by replacement of a cable system component has been analyzed with the same procedure as for PM by rehabilitation. Figures 6.61 and 6.62 show the results of the functions for the breakdown voltage and the failure rate. As expected, the water-tree length determines these characteristics.

Figure 6.63 shows the resulting model for failure rate when the XLPE cable component undergoes PM by replacement.

6.6.10 The Effect of Different PM Plans

In the previous sections, the impact of PM by either rehabilitation or replacement has been modeled. However, one effect that has not yet been introduced is when to apply the PM activity.

The point in time when PM is applied is crucial to the maintenance planning and strategy definitions. These issues are considered in detail in the next section of this chapter. However, in this section the various effects of PM have been analyzed in respect to when the PM effort is applied. The results have been shown for the Cable Application Study. A specific maintenance plan has also been defined and used for the analysis in the next chapter.

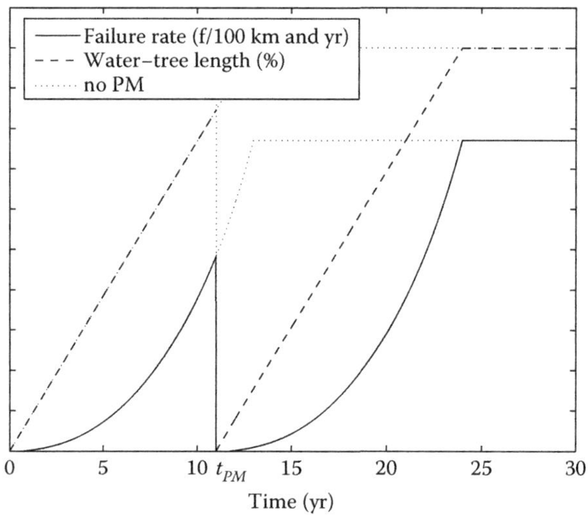

FIGURE 6.62
The effect of the replacement of a cable system component on failure rate and water-tree length.

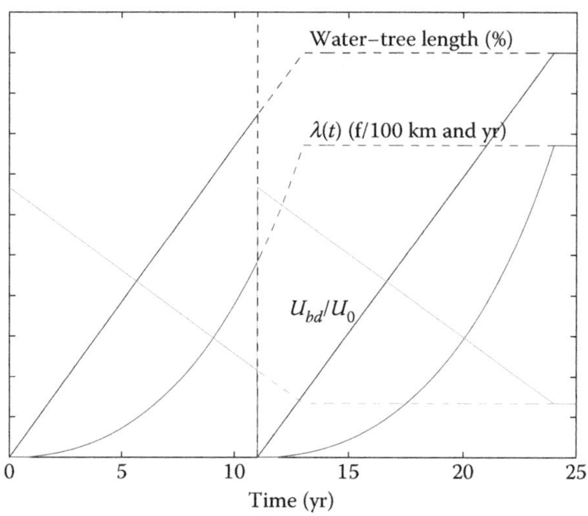

FIGURE 6.63
Resulting failure rate model with the effect of PM by replacement.

6.6.10.1 The Effect of Applying Rehabilitation at Different Times

Figure 6.64 demonstrates the impact of PM on the breakdown voltage and the failure rate function depending on when the PM activity is applied. The results are shown for PM applied within the period 6–13 years.

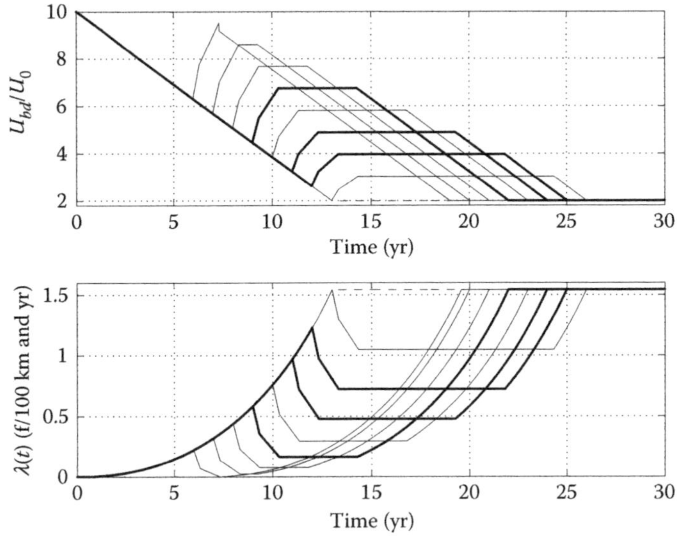

FIGURE 6.64
Impact on the breakdown voltage and failure rate of applying PM at different times.

It can be seen that the resulting benefit on failure rate that is gained by the PM action varies according to when the PM measure is applied. If it is applied too early or too late, then the benefit gained can be limited. This has been highlighted by looking at the functions when PM is applied after 9, 11, and 12 years. These three PM occasions are spread within the period where PM has a relatively significant effect on the failure rate.

In the Cable Application Study, a maintenance plan where PM is applied on three occasions at years 9, 11, and 12 has been considered in the following analysis of PM times.

Further analysis focuses on these three PM occasions. Figures 6.65 and 6.66 illustrate the results for the breakdown voltage and failure rate when PM is applied in these three different years. The figures show the different effects of the PM activity.

6.6.10.2 The Effect of Applying Replacement at Different Times

In the same way as for the rehabilitation, the effect of applying replacement at different times has also been analyzed. Figure 6.67 shows the results for the cable system component. Furthermore, the plan for applying PM at years 9, 11, and 12 has been analyzed. The results from this analysis are presented in Figure 6.68 for the breakdown voltage and in Figure 6.69 for the failure rate.

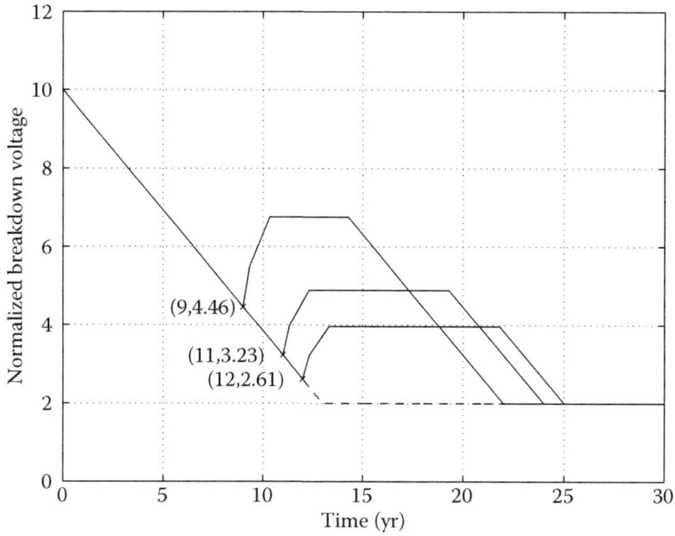

FIGURE 6.65
Effect of rehabilitation of a cable system component on the breakdown voltage.

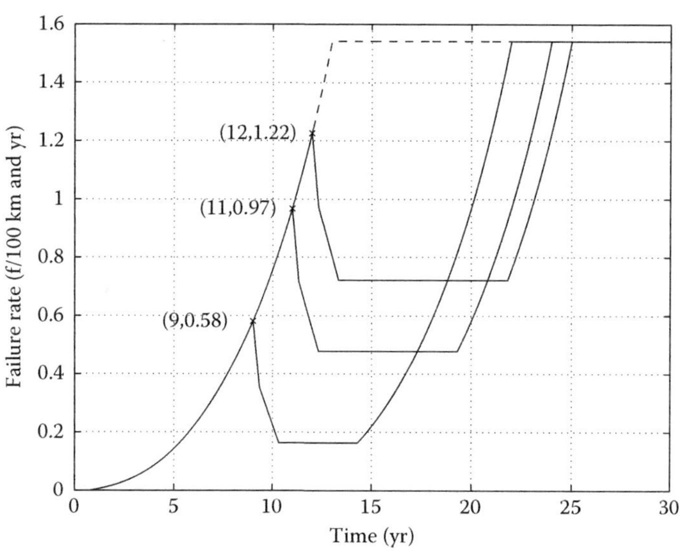

FIGURE 6.66
Effect of rehabilitation of a cable system component on the failure rate.

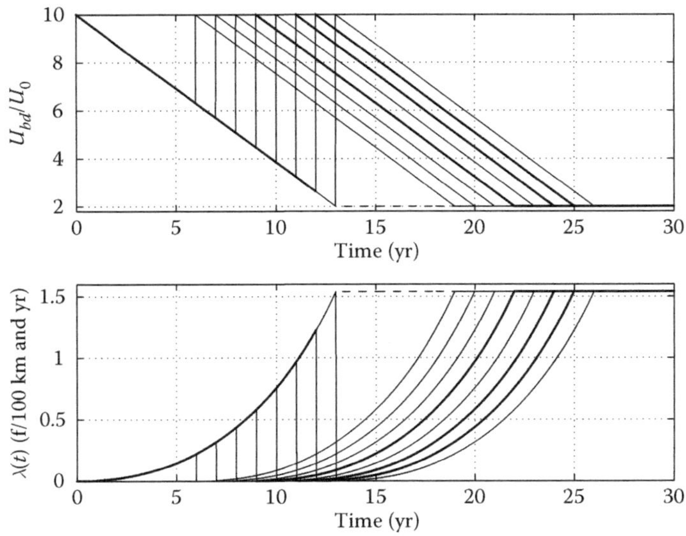

FIGURE 6.67
Effect on failure rate caused by cable system component replacement at different years.

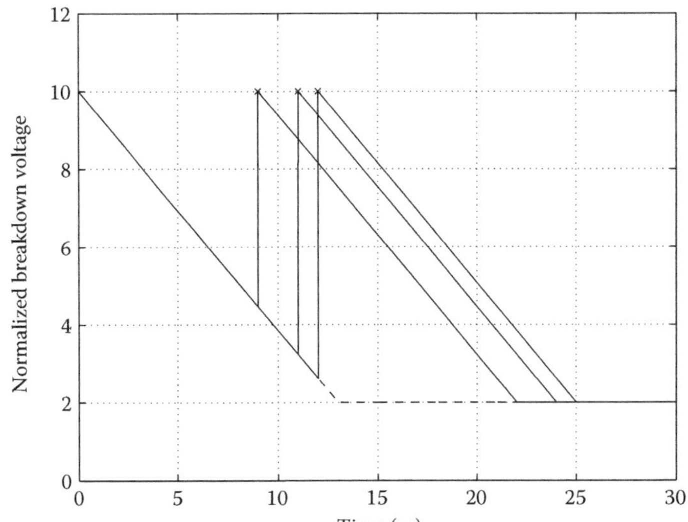

FIGURE 6.68
Effect on the breakdown voltage function caused by cable system component replacement at three different points in time.

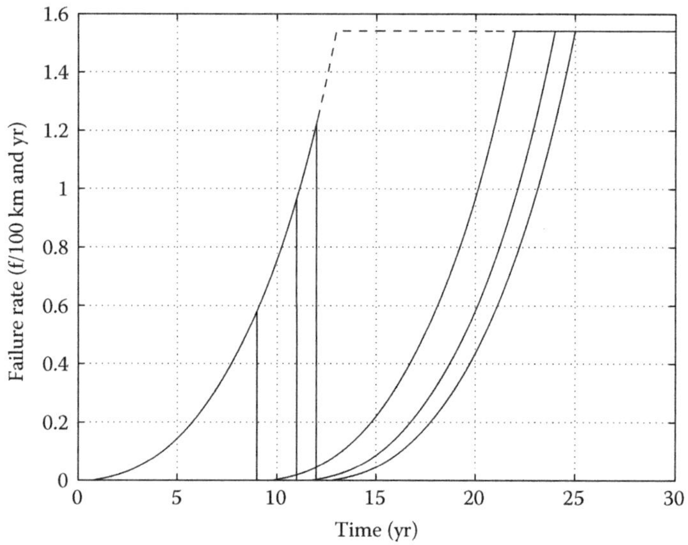

FIGURE 6.69
Effect on the failure rate function caused by cable system component replacement at three different points in time.

6.6.10.3 Modeling Corrective Maintenance

The work presented in this book focuses on the modeling of PM. However, to capture the effect of PM and to enable the defining of a PM strategy, the alternative to PM (which is to correct failures by corrective maintenance or CM) should be addressed in the model.

One of the fundamental differences between PM and CM is that CM is an activity undertaken as a result of a random process the occurrences of failures. This is contrary to PM, which is performed at certain defined points in time. Traditionally, these time points were at predefined intervals. This work suggests, however, that CM measures should be related to the actual condition and reliability of the components.

As the main focus of this work is not on developing theory for modeling CM, the following simple assumptions have been made. It has been assumed that CM restores the component after failure, and that the component behavior after the CM activity is the same as the time just before the failure occurred.

CM will be included in the model based on this assumption without introducing any more complications into the model. However, the different effects of CM should be discussed along with the basis for making the assumption above.

The alternative effects of CM are that it either improves the cable behavior, reduces it, or restores it to what it was before. Consequently, the assumption made follows the last of these three.

It could be argued that one or another assumption is more possible, however, investigating this falls outside the scope of this work. Briefly, it could be said that if the cable is assumed to be improved by CM, then one possible model would be to assume, in the same way as for replacement, that the cable would then become as good as new. Instead of assuming there would be no effect on the component behavior, this would give an optimistic model showing that CM is as good as replacement, which seems unreasonable. On the other hand, if it is assumed that the CM worsens the component behavior after the restoration, then a very pessimistic view of the effect of CM is captured. For some circumstances, it is arguable that the best effect is not to do any PM at all, such as shown in Section 6.3, where it can be argued that an unmaintained cable has the longest lifetime. However generally, maintenance has been demonstrated to have a positive effect on component lifetime, and therefore there is no reason not to incorporate such an aspect into the limited scope for modeling CM.

The conclusion is consequently that the assumption made is valid for a simple model of CM that would produce results with a pessimistic rather than optimist outcome (compared with if the behavior was improved by the CM activity).

There is one further aspect of importance to consider that the different component types in the distribution system can be as either (i) discrete components such as transformers, breakers, and so on or (ii) lines and cables which are considered in units of length. This difference in component types could impact on the above statement about the effect of CM. For CM of feeders (lines or cables), it is most probable that just a part of the length is effected by the CM measure. Therefore, it would be reasonable not to assume that the component is as good as new after the CM. However, considering for example a breaker, it is more reasonable to assume that the component would behave like new after a CM measure. The conclusion is consequently that the treatment by CM should be evaluated depending on the type of component that is considered.

6.6.11 Comparison between PM Methods

In this Section relationships to support functions for failure rate due to failures caused by water-tree growth in a cable system component has been deduced. For the Cable Application Study, these functions include the effect of applying PM either by rehabilitation (injection of silicon PM_{si}) or replacement (PM_{rp}) of cable. CM has also been considered leading to a restoration result where there is no effect on the failure rate.

Tables 6.39 through 6.41 present the results for the three following resulting functions when:

1. Water-tree growth in the cable is left as it is, as an aging process denoted by $\lambda_{old}(t)$.
2. PM is applied by rehabilitation and denoted by $\lambda_{PM_{si}}(t)$.
3. PM is applied by replacement and denoted by $\lambda_{PM_{rp}}(t)$.

Note that the values in the tables are presented for integer values of time (selected function values for complete years).

The first function refers to the aging process in the cable resulting from water-tree effects impacting on the failure rate with no impact of maintenance. The two following functions represent the results of applying PM either by rehabilitation or replacement. The CM as described earlier would follow the first of the three functions, that is, have no impact on the failure rate.

Figure 6.70 compares the two different PM activities that have been studied. Three different characteristics for the failure rate function have been shown: (i) without maintenance, (ii) with maintenance by rehabilitation, and (iii) maintenance by replacement. The results are shown for PM applied after 11 years. The three functions have similar initial and final behaviors. When PM is applied, the failure rate either drops down to an initial growth rate for the replacement case or shows a decreased drop in the rehabilitation case.

It is understood that benefit by rehabilitation would change with the time of maintenance application, and furthermore that there is a period after Year 6 and before Year 13, within which a benefit would be gained. Figure 6.71 shows the results when the time for the maintenance activity is changed between Years 8 and 13. On the one hand, it can be seen that the benefit for the rehabilitation is not significant if applied too early or too late. On the other hand, it can be seen that the benefit for the replacement is proportional to when the measure is undertaken, since the effect is always that the failure rate is set to zero. This issue of comparing the effect of different maintenance methods is further analyzed in Section 6.7, the system and cost benefit analysis.

This summary presents the final results of this chapter, which serve as input data for the proceeding analysis. In the next chapter, the model developed ($\lambda(t, PM)$) has been used to predict the behavior of an average cable as one component in a distribution system.

In the validation of the model implemented here, some calculations had to be made by hand. Tables 6.39 through 6.41 present time parameters which determine the deduction of the failure rate functions due to rehabilitation. For correctness of the results, the following should be valid:

$$\begin{aligned}
\lambda_{PM_{si}}(t_1) &\Leftrightarrow \lambda_{old}(t_{e1}) \\
\lambda_{PM_{si}}(t_2) &\Leftrightarrow \lambda_{old}(t_{e2}) \\
\lambda_{PM_{si}}(t_3) &\Leftrightarrow \lambda_{old}(t_{e2}) \Leftrightarrow \lambda_{PM_{rp}}(t_3).
\end{aligned} \tag{6.23}$$

TABLE 6.39

Result for the $\lambda(t, PM)$ Model of a Cable System Component with PM Applied at Year 9

Time [yr]	$\lambda_{old}(t)$ [f/100 km and yr]	$\lambda_{PM_{si}}(t, t_{PM} = 9)$ [f/100 km and yr]	$\lambda_{PM_{rp}}(t, t_{PM} = 9)$ [f/100 km and yr]
0	0.0000	0.0000	0.0000
1	0.0030	0.0030	0.0030
2	0.0188	0.0188	0.0188
3	0.0453	0.0453	0.0453
4	0.0860	0.0860	0.0860
5	0.1437	0.1437	0.1437
6	0.2200	0.2200	0.2200
7	0.3168	0.3168	0.3168
8	0.4361	0.4361	0.4361
9*	0.5809	0.5809	0.0000
10	0.7558	0.2259	0.0030
11	0.9676	0.1631	0.0188
12	1.2254	0.1631	0.0453
13	1.5418	0.1631	0.0860
14	1.5418	0.1631	0.1437
15	1.5418	0.2200	0.2200
16	1.5418	0.3168	0.3168
17	1.5418	0.4361	0.4361
18	1.5418	0.5809	0.5809
19	1.5418	0.7558	0.7558
20	1.5418	0.9676	0.9676
21	1.5418	1.2254	1.2254
22*	1.5418	1.5418	1.5418
23	1.5418	1.5418	1.5418
24	1.5418	1.5418	1.5418
25	1.5418	1.5418	1.5418
26	1.5418	1.5418	1.5418
27	1.5418	1.5418	1.5418
28	1.5418	1.5418	1.5418
29	1.5418	1.5418	1.5418
30	1.5418	1.5418	1.5418
Special time points: ______			
$t_{e2} = 5.28$	0.1631	0.1631	0.1631
$t_{e1} = 7.33$	0.3535	0.3535	0.3535
$t_1 = 9.33$	0.6350	0.3535	0.0000
$t_2 = 10.33$	0.8213	0.1631	0.0073
$t_3 = 14.28$	1.5418	0.1631	0.1631

TABLE 6.40

Result for the $\lambda(t, PM)$ Model for a Cable System Component with PM Applied at Year 11

Time [yr]	$\lambda_{old}(t)$ [f/100 km and yr]	$\lambda_{PM_{si}}(t, t_{PM} = 11)$ [f/100 km and yr]	$\lambda_{PM_{rp}}(t, t_{PM} = 11)$ [f/100 km and yr]
0	0.0000	0.0000	0.0000
1	0.0030	0.0030	0.0030
2	0.0188	0.0188	0.0188
3	0.0453	0.0453	0.0453
4	0.0860	0.0860	0.0860
5	0.1437	0.1437	0.1437
6	0.2200	0.2200	0.2200
7	0.3168	0.3168	0.3168
8	0.4361	0.4361	0.4361
9	0.5809	0.5809	0.5809
10	0.7558	0.7558	0.7558
11*	0.9676	0.9676	0.0000
12	1.2254	0.5568	0.0030
13	1.5418	0.4780	0.0189
14	1.5418	0.4780	0.0453
15	1.5418	0.4780	0.0860
16	1.5418	0.4780	0.1437
17	1.5418	0.4780	0.2200
18	1.5418	0.4780	0.3168
19	1.5418	0.4780	0.4361
20	1.5418	0.5809	0.5809
21	1.5418	0.7558	0.7558
22	1.5418	0.9676	0.9676
23	1.5418	1.2254	1.2254
24*	1.5418	1.5418	1.5418
25	1.5418	1.5418	1.5418
26	1.5418	1.5418	1.5418
27	1.5418	1.5418	1.5418
28	1.5418	1.5418	1.5418
29	1.5418	1.5418	1.5418
30	1.5418	1.5418	1.5418
Special time points: ______			
$t_{e2} = 8.31$	0.4780	0.4780	0.4780
$t_{e1} = 9.79$	0.7163	0.7163	0.7163
$t_1 = 11.33$	1.0471	0.7163	0.0000
$t_2 = 12.33$	1.3227	0.4780	0.0073
$t_3 = 19.31$	1.5418	0.4780	0.4780

TABLE 6.41

Result for the $\lambda(t, PM)$ Model of a Cable System Component with PM Applied at Year 12

Time [yr]	$\lambda_{old}(t)$ [f/100 km and yr]	$\lambda_{PM_{si}}(t, t_{PM} = 12)$ [f/100 km and yr]	$\lambda_{PM_{rp}}(t, t_{PM} = 12)$ [f/100 km and yr]
0	0.0000	0.0000	0.0000
1	0.0030	0.0030	0.0030
2	0.0188	0.0188	0.0188
3	0.0453	0.0453	0.0453
4	0.0860	0.0860	0.0860
5	0.1437	0.1437	0.1437
6	0.2200	0.2200	0.2200
7	0.3168	0.3168	0.3168
8	0.4361	0.4361	0.4361
9	0.5809	0.5809	0.5809
10	0.7558	0.7558	0.7558
11	0.9676	0.9676	0.9676
12*	1.2254	1.2254	0.0000
13	1.5418	0.8045	0.0030
14	1.5418	0.7219	0.0188
15	1.5418	0.7219	0.0453
16	1.5418	0.7219	0.0860
17	1.5418	0.7219	0.1437
18	1.5418	0.7219	0.2200
19	1.5418	0.7219	0.3168
20	1.5418	0.7219	0.4361
21	1.5418	0.7219	0.5809
22	1.5418	0.7558	0.7558
23	1.5418	0.9676	0.9676
24	1.5418	1.2254	1.2254
25*	1.5418	1.5418	1.5418
26	1.5418	1.5418	1.5418
27	1.5418	1.5418	1.5418
28	1.5418	1.5418	1.5418
29	1.5418	1.5418	1.5418
30	1.5418	1.5418	1.5418
Special time points:			
$t_{e2} = 9.82$	0.7219	0.7219	0.7219
$t_{e1} = 11.02$	0.9723	0.9723	0.9723
$t_1 = 12.33$	1.3227	0.9723	0.0000
$t_2 = 13.33$	1.5418	0.7219	0.0073
$t_3 = 21.82$	1.5418	0.7219	0.7219

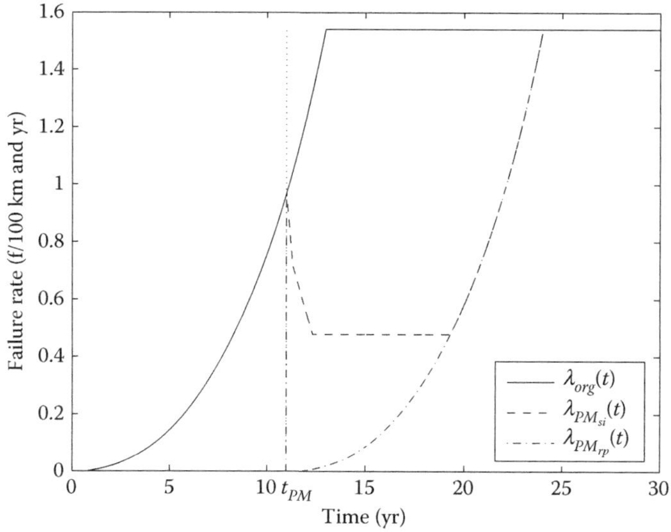

FIGURE 6.70
Benefit to failure rate of PM by either replacement or rehabilitation of the cable component.

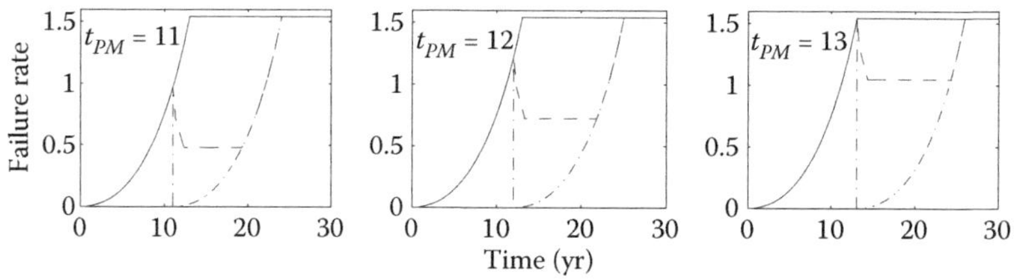

FIGURE 6.71
Comparison of the benefit to failure rate of PM by either replacement or rehabilitation of the cable component.

It is interesting to note that the last of these relations (Equation 6.23) states that the failure rate function due to rehabilitation intersects the failure rate function due to replacement at a specific point t_3. This time parameter varies depending on when the PM action is applied.

The following assumption was made in the model: $\max(\lambda_{old}(t)) = \lambda_{old}(t = 13.00)$. This leads to: $\max(\lambda_{PM_{rp}}(t)) = \lambda_{PM_{rp}}(t = 13 + t_{PM})$. Since the rehabilitation process results in a failure rate that terminates into the same function as for the replacement situation, then the maximum value is equal and $\max(\lambda_{PM_{si}}(t)) = \lambda_{PM_{si}}(t = 13 + t_{PM})$. The maximum values for the failure rate function should therefore for the cases with ($t_{PM} = 9, 11$, and 12) be $t = 22, 24$, and 25, respectively, which is supported by the results presented in Tables 6.39 through 6.41.

6.6.12 Remarks about the Implementation of the Model

The model developed for the failure rate effected by PM has been implemented in the Cable Application Study. This has been done using the programming tool MATLAB, which was introduced in Section 5.3. A summary of the input data files and used is provided in Appendix.

6.7 System Analyses for Various Maintenance Strategies

This section relates to the RCM procedures of estimation of composite failure rate, sensitivity analysis, and cost and benefit analysis as presented in Table 4.1. The objective of the work relates to defining effective maintenance strategies focusing on PM and reliability. Therefore, the scope of the financial analysis is the comparison of PM methods including the possibility of performing no PM, that is, only corrective maintenance (CM).

6.7.1 Introduction

The previous section presented procedures developed for modeling component behavior that define a qualitative relationship between PM and reliability performance. Furthermore, this was illustrated by studying a specific component the XLPE cable component affected by water treeing. However, this model relates to one failure cause only for the cable component. Consequently, the result from Section 6.6 provides a model of the failure rate behavior of one component (XLPE cable) and one cause of failure (water treeing).

In this chapter, the model developed for the failure rate has been extended to include a system perspective. The purpose here is to achieve a model that can be used to analyze the effect of preventing failures in the included components at the system level. Furthermore, the aim has been to define effective maintenance strategies, which means that system analysis should also consider when it is the best time for performing the PM activities, and on which components. Due to limitations in cost and system requirements,

PM is applied to certain components at a time, and within a given time period it is not automatic that all components will be exposed to PM.

The above discussion has introduced several complications to the analysis of the effect of PM on the system. Consequently, before a system effect analysis can be performed, there are several issues that must be solved, and two of these can be summarized as follows. First, the component has to be modeled as part of a system which can be done by introducing the concept of an average component. Second, the composite failure rate has to be modeled for an average component, which is the failure rate due to all causes of failures.

Similar to as in the previous chapter, the theory developed is presented by using an application study. To enable the system effect analysis to be performed, input data are required for both a system as well as its included components.

In earlier sections, a detailed analysis of the Birka system was presented by both analysis of the causes of failures for the 11 kV cable component (see Section 6.3), and by reliability analysis (see Chapter 9). Consequently, the analysis of the Birka system performed together with the model for the cable component behavior developed in Section 6.6 constitutes the required input data for the analysis of the system effects and maintenance strategies in this chapter.

The following sections present all these issues in greater detail by on the one hand introducing the concepts of maintenance strategies, average cables, and composite failure rate, and on the other hand by illustrating the application to the cable component as part of the Birka system and with results for reliability indices from RADPOW. Finally, a cost and benefit analysis is performed to achieve the solution that is the cost-effective maintenance strategy.

6.7.2 Input Data for the System Effect Analysis

6.7.2.1 Birka Liljeholmen System

Both the Birka and Liljeholmen 11 kV stations have been presented and analyzed earlier, first by a study of the statistics and practices (in Section 6.3), and second by a system effect analysis with Approach I (in Section 6.5). The focus for the Birka system analysis in this chapter is on the effect of PM applied to the 11 kV cable component. Furthermore, these 11 kV cables were represented in the reliability network as being of the one average cable type and laid in parallel.

The RADPOW input data for the Birka system analysis are the same as those used for the system analysis presented earlier. The component reliability input data for analyzing the system with RADPOW are presented in Section 6.5.3.5 and Tables 6.28 and 6.29. These data represent the base case for the Birka system, which is the data set that is being varied for different cases in the sensitivity analysis. The input reliability data that are effected in the system analysis for the Cable Application Study are the failure rate for the average 11 kV cable. The cable used here (LH11) is defined by the average failure rate and average length as follows:

$$\begin{array}{lll} \lambda_{LH11} & = 0.10069 & [\text{f/yr}] \quad \text{with} \\ l_{av} & = 8.10584 & [\text{km}] \end{array} \tag{6.24}$$

Furthermore, Table 6.42 presents the resulting output data from the RADPOW analysis for the base case. This provides reliability load point indices as well as different system performance indices. In the Cable Application Study, the impact of PM on the 11 kV cable has been analyzed. However, for the Birka system, there is only one load point at this voltage level, called Load Point 1, Lpl or LH11, since it is this load point that is connected to the Liljeholmen Station (LH).

6.7.2.2 Failure Rate Functions for XLPE Cable Component

Failure rate functions for an XLPE cable component were modeled in Section 6.6. Furthermore, functions for the PM regime applied after 9, 11, and 12 years were defined. Function values for the beginning of each year are presented in Tables 6.39 and 6.41. These tables present the results for three different failure rate functions, first due to aging ($\lambda_{old}(t)$), second and third due to the impact of PM using either the rehabilitation method ($\lambda_{PMsi}(t, t_{PM})$) or the replacement method ($\lambda_{PMrp}(t, t_{PM})$).

Figure 6.72 illustrates these failure rate functions with the three PM measures: $t_{PM1} = 9$, $t_{PM2} = 11$, and $t_{PM3} = 12$. For all these functions, the failure rate lies in the interval [0, 1.5418] f/100 km and yr = [0, 0.015418] f/yr and km.

Note that these failure rate functions refer to failures due only to water treeing. Therefore, a general description of the failure rate functions of $\lambda_{wt}(t)$ [f/yr] has been defined as follows. The input data for the Birka system analysis presented in Section 6.7.2.1 used the unit for failure rate of failures per year. Moreover, the average cable length for the 11 kV cable shown in Equation 6.25 is 8.10584 km. If the failure rate functions discussed earlier are transformed into an equivalent unit, the following corresponding interval is achieved:

$$\lambda_{wt}(t) \subseteq [0, 0.12579] \quad [\text{f/yr}] \tag{6.25}$$

where $0.12579 \approx (1.5418 \cdot 8.1584)/100$. This failure rate represents an XLPE cable that is affected by water treeing and has a length of about 8.106 km.

6.7.2.3 Validation of the Combination of Input Data

Two different sources of data and relationships have been deduced which both relate to the cable component behavior. First of all, a model of the failure rate characteristics due to a specific failure cause and a model of PM for

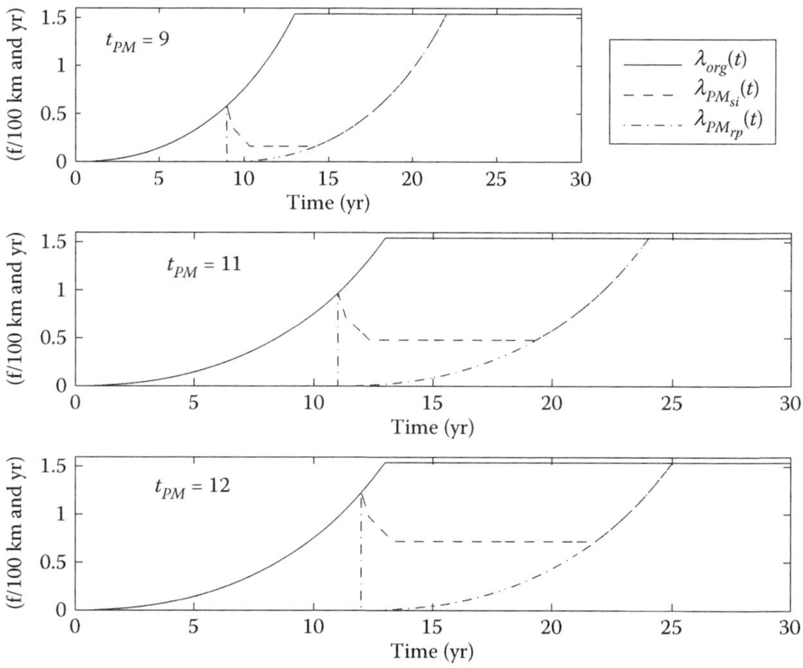

FIGURE 6.72
Resulting failure rate functions for the cable component for three maintenance plans: $t_{PM1} = 9$, $t_{PM2} = 11$, and $t_{PM3} = 12$.

the same have been modeled. Second, a system cable component (the average 11 kV cable) has been analyzed in great detail for the Birka system. The combination of these two modeling efforts results in realizing a complete RCM study. However before realizing this, it must be clear that the data are compatible, which is discussed in this section. The purpose of this is to justify that the output result not only provides an illustration but also a relevant analysis. The input data for the system effect analysis comes from the following:

1. The failure rate model for a component affected by water treeing, based on Svel data: λ_{wt} [0,0.015418] [f/yr,km] (from Chapter 10).
2. Birka system 11kV cable based on Birka data: $\lambda_{LH11} \approx 0.0124$ [f/yr,km] (from Equation 6.1).

Consequently, the contribution of water-treeing failures to total failures corresponds to a failure rate increase of more than about 100%. To verify whether this is reasonable or not, the two different data sources should be studied in greater depth.

1. *Svel report data analysis*: This data refers to XLPE cables for 12 kV and higher. Most quantities refer to 12 and 24 kV, and further details are not provided in the report.

   The data used from the Svel report are data for the population of cables without an extruded layer put into operation 1970–1975, and have a total length 1800 km. Within the years 1975–1988, these cables fail 175 times $= 175/(14 \cdot 1800) \approx 0.00694$ f/yr and km; or for the total population of cables with an aging time of 13 years (cable put into operation 1970, which includes failures occurring between 1970 and 1983), $114/(13 \cdot 1,800) \approx 0.0045$ f/yr and km.

   Furthermore, the Svel report indicates that an XLPE cable with an extruded layer should have about 0.0003 f/yr and km, and that the average failure rate for XLPE cables in Sweden (mixture of XLPE types) is about 0.003 f/yr and km.
2. *11 kV data*: The Birka 11 kV data includes a mixture of cable types and ages with the failure rate 0.01242 f/yr and km. Other sources of cable statistics (this is discussed in the report Causes of failures) show values of 0.05 f/yr and km in Norway (1967–1983, 1–380 kV), Svel 0.03 f/yr and km (11 kV 1995–1997), 0.02 f/yr and km mass-impregnated cable (Norway).

   From the above comparison of data, it can be concluded that the failure rate data for the 11 kV cable are higher than the data for the general XLPE cable not exposed to water trees. This result is expected since more failures tend to occur at lower voltage levels. However, the question was whether the relationship between $\lambda_{wt}(t)$ and λ_{LH11} was reasonable. It is difficult to answer this, but the observations made indicate that the relationship is reasonable and that the difference was expected to be larger.

6.7.2.3.1 Conclusion

The conclusion is therefore that it is reasonable to assume that the two data sources agree for time zero, which is before the water trees have any impact on the failure rate. This model is supported by the interviews and discussions with maintenance personnel about the 11 kV Birka cables where it is not identified that water-tree growth is an actual problem at this point.

6.7.2.3.2 Remarks

A further comment about the input data relates to the detailed knowledge gained about failure causes for the Birka 11 kV cables. The statistical data analysis of Birka data shows the result that the maximal contribution from water trees (failures due to material and method) is 14%.

This figure (14% of 0.1007 f/yr) equals 0.01410 f/yr which implies 0.00174 f/yr and km. It is interesting to compare this number with the failure rate model. It can be seen that it would correspond to the time between 0 and 1 year of water-tree growth. Consequently, the assumption to superpose the two data sources at time 0 is validated.

6.7.2.4 Loss Parameter for Supporting Maintenance Strategy

The linear loss parameter ($\Delta''\epsilon$) was introduced in Section 6.4. Furthermore, in Section 6.6 it was identified to be useful for defining the condition of the cable that in turn was transformed into a relation for the failure rate of the cable. The benefit with using this parameter is that it can be measured without destroying the sample material, which is not the case for the alternative method of breakdown strength.

The cable condition can therefore be measured by using the loss parameter. Results from such measurements enable distinguishing between those cables that have a better or a worse condition than an average cable. Moreover, this would support obtaining:

- A ranking of cables based on cable condition, which would define which cables to apply the PM activity
- A criteria for when a PM activity should be applied, for example, $(\Delta''\epsilon) < 0.1$, and if it is reasonable at any one time to delay the PM activity

The conclusion is that the measurements of the loss parameter can be used as input data for a maintenance strategy. This can be done by using historical information about the component up to the decision time. These data can support the two decisions, first whether PM should be applied based on a criteria and second the order in which to maintain the components.

6.7.3 Concepts for the System Effect of Maintenance Strategies

6.7.3.1 Composite Failure Rate

As discussed earlier, component failures occur due to different causes of failures, which in turn could be effected by PM. The causes of failures in the cable component have been analyzed in detail, and a result from these studies is the development of a model that defines the variation of the failure rate function over time and with the PM method. This model refers to a cable type that is prone to fail due to water treeing and the PM methods analyzed to prevent this specific cause of failure. However, the cable is affected by other types of failures, for example, due to corrosion, which also contributes to the failure rate of the component. Consequently, a component failure rate is made up of the contribution from several causes of failures, and the total failure rate means a composite rate of those contributions.

A next step in the Cable Application Study is therefore to extend the model developed for failure rate in order to represent the composite failure rate. The input data required to obtain such a model for failure rate were presented and analyzed in Section 6.7.2. Furthermore, it was identified that these data originate from two data sources: first from the Birka system and specifically data for the 11 kV cable component, and second from the failure rate function developed for the water-tree contribution to the XLPE cable.

The corresponding failure rate data are: 0.10069 f/yr, as defined in Equation 6.24, and the interval of [0, 0.12579] f/yr as defined in Equation 6.25. The first refers to the Birka cable data for the average failure rate of an 11 kV cable (failures due to all failure causes). The second refers to the XLPE cable data for a failure rate function due to the failure caused by water treeing.

To enable a system effect analysis of the impact of PM to be performed, the following assumptions have been made:

- The average 11 kV cable in the Birka system is affected by water treeing.
- At time 0, the failure rate due to all failure causes equals the input data for the Birka base system (0.10069 f/yr).
- From time 0 until the end of the interval, the composite failure rate is made up of the sum of the contribution from the water-tree failure rate effect and the nominal value which represents the effects from other causes of failures.

These assumptions mean that the failure rate model developed is superposed on the Birka system data. A discussion about the justification of such an approximation was made in Section 6.7.2.3. The resulting composite failure rate function for the Cable Application Study (for the average 11 kV cable in the Birka system) is therefore given by

$$\lambda_{tot}(t) = \lambda_{LH11} + \lambda_{wt}(t) \quad [\text{f/yr}] \quad : \quad t_0 \leq t \leq T$$

$$\text{where} \quad \lambda_{wt}(t) \subseteq \begin{cases} \lambda_{old}(t) & \text{if} \quad \text{no} \quad \text{PM} \\ \lambda_{PM_{si}}(t) & \text{if} \quad PM_{si} \\ \lambda_{PM_{rp}}(t) & \text{if} \quad PM_{rp}. \end{cases} \tag{6.26}$$

The corresponding interval for the function is given by

$$\lambda_{tot}(t) \subseteq [0.10069, 0.22650] \quad [\text{f/yr}]. \tag{6.27}$$

This interval suggests that the composite failure rate for $\lambda_{tot}(t)$ increases with somewhat more than 100% due to the impact of water-tree failures. This characteristic is illustrated in Figure 6.73. This figure shows three failure rate functions for the Cable Application Study. The first is the composite failure rate function ($\lambda_{tot}(i)$), the second is the failure rate function due to water treeing ($\lambda_{wt}(i)$), and the third is the nominal value for the Birka system average

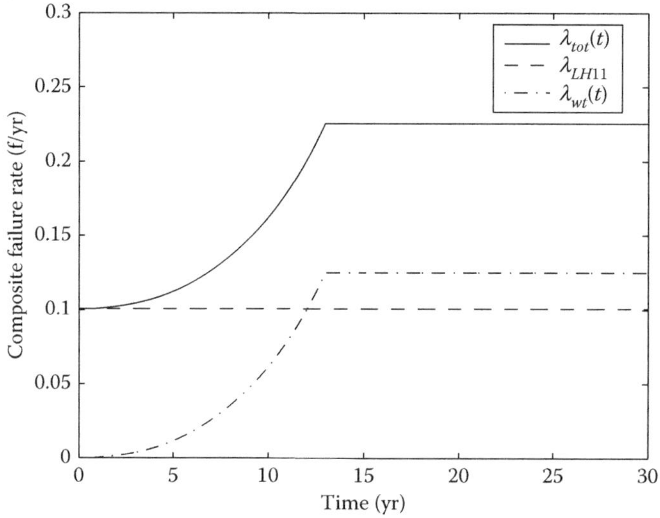

FIGURE 6.73
Illustration of the composite failure rate for the 11 kV average cable in the Birka system affected by water treeing.

11 kV cable (λ_{LH11}). The conclusion is that in the model of the composite failure rate for the 1 kV cable in the Birka system, that the failure cause water treeing makes a significant contribution to failure.

6.7.3.2 Maintenance Strategy

In previous analysis, the system has been reduced by (i) focusing the analysis on those included components that have a significant impact on the system reliability and could be improved by PM, and (ii) the failure causes that could be affected by PM for those components have been identified.

The maintenance strategy for the system aims to find suitable and effective PM activities to reduce those failure causes which would in turn result in finding the best solution for system reliability.

To create this strategy for a certain time period (e.g., $T \subseteq [O, 30]$), the following questions have to be answered for each failure cause that is related to a specific component or component type such as XLPE cable affected by treeing:

1. What type of PM method should be applied?
2. With what frequency should it be applied?
3. At what points in time should it be applied?
4. What proportion of the component type will undergo PM each time?

TABLE 6.42

Evaluated Reliability Indices for the Birka System Base Case

Load point	λ [f/yr]	U [h/yr]	r [h/f]	L [kW]	LOE [kWh/yr]
LH11	0.334494	0.506714	1.51487	768.974	389.65
HD	0.111307	0.133388	1.19837	22999.9	3067.9
SJ	0.110706	0.150543	1.35984	0.8	0.120434
SAIFI = 0.115491 [int/yr, cust.]			SAIDI = 0.140386 [h/yr, cust.]		
CAIDI = 1.21556 [h/int.]			AENS = 0.144988 [kWh/yr, cust.]		

The last of these questions is very much related to the second. Obviously, if it is possible to maintain all system components at one time, it is probably only necessary to apply one PM activity during the period. However for a power delivery system in operation, this would not be possible, since the system reliability for the supply has to fulfill certain levels, and it would also be very costly to undertake all the investments at any one point in time instead of spreading them over a period of years. Consequently, a sound maintenance strategy would divide the PM activities over several points in time for different components.

For the Cable Application Study as presented in Section 6.6, a plan of PM activities was suggested with three application occasion times 9, 11, and 12 years after coming into operation. To complete the maintenance strategy for the Cable Application Study, there is one further issue to decide, and that is, the proportion of the XLPE cable components that should be treated to prevent water-tree failures at each maintenance activity.

The Cable Application Study results are presented for two different strategies which are presented in Table 6.43. These have been chosen to capture a reasonable spread, that is, with a total of either 30% or 90% of cables undergoing PM. Furthermore, equal time points have been chosen to provide results for comparisons between the two PM methods. It should, however, be recognized that the methodology as well as the implementation supports the analysis of the whole range of possible strategies. In practice, a range of sensitivity studies would be conducted to assess all possible and practical alternatives.

Finally, some general remarks about the choice of maintenance strategy have been presented. An effective maintenance strategy requires that the benefit gained must justify the expenditure made to achieve it. Furthermore, a limited capital expenditure per year means that the investments must be spread over a number of years. Consequently, the PM will be spread over a number of years, with the proportion of components maintained being less than 100%. Broadly, the choice of strategy depends on how much money there is to spend.

TABLE 6.43

Two Maintenance Strategies for the Cable Component Application Study with Either Rehabilitation $PM_{si,j}$ or Replacement $PM_{rp,j}$

Measure	Year	Proportion
Strategy 1 (S_1)		
$PM_{i,1}$	$t_{PM1} = 9$	$s_1 = 0.1$
$PM_{i,2}$	$t_{PM2} = 11$	$s_2 = 0.1$
$PM_{i,3}$	$t_{PM3} = 12$	$s_3 = 0.1$
Strategy 2 (S_2)		
$PM_{i,1}$	$t_{PM1} = 9$	$s_1 = 0.3$
$PM_{i,2}$	$t_{PM2} = 11$	$s_2 = 0.3$
$PM_{i,3}$	$t_{PM3} = 12$	$s_3 = 0.3$

The first strategy implies that a total of 30% of the component type undergoes PM. This situation could correspond to where a maintenance strategy for a certain time period is constricted by a limit or a maximum number of components to be affected by PM. Such a limit would primarily be based on restrictions on reliability performance rather than on cost, since the latter would imply restrictions on when to make the PM. It is assumed that the maintenance is applied on three occasions. Consequently, the maintenance strategy should distribute the 30% of cables over three PM times. Table 6.44 demonstrates the eight different maintenance strategies that fulfill those requirements.

TABLE 6.44

Different Maintenance Strategies with Preventive Maintenance Applied on Three Occasions to a Total of 30% of the Components

Strategy	s_1	s_2	s_3
1[a]	0.1	0.1	0.1
2	0.2	0.1	0
3	0.2	0	0.1
4	0.3	0	0
5	0	0.2	0.1
6	0	0.1	0.2
7	0	0.3	0
8	0	0	0.3

[a]Corresponds to Strategy S_1.

6.7.3.3 Average Component Composite Failure Rate

Previous sections have identified the need for and showed the definitions of: (i) composite failure rates for the components (presented in Section 6.7.3.1) and (ii) a maintenance strategy for the system (presented in Section 6.7.3.2). Both these are required to support a system analysis of the effect by PM. However, one further concept has to be introduced before the system reliability analysis can be performed and that is, the focus of this section.

As discussed earlier, a general maintenance strategy will result in less than all the components of a certain type undergoing PM at the same time or by the same method. Moreover, some components might not undergo PM at all during the time period for which the strategy corresponds. This situation creates different component behavior over time for components that have the same original conditions. In other words, the failure rate functions for components with similar failure rate characteristics may differ with time since they are subject to differing PM methods. Consider, for example, three XLPE cables that are exposed to water treeing. Assume that one of these undergoes PM year 9 by rehabilitation of the cable, the next cable at year 11 by replacement, and the third cable has no PM at all, then the maintenance strategy would result in three different failure rate functions for this one component type. Furthermore, the situation could be that 10% of the XLPE cables follow the first pattern, and so on. Consequently, it can be acknowledged that the complexity of the maintenance issue has become even greater. To capture this complication with a mixture of different failure rate functions for components with similar component behavior, the concept of average component has been introduced.

An average component is one that represents all the components of that type and their different behaviors due to a specific maintenance strategy. For the above example with the three components, the average component failure rate would be the resulting failure rate function when these different failure functions have been combined in a way that takes into account the type of PM method, the time for the activity, the number of components, and so on.

To demonstrate the different failure rate functions, the following example has been used. Consider a system with cable components that are exposed to PM by rehabilitation at three different points in time, being 9, 11, and 12 years after coming into service. Furthermore, assume that 10% of the cables undergo PM by rehabilitation at each time. This maintenance strategy corresponds to S1 as defined in Table 6.43. This situation results in four different failure rate functions which represent the component behavior for the four different classes of cables, being those rehabilitated at years 9, 11, or 12, or not exposed to PM by rehabilitation. The first three of these correspond to a failure rate function that is valid for 10% of the cables each. The last cable class is represented by 70% of the cables. Figure 6.74 illustrates the concept of the average cable for this example, that is, where 10%

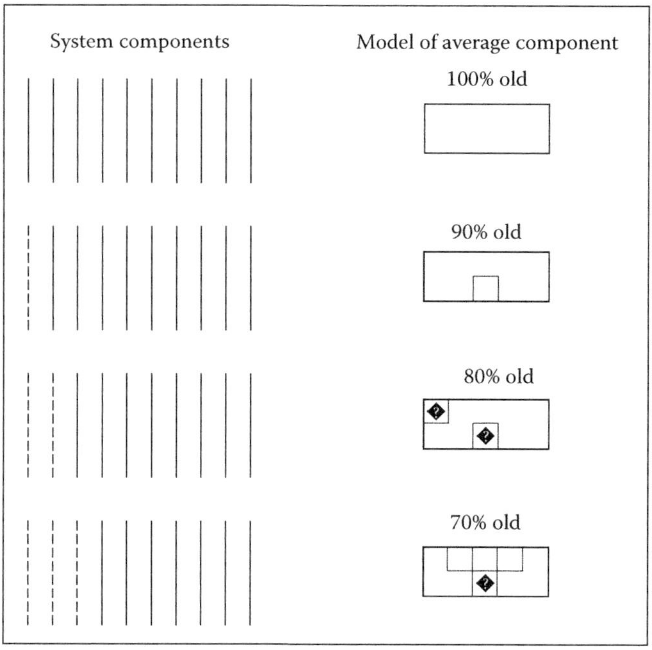

FIGURE 6.74
Representation of the average component.

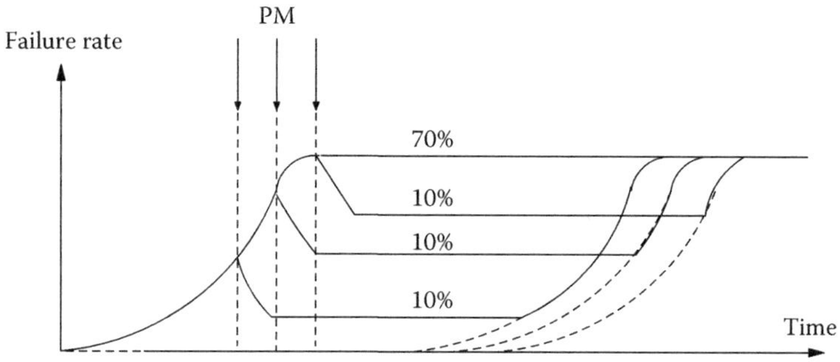

FIGURE 6.75
Illustration of different failure rate functions due to a maintenance strategy.

of the components are replaced each time. The resulting model for the average component then implies a 70% "old component." Furthermore, Figure 6.75 illustrates the resulting four different failure rate functions.

However, the aim here is to define one failure rate function that represents the average behavior of the component type. An average failure rate function ($\lambda_{av}(t)$) for this example and with time t: $0 \rightarrow 30$ is made stepwise of the following different functions:

$$\begin{aligned}
\lambda_{av}(0) &= \lambda_{old}(t)\\
\lambda_{av}(1) &= \lambda_{old}(t)\\
\ldots &= \lambda_{old}(t)\\
\lambda_{av}(9) &= 0.9\cdot\lambda_{old}(t)+0.1\cdot\lambda_{PM1}(t),\\
\ldots &= 0.9\cdot\lambda_{old}(t)+0.1\cdot\lambda_{PM1}(t),\\
\lambda_{av}(11) &= 0.8\cdot\lambda_{old}(t)+0.1\cdot\lambda_{PM1}(t)+0.1\cdot\lambda_{PM2}(t),\\
\lambda_{av}(12) &= 0.7\cdot\lambda_{old}(t)+0.1\cdot\lambda_{PM1}(t)+0.1\cdot\lambda_{PM2}(t)+0.1\cdot\lambda_{PM3}(t),\\
\ldots &= 0.7\cdot\lambda_{old}(t)+0.1\cdot\lambda_{PM1}(t)+0.1\cdot\lambda_{PM2}(t)+0.1\cdot\lambda_{PM3}(t),\text{ and}\\
\lambda_{av}(30) &= 0.7\cdot\lambda_{old}(t)+0.1\cdot\lambda_{PM1}(t)+0.1\cdot\lambda_{PM2}(t)+0.1\cdot\lambda_{PM3}(t),
\end{aligned}\tag{6.28}$$

where $\lambda_{old}(t)$ refers to the components that do not undergo PM and $\lambda_{PMi}(t)$ refers to the components undergoing PM at times $i=1,2,3$.

This equation can be rewritten in a more compact form as follows: $\lambda_{av}(t)=$

$$\begin{cases}
\lambda_{old}(t) & : \quad 0\leq t\leq 9\\
0.9\cdot\lambda_{old}(t)+0.1\cdot\lambda_{PM1}(t) & : \quad 9\leq t\leq 11\\
0.8\cdot\lambda_{old}(t)+0.1\cdot\lambda_{PM1}(t)+0.1\cdot\lambda_{PM2}(t) & : \quad 11\leq t\leq 12\\
0.7\cdot\lambda_{old}(t)+0.1\cdot\lambda_{PM1}(t)+0.1\cdot\lambda_{PM2}(t)+0.1\cdot\lambda_{PM3}(t) & : \quad 12\leq t\leq 30
\end{cases}\tag{6.29}$$

6.7.4 Modeling Composite Failure Rate for the Average Cable Component

6.7.4.1 Resulting Composite Failure Rate Model

A composite failure rate function for an average 11 kV cable component can be formulated based on those concepts and results that were presented in Section 6.7.3. This model for the component failure rate constitutes an extension of the Cable Application Study that enables the system reliability analysis for analyzing the effect of different maintenance strategies.

The composite failure rate for an average 11 kV cable component has been defined in Equation 6.26 as follows:

$$\lambda_{tot}(t)=\lambda_{LH11}+\lambda_{wt}(t)\quad [\text{f/yr}] \quad : \quad t_0\leq t\leq T$$

$$\text{where}\quad \lambda_{wt}(t)\subseteq\begin{cases}
\lambda_{old}(t) & \text{if no PM}\\
\lambda_{PM_{si}}(t) & \text{if } PM_{si}\\
\lambda_{PM_{rp}}(t) & \text{if } PM_{rp}.
\end{cases}\tag{6.30}$$

Two maintenance strategies denoted S_1 and S_2 were defined in Table 6.43. For both of these, PM is applied at three points in time: 9, 11, and 12 years after coming into service. The proportion of cables undergoing PM at each time is 10% for S_1 and 30% for S_2.

Finally, the average component failure rate was introduced and Equation 6.29 illustrated how it can be evaluated.

The composite failure rate for the average 11 kV cable can be evaluated based on these three concepts with corresponding results. The following is a general presentation of the resulting equation:

$$\lambda_{atot}(t) = \lambda_{LH11} + \lambda_{awt}(t) \quad : \quad t_0 \leq t \leq T \quad [\text{f/yr}] \tag{6.31}$$

$$\lambda_{atot}(t, \text{no PM}) = \lambda_{LH11} + \lambda_{old}(t) \cdot (l_{av}/100) \quad : \quad t_0 \leq t \leq T[\text{f/yr}] \tag{6.32}$$

$$\lambda_{atot}(t, PM_{i3}, S) = \lambda_{LH11} + (l_{av}/100) \cdot \begin{cases} \lambda_{old}(t) & : \; t_0 \leq t \leq t_{PM1} \\ (1 - s_1) \cdot \lambda_{old}(t) + s_1 \cdot \lambda_{PM_{i,1}}(t) & : \; t_{PM1} \leq t \leq t_{PM2} \\ (1 - (s_1 + s_2)) \cdot \lambda_{old}(t) + s_1 \cdot \lambda_{PM_{i,1}}(t) + & : \; t_{PM2} \leq t \leq t_{PM3} \\ s_2 \cdot \lambda_{PM_{i,2}}(t) & : \\ (1 - (s_1 + s_2 + s_3)) \cdot \lambda_{old}(t) + s_1 \cdot \lambda_{PM_{i,1}}(t) + & : \; t_{PM3} \leq t \leq T \\ + s_2 \cdot \lambda_{PM_{i,2}}(t) + s_3 \cdot \lambda_{PM_{i,3}}(t) & : \end{cases} \tag{6.33}$$

6.7.4.2 Analysis of the Effect of Different Strategies

Maintenance strategies have been discussed earlier, and in Section 6.7.3.2, there were two strategies defined for demonstrating purposes, presented in Table 6.2.

These two different strategies have been studied in detail for the Cable Application Study. Furthermore, this includes the analyses of PM featuring both rehabilitation and replacement methods. Consequently, there are three different maintenance methods being represented, each with a failure rate function as follows.

- No PM, where the components are only affected by CM, $\lambda_{atot}(t)$ $t_0 \leq t \leq T$ [f/yr].
- PM by rehabilitation, $\lambda_{atot}(t, PM_s i)$ for a specific strategy $S : t_0 \leq t \leq T$ [f/yr].
- PM by replacement, $\lambda_{atot}(t, PM_r p)$ for a specific strategy $S : t_0 \leq t \leq T$ [f/yr].

6.7.4.3 Comparison of Maintenance Methods

A comparison of these different maintenance methods has been done. Figure 6.76 demonstrates the results for the Cable Application Study. This figure illustrates the characteristics of the three failure rate functions presented earlier. Furthermore, this figure shows the results for the situation where the

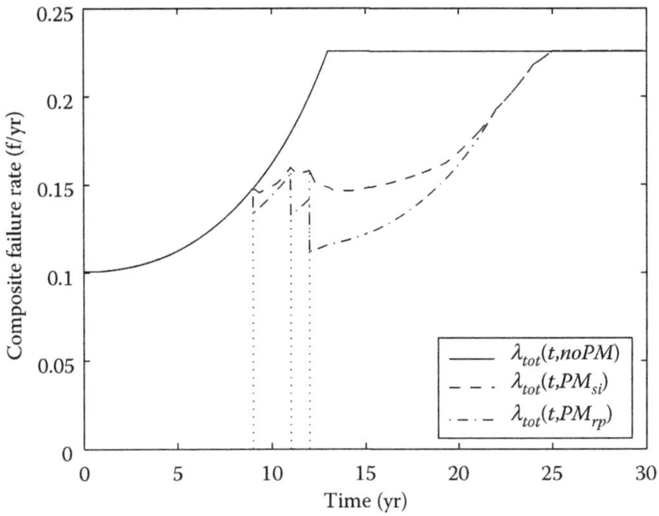

FIGURE 6.76
Comparison of failure rate functions for different maintenance methods, showing results for the situation with the 11 kV average cable component in the Birka system which is affected by water treeing and exposed to the maintenance strategy S_2.

11 kV average cable in the Birka system is exposed to strategy 82 (with 30% of the cables rehabilitated at years 9, 11, and 12). A significant result is that PM for reducing failures is the most effective method. Moreover, the result indicates that the lowest failure rate function is obtained for the situation where PM is performed by replacement.

6.7.4.4 Comparing the Two Maintenance Strategies

In the next step of the analysis, the effect of different strategies for maintenance has been analyzed. For the Cable Application Study, the previously defined strategies S1 and S2 have been studied in detail.

Figure 6.77 shows the effect of two different strategies for PM on the composite failure rate. With S1, 10% of the cables are rehabilitated on the three maintenance occasions, and with S2, 30% of the cables are rehabilitated on the three maintenance occasions. This means that with the first strategy a total of 30% of the total population of cables are changed within the period compared with 90% for the second. Consequently, the figure shows a span for the probable effect of the maintenance strategy that is the benefit in the composite component failure rate.

Figure 6.78 shows the results for the same analysis when PM is applied by replacement instead. A similar behavior can be seen in cable rehabilitation, however, the difference in benefit for the different strategies is even more significant.

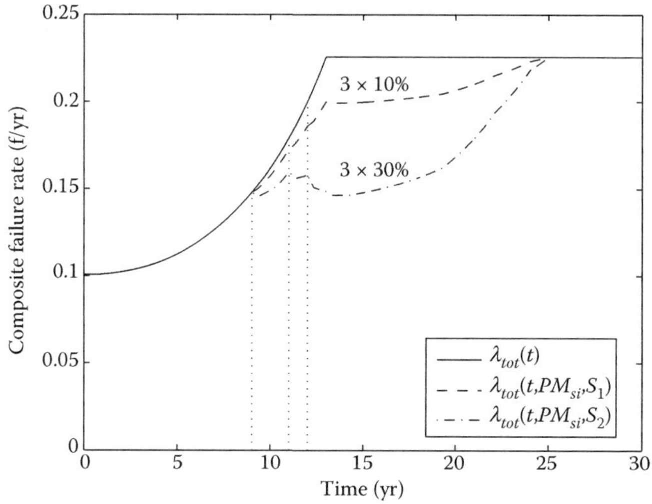

FIGURE 6.77
Effect of PM_{si} using different strategies for the 11 kV average cable in the Birka system.

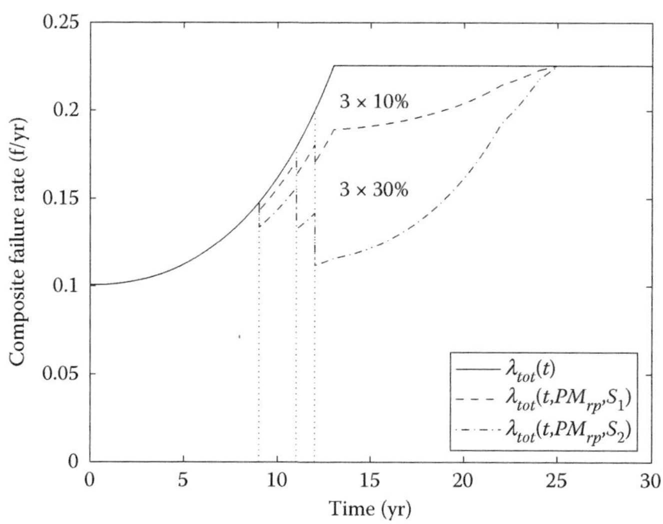

FIGURE 6.78
Effect of PM_{rp} using different strategies for the 11 kV average cable in the Birka system.

6.7.4.5 The Effect of Different Strategies

Detailed analyses have been made for the different strategies. In strategy S_2, a total of 90% of cables are maintained with equal amounts at three points in time (30% at years 9, 11, and 12). If it is assumed instead that the constraint for

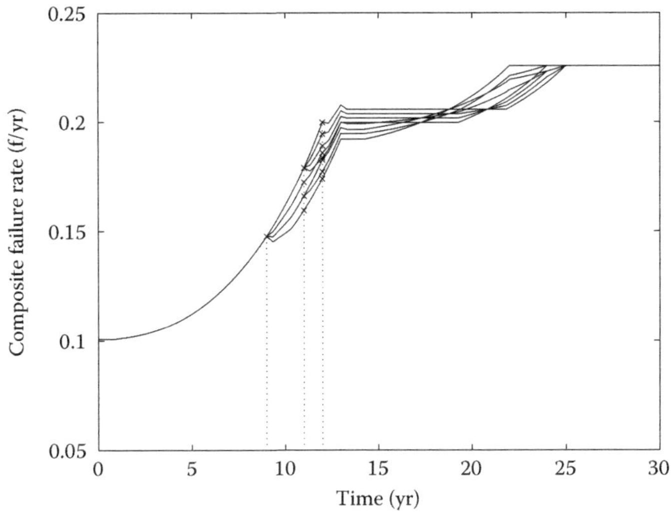

FIGURE 6.79
Effect of different strategies for rehabilitating 30% of cables on three occasions (variations of S_2).

the maintenance strategy is that the amount of cables maintained constitute 90% of the total, then numerous different possible strategies are obtained. This aspect of constraint for maintenance plans was previously discussed in Section 6.7.3.2. Figure 6.79 illustrates the effect of different strategies when rehabilitation is applied that sum up to a total of 90% of cables maintained. Results from a similar analysis when replacement is applied are displayed in Figure 6.80. This figure suggests that an optimal solution exists, for example, where the composite failure rate is limited by a criteria for the upper level, and the optimal solution is the maintenance strategy for which the failure rate reaches that level last in time.

A similar analysis has been made for the first PM strategy (S_1). This strategy implies that a total of 30% of cables are maintained at three points in time. As presented earlier in Table 6.44, this constraint results in eight different cases for dividing the PM measures. Figure 6.81 displays these possible maintenance strategies. It can be seen that the effect of applying the PM method by rehabilitation is different depending on when it is applied. For example, the extremes, that is, either performing all PM at the first or last point in time for maintenance show that the benefit in reliability would increase in the earlier one. In other words the greatest benefit to failure rate is gained if the rehabilitation effort is concentrated to the first maintenance occasion. However, it is important to note that the cost is dependent on when the investment in maintenance is made, which is further analyzed in Section 12.6. Similar analysis of when PM involves replacement is presented in Figure 6.82. Both these results indicate that the time aspect is important when determining an effective maintenance plan.

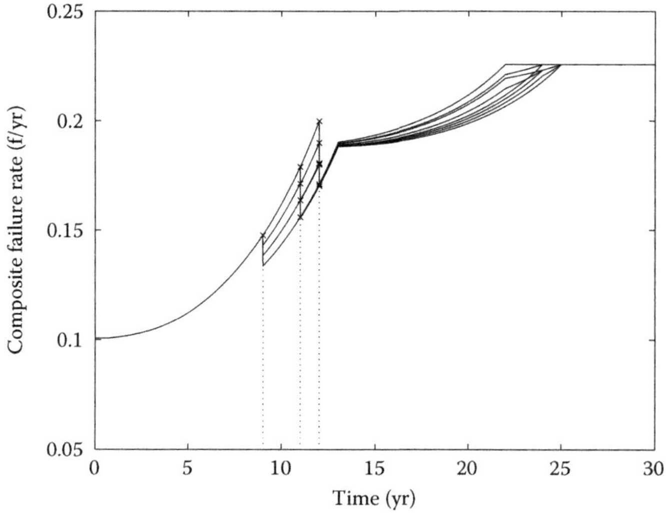

FIGURE 6.80
Effect of different strategies for replacing 30% of cables on three occasions (variations of S_2).

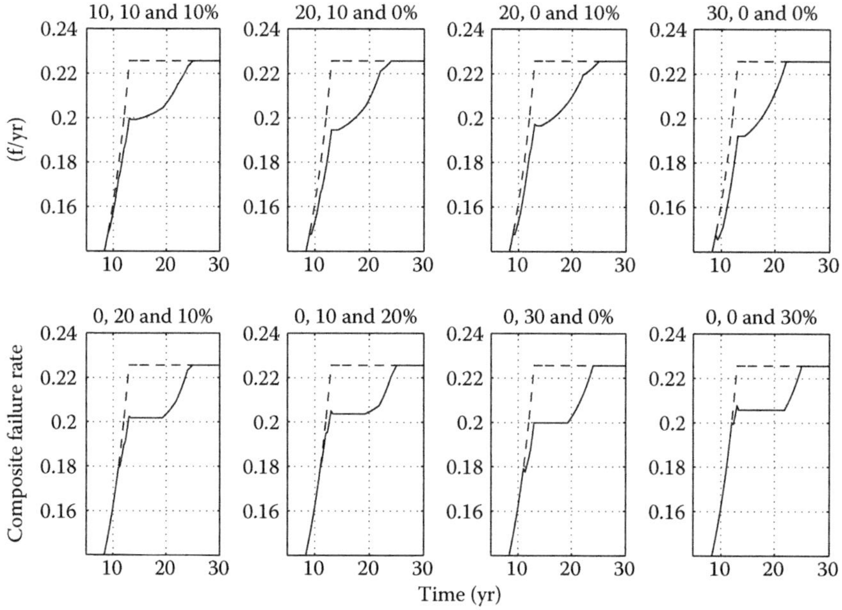

FIGURE 6.81
Effect of different strategies for rehabilitating 10% of cables on three occasions (variations of S_1).

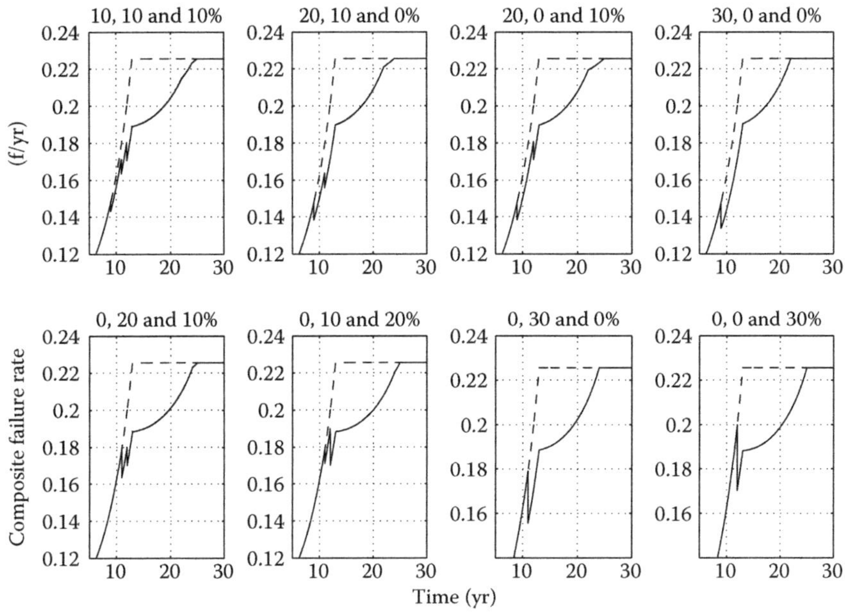

FIGURE 6.82
Effect of different strategies for replacing 10% of cables on three occasions (variations of S_1).

6.7.4.6 Resulting Failure Rate Functions as Input Data for the System Analysis

Tables 6.45 and 6.46 present selected values for each year for the resulting composite failure rate functions for an average 11 kV cable component in the Birka system. Furthermore, these correspond to strategies S_1 and S_2, respectively. These data demonstrate reliability input data for the system effect analysis with RADPOW.

The results have been validated by comparisons with corresponding manually calculated values using Equation 6.33. The selected values have been presented below. The conclusion is that the results for the implemented equations are correct.

$$t = 9 \Leftrightarrow t_{PM1} \Rightarrow$$

$$\lambda_{atot}(9, \text{no PM}) = \lambda_{LH11} + \lambda_{old}(9) \cdot (l_{av}/100) =$$

$$0.10069 + 0.5809 \cdot (8.10584/100) = 0.14778,$$

$$\lambda_{atot}(9, PM_{si3}, S) = \lambda_{LH11} + (l_{av}/100) \cdot (0.9 \cdot \lambda_{old}(9) + 0.1 \cdot \lambda_{PM_{si,1}}(9)) =$$

$$= 0.10069 + (8.10584/100) \cdot (0.5809) = 0.14778,$$

TABLE 6.45

Results for the Composite Failure Rate Function ($\lambda_{atot}(t)$) for the 11 kV Average Cable in the Birka System Affected by Water Treeing and Subject to Preventive Maintenance Strategy S_1

Time [yr]	$\lambda_{atot}(t, \text{no PM})$ [f/yr]	$\lambda_{atot}(t, PM_{si}, S_1)$ [f/yr]	$\lambda_{atot}(t, PM_{rp}, S_1)$ [f/yr]
0	0.10069	0.10069	0.10069
1	0.10094	0.10094	0.10094
2	0.10222	0.10222	0.10222
3	0.10436	0.10436	0.10436
4	0.10767	0.10767	0.10767
5	0.11234	0.11234	0.11234
6	0.11853	0.11853	0.11853
7	0.12637	0.12637	0.12637
8	0.13604	0.13604	0.13604
9*	0.14778	0.14778	0.14307
10	0.16196	0.15767	0.15586
11*	0.17913	0.17261	0.16359
12*	0.20002	0.18599	0.17062
13	0.22567	0.19990	0.18905
14	0.22567	0.19923	0.18986
15	0.22567	0.19969	0.19103
16	0.22567	0.20047	0.19261
17	0.22567	0.20144	0.19466
18	0.22567	0.20261	0.19724
19	0.22567	0.20403	0.20041
20	0.22567	0.20658	0.20427
21	0.22567	0.21009	0.20895
22	0.22567	0.21465	0.21465
23	0.22567	0.21845	0.21845
24	0.22567	0.22311	0.22311
25	0.22567	0.22567	0.22567
26	0.22567	0.22567	0.22567
27	0.22567	0.22567	0.22567
28	0.22567	0.22567	0.22567
29	0.22567	0.22567	0.22567
30	0.22567	0.22567	0.22567
Special time point:			
$t_3 = 21.82$	0.22567	0.21353	0.21353

Note: Integer values are listed that provide input data for reliability system analysis (RADPOW) and cost benefit analysis.

TABLE 6.46

Results for the Composite Failure Rate Function ($\lambda_{atot}(t)$) for the 11 kV Average Cable in the Birka system Affected by Water Treeing and Subject to PM Strategy S_2

Time [yr]	$\lambda_{atot}(t, \text{no PM})$ [f/yr]	$\lambda_{atot}(t, PM_{si}, S_2)$ [f/yr]	$\lambda_{atot}(t, PM_{rp}, S_2)$ [f/yr]
0	0.10069	0.10069	0.10069
1	0.10094	0.10094	0.10094
2	0.10222	0.10222	0.10222
3	0.10436	0.10436	0.10436
4	0.10767	0.10767	0.10767
5	0.11234	0.11234	0.11234
6	0.11853	0.11853	0.11853
7	0.12637	0.12637	0.12637
8	0.13604	0.13604	0.13604
9*	0.14778	0.14778	0.13365
10	0.16196	0.14908	0.14365
11*	0.17913	0.15956	0.13252
12*	0.20002	0.15793	0.11180
13	0.22567	0.14835	0.11581
14	0.22567	0.14634	0.11824
15	0.22567	0.14772	0.12174
16	0.22567	0.15007	0.12648
17	0.22567	0.15297	0.13264
18	0.22567	0.15650	0.14037
19	0.22567	0.16075	0.14988
20	0.22567	0.16840	0.16145
21	0.22567	0.17893	0.17550
22	0.22567	0.19260	0.19260
23	0.22567	0.20401	0.20401
24	0.22567	0.21798	0.21798
25	0.22567	0.22567	0.22567
26	0.22567	0.22567	0.22567
27	0.22567	0.22567	0.22567
28	0.22567	0.22567	0.22567
29	0.22567	0.22567	0.22567
30	0.22567	0.22567	0.22567
Special time point:			
$t_3 = 21.82$	0.22567	0.18926	0.18926

Note: Integer values are listed that provide input data for reliability system analysis (RADPOW) and cost benefit analysis.

$\lambda_{atot}(9, PM_{rp3}, S) = \lambda_{LH11} + (l_{av}/100) \cdot (0.9 \cdot \lambda_{old}(9) + 0.1 \cdot \lambda_{PM_{rp,1}}(9)) =$

$= 0.10069 + (8.10584/100) \cdot (0.9 \cdot 0.5809) = 0.14307,$

$t = 22 \Leftrightarrow t_{PM2} \Rightarrow$

$\lambda_{atot}(22, \text{no PM}) = \lambda_{LH11} + \lambda_{old}(22) \cdot (l_{av}/100) =$

$0.10069 + 1.5418 \cdot (8.10584/100) = 0.22567,$

$\lambda_{atot}(22, PM_{si3}, S) = \lambda_{LH11} + (l_{av}/100) \cdot (0.7 \cdot \lambda_{old}(22) + 0.1 \cdot \lambda_{PM_{si,1}}(22)) +$

$0.1 \cdot \lambda_{PM_{si,2}}(22)) + 0.1 \cdot \lambda_{PM_{si,3}}(22)) == 0.10069 + (8.10584/100) \cdot$

$\cdot (0.7 \cdot 1.5418 + 0.1 \cdot 1.5418 + 0.1 \cdot 0.9676 + 0.1 \cdot 0.7558) = 0.21464,$ and

$\lambda_{atot}(22, PM_{rp3}, S) = \lambda_{LH11} + (l_{av}/100) \cdot (0.7 \cdot \lambda_{old}(22) + 0.1 \cdot \lambda_{PM_{rp,1}}(22)) +$

$0.1 \cdot \lambda_{PM_{rp,2}}(22)) + 0.1 \cdot \lambda_{PM_{rp,3}}(22)) == 0.10069 + (8.10584/100) \cdot$

$\cdot (0.7 \cdot 1.5418 + 0.1 \cdot 1.5418 + 0.1 \cdot 0.9676 + 0.1 \cdot 0.7558) = 0.21464.$

6.7.5 System Reliability Analysis Comparing PM Strategies

This section shifts the analysis from the component to the system level and focuses on the evaluation of reliability for which the tool RADPOW has been used. RADPOW provides reliability indices for load points and system reliability indices. The tool was briefly introduced in Section 3.8.

The objective here is to define the effect on system reliability by PM on selected components in the system. Furthermore, this should be made in quantitative terms. This implies that the Cable Application Study should be extended by performing a reliability analysis where the defined model for failure rate that is the composite failure rate for an average component as defined in Equations 6.32 and 6.33 that is the component input data.

RADPOW has been run repeated times to simulate the failure rate change over time. The previously defined two maintenance strategies (S_1 and S_2) have been analyzed. Input data for the corresponding failure rate functions for these two maintenance strategies were presented in Tables 6.45 and 6.46. This section presents the resulting output, which is the load point reliability indices and system performance indices. MATLAB has been used as a tool for the analysis of the data.

The output is reliability indices varying in time depending on the changed characteristics of the failure rate, and specifically the contribution of the failure causes due to water trees.

6.7.5.1 Effect on Reliability Indices Due to Water-Tree Growth

Figure 6.83 presents the resulting functions for the Birka system reliability indices when the 11 kV cable components have an increased failure rate

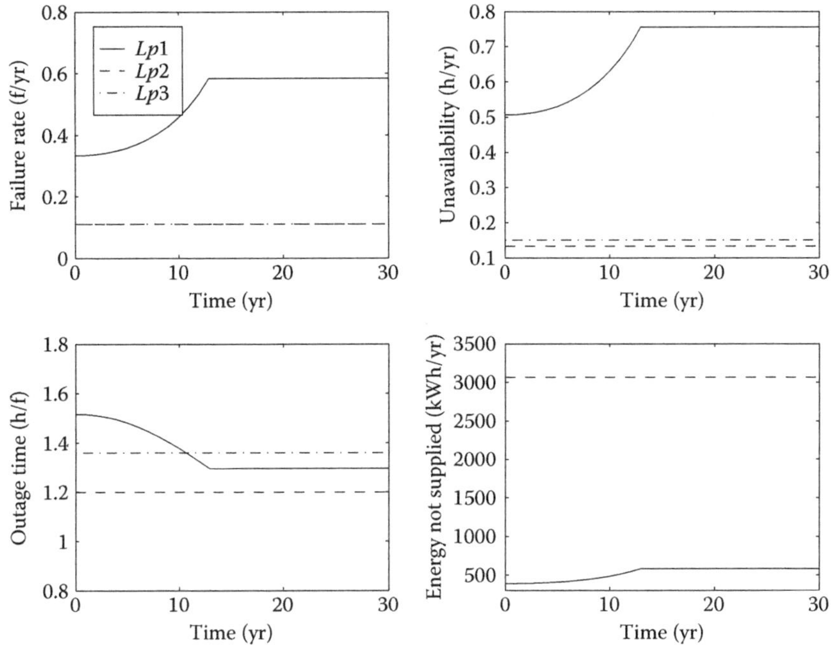

FIGURE 6.83
Load point indices for the Birka system with 11 kV cables affected by water-tree growth. Note that load point *Lp*1 refers to the only customer using 11 kV level (*LH*11).

function due to water-tree growth. The system includes three load points where the first is on the 11 kV level but the other two are on the 33 kV voltage level. However, the model for failure rate change over time only relates to the failures for the 11 kV cable, and failures for other components are assumed constant in time. Consequently, load points 2 and 3 (i.e., SJ and HD) have reliability functions that are constant in time and are therefore not considered in the following analysis of the effect of PM.

The first graph shows the failure rate function. The effect on the failure rate due to water-tree growth has been discussed in earlier sections, however, then the focus was on either the effect on the individual cable component or as a model for the average 11 kV cable in the Birka system. In this analysis, the 11 kV component has been put into a system context, and the effect on the system reliability analyzed. The failure rate functions refer to different load points in the system, and as earlier stated, it is for Load Point 1 that the impact of the included failure rate function can be seen. It can also be seen that the failure rate for Load Point 1 varies from about 0.3345 to 0.5844 failures per year, which implies an increase of about 75%. However, the customer (load point) is not only affected by the failures occurring but also of the duration that these appear (the outage time) as well as the amount of energy

TABLE 6.47

Boundary Values for the Reliability Indices for Load Point 1 (*LH*11) Affected by Water Treeing

Indices	$t = t_0$ yr	$t = T$ yr	Increase [%]
λ_{Lp1}	0.33449 f/yr	0.58445 f/yr	+75
U_{Lp1}	0.50671 h/yr	0.75667 h/yr	+49
r_{Lp1}	1.51487 h	1.29467 h	−17
LOE_{Lp1}	389.650 kWh/yr	581.739 kWh/yr	+49

that is not supplied. Indices to capture these parameters are illustrated in the following plotting. It can be seen that the characteristic of the failure rate function (increasing due to water tree growth) dominates in the functions for unavailability and loss of energy, compared with the contribution from the outage time ($U = \lambda r, LOE = LU$). This is explained by the restoration time being equal for both lengths of water trees, and consequently is constant in time in this context; and this should be compared with the failure rate with a significant increase in time. Furthermore, it can be seen that the outage time is the only one of the functions that is decreasing due to the water-tree effect, meaning that the average restoration time when a failure occurs is shorter. This is because the contribution to restoration time from the 11 kV cables is comparatively small compared with the other components in the system; for example, repair time is 6 h for an 11-kV cable compared with 48 h for a 33-kV cable. For the energy not-supplied, it can be seen that the load points represent different types of customers with a wide range of energy requirements. The third load point (the subway) has such a low consumption per customer so it is not seen in the figure.

Table 6.47 summarizes some results for the load point reliability indices shown in Figure 6.12. This presents the boundary values for the load point indices at Load Point 1, being the values at time 0 (t_0) and at the end of the period (T). These numbers correspond to the reliability indices: maximum and minimum values. As stated earlier, for all of the indices except for the outage time, the maximum value is reached as a result of the maximum contribution to failure rate due to the water trees. Moreover, the last column in the same table presents the individual percentage increases for the different indices. For the maximum value of the failure rate function, the increase from the initial value at time 0 is about 75%. For the unavailability, the corresponding increase is 49%, which is the same increase that appears for the loss of supplied energy. On the other hand, the restoration time decreases (−17%). These results show a significant impact on the load point reliability indices due to the failures caused by water trees. Consequently, it is of great interest to analyze how these percentages could be reduced by PM. Methods to solve this are analyzed further in the following section.

6.7.5.2 Impact of PM on Reliability Indices

Resulting reliability indices for Load Point 1 have been analyzed for different choices of maintenance strategies where PM has been applied to reduce failures due to water trees. Moreover, the PM is performed using either one of the methods of rehabilitation or replacement of 11 kV cables.

Figures 6.84 and 6.85 demonstrate the result for the two strategies S_1 and S_2. These results demonstrate the impact of a maintenance strategy where maintenance is applied on three occasions, years 9, 11, and 12, and at each time with the proportion of either 10% of the cables maintained (S_1) or 30% (S_2). Consequently, 30% (or 90%) of the 11 kV cables are maintained during the time period by either replacement or rehabilitation. The results show a significant effect of PM in reducing the failure rate. Furthermore, this leads to reductions in the unavailability and the energy not-supplied.

Energy not-supplied is a useful load point index that relates the unavailability with the energy supply demand. Table 6.48 presents results for Load Point 1 subject to either one of the maintenance strategies S_1 or S_2. These data show that the total energy not-supplied during the studied period of 30 years is 16.14 MWh when no PM is applied. Furthermore, for strategy S_1, this number is reduced to 15.75 MWh when subject to the silicon injection

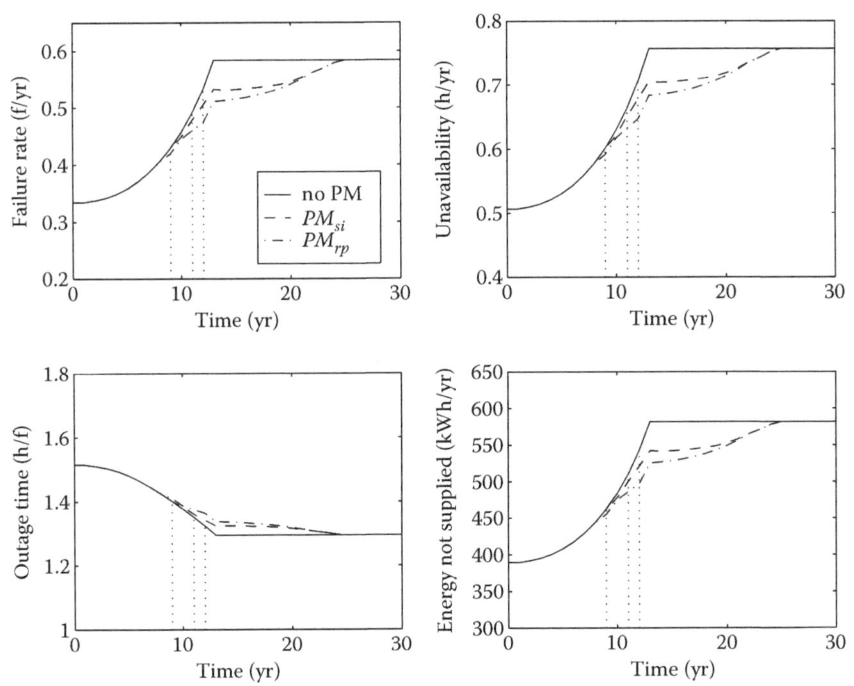

FIGURE 6.84
Effect on reliability indices for Load Point 1 (*LH*11) due to different PM methods. PM applied with strategy S_1, at years 9, 11, and 12 with 10% of the cables subject to PM on each occasion.

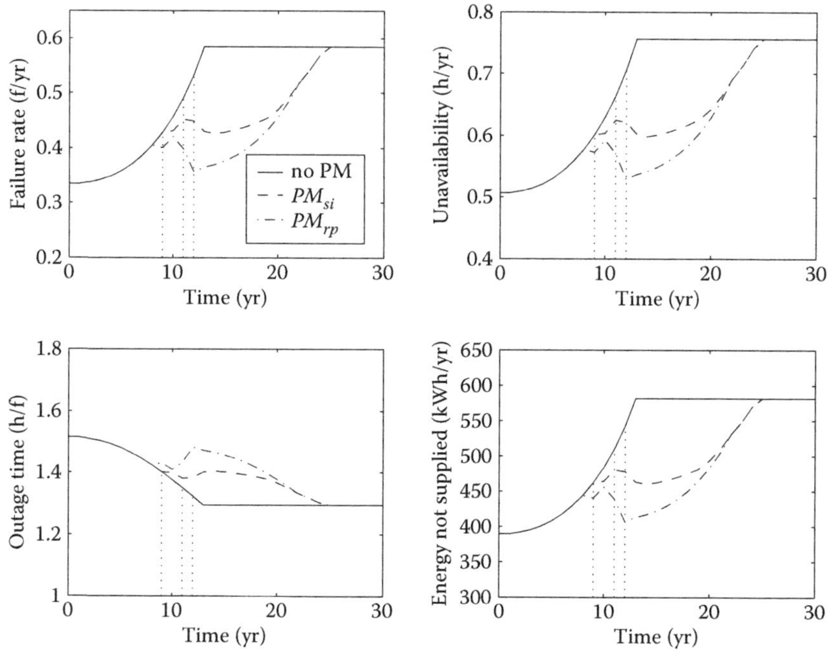

FIGURE 6.85
Effect on reliability indices for Load Point 1 (*LH*11) due to different PM methods. PM applied with strategy S_2, at years 9, 11, and 12 with 30% of the cables subject to PM on each occasion.

method (rehabilitation) and to 15.61 MWh for the replacement method. For strategy S_2, the equivalent numbers are 14.97 MWh and 14.57 MWh, respectively, that is with a total of 90% of 11 cables applied to PM. Consequently, the PM methods analyzed can reduce the energy not-supplied at Load Point 1 by up to around 2 MWh. This benefit can be valued and depends on the cost of the customer interruption. This aspect is analyzed further in Section 6.7.6.

Furthermore, the impact on the system performance indices has been analyzed. Figures 6.86 and 6.87 present the results for four different system indices. As can be seen from the figures, the characteristics follow the behavior of the load point indices for Load Point 1. This is an expected result, since the load point failure for this system is dominated by failures at the 11 kV level and especially by the significant contribution from the failures due to water-tree growth.

The reduction in the reliability functions leads to improved system behavior. However, the figures demonstrate that this benefit is reducing with time, and after a certain time the functions reach the same function values as for the system not affected by maintenance. Both for the rehabilitation, the failure rate reaches the maximum value at year 25. However, the replacement method provides better reliability after PM until about year 22 ($t_3 = 21.82$ years) for both of the strategies.

TABLE 6.48

Results for Energy Not-Supplied for the Cable Application Study with Different Maintenance Methods and Two PM Strategies S_1 and S_2

Time [yr]	$LOE_{Lp1}(t, \text{no PM})$ [kWh/yr]	$LOE_{Lp1}(t, PM_{si}, S)$ [kWh/yr]		$LOE_{Lp1}(t, PM_{rp}, S)$ [kWh/yr]	
		S_1	S_2	S_1	S_2
0	389.65	389.65	389.65	389.65	389.65
1	389.65	389.65	389.65	389.65	389.65
2	392.00	392.00	392.00	392.00	392.00
3	395.29	395.29	395.29	395.29	395.29
4	400.39	400.39	400.39	400.39	400.39
5	407.57	407.57	407.57	407.57	407.57
6	417.09	417.09	417.09	417.09	417.09
7	429.14	429.14	429.14	429.14	429.14
8	444.02	444.02	444.02	444.02	444.02
9*	462.07	462.07	462.07	454.83	440.34
10	483.88	477.28	464.07	474.50	455.72
11*	510.29	500.26	480.20	486.40	438.60
12*	542.41	520.84	477.70	497.20	406.74
13	581.86	542.22	462.95	525.54	412.90
14	581.86	541.20	459.86	526.80	416.64
15	581.86	541.91	461.98	528.60	422.02
16	581.86	543.11	465.60	531.02	429.31
17	581.86	544.60	470.05	534.20	438.79
18	581.86	546.40	475.48	538.14	450.68
19	581.86	548.58	482.02	543.01	465.30
20	581.86	552.50	493.78	548.95	483.09
21	581.86	557.90	509.98	556.15	504.70
22	581.86	564.91	531.00	564.91	531.00
23	581.86	570.76	548.55	570.76	548.55
24	581.86	577.93	570.04	577.93	570.04
25	581.86	581.86	581.86	581.86	581.86
⋮	581.86	581.86	581.86	581.86	581.86
30	581.86	581.86	581.86	581.86	581.86
Total: [MWh]	16.14	15.75	14.97	15.61	14.57

6.7.5.3 Which PM Method Is the Best?

A fundamental question for an RCM plan and the actual question raised at this point in the Cable Application Study is the following:

Which PM method is the best?

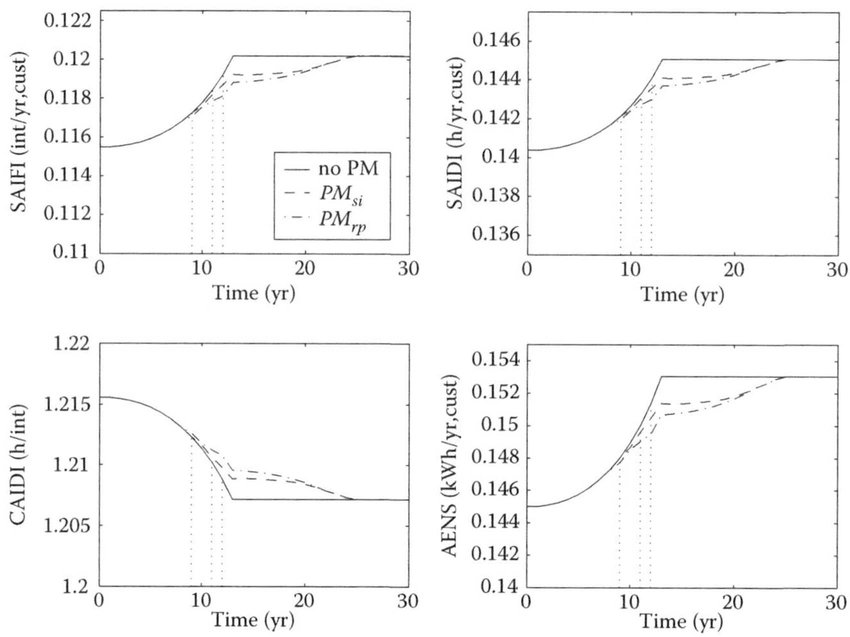

FIGURE 6.86
Effect on system performance indices due to PM methods to reduce water-tree failures. PM applied with strategy S_1 at years 9, 11, and 12 with 10% of cables subject to PM on each occasion.

For the Cable Application Study, this question relates to the choice between no PM, or PM by either rehabilitation or replacement of the 11 kV cables in the Birka system.

The system reliability effect analysis, which takes into account the impact of maintenance methods, should provide results that support the decision. However, there is no obvious answer to the question. The optimal choice of method would depend on the specific requirements. However, one solution is to evaluate expected reliability indices over the time period that is an average value for a load point as follows:

$$
\begin{aligned}
\lambda_{av,Lp} &= \frac{\sum_{t=t_0}^{T} \lambda_{Lp}(t,PM)}{(T-t_0)} \\
U_{av,Lp} &= \frac{\sum_{t=t_0}^{T} U_{Lp}(t,PM)}{(T-t_0)} \\
r_{av,Lp} &= \frac{\sum_{t=t_0}^{T} r_{Lp}(t,PM)}{(T-t_0)} \\
LOE_{av,lp} &= \sum_{t=t_0}^{T} LOE_{Lp}(t,PM).
\end{aligned}
\tag{6.34}
$$

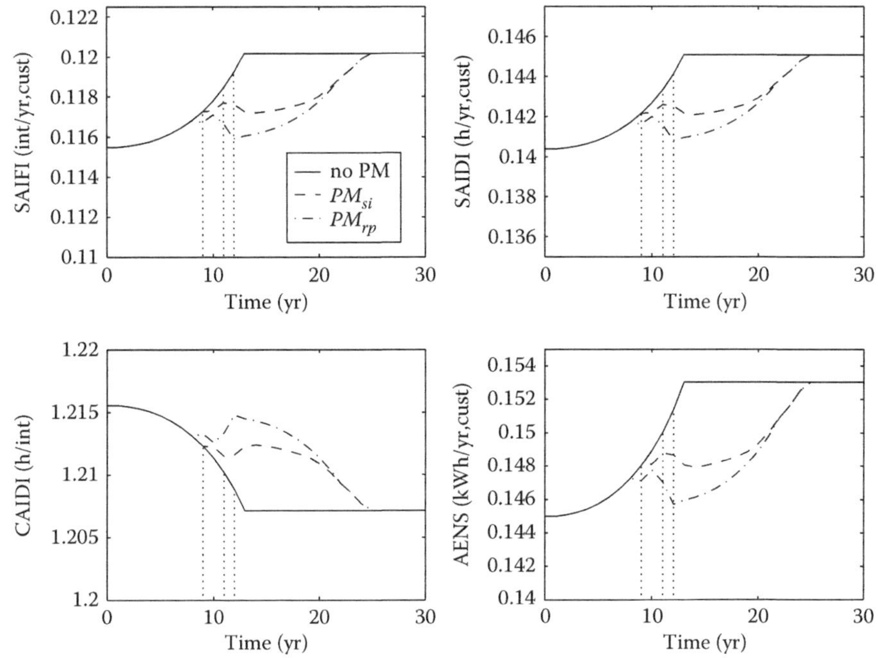

FIGURE 6.87
Effect on system performance indices due to PM methods to reduce water-tree failures. PM applied with strategy S_2 at years 9, 11, and 12 with 30% of cables subject to PM on each occasion.

The first three average values relate to the basic reliability performance indices, and the last to a reliability index that includes the effect on customer supply.

These indices have been evaluated for the Cable Application Study. Table 6.49 summarizes the result. The results show consistently that the best reliability is gained with PM by replacement and with as much as possible of the component subject to PM (strategy 82). For example, energy not supplied is reduced by about 1 MWh with this solution. Furthermore, this is a consequence of the "gain" in the lowest unavailability, which in this example is from 0.69 to 0.63 h/yr.

6.7.6 Cost and Benefit Analysis Comparing Maintenance Strategies

This section introduces economic analysis, defines cost factors of special interest for the RCAM method. It develops models for the evaluation of cost functions for the same and finally performs a total cost analysis. Moreover, all these theories are exemplified for the Cable Application Study. The section therefore also includes a treatment of required input data to support the methods defined for the cost evaluation.

TABLE 6.49

Comparing Maintenance Methods Based on Reliability Indices with Results for the Cable Application Study

		Average Values for Different Maintenance Methods				
Reliability			PM_{si}		PM_{rp}	
Factor	**Unit**	**CM**	S_1	S_2	S_1	S_2
$\lambda_{av,Lp1}$	[f/yr]	0.52154	0.50470	0.47101	0.49890	0.45363
$U_{av,Lp1}$	[h/yr]	0.69950	0.68266	0.64897	0.67687	0.63159
$r_{av,Lp1}$	[h/f]	1.40374	1.41318	1.43738	1.41710	1.45511
$LOLE_{av,Lp1}$	[MWh/yr]	16.14	15.75	14.97	15.61	14.57

Chapter 3 and Section 3.7 present some fundamental techniques for cost analysis that are used in this section.

The nomenclature used for the cost analysis has been chosen to where possible agree with those defined in the reference books [139] and [140].

6.7.6.1 Introduction to Cost Analysis

This section introduces the aspect of cost into the methodology development. Thereby the RCM analysis reaches its final step which relates the benefit in cost due to reliability and PM. The motivation for any PM strategy must include financial considerations. These should justify that the cost for applying the PM measure should be less than not taking any action, that is, CM. Therefore, a very important issue is to compare the costs associated with different PM methods as well as only CM, which is the focus for the financial analysis presented in this section. The problem being investigated is how to relate reliability with PM. Furthermore, it has been identified that this can be made by analyzing component failures and how to prevent them. The introductory general question is therefore:

What kind of costs are associated with failures?

There are several costs that can be related to the effect of interruption of supply. One way of grouping these is by distinguishing between cost due to failure and cost due to preventing failure. The first of these two categories is related to the cost of restoring failure, for example, repairs, penalty costs to the affected customers, or losses in revenue due to reductions for nondelivered energy. The second category relates to the costs of the PM actions such as, for example, replacement of the cable. Another way of grouping the same costs is by distinguishing between the parties utility costs or customer costs. On the one hand, the cost for the restoration of failure cost of failure and the cost for the preventive action to combat the failure cost of PM are both related to the utility. On the other hand, the cost for a supply interruption affects the customer, who will suffer from unavailability and, for example, then be compensated for via a penalty cost for the cost of interruption. Finally, with the focus on the reliability, the costs can be grouped into either reliability costs (those costs for investment needed to achieve a certain level of reliability, for example, PM actions) or reliability value (the benefit derived by the customers and society). The first would include the balance between no PM (which is CM) and PM actions. The concepts of reliability cost and reliability worth (value) are discussed further in [79].

The cost analysis that follows relates to the modeling and the evaluation of the following.

1. The cost of failure (C_f)
2. The cost of PM (C_{PM})
3. The cost of interruption (C_{int}).

The optimal maintenance method and PM strategy is the solution that minimizes the total cost, which is the sum of these three costs that relates to both the utility and the customer cost.

It is interesting to note that some of these above introduced costs are related to the component failure itself, and some to the system effect of the failure. Therefore, the reliability input data for the cost and benefit analysis are both from the component and system levels. The input data required for the Cable Application Study are presented in the following section.

6.7.6.2 Input Data for the Cost Benefit Analysis

The input data required for the cost and benefit analysis can be grouped into three types: (i) the cost specific data (e.g., the cost of repair), (ii) the reliability data for the system being analyzed both for the component level (failure rate for average component) and for the system level (loss of energy for the affected load point), and (iii) the economic factors (e.g., interest rates). In the following paragraphs, these data are presented as input data for the Cable Application Study, that is, the Birka system.

The costs of different maintenance activities for the Birka System have been analyzed, and specifically costs related to faults in the 11 kV underground cables analyzed as presented in the report [112]. A result from this analysis is the average cost for the repair of cable, which is SEK 20,000–40,000 depending on cable length, laying conditions, and related factors. Another interesting figure to keep in mind is MSEK 5 that Birka service receive per month for handling the maintenance of Birka Nats network (not only the Stockholm network). Moreover, the cost of cable replacement (for two parallel cables) costs about SEK 2000/m, where the cost for the cable itself is a small part of the whole cost. The largest contribution to the cost relates to the laying.

The SINTEF study presents cost data for the two PM methods of rehabilitation and replacement as shown in Table 6.15. From the values in the table, the following average costs per meter cable are obtained: NKR 294/m and 261/m for the rehabilitation, and NKR 452/m and 569/m, respectively, for the replacement. One Norwegian crown (NKR) is about 1.5 Swedish crowns (SEK) and the resulting average values in SEK are 416 and 766. Consequently, the cost for rehabilitation is about 54% of the cost for replacement.

It can be seen that the data from the SINTEF study show much lower costs than the Birka study. This is probably due to the Birka system being analyzed to represent an urban network which implies many complications for the cable laying, resulting in high costs. Based on this observation and the data above, the following assumptions are made for the input cost data for the analysis: C_{PMsi} = SEK 1000/m for silicon rehabilitation, and C_{PMrv} = SEK 2000/m for replacement. Consequently, it is assumed that the rehabilitation method costs half as much as the replacement method.

In the Cable Application Study, two different types of PM methods have been analyzed (rehabilitation and replacement of cable), along with the situation without any PM. Therefore, there are three different sets of input data for the system being analyzed, referred to as no PM, PM_{rp}, and PM_{si}. Furthermore, these maintenance methods have been analyzed for two different strategies (S_1 and S_2) as defined in Table 6.43. Moreover, the component being affected by the PM measure is an average cable of the 11 kV underground cable referred to as LH11. As defined in Equation 6.24, the length of this cable is 8.10584 km.

The component input data used for the cost analysis are represented by the composite failure rate function which is $\lambda_{atot}(t, noPM)$, $\lambda_{atot}(t,\ PM_si,\ S)$, and $\lambda_{atot}(t,\ PM_rp,\ S)$ [f/yr]. Tables 6.45 and 6.46 present annual values for these failure rate functions for the different maintenance methods and strategies analyzed.

The system input data used for the cost analysis are represented by the energy not-supplied function, that is, $LOE_{Lp1}(t, \text{no PM})$, $LOE_{lp1}(t, PM_{si}, S)$, and $LOE_{lp1}(t, PM_{rp}, S)$[kWh/yr]. Table 6.48 presents annual values for these functions, which are output values from the reliability analysis using RADPOW.

Finally to complete the input data required for the cost analysis, the economic evaluation factors need to be defined. These include, for example, the rates on capital and the cost of interruption, plus related costs. These kinds of data are somewhat more subjective than those discussed earlier, and also sensitive to changes on the financial market. Therefore, it has been decided to use the same numerical values for these factors as those used in a recent economic analysis at a Swedish utility, Vattenfall, and presented in [141]. The interruption cost is discussed further when modeling the cost of interruption.

TABLE 6.50

Input Data for the Cost and Benefit Analysis of the Cable Application Study

Parameter	Notation	Value	Source
Repair cost	c_f	30,000 SEK/f	[112,116]
Replacement cost	$c_{PM_{rp}}$	$2 \cdot 10^6$ [SEK/km]	[112,116]
Rehabilitation cost	$c_{PM_{si}}$	$1 \cdot 10^6$ [SEK/km]	[112,116,127]
Cost for interruption	c_{int}	43 [SEK/kWh]	[141,142]
Inflation	d_1	0.02, i.e., 2%	[141,143]
Interest rate	d_2	0.07, i.e., 7%	[141,143]

Table 6.50 summarizes the specifically introduced input data for cost and benefit analysis of the Cable Application Study.

6.7.6.3 Modeling the Cost of Failure

The cost for restoration of supply after failure from a utility perspective can be expressed as the product of the failure rate and the cost for restoration. A general formula for the annual cost of restoration, referred to here as cost of failure, can consequently be defined with the simple relation:

$$C_f(t) = \lambda(t) \cdot c_f(t) \quad : \quad t_0 \leq t \leq T \quad [\text{SEK/yr}] \tag{6.35}$$

where $\lambda(t)$ is the failure rate [f/yr] and $c_f(t)$ is the cost of failure [SEK/f].

The failure rate function which depends on the choice of maintenance method and strategy has been analyzed previously in great detail. The input data functions for the Cable Application Study are as introduced in Section 6.7.6.2: $\lambda_{atot}(t, \text{no PM})$, $\lambda_{atot}(t, PM_{si}, S)$, and $\lambda_{atot}(t, PM_{rp}, S)$.

For defining the cost function, it is desirable to predict how the cost of failure could change in the future. From the input data (see Section 6.7.6.2), a fixed cost for repair is known. Assume that this cost of a failure changes with time ($c_f(t)$ during the period $t = 0, \ldots, n$), and increases with a rate of d_1 (e.g., due to inflation) each year, then the resulting cost per failure for each year is then defined by a geometric series [43] Consequently, the cost of failure function can be defined by the following equation:

$$c_f(t) = c_f(t = 0) \cdot (1 + d_1)^t \quad : \quad t_0 \leq t \leq T \quad [\text{SEK/f}] \tag{6.36}$$

where $c_f(t = 0)$ is the cost of failure (e.g., the repair cost provided by present data in [SEK/f], and d_1 is the inflation rate, e.g., 2%)$\Rightarrow \quad d_1 = 0.02$.

Figure 6.88 illustrates a function for the cost per failure, when the cost is increased with the inflation rate of 2% over a 30-year period. For this situation, the cost is increased by the factor $(1.02)^{30} \approx 1.81$, that is, by 80%, and the cost of failure increases from SEK 30,000 to 54,000.

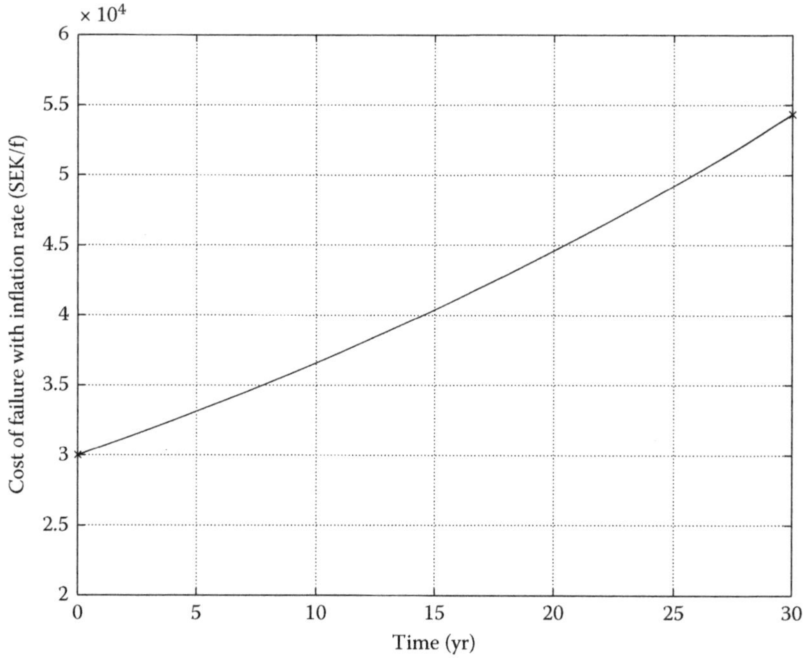

FIGURE 6.88
Cost of failure increases due to an inflation rate of 2%.

The cost of failure with CM means simply that the maintenance activity is applied when failure occurs. As a result from the earlier analysis and assumptions, the CM results in no effect on the failure rate function. Therefore, the failure rate function corresponding to the CM case is $\lambda_{tot}(t, \text{no PM})$. Based on Equations 6.35 and 6.36, the resulting function for cost of failure with the CM method can therefore be expressed as follows:

$$CCM_f(t) = \lambda_{atot}(t, \text{no PM}) \cdot c_f \cdot (1 + d_1)^t \quad [\text{SEK/yr}] \tag{6.37}$$

The cost of failure functions when PM is applied can be defined in a similar way as for the CM method. However, the characteristic of the failure rate functions depends on the choice of PM strategy (S). In the Cable Application Study with the two different PM methods of rehabilitation (PM_{si}) and replacement (PM_{rp}), the resulting cost of failure functions is expressed as follows:

$$CPM_{f,si}(t, S) = \lambda_{atot}(t, PM_{si}) \cdot c_f \cdot (1 + d_1)^t \quad [\text{SEK/yr}] \tag{6.38}$$

and

$$CPM_{f,rp}(t, S) = \lambda_{atot}(t, PM_{rp}) \cdot c_f \cdot (1 + d_1)^t \quad [\text{SEK/yr}]. \tag{6.39}$$

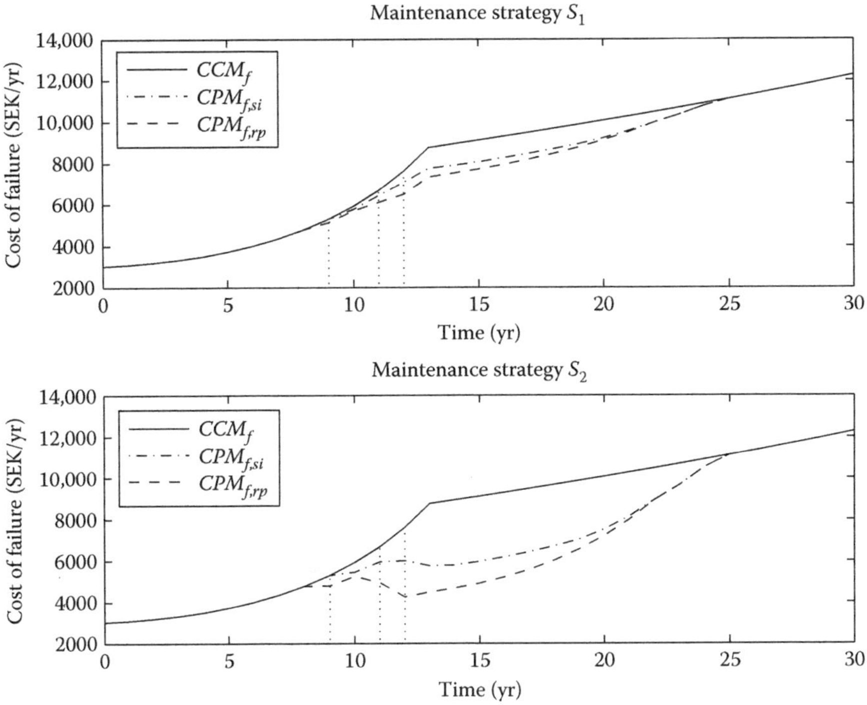

FIGURE 6.89
Cost of failure function for three different maintenance methods and two PM strategies, using the results from the Cable Application Study.

The functions for cost of failure defined above represent three different maintenance methods. The effect of these different methods has been analyzed for the Cable Application Study.

Figure 6.89 shows yearly cost results for the three different cost functions of $CCM_f(t)$, $CPM_{f,si}(t, S)$, and $CPM_{f,rp}(t, S)$. Furthermore, the results are shown for two PM strategies, where PM is applied at years 9, 11, and 12, and with 10% (S_1) or 30% (S_2) of the cables subject to PM on each occasion. The trend in the increase of cost due to the interest rate can be seen as well as the rise due to PM activities. Furthermore, it can be seen that the costs for PM methods are consistently higher than for the CM method. Table 6.51 presents the resulting values for the cost of failure function. These values demonstrate the reduction in the cost of failure due to PM. This decrease depends on the selected strategy; where in S_1 a total of 30% of the 11 kV cables are maintained, and in S_2 a total of 90% are. These strategies, for the replacement method regime, give a benefit within the interval SEK 20,000–50,000. The overall conclusion from these results is that the lowest cost of failure is gained with PM by replacement. This is due to both the lowest total cost over the whole period, as well as for each year during the period.

TABLE 6.51

Results for the Cost of Failure per Year from the Cable Application Study, by Different Maintenance Methods and Two PM Strategies

		$CPM_{f,si}(t, S)$ [SEK/yr]		$CPM_{f,rp}(t, S)$ [SEK/yr]	
Time [yr]	$CCM_f(t)$ [SEK/yr]	S_1	S_2	S_1	S_2
0	3021	3021	3021	3021	3021
1	3089	3089	3089	3089	3089
2	3191	3191	3191	3191	3191
3	3323	3323	3323	3323	3323
4	3496	3496	3496	3496	3496
5	3721	3721	3721	3721	3721
6	4004	4004	4004	4004	4004
7	4355	4355	4355	4355	4355
8	4782	4782	4782	4782	4782
9*	5298	5298	5298	5129	4792
10	5923	5766	5452	5700	5253
11*	6682	6438	5952	6102	4943
12*	7610	7077	6009	6491	4254
13	8758	7758	5757	7337	4495
14	8933	7886	5793	7516	4681
15	9112	8063	5964	7713	4915
16	9294	8256	6181	7932	5209
17	9480	8462	6426	8177	5572
18	9669	8681	6705	8451	6015
19	9863	8917	7025	8759	6550
20	10,060	9209	7507	9106	7197
21	10,261	9553	8136	9501	7980
22	10,466	9955	8932	9955	8932
23	10,676	10,334	9651	10,334	9651
24	10,889	10,766	10,518	10,766	10,518
25	11,107	11,107	11,107	11,107	11,107
26	11,329	11,329	11,329	11,329	11,329
27	11,556	11,556	11,556	11,556	11,556
28	11,787	11,787	11,787	11,787	11,787
29	12,023	12,023	12,023	12,023	12,023
30	12,263	12,263	12,263	12,263	12,263
Total: [MSEK]	0.24602	0.23546	0.21435	0.23202	0.20400

Another way of presenting the cost is by cumulative values. This involves a cost function that adds the cost for each year to that year. Figure 6.90 shows the cumulative cost functions for a similar situation to what was presented above. These results illustrate what the maximum and minimum costs in

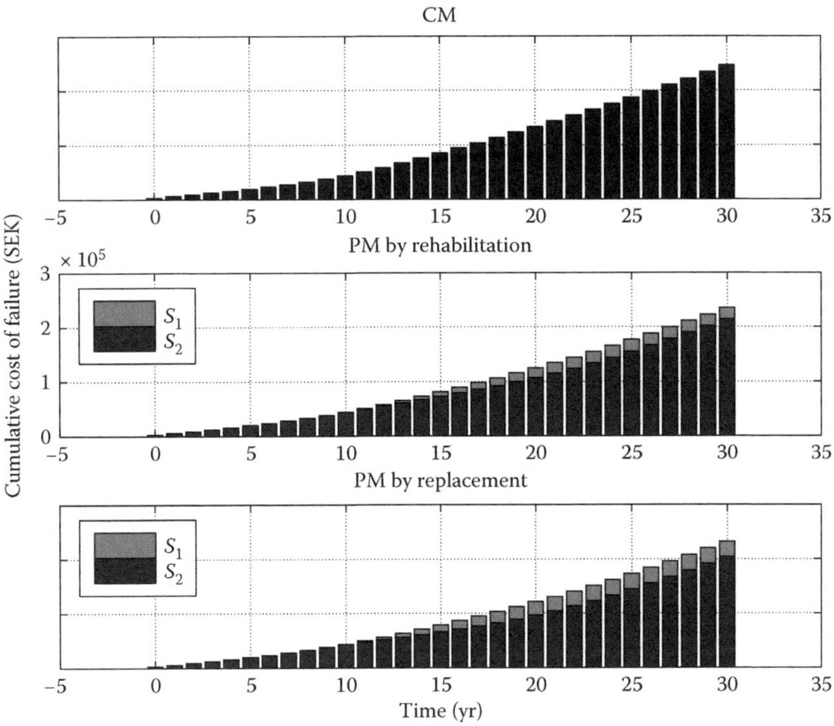

FIGURE 6.90
Cumulative cost of failure function for three different maintenance methods and two PM strategies, using the results from the Cable Application Study.

respect to failure are each year for the CM method and the PM_{rp} method, respectively. This implies that in this application study comparing the combinations of two maintenance strategies, that the most cost- effective solution is to perform PM using the replacement method.

The results for the Cable Application Study have been validated by comparisons with corresponding manually calculated values using Equations 6.37 through 6.39. The conclusion is that the results presented from the equations implemented are correct. The values selected have been presented below.

$$\text{Input data} \quad c_f = 30000 \quad \text{SEK/f}, d_1 = 0.02$$

$$\text{Strategy} \quad S_1 \quad \text{applies:}$$

$$s_1 = s_2 = s_3 = 0.1, t_{PM1} = 9, t_{PM1} = 11, t_{PM1} = 12$$

$$\Rightarrow CCM_f(t = 0) = \lambda_{atot}(t = 0, \text{no PM}) \cdot 30000 \cdot 1$$

$$= 0.10069 \cdot 30000 \approx 3021 \quad \text{SEK/yr}$$

$$\Rightarrow CPM_{f,si}(t=9,S) = \lambda_{atot}(t=9, PM_{si}, S_1) \cdot 30000 \cdot (1.02)^9$$

$$= 0.14778 \cdot 30000 \cdot 1.19509 \approx 5298 \quad \text{SEK/yr}$$

$$\Rightarrow CPM_{f,rp}(t=9,S) = \lambda_{atot}(t=9, PM_{rp}, S_1) \cdot 30000 \cdot (1.02)^9$$

$$= 0.14307 \cdot 30000 \cdot 1.19509 \approx 5129 \quad \text{SEK/yr}$$

Strategy S_2 applies:

$$s_1 = s_2 = s_3 = 0.3, t_{PM1} = 9, t_{PM1} = 11, t_{PM1} = 12$$

$$\Rightarrow CPM_{f,si}(t=12,S) = \lambda_{atot}(t=12, PM_{si}, S_2) \cdot 30000 \cdot (1.02)^{12}$$

$$= 0.15793 \cdot 30000 \cdot 1.26824 \approx 6009 \quad \text{SEK/yr}$$

$$\Rightarrow CPM_{f,rp}(t=12,S) = \lambda_{atot}(t=12, PM_{rp}, S_2) \cdot 30000 \cdot (1.02)^{12}$$

$$= 0.11180 \cdot 30000 \cdot 1.26824 \approx 4254 \quad \text{SEK/yr}$$

6.7.6.4 Modeling the Cost of PM

The cost for a PM activity is a one-time cost that in economic values would be spread by depreciation. One common way to consider the cost of depreciation economically is to assume a constant amount for each year, being the sum of the cost of the interest rate and the depreciation [52]. This method is called an annualized method and was presented in Section 3.7. It would, however, be very difficult to show the benefit in a PM activity if not only the cost for the activity itself but also an interest on the capital investment would be required. Furthermore, in today's electricity market, it is common that the utility draws up a contract with a fixed cost for a future PM measure. Both these observations indicate that a good model for the cost of PM is annual depreciation based on the fixed cost of the PM, which has been used in the rest of the analysis.

Two strategies for the PM have been analyzed for the Cable Application Study, being S_1 and S_2. Both these mean that PM is applied three times within the study period $t: 0 \rightarrow 30$ years. Consequently, these strategies result in three different depreciation times for the spread of the investment costs required for PM. This depreciation time would be the time until the end of the study period and begin at the maintenance point, expressed as $n = T - t_{PMj} + 1$ for $j = 1, 2, 3$.

However, the cost of PM for a cable component depends on the length of cable being measured. In the Cable Application Study, the 11 kV cable being analyzed had an average length (l_{av}) of about 8.106 km. Furthermore, the total PM cost depends on the percentage of the components that are exposed to the PM routines. This was earlier defined in the maintenance strategy, for example, in Strategy S_2, with $s_1 = s_2 = s_3 = 30\%$ of components subject to PM each occasion.

Assuming that the investment costs for PM are either $c_{PM_{si}}$ for the rehabilitation method, or $c_{PM_{rp}}$ for the replacement method, then the resulting characteristic for the PM cost functions can then be defined by the following equations:

$$\begin{aligned} C_{PM_{si,j}} &= c_{PM_{si}} \cdot s_j \cdot l_{av} \quad j = 1, 2, 3, \ldots \text{[SEK]} \\ C_{PM_{rp,j}} &= c_{PM_{rp}} \cdot s_j \cdot l_{av} \quad j = 1, 2, 3, \ldots \text{[SEK]}, \end{aligned} \tag{6.40}$$

where

- $c_{PM_{si}}$ is the cost for PM by rehabilitation [SEK/km]
- $c_{PM_{rp}}$ is the cost for PM by replacement [SEK/km]
- s_j is the proportion of cables affected by PM at time t_{PMj}
- l_{av} is the average length of a cable being affected by PM

The annual cost (in SEK/yr) for the Cable Application Study with PM applied three times with either rehabilitation (PM_{si}) or replacement (PM_{rp}) can then be formulated as follows:

$$CPM_{PM_{si}}(t, S) \quad \text{[SEK/yr]} \quad = \\ \begin{cases} 0 & : \; t_0 \le t \le t_{PM1} \\ \frac{C_{PMsi,1}}{(T - t_{PM1} + 1)} & : \; t_{PM1} \le t \le t_{PM2} \\ \frac{C_{PMsi,1}}{(T - t_{PM1} + 1)} + \frac{C_{PM_{si,2}}}{(T - t_{PM2} + 1)} & : \; t_{PM2} \le t \le t_{PM3} \\ \frac{C_{PMsi,1}}{(T - t_{PM1} + 1)} + \frac{C_{PM_{si,2}}}{(T - t_{PM2} + 1)} + \frac{C_{PM_{si,3}}}{(T - t_{PM3} + 1)} & : \; t_{PM2} \le t \le T \end{cases} \tag{6.41}$$

and

$$CPM_{PM_{rp}}(t, S) \quad \text{[SEK/yr]} \quad = \\ \begin{cases} 0 & : \; t_0 \le t \le t_{PM1} \\ \frac{C_{PMrp,1}}{(T - t_{PM1} + 1)} & : \; t_{PM1} \le t \le t_{PM2} \\ \frac{C_{PMrp,1}}{(T - t_{PM1} + 1)} + \frac{C_{PM_{rp,2}}}{(T - t_{PM2} + 1)} & : \; t_{PM2} \le t \le t_{PM3} \\ \frac{C_{PMrp,1}}{(T - t_{PM1} + 1)} + \frac{C_{PM_{rp,2}}}{(T - t_{PM2} + 1)} + \frac{C_{PM_{rp,3}}}{(T - t_{PM3} + 1)} & : \; t_{PM2} \le t \le T \end{cases} \tag{6.42}$$

Figure 6.91 illustrates the cost functions due to PM, being $CPM_{PM_{si}}(t, S)$ and $CPM_{PM_{rp}}(t, S)$. The results are shown for the Cable Application Study and for the two PM strategies (S_1 and S_2). The three different cost levels due to the depreciation of the PM investment costs can be seen. As expected, these differ by a factor of 3 (the difference in components affected by PM). Furthermore, Table 6.52 presents annual values for the Cable Application Study with the cost of the three PM measures distributed yearly. The results are presented for the application of the two strategies S_1 and S_2. It can be seen that the cost of PM significantly increases with more components being

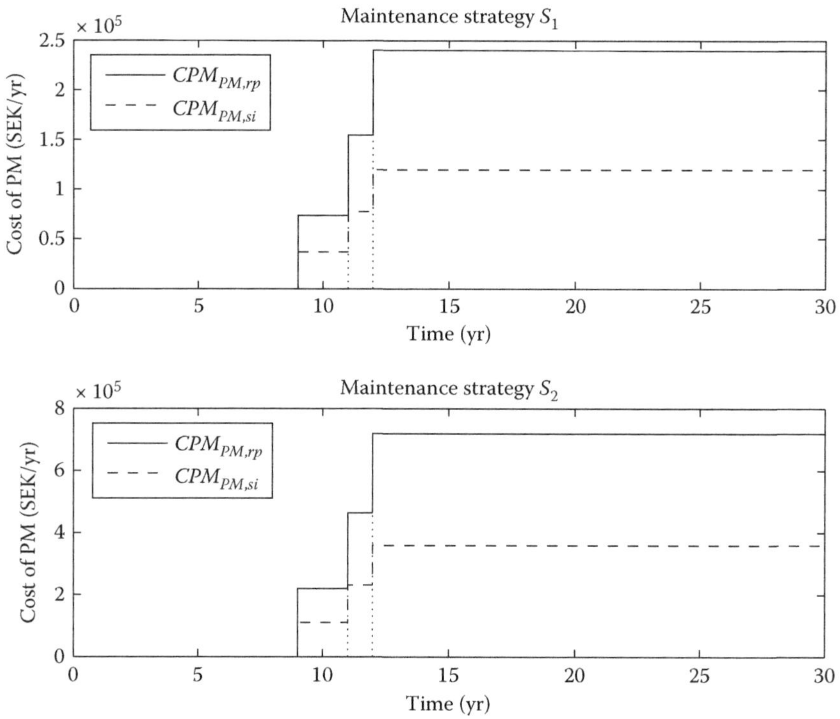

FIGURE 6.91
Cost of the PM function for three different maintenance methods and two PM strategies, showing results for the Cable Application Study.

affected by the maintenance action, which is expected for the underground cable component.

The corresponding cumulative cost functions are illustrated in Figure 6.92.

The results for the Cable Application Study have been validated by comparisons with corresponding manually calculated values using Equation 6.40 or 6.42. The conclusion is that the results for the implemented equations are correct. The values selected have been presented below.

Input data: $T = 30$ years, $l_{av} = 8.10584$ km, $c_{si} =$ SEK1,000,000

Strategy S_1 applies:

$s_1 = s_2 = s_3 = 0.1, t_{PM1} = 9, t_{PM1} = 11, t_{PM1} = 12$

$\Rightarrow C_{PM_{si,1}} = c_{PM_{si}} \cdot l_{av} \cdot s_1 \approx 1{,}000{,}000 \cdot 8.106 \cdot 0.1 \approx \text{SEK}810{,}584$

$s_3 = s_2 = s_1 \Rightarrow C_{PM_{si}2} = C_{PM_{si}3} = \text{SEK}810{,}600$

TABLE 6.52

Results for the Cost of PM per Year in the Cable Application Study with Different Maintenance Methods and Two Strategies (S_1 and S_2)

Time [yr]	$CPM_{PM_{si}}(t, S)$ [SEK/yr]		$CPM_{PM_{rp}}(t, S)$ [SEK/yr]	
	S_1	S_2	S_1	S_2
0	0	0	0	0
⋮	0	0	0	0
8	0	0	0	0
9*	36,845	110,534	73,689	221,068
10	36,845	110,534	73,689	221,068
11*	77,374	232,121	154,748	464,244
12*	120,036	360,109	240,073	720,218
13	120,036	360,109	240,073	720,218
⋮	120,036	360,109	240,073	720,218
30	120,036	360,109	240,073	720,218
Total: [MSEK]	2.43175	7.29526	4.86350	14.59052

Note: The results are shown with the cost of PM activity distributed by depreciation without any rate.

$$\Rightarrow CPM_{PM_{si}}(t = 9, S) = \frac{C_{PM_{si,1}}}{(T - t_{PM1} + 1)} = \frac{810584}{22} \approx 36,845\text{SEK/yr}$$

$$\Rightarrow CPM_{PM_{si}}(t = 11, S) = CPM_{PM_{si}}(t = 9, S) + \frac{C_{PM_{si,2}}}{(T - t_{PM2} + 1)} = 36,845$$

$$+ \frac{810,584}{20} \approx 36,845 + 40,529 \approx 77,374\text{SEK/yr}$$

$$\Rightarrow CPM_{PM_{si}}(t = 12, S) = CPM_{PM_{si}}(t = 11, S) + \frac{C_{PM_{si,3}}}{(T - t_{PM3} + 1)}$$

$$= 81,060 + \frac{810,584}{20} \approx 77,374 + 42,662 \approx 120,036\text{SEK/yr}$$

$$TCPM_{PM_{si}} = 2 \cdot CPM_{PM_{si}}(t = 9, S) + CPM_{PM_{si}}(t = 11, S) + 19 \cdot$$
$$CPM_{PM_{si}}(t = 12, S) \approx \text{SEK}2,431,800 \Leftrightarrow 3 \cdot \text{SEK}810,600.$$

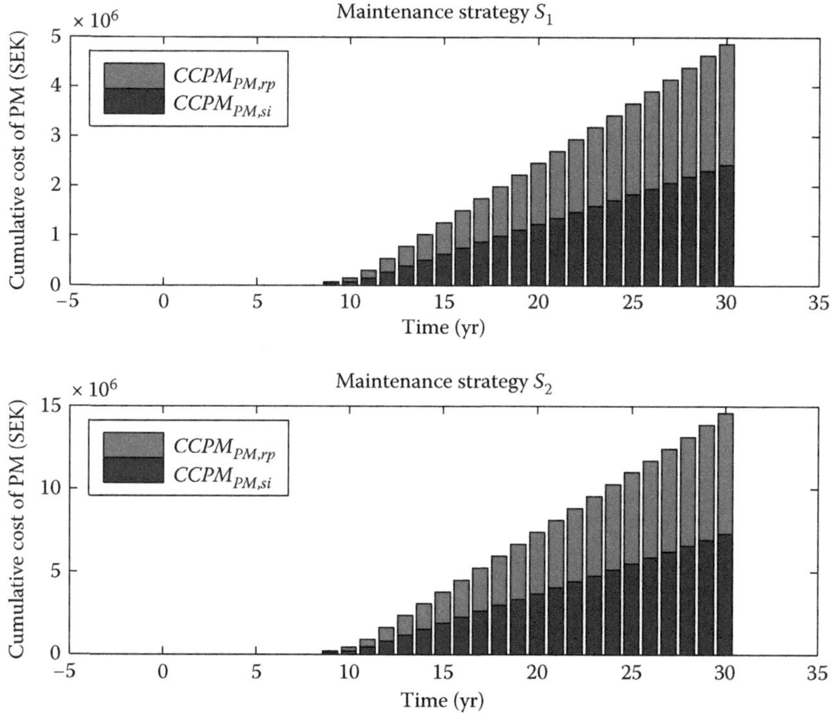

FIGURE 6.92
Cumulative cost of the PM function for three different maintenance methods and two PM strategies, showing results for the Cable Application Study.

6.7.6.5 Modeling the Cost of Interruption

As previously discussed, the cost of interruption is related to the customer effect of the availability of supply. This cost is related to the reliability value. It is a difficult matter to define this cost since it is based on subjective judgment, for example, how many interruptions that a specific customer determines is acceptable during a certain period of time. Investigations have been made to define this cost, based on questionnaires sent to customers who are divided into different customer categories. Different methods to consider customer interruption costs are evaluated in [144]. Results from a comprehensive study made in Sweden are presented in [142]. This same study was based on a questionnaire sent to 4000 customers with a reply frequency of 57%. The Swedish utilities (including e.g., Vattenfall and Birka Nät) use the results from this report when evaluating the cost of interruptions. As input data for this analysis, the cost of interruption (c_{int}) in [SEK/kWh] as defined in Table 6.50 is the value currently used by Vattenfall.

One reliability index for measuring the customer effect is the energy not-supplied (*LOE*). A change in this index can be used to define the benefit in reliability value. In the Cable Application Study, this function has been evaluated as part of the previously presented reliability analysis of the Birka system. Note that this index is evaluated for each load point, where for the Birka system Lp_1 is where the 11 kV cable components are connected.

These reliability functions together with the cost of interruption constitute the input data for the model defined below for the cost of interruption function. By assuming that the cost for interruptions change in time in similar ways to the cost of failures and increase with the inflation rate of 2% over a 30-year period, then the cost of interruption with CM can then be formulated as follows:

$$CCM_{Lp1,int}(t) = LOE_{Lp1}(t, \text{no PM}) \cdot c_{int} \cdot (1+d_1)^t \quad t_0 \leq t \leq T \quad [\text{SEK/yr}] \tag{6.43}$$

where $LOE_{Lp1}(t)$ is the energy not-supplied function at load point $Lp1$ in [kWh/yr], c_{int} is the customer cost of interruption [SEK/kWh], and d_1 is the inflation rate $2\% \Rightarrow d_1 = 0.02$.

The cost of interruption functions when PM is applied is defined in similar way as the CM as follows:

$$CPM_{int,PM_{si}}(t,S) = LOE_{Lp1}(t, PM_{si}, S) \cdot c_{int} \cdot (1+d_1)^t \quad [\text{SEK/yr}] \tag{6.44}$$

and

$$CPM_{int,PM_{rp}}(t,S) = LOE_{Lp1}(t, PM_{rp}, S) \cdot c_{int} \cdot (1+d_1)^t \quad [\text{SEK/yr}]. \tag{6.45}$$

The resulting cost functions for interruption are shown in Figures 6.93 and 6.94. The results show that the lowest cost of interruption is gained with PM by replacement. Table 6.53 presents the results for the Cable Application Study with the cost of the interruption distributed yearly.

Results for the Cable Application Study have been validated by comparison with corresponding manually calculated values using Equations 6.43 to 6.45. The conclusion is that the results from the implemented equations are correct. The values selected have been presented below.

$$T = 30 \quad \text{year}, c_{int} = 43 \quad \text{SEK/kWh}$$

$$\text{Strategy} \quad S_1 \quad \text{applies:}$$

$$s_1 = s_2 = s_3 = 0.1, t_{PM1} = 9, t_{PM1} = 11, t_{PM1} = 12$$

$$\Rightarrow CCM_{int}(t=0) = LOE_{Lp1}(t=0, \text{ no PM}) \cdot 43 \cdot 1$$

$$= 389.65 \cdot 43 \approx 16{,}755 \quad \text{SEK/yr}$$

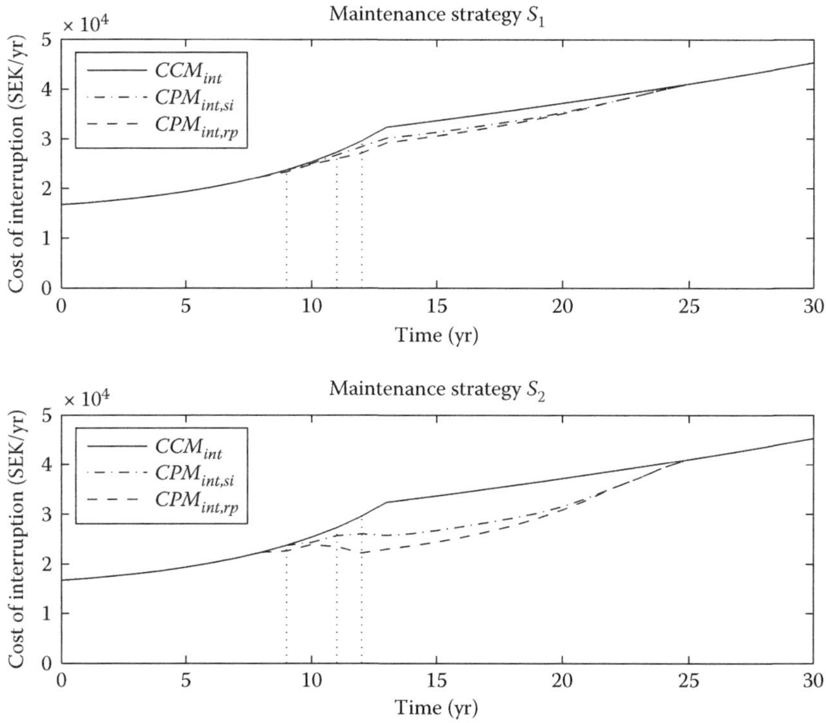

FIGURE 6.93
Cost of the interruption function for three different maintenance methods and two PM strategies, using results for the Cable Application Study.

$$\Rightarrow CPM_{int,PM_{si}}(t = 9, S) = LOE_{Lp1}(t = 9, PM_{si}, S_1) \cdot 43 \cdot (1.02)^9$$

$$\approx 462.07 \cdot 43 \cdot 1.19509 \approx 23,745 \quad \text{SEK/yr}$$

$$\Rightarrow CPM_{int,PM_{rp}}(t = 9, S) = LOE_{Lp1}(t = 9, PM_{si}, S_1) \cdot 43 \cdot (1.02)^9$$

$$\approx 454.83 \cdot 43 \cdot 1.19509 \approx 23,373 \quad \text{SEK/yr.}$$

6.7.6.6 Evaluation and Comparison of Total Annualized Costs

Previous sections have defined cost functions for the various situations with different types of PM methods as well as with CM. Furthermore, the impact of different PM strategies has been modeled and analyzed. The three different cost functions that have been analyzed are: cost of failure, cost of PM, and cost of interruption.

The aim was to compare these maintenance methods and strategies and to define which of the same that would give the most cost-effective solution, that is, result in a cost-effective maintenance plan. Consequently, this completes the RCM plan.

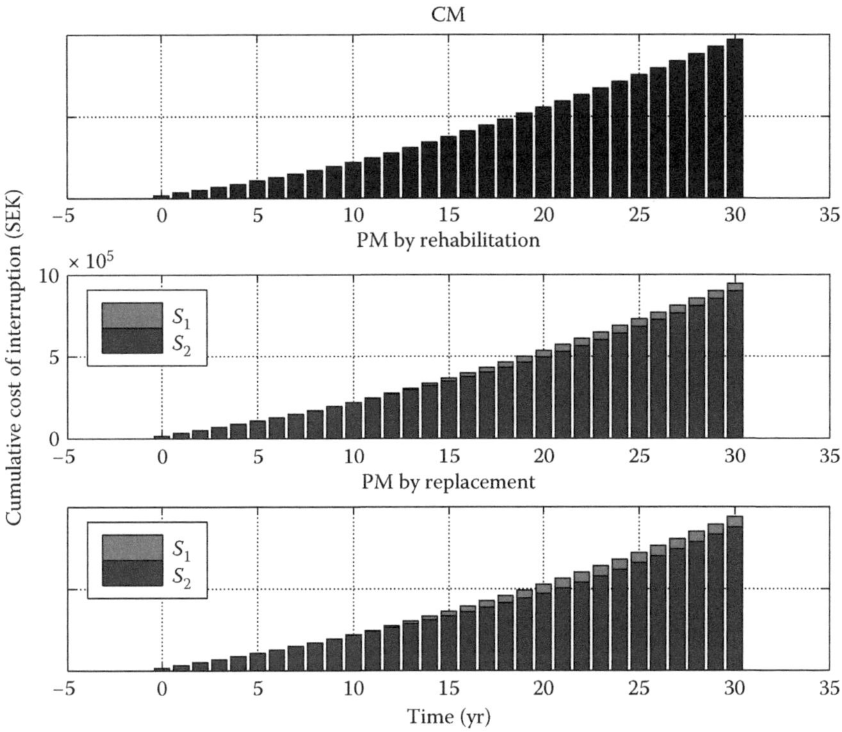

FIGURE 6.94
Cumulative cost of the interruption function for three different maintenance methods and two PM strategies, showing results for the Cable Application Study.

One way for comparing these maintenance methods and different strategies is by analyzing the annual costs. It is interesting to compare these annual costs both for each year (e.g., if there is an upper limit for yearly expenses) and as a total cost for the period. The latter is possibly an assessment of the total required resources. However, as identified in Section 6.7.6, payments made at different points in time are not comparable, since money today has a greater value than the same amount of money in the future. It was also identified that it is cost effective to postpone a PM activity if possible. Consequently, annual cost values are of interest but should be evaluated as present values.

In the Cable Application Study, the total annual costs can be formulated for the three different maintenance methods (CM and PM by either rehabilitation or replacement). Furthermore, the PM methods have been analyzed for two different strategies: S_1 and S_2. Moreover, cost functions have been analyzed for three different costs defined as failure, PM, and interruption. This results in eight different cost functions that can be summarized into the three maintenance methods as follows.

TABLE 6.53

Results for the Cost of Interruption per Year for the Cable Application Study with Different Maintenance Methods and Two Strategies (S_1 and S_2)

Time [yr]	$CCM_{int}(t)$ [SEK/yr]	$CPM_{int,si}(t,S)$ [SEK/yr]		$CPM_{int,rp}(t,S)$ [SEK/yr]	
		S_1	S_2	S_1	S_2
0	16,755	16,755	16,755	16,755	16,755
1	17,090	17,090	17,090	17,090	17,090
2	17,537	17,537	17,537	17,537	17,537
3	18,038	18,038	18,038	18,038	18,038
4	18,636	18,636	18,636	18,636	18,636
5	19,349	19,349	19,349	19,349	19,349
6	20,197	20,197	20,197	20,197	20,197
7	21,197	21,197	21,197	21,197	21,197
8	22,370	22,370	22,370	22,370	22,370
9*	23,745	23,745	23,745	23,373	22,629
10	25,363	25,018	24,325	24,872	23,887
11*	27,282	26,746	25,673	26,005	23,450
12*	29,580	28,403	26,050	27,114	22,181
13	32,366	30,161	25,751	29,233	22,968
14	33,013	30,706	26,091	29,889	23,639
15	33,674	31,361	26,736	30,591	24,423
16	34,347	32,059	27,484	31,346	25,342
17	35,034	32,790	28,302	32,163	26,419
18	35,735	33,557	29,202	33,050	27,678
19	36,449	34,365	30,195	34,016	29,148
20	37,179	35,303	31,551	35,076	30,868
21	37,922	36,360	33,237	36,246	32,893
22	38,680	37,554	35,299	37,554	35,299
23	39,454	38,701	37,195	38,701	37,195
24	40,243	39,971	39,425	39,971	39,425
25	41,048	41,048	41,048	41,048	41,048
26	41,869	41,869	41,869	41,869	41,869
27	42,706	42,706	42,706	42,706	42,706
28	43,560	43,560	43,560	43,560	43,560
29	44,432	44,432	44,432	44,432	44,432
30	45,320	45,320	45,320	45,320	45,320
Total: [MSEK]	0.97017	0.94691	0.900037	0.93930	0.87755

TABLE 6.54
Sum of the Total Annual Cost for Different Maintenance Methods for the Cable Application Study and PM Strategies S_1 and S_2

Cost Factors	Cost of Different Maintenance Methods [MSEK]				
	TCCM	$TCPM_{si}$		$TCPM_{rp}$	
		S_1	S_2	S_1	S_2
C_f	0.24602	0.23547	0.21435	0.23206	0.20400
C_{PM}	–	2.43175	7.29526	4.86350	14.59052
C_{int}	0.97017	0.94691	0.90037	0.93930	0.87755
Total	1.21620	3.61413	8.40998	6.03483	15.67207

Total annual cost of CM:

$$TCCM(t) = CCM_f(t) + CCM_{int}(t): \quad t_0 \leq t \leq T \quad [\text{SEK/yr}] \tag{6.46}$$

Total annual cost of PM by rehabilitation:

$$\begin{aligned} &TCPM_{si}(t,S) = \\ &CPM_{f,si}(t,S) + CPM_{PM,si}(t,S) + CPM_{int,si}(t,S) \quad [\text{SEK/yr}] \end{aligned} \tag{6.47}$$

Total annual cost of PM by replacement:

$$\begin{aligned} &TCPM_{rp}(t,S) = \\ &CPM_{f,rp}(t,S) + CPM_{PM,rp}(t,S) + CPM_{int,rp}(t,S) \quad [\text{SEK/yr}] \end{aligned} \tag{6.48}$$

Figures 6.95 and 6.96 show these total annual cost functions for the Cable Application Study and for the two different analyzed strategies. It can be seen clearly that the dominant cost values are those for PM. However, the figures also stress the important factor of the distribution of the costs over time, where for example the significant rise in annual cost due to PM measures occurs after about 11 years. Table 6.54 presents the total annual values for this study. This table shows that for both of the strategies PM implies a reduction in the cost of failures and the cost of interruption. However, the cost of the PM activity itself (rehabilitation or replacement of underground 11 kV cable) is extensive. The cost analysis therefore indicates that the cost-effective solution found by comparing the total effect on these cost factors is "not to perform PM." However, this analysis does not take into account the effect of the point in time when the PM is made, which is analyzed in detail in the next section.

The results could also be analyzed regarding the sensitivity of the parameters. When comparing the resulting cost functions, it is understood that the cost of failure and the cost of interruption include an economic factor (the

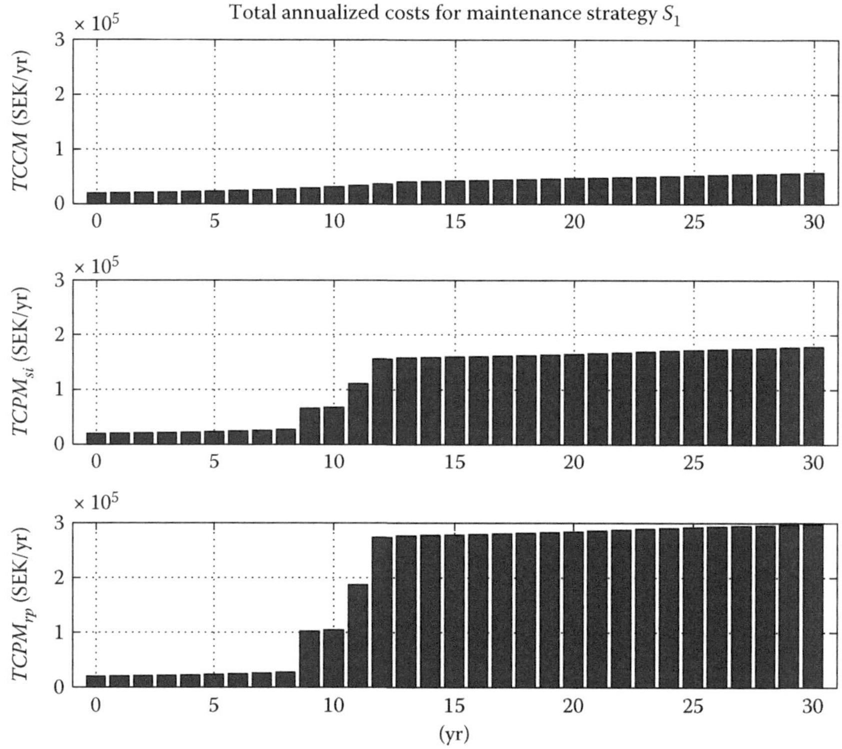

FIGURE 6.95
Annual total costs for Strategy S_1 showing the results for the Cable Application Study.

rate d_1), which is not included in the cost of PM. This means that one way of changing the balance between total costs of the CM and PM methods is to adjust this factor. The rate $d_1 = 2\%$ is used as input data for the Cable Application Study. Figure 6.97 illustrates the effect of using a rate of 10% instead. These results indicate a significant change in the balance between the cost of CM and PM. Consequently, the conclusion is that it is possible to obtain an equilibrium level between CM and PM. Whether this would be expected economically is, however, another matter.

Nonetheless, it is important to recognize that the maintenance of the cable component will be expensive since it involves extensive changes, for example, by injection or replacement. For other types of components, this situation would be different, for example, a circuit breaker could be maintained by less complicated and less expensive methods, and would not be replaced as a whole entity. Therefore, it is understood that the PM benefit for other types of components could be more easily seen than for the cable component. This observation is important for not drawing incorrect conclusions from these results.

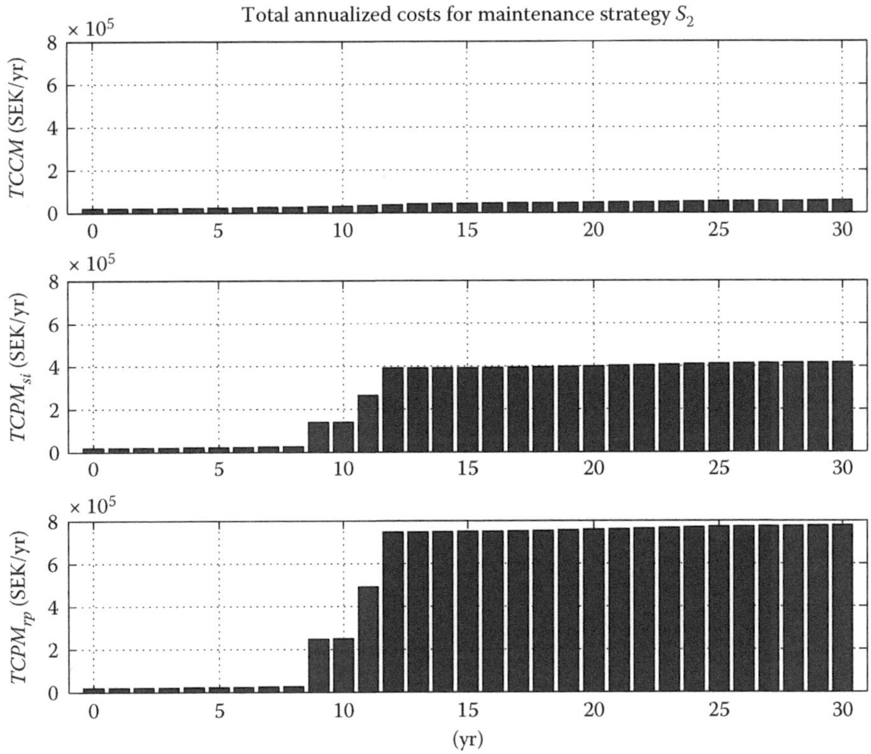

FIGURE 6.96
Annual total costs for Strategy S_2 showing the results for the Cable Application Study.

6.7.6.7 Present Value Analysis

Total annual cost values for the different maintenance methods were evaluated in the previous section. However, it was also identified that for an accurate comparison it is necessary to evaluate present values. The next step in the analysis is therefore to evaluate present values for these annual costs. The cost functions have been evaluated as present values according to the technique presented earlier in Section 3.7.

Present values can be formulated as the sum of discounted values for payments at different points in time as follows:

$$TCPV = \sum_{n=t_0}^{T} C(n) \cdot PV_f(n, d_2) : \quad t_0 \leq n \leq T \quad \text{[SEK]} \tag{6.49}$$

where n is the time in numbers of years after the present time for the payment $C(n)$, d is the discount rate, $PV_f(n, d_2)$ is the preset value factor as defined in Equation 3.13.

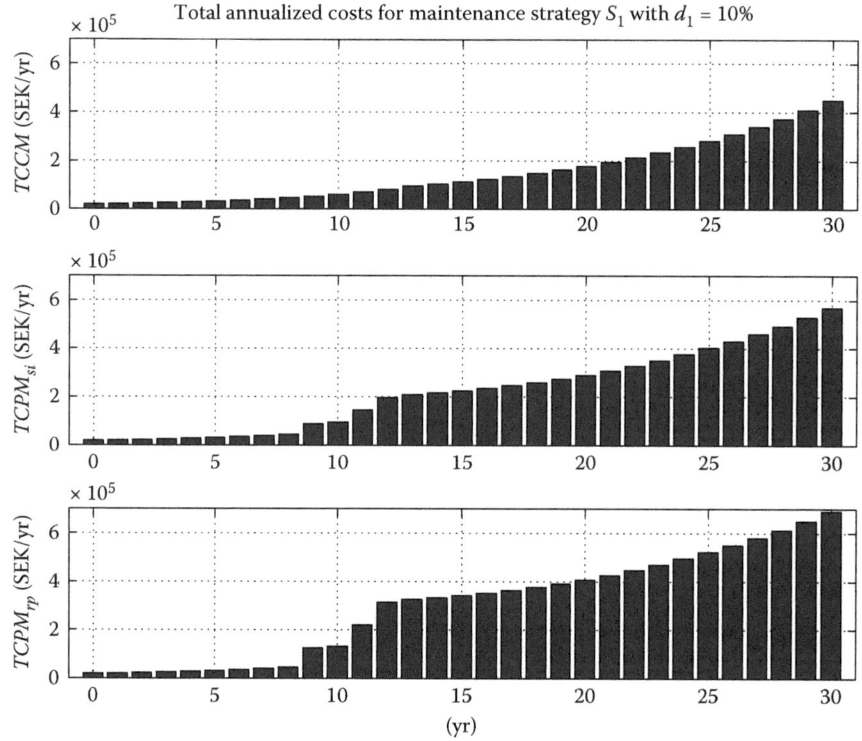

FIGURE 6.97
Effect of change in economic variables. Annual total costs for Strategy S_1 showing the results for the Cable Application Study.

For the Cable Application Study, the following present values can then be defined in a similar way.

Present value for the total annual cost of CM:

$$TCCMPV = \sum_{t=t_0}^{T} TCCM(t) \cdot PV_f(t, d_2) \quad : \quad t_0 \leq t \leq T \quad [\text{SEK}] \qquad (6.50)$$

Present value for the total annual cost of PM by rehabilitation:

$$TCPM_{si}PV = \sum_{t=t_0}^{T} TCPM_{si}(t, S) \cdot PV_f(t, d_2) \quad : \quad t_0 \leq t \leq T \quad [\text{SEK}] \qquad (6.51)$$

Present value for the total annual cost of PM by replacement:

$$TCPM_{rp}PV = \sum_{t=t_0}^{T} TCPM_{rp}(t, S) \cdot PV_f(t, d_2) \quad : \quad t_0 \leq t \leq T \quad [\text{SEK}] \qquad (6.52)$$

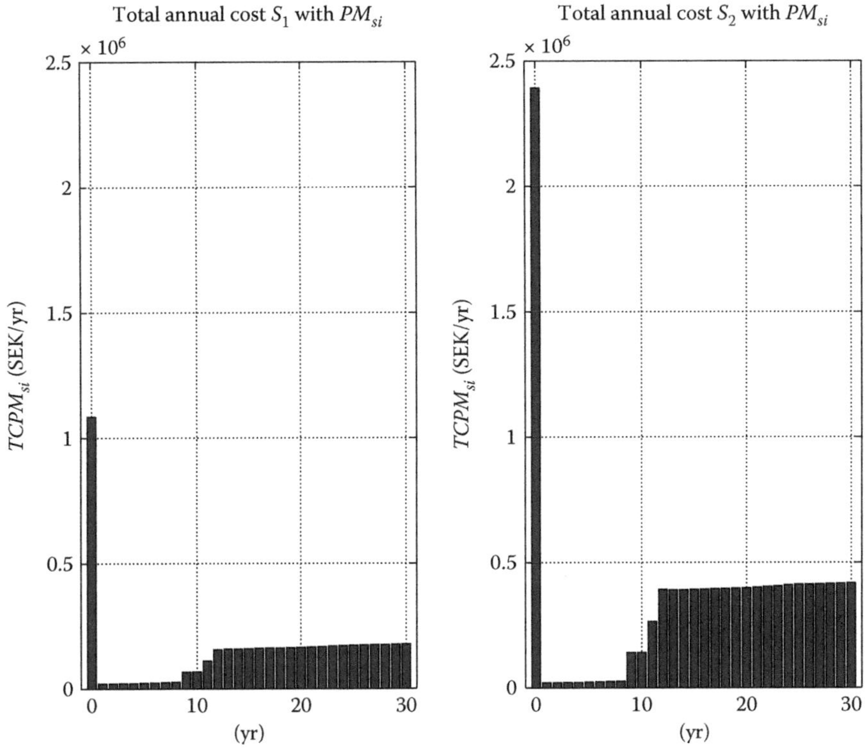

FIGURE 6.98
Present value analysis illustrating the impact of different maintenance strategies when applying PM by rehabilitation, showing the results for the Cable Application Study. Note that the first bar shows the present value which is the sum of the annual costs (the other bars).

Figures 6.98 and 6.99 illustrate the procedure for present value analysis and show the results for the Cable Application Study. These figures present the present value as the sum of the total annual costs which are each individually discounted with a certain rate. For the results presented, the rate is 7%, which equals the interest rate presented in Table 6.50. Furthermore, the figures illustrate the impact of different maintenance strategies, where S_2 implies a significantly higher present value than S_1, which is due to the greater required investment in PM.

Figure 6.100 displays the total present values for the three alternative maintenance methods and the two different strategies. The results for all cases show that the PM methods result in the highest present values. This indicates that the cost-effective solution is to "not-perform PM," in other words apply CM only, as is shown by the first method. However, it should be emphasized that the outcome is totally dependent on the conditions created by the input data. If, for example, the discount rate was changed, then different results

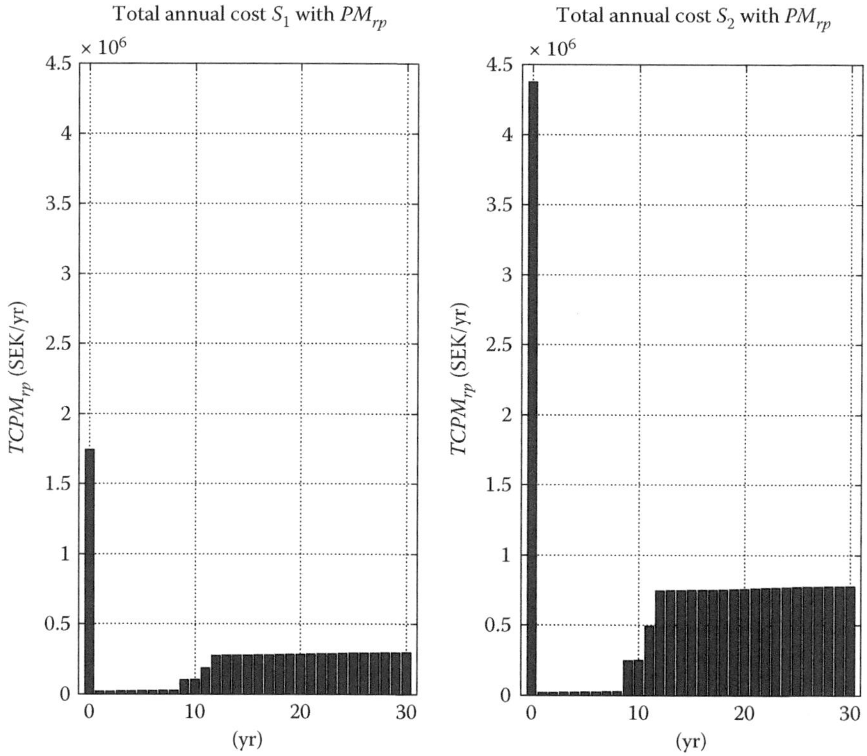

FIGURE 6.99
Present value analysis illustrating the impact of different maintenance strategies when applying PM by replacement, showing the results for the Cable Application Study. Note that the first bar shows the present value which is the sum of the annual costs (the other bars).

would be obtained. This effect can also be seen from the figure comparing the results for Strategy S_1 with two different discount rates of: (i) zero, which implies that the present value equals the direct sum of the annualized values, and (ii) 7% as previously shown. Two interesting results can be observed here. The first is that the discount rate has a significant impact on the total present value, and second, that the difference between the payments required for CM and PM can be reduced. This latter observation leads to the conclusion that there would exist an optimal point when it is equally beneficial to apply PM as CM, and moreover that there would be a solution where PM would be preferable to CM. Analysis of the impact of different rates has been made and the point for which it is equally beneficial to use either CM or PM has been identified. Table 6.55 summarizes some resulting values for the present value analysis. A point where all methods result in the same cost is where the cost of failure $c_f = 30{,}000$ SEK/f. Although this equilibrium point $d_2 = 30{,}000\%$ seems unreasonable it is theoretically possible. However, the

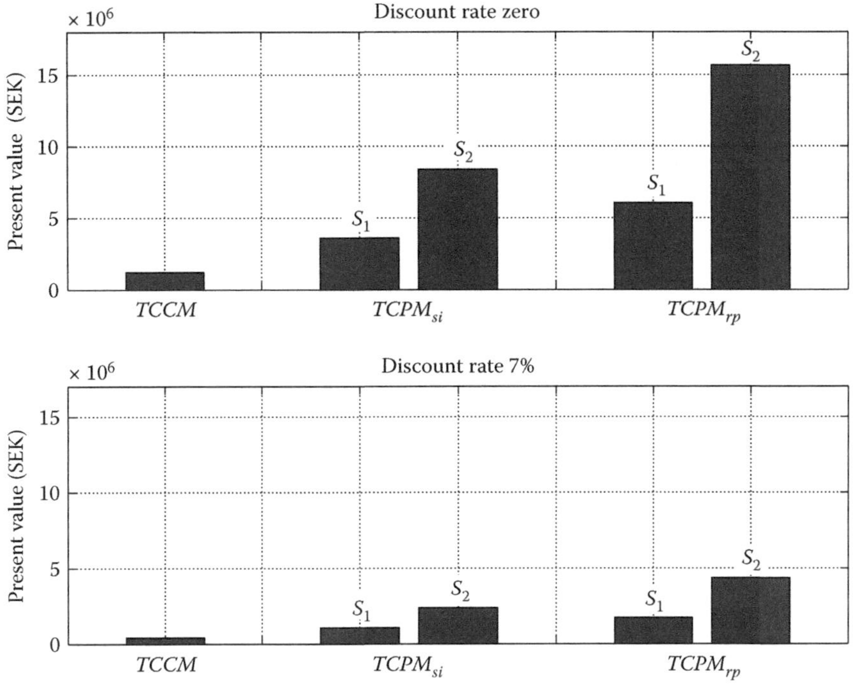

FIGURE 6.100
Summary of present value analysis for the Cable Application Study: comparisons of two PM strategies (S_1 and S_2) and two discount rates.

TABLE 6.55
Present Value Analysis with the Effect of Different Discount Rates Showing the Results for the Cable Application Study and Strategies S_1 and S_2

Discount Rate %	Present Value of Maintenance Methods [MSEK]				
	TCCMPV	$TCPM_{si}PV$		$TCPM_{rp}PV$	
		S_1	S_2	S_1	S_2
0	1.21620	3.61413	8.40998	6.03483	15.67207
7	0.43058	1.08443	2.39213	1.74508	4.37441
10	0.31003	0.71177	1.51525	1.11770	2.73303

significant result is that it is possible to find an optimal balance between CM and PM based on cost considerations.

6.7.7 General Remarks about the Implementation

The RCAM analysis includes several sensitivity studies of the distribution systems analyzed, involving repeated analyses using different sets of input data. To support this process, some simple routines for creating output data tables have been implemented in RADPOW, though the output data has almost exclusively been handled using MATLAB.

Graphical representation has been used for the analysis of the output from RADPOW (RADPOW output consists of a series of reliability indices for different input values such as failure rates for components). This is supported by MATLAB and several program routines have been developed for that purpose.

The following list summarizes the different computer routines that have been developed and implemented in the program, as well as the various steps featured in the RCAM Case Study for Electrical Distribution systems.

1. Modeling of general life distribution functions
2. Evaluation of cable component behavior
   a. Modeling the failure rate function over time $\lambda(t)$
   b. Modeling cable component parameter behavior $\lambda(t, PM)$
3. System effect analysis
   a. Estimation of the composite failure rate for a cable component
   b. Evaluation of reliability indices
      i. Repeated running of RADPOW
      ii. Translation of results into functions in time
   c. Analysis of the effect of different strategies and maintenance methods
4. Cost and benefit analysis
   a. Modeling cost functions
   b. Annual cost and present value analysis

6.8 Summary of the RCAM Case Study Distribution System

An overall conclusion from this description of the RCAM Case Study of Electrical Distribution system is that it is an extensive effort required to accomplish such an analysis.

To enable the carrying out of an RCAM analysis, comprehensive input data and knowledge were required, specifically about the cable component behavior. Furthermore, it can be identified that input data for the cable component behavior were obtained primarily from two different types of sources: (i) data based on experience, which were used for predicting the failure rate characteristic, and (ii) knowledge about the cable condition by modeling the component behavior. Furthermore, it can be seen that some of the steps in

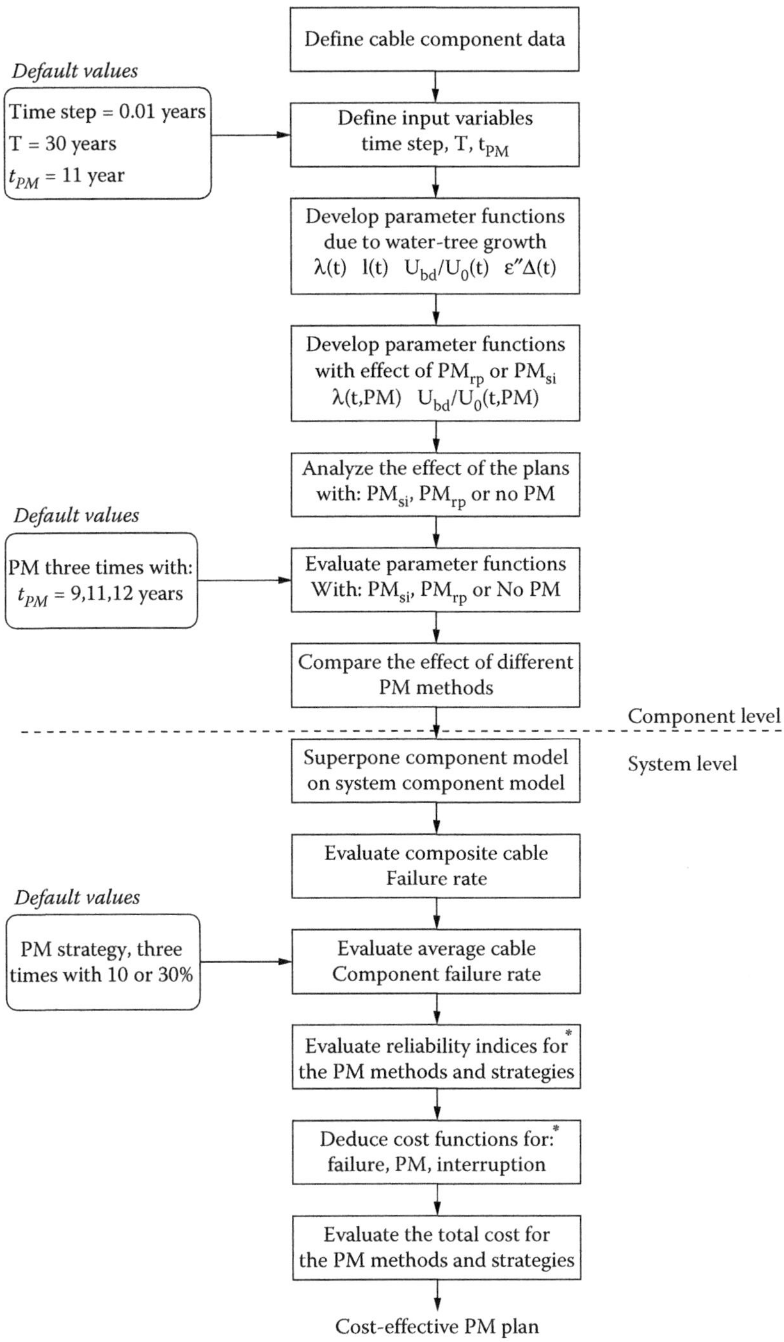

FIGURE 6.101
RCAM Implemented for the Distribution System Case. Approach II with analysis of one component, one cause of failure, and two alternative PM methods. Implementation of the systematic process has been made in MATLAB and the resulting program uses the results from RADPOW for the reliability analysis in the steps noted with *.

the approach are the same as those used as input data in the framework. This is an interesting result since it reflects that one of the greatest challenges and difficulties when applying RCM is to define the data required to support the model, not only because of difficulties in obtaining these, but because it is simply part of the analysis to define suitable data (e.g., choosing maintenance frequency).

It is important to recognize that first, a model for the component failure rate characteristic is defined and then second, this is implemented for a PM strategy. The system analysis is consequently an analysis of the effect of implementing the model developed for different PM strategies.

Figure 6.101 illustrates the resulting logic for the Case Study, which has been implemented and validated. This figure includes the different procedures in the RCAM approach as well as the systematic process for analyzing the included components and causes of failures in the system. Furthermore, this figure shows the logic of the model for the functional relationship between failure rate and PM. The choice of input variables that was used for the results presented are shown on the left-hand side of this figure.

7

RCAM Case Wind Power Systems

7.1 Introduction

This chapter gives a summary presentation of a selection of RCAM (Reliability-Centered Asset Maintenance) studies for wind power systems. The first study involves a comprehensive survey of failure statistics on wind turbines (WT), one of the first published in the world. Second, main results from an RCM (Reliability-Centered Maintenance) analysis following the RCAM approach are presented. This study involves performing an FMEA (failure mode and effect analysis) and comparing two different WTs and shows detailed results identifying underlying causes of failures for the main components in a wind turbine. Finally, results are presented from comparing different maintenance plans using life-cycle cost (LCC) analysis methods.

7.2 Failure Statistics of Wind Power Systems

7.2.1 The Wind Power System

The main function of a wind power system is to transform kinetic energy in the wind into electric energy. Wind energy forces an aerodynamic rotor to turn. The wind energy is thus transformed into mechanical energy. Mechanical energy in wind blades' slow turning rotor shaft is geared up to a high-speed shaft (HSS) which is connected to a generator. Inside the generator, the rotational mechanical energy is transformed into electrical energy. The electric power output is then connected to the grid.

7.2.2 Introduction

The fast expansion of the wind power market came with some initial problems. The new designs were not always fully tested, and the designed 20-year lifetime was typically never achieved before the next generation of turbines was erected. This case study presents results from an investigation of failure statistics which are described in detail in [145].

The study included statistics from four sources, that is, two separate sources from Sweden, one from Finland, and one from Germany. Statistics reveal reliability performance of the different components within the wind turbine. The gearbox is the most critical because downtime, per failure, is high compared to the other components in the wind power turbine. The statistical data for larger turbines also show trends toward higher, ever increasing failure frequency when compared to small turbines, which have a decreasing failure rate over the operational years.

7.2.3 Approach for the Survey of Failures

The survey uses a quantitative approach and a qualitative approach for the statistical research. This work's main findings on component lifetime and failure rate are statistical data which have been analyzed with measurements used in reliability theory. The findings were then supported by information from articles, books, field trips, and interviews related to failures of wind power systems.

The quantitative analysis is based on the statistical data from Sweden, Finland, and Germany [146–149].

7.2.4 Survey of Failures for Wind Power Turbines

The study of wind power turbine statistics gives knowledge of reliability performance. Operational data will verify the correctness of the predicted system lifetime. Statistical data analysis may result in component redesign or a changed maintenance schedule. In the prestudy, it became evident that statistical data on failures were difficult to find. This was for several reasons, for example, no statistical data were collected, wind turbine manufacturers refused to reveal data, data from different designs could not be compared, or data retrieval was too expensive to access. These restrictions led to the following decisions concerning data for the survey:

1. Data need to be comparable.
2. Data need to be within the same or similar time span.
3. Data need to be reliable.

7.2.4.1 Sources for Statistical Data

The statistical data used for this survey stem from the same region, that is, northern Europe, and comes from four different sources, which obey these restrictions. Source 1 [146] and 2 [147] stem from Sweden and partly overlap each other. Source 3 [148] comes from Finnish statistics and Source 4 [149] is based on German statistics. The statistics from Source 3 and 4 were only used as reference to see if the results were similar. The background data for the statistics from Source 3 and 4 were not accessible for this research, and therefore no deeper study was performed for the Finnish and German statistics. In the concluding paragraph, some statistical values from these sources are shown, just to give a hint of what similar values look like in the other countries.

1. *Source 1—Swedish annual report*: Driftuppföljning av vindkraftverk, årsrapport (wind power operations, annual report), Sweden: An annually published report from Elforsk concerns statistical data of performance, failures, and downtimes for almost all wind power plants situated in Sweden.
2. *Source 2—Swedish database*: Felanalys (failure analysis), Sweden: A database of all reported failures in Sweden since 1989 is maintained by Vattenfall Power Consultant (former Swedpower AB). The database was created in 1997 and also reports on failures dating as far back as 1989. This database is also a part of the source for Driftuppföljning av vindkraft, årsrapport, but it contains more information about the failures compared to the published annual report.

7.2.4.2 Failure Statistics

Operational statistics from wind power plants are regularly collected by the control unit inside the turbine. Today, most turbines are fitted with equipment that makes it possible to collect the data remotely, via modem or internet. The data from this work's accessible sources are assembled by governmental authorities or research groups. Special research program participants or subsidized companies are obligated to report not only performance but also turbine downtime and failures to authorities or research groups in their respective countries.

7.2.4.3 General Procedure to Restore a Failure

In this work failure is defined as the termination of the ability of an item to perform a required function. A typical procedure for handling the failure can be summarized as follows [14,150]:

A failure occurs inside the wind turbine, for example, the gearbox fails. The control unit inside the turbine registers the failure directly or the consequence of the failure, and responds according to the failure type. In case of safety hazard or major damage, the turbine is shut down.

If the unit is remotely monitored, an alarm is sent to the wind power plant operators. Often the wind turbine can be restarted because many alarms are not crucial. If the failure of the turbine is severe, a visual inspection of the turbine has to be made. This inspection can be performed by the operators or by locally authorized personnel. If a major failure has occurred, maintenance and repair personnel have to be contacted to repair or replace the damaged parts. When a major failure has occurred, a report is filed which describes the parts involved, possible failure causes, and downtime related to the failure. The report is maintained and transformed into databases.

The failure reports are designed with check boxes and use similar terminology in Sweden, Finland, and Germany.

7.2.4.4 Statistics from Sweden

Statistical data from the wind power systems in Sweden are collected by Vattenfall Power Consultant, which in turn performs this task on behalf of Elforsk (Table 7.1). Every year, Elforsk publishes a report on the performance of the turbines in Sweden, including values of downtimes for different component failures [146].

Almost all Swedish wind turbines are connected to a system with automatic monitoring of turbine performance. Failure reports are not automated but are handed in as reports and are then compiled in a database. The database contains information about the production and downtimes as well as failure reports dating back to 1989.

The following figures present an overview of failures and downtimes for the period 2000–2004. Table 7.2 presents the amount of installed turbines during this period. The number of turbines has changed over the years as more turbines are installed.

7.2.4.5 Failure Frequency

Figure 7.1 shows the percentage breakdown of failures that occurred during the years 2000–2004. This distribution of failures was calculated from the statistics in Source 2 [147]. Most failures were linked to the electric system

TABLE 7.1

Overview of Sources for the Statistical Survey

Name of Source	Source 1 [9]: Driftuppföljning av vindkraftverk, Arsrapport	Source 2 [10]: Felanalys
Country	Sweden	Sweden
Time span of data in survey	1997–2004	1989–2005
Number of turbines in survey (2005)	723	786
The data coverage of all turbines in the country	>95%	>95%
Accumulated number of reported failures	1658	1658

Source: Ribrant, J. and Bertling, L., Survey of failures in wind power systems with a focus on Swedish wind power plants, 1997–2005, *IEEE Transactions on Energy Conversion*, **22**(1), 167–173 © 2007 IEEE [145].

TABLE 7.2

Total Amount of Installed Turbines Used in Survey in Sweden

Year	2000	2001	2002	2003	2004	2000–2004 (Average)
Number of turbines in survey	527	570	620	682	723	624.5

Source: Ribrant, J. and Bertling, L., Survey of failures in wind power systems with a focus on Swedish wind power plants, 1997–2005, *IEEE Transactions on Energy Conversion*, **22**(1), 167–173 © 2007 IEEE [145].

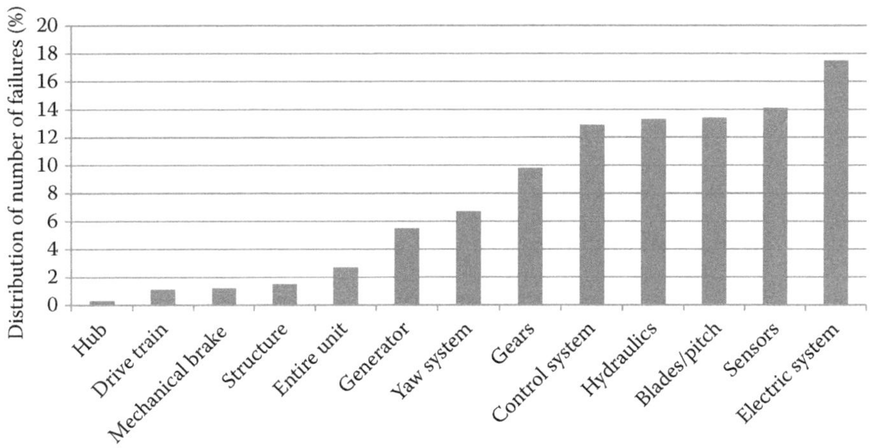

FIGURE 7.1

Distribution of number of failures for Swedish wind power plants between 2000 and 2004. (Data from Ribrant, J. and Bertling, L., Survey of failures in wind power systems with a focus on Swedish wind power plants, 1997–2005, *IEEE Transactions on Energy Conversion*, **22**(1), 167–173 © 2007 IEEE [145].)

followed by sensors and blades/pitch components. Table 7.3 overviews the values for frequency of failures.

7.2.4.6 Downtime

Figure 7.2 shows the distribution of downtime per component in Sweden between the years 2000 and 2004. Data for this figure were calculated using the values taken from Elforsk's yearly published report [146]. The results show that the most troublesome component is the gearbox, closely followed by the control system, and the electric system. This means that the gearbox has the longest downtime compared to the other components. Table 7.3 presents an overview of the values for frequency of failures in Sweden.

Table 7.3 also shows annual failure frequencies and turbine downtimes. These calculations take into account that some turbines use neither

TABLE 7.3

Downtimes and Failure Frequencies for Components in the Swedish Wind Power Plants 2000–2004

Component	Entire Unit	Structure	Yaw System	Hydraulics	Mechanical Brakes	Gears	Sensors	Drive Train	Control System	Electric System	Generator	Blades/ Pitch	Hub	Total
Total downtime per component, 2000–2004 (hours)	2631	1874	20,754	6918	1881	30,286	8357	3788	28,620	22,395	13,906	14,743	50	156,202
Average downtime per year (hours/year)	526	375	4151	1384	376	6057	1671	758	5724	4479	2781	2949	10	31,240
Average downtime per year per turbine (hours)	0.8	0.6	6.6	2.6	0.6	11.6	2.7	1.2	9.2	7.2	4.5	4.7	0.0	52.4
Distribution of downtime, 2000–2004 (%)	1.7	1.2	13.3	4.4	1.2	19.4	5.4	2.4	18.3	14.3	8.9	9.4	0.0	100.0
Total number of failures per component, 2000–2004	33	18	80	160	15	118	169	13	155	210	66	161	4	1202
Average number of failures per year	6.6	3.6	16.0	32.0	3.0	23.6	33.8	2.6	31.0	42.0	13.2	32.2	0.8	240.4
Average number of failures per year per turbine	0.011	0.006	0.026	0.061	0.005	0.045	0.054	0.004	0.050	0.067	0.021	0.052	0.001	0.402
Distribution of failures, 2000–2004 (%)	2.7	1.5	6.7	13.3	1.2	9.8	14.1	1.1	12.9	17.5	5.5	13.4	0.3	100.0
Average downtime per failure, 2000–2004 (hours)	79.7	104.1	259.4	43.2	125.4	256.7	49.4	291.4	184.6	106.6	210.7	91.6	12.5	130.0

Source: Ribrant, J. and Bertling, L., Survey of failures in wind power systems with a focus on Swedish wind power plants, 1997–2005, *IEEE Transactions on Energy Conversion*, **22**(1), 167–173 © 2007 IEEE [145].

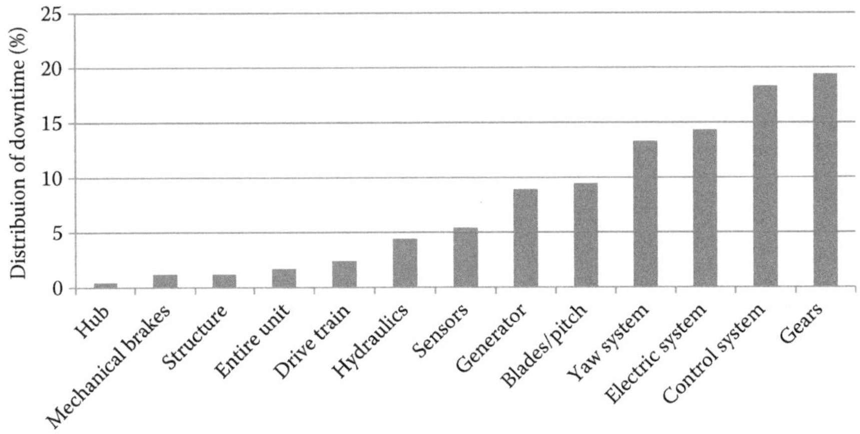

FIGURE 7.2
Percentage of downtime per component in Sweden between 2000 and 2004. (Data from Ribrant, J. and Bertling, L., Survey of failures in wind power systems with a focus on Swedish wind power plants, 1997–2005, *IEEE Transactions on Energy Conversion*, **22**(1), 167–173 © 2007 IEEE [145].)

hydraulics nor gearboxes. A suitable measurement of the failure's severity considers component failure frequency and average downtime per failure. If, for example, a component fails often but has a very short downtime, this does not noticeably disturb the production compared to a seldom occurring failure that has a long downtime.

The most important finding from the values in this table is that a typical turbine in Sweden has a failure 0.402 times a year and the mean downtime for each failure is 130 hours a year. The most critical components are drive train, gearbox, and yaw system which take 250–290 hours to repair.

7.2.4.7 Failures Versus Operational Year

Another way of examining the data is to investigate the relationship between the failure number and the turbine age, that is, operational year. The database source [147] contains enough information so that the calculation of failures per operational year is possible. Figure 7.3 shows the total number of failures per operational year, together with the number of turbines that have "survived" respective operational years. Failures for turbines with up to 19 years of operation have been recorded.

Since so many turbines have been installed within the last 8 years, a recalculation of failure frequency has been made. In Figure 7.4, the annual failure rate per turbine versus operational year is plotted using recalculated values of the failure frequency [147]. This figure gives a better view of the

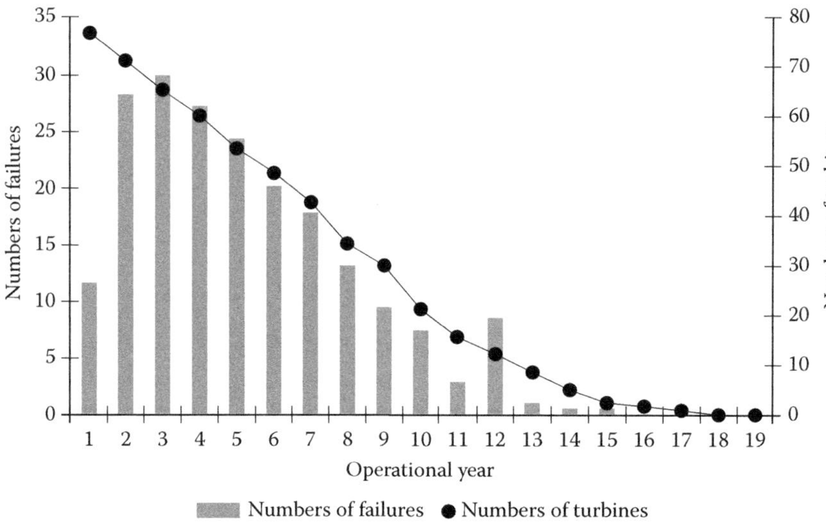

FIGURE 7.3
Number of failures versus operational age. (Adapted from Ribrant, J. and Bertling, L., Survey of failures in wind power systems with a focus on Swedish wind power plants, 1997–2005, *IEEE Transactions on Energy Conversion*, **22**(1), 167–173 © 2007 IEEE [145].)

development of the failures since it takes into account the population, that is, the number of turbines within every operational year.

The number of failures in the first operational year is much lower than the second year. A reason for this could be that not all failures are reported during the run-in period of a new turbine. For the following years, the failure rate is constant until about 5 years of operation, when the failure rate seems to drop. In year 12, there is a peak, and there seems to be an upward trend in the following years towards year 19, but for the last two bars one must also consider the small amount of data and the diminishing population of turbines. The failures for years 18 and 19 constitute failures from only a few surviving turbines.

A breakdown of the failures and the failure rate into groups of rated power of <500, 500–1000, and >1000 kW was made to give a more detailed overview of how failures distribute among power groups. Results from these calculations are shown in Figures 7.5 and 7.6.

Figure 7.6 shows the different power groups' trends and shows that turbines rated below 500 and 500–1000 kW follow the same pattern of a slight increase in failures during the first 3 years of operation and an annual failure rate decrease after 5 years. The most remarkable finding is that the turbines rated above 1000 kW have a higher annual failure rate than smaller turbines. The second important finding is that for the turbines rated above

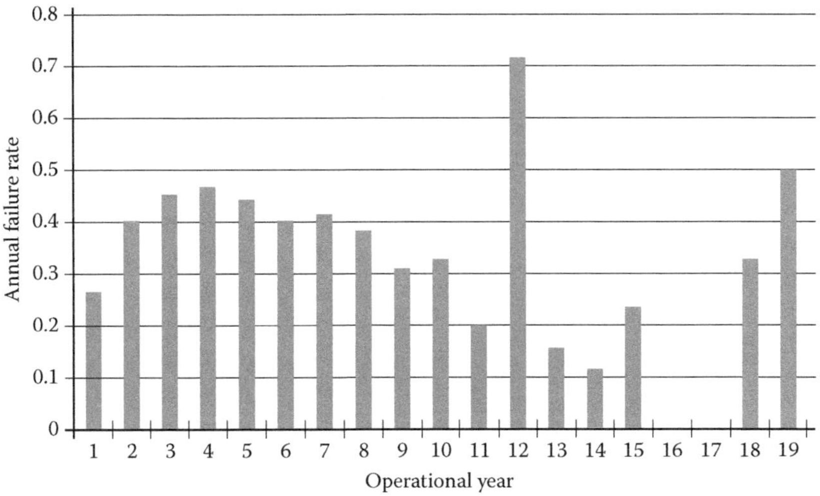

FIGURE 7.4
Annual failure rate per turbine versus operational year. (Adapted from Ribrant, J. and Bertling, L., Survey of failures in wind power systems with a focus on Swedish wind power plants, 1997–2005, *IEEE Transactions on Energy Conversion*, **22**(1), 167–173 © 2007 IEEE [145].)

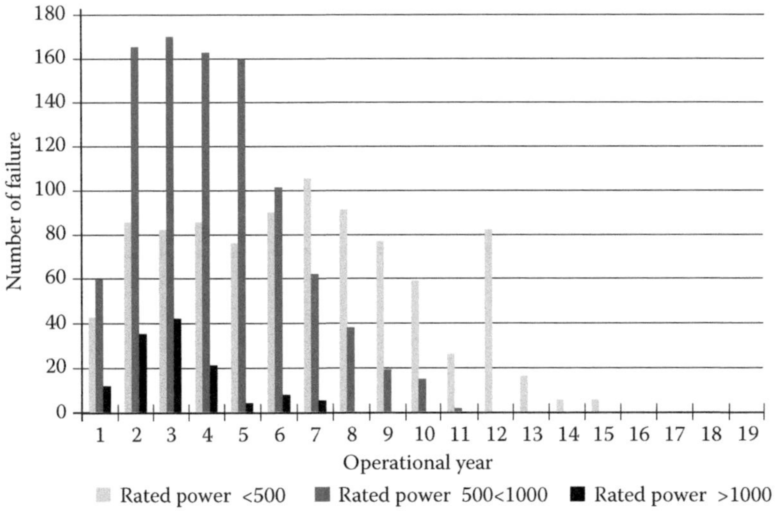

FIGURE 7.5
Number of failures in respective rated power group versus operational year. (Adapted from Ribrant, J. and Bertling, L., Survey of failures in wind power systems with a focus on Swedish wind power plants, 1997–2005, *IEEE Transactions on Energy Conversion*, **22**(1), 167–173 © 2007 IEEE [145].)

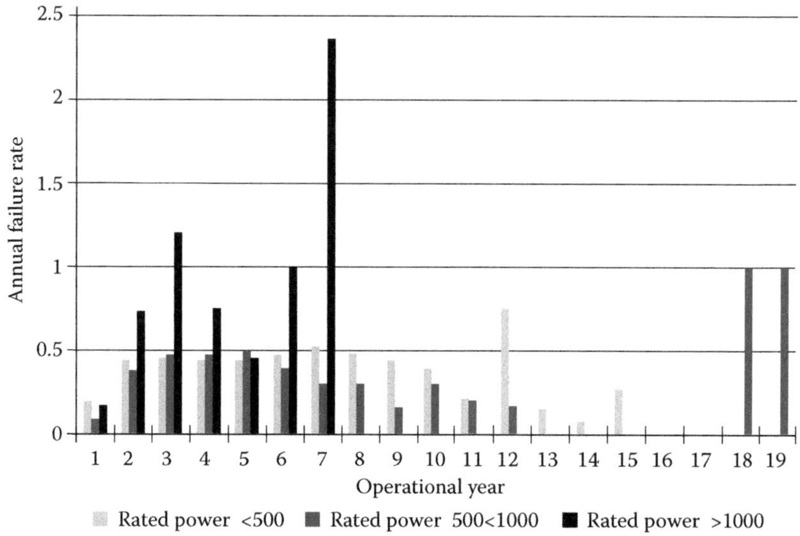

FIGURE 7.6
Failure rate in respective rated power group versus operational year. (Adapted from Ribrant, J. and Bertling, L., Survey of failures in wind power systems with a focus on Swedish wind power plants, 1997–2005, *IEEE Transactions on Energy Conversion*, 22(1), 167–173 © 2007 IEEE [145].)

1000 kW, the failure rate appears to increase, rather than decrease or remain constant.

7.2.4.8 Gearboxes—A Deeper Study

In this work, it has been shown that the gearbox is a troublesome component within the wind power turbine. For a typical turbine, 20% of the downtime is due to gearbox failures and an average gearbox failure takes about 256 hours to repair. Since the gearbox has been identified as critical for the availability of the turbine, some more statistics concerning gearboxes in Swedish power plants are presented.

Table 7.4 shows failures for gearboxes between 1997 and 2004. This table includes all gearbox failures that have occurred since the start of the database. In some cases, the failure statistics are reported and put into the database much later than the report is printed, hence there might be some small deviance between the officially published report from Elforsk [146] and the figures found in the database at Vattenfall Power Consultant [147]. These small differences can be found in the amount of hours in reported downtime but this does not affect the overall conclusion in this study that gearboxes constitute a great part of the total downtime.

TABLE 7.4

Overview of Data for Gearbox Failures in Sweden between 1997 and 2004

Year	1997	1998	1999	2000	2001	2002	2003	2004	1997–2004
Number of failures	21	41	52	26	30	42	13	7	232
Total downtime (hours)	4031	2518	5061	6172	5228	12,589	3987	2309	41,895
Average downtime per failure (hours)	192	61	97	237	174	300	307	330	181
Percentage of total downtime (%)	9.4	5.3	7.3	15.5	13.6	33.5	14.8	17.4	14.6

Source: Ribrant, J. and Bertling, L., Survey of failures in wind power systems with a focus on Swedish wind power plants, 1997–2005, *IEEE Transactions on Energy Conversion*, **22**(1), 167–173 © 2007 IEEE [145].

The most important finding from the overview shown in Table 7.4 is that the amount of gearbox failures, for the most recent years, tends to decrease, while the average downtime per failure increases.

The distribution of the failures according to the brands is a way of finding out if some types of turbines fail more than others. Table 7.5 shows the distribution of gearbox failures for wind turbines greater than 490 kW. The limit of 490 kW was chosen since this includes a large group of Wind World turbines. As turbine development continues, so does the power output. While still available, smaller turbines are decreasing as new ones are installed.

It does not look like any type of turbine is worse than the other, considering that the total amount of available data is small. Two types of turbines do stand out more than the others. The reason for this has not been investigated. The average number of gearbox failures per turbine for turbines above 490 kW is 0.18.

Table 7.5 also contains an Enercon E40 turbine. This turbine is a gearless turbine and has a design different from the other turbines. Instead of the gearbox, it utilizes a direct driven synchronous ring generator. Although direct drive has no cost or weight advantage over conventional geared systems, it has a simpler power train than the gearbox/high-speed generator combination, and may be more reliable [151] especially the PMG (Permanent Magnet Generator) design type. For some reasons, this failure was reported as a gearbox problem.

In the end, gearboxes are not all that different from each other. There might be differences in design, but when it comes to the basics, they are similar. A gear is a toothed wheel that works with other toothed wheels to alter the relationship between the speed of the rotor and the speed of the driven parts,

TABLE 7.5

Distribution of Gearbox Failures for Turbines with a Power of more than 490 kW

Brands	Power [kW]	Number of Failures [*n*]	Total Downtime [h]	Average Downtime [h]	Total Number of Installed Turbines (2005) [*n*]	Average Number of Gearbox Failures Per Turbine [*n*]
Bonus	600	6	654	109	22	0.17
Enercon	500	1	31	31	31	0.03
GE Wind Energy/ EnronWind	1425	16	4519	283	7	2.29
Neg-Micon	600	5	1247	250	3	1.67
Neg-Micon	750	9	1329	148	20	0.45
Nordic	1000	1	224	224	4	0.25
Vestas	500	5	570	114	23	0.22
Vestas	600	19	7887	415	56	0.34
Vestas	660	19	7151	376	79	0.24
Vestas	850	3	370	124	93	0.03
Vestas	1500	6	5138	857	16	0.37
WindWorld	490	11	3517	320	16	0.69
WindWorld	500	6	1174	196	27	0.22
WindWorld	600	2	940	470	43	0.05
All turbines	–	109	34,751	319	593	0.18

Source: Ribrant, J. and Bertling, L., Survey of failures in wind power systems with a focus on Swedish wind power plants, 1997–2005, *IEEE Transactions on Energy Conversion*, **22**(1), 167–173 © 2007 IEEE. [145]; *Felanalys*, Database of failures for Swedish wind power turbines 1997–2005, used with permission from Vattenfall Power Consultant and the Swedish Energy Authority, 2005. [147]

TABLE 7.6

Type of Gearbox Failure

Types of Reported Failure Codes	Component	Number of Failures	Average Downtime (hours)	Number of Failures, Cause: B1	Average Downtime, Cause: B1 (hours)
I-1	Bearings	41	562	36	601
I-2	Gearwheels	3	272	2	379
I-3	Shaft	0	0	0	0
I-4	Sealing	8	52	4	30
I-5	Oil system	13	26	5	36
I-other	Not specified	44	230	19	299

Source: Ribrant, J. and Bertling, L., Survey of failures in wind power systems with a focus on Swedish wind power plants, 1997–2005, *IEEE Transactions on Energy Conversion*, **22**(1), 167–173 © 2007 IEEE [145].

such as the generator. A gearbox is a set of interlocking gears in some kind of casing, usually cast iron [152]. They all have a containing case, lubricant system, and wheels held in place by bearings.

Table 7.6 shows the failed subcomponent within the gearbox for turbines with a rated power of 490 kW or more. This data was calculated from information given in the database source [147]. Each type of failure has a specific code assigned to it on the report form. The letter I stands for gearbox and the subsequent number is the code for what part of the gearbox that failed, for example, $I - 1$ is a general failure to the gearbox.

The last row in Table 7.6 represents unspecified failures. This means that they have been filed under gearbox failure but no subcomponent has been specified. It is possible to examine additional text added in the database of failures, and see at a quick glance that about half the unspecified gear failures correspond to serious failures that resulted in a replacement of the whole gearbox. Most I1-failures, bearing failures, demanded a total change of the gearbox or gearbox bearings. The last two columns in Table 7.6 show category B1-failures. When reporting the failed component, it was also possible to put in the cause. A B1-failure is a failure caused by wear. The majority of the failures to the gearbox are caused by wear and therefore they are shown here explicitly.

7.2.5 Comparison among Sweden, Finland, and Germany

A comparison of the three countries is possible to some extent. The available statistics from Finland [148] and Germany [149] were used to find representative values for downtime and amount of failures for each component. One can see trends and similarities among the three different countries. A compilation of the most interesting values is presented in Table 7.7.

TABLE 7.7
Summary of Statistical Findings

Country	Sweden	Finland [13]	Germany [14]
Average number of failures per turbine	0.402 times a year	1.38 times a year	2.38 times a year
Average downtime per year	52 hours per year	237 hours per year	149 hours per year
Average downtime per failure	170 hours per failure	172 hours per failure	62.6 hours per failure
Most number of failures	1. Electrical system 2. Sensors 3. Blades/pitch	1. Hydraulics 2. Blades/pitch 3. Gears	1. Electrical system 2. Control system 3. Hydraulics, sensors
Most amount of downtime	1. Gears 2. Control system 3. Electrical system	1. Gears 2. Blades/pitch 3. Hydraulics	1. Generators 2. Gears 3. Drive train
Longest downtime per failure	1. Drive train 2. Yaw system 3. Gears	1. Gears 2. Blades/pitch 3. Structure	1. Generators 2. Gears 3. Drive train

Source: Ribrant, J. and Bertling, L., Survey of failures in wind power systems with a focus on Swedish wind power plants, 1997–2005, *IEEE Transactions on Energy Conversion*, **22**(1), 167–173 © 2007 IEEE [145].

It is possible to say that the downtime of the failures is similar. The average downtime is less in Germany compared to Sweden and Finland and that could be explained by a better and nearer service organization, but overall the average downtime for a failure is as high as 62–172 hours, that is, 2–7 days. According to the findings, each turbine needs to be attended for repair at least once a year, that is, 0.4–2.38 times a year. One must also consider that in this survey only the major incidents are reported and the downtime for scheduled maintenance is omitted. If the wind power is to be competitive, the downtime needs to be shortened and visits to the turbine should be kept to a minimum (Table 7.8).

The gears and the drive train are the components that demand the longest downtime per failure. The reason for this is that they are big and cumbersome to replace, and replacement involves equipment such as cranes and so on. Since drive train and gearboxes seldom fail, one reason for the long downtime could be that spare parts need to be ordered which could prolong the time for repair.

The study of failures has also shown that the gearbox is critical to the availability of the wind turbine. Most of the gearbox failures are caused by wear on the mechanical parts.

TABLE 7.8
Result of Statistical Findings for Sweden

Country	Sweden
Average number of failures	0.402 times a year
Average downtime	52 hours per year; 170 hours per failure
Most number of failures	Electrical system
	Sensors
	Blades/pitch
Most amount of downtime	Gears
	Control system
	Electrical system
Longest downtime per failure	Drive train
	Yaw system
	Gears
Important findings from the failures statistics	Turbines below 1 MW show similar failure rate trends, with a small increase for the first 3 years of operations and after 5 years it decreases
	Turbines above 1 MW show an increasing failure rate over the operational years.
Important findings concerning gears	The amount of failures has decreased for the most recent years
	The downtime for gearbox failures has increased in recent years
	The majority of the gearbox failures are caused by wear
	No link between brand of turbine and amount of gearbox failures can be proved since the amount of available data is small, but two types of turbines stand out compared to the others

Source: Ribrant, J. and Bertling, L., Survey of failures in wind power systems with a focus on Swedish wind power plants, 1997–2005, *IEEE Transactions on Energy Conversion*, **22**(1), 167–173 © 2007 IEEE [145].

7.3 RCAM Implemented for Wind Power Turbines

7.3.1 Introduction

The RCM analysis presented in this section is an essential part in the implementation of RCAM. Its purpose in the context of RCAM is, by means of an FMEA, to reveal the components, the failure modes as well as the major underlying failure causes that are most relevant for the system reliability and availability, and to identify suitable preventive measures. In this way, the RCM study forms the basis of RCAM; it ensures to focus the subsequent development and application of mathematical models on the practically relevant items and failures.

The motivation of focusing the RCM analysis on selected components is directly supported from failure statistics performed within the case study. Summarizing failure data from over 750 wind turbines in Sweden in the period 1989–2005, it reveals that about 75% of the annual downtime is caused by only 15% of the failures. A similar result has been obtained from wind turbine failure data from the German WMEP program as presented in [153].

The RCM analysis has been carried out within a work group involving a wind turbine owner and operator, a maintenance service provider, a provider of condition-monitoring services and wind turbine component supplier as well as researchers at academia. The case study is discussed in detail in [15]. The study forms the basis for the development of quantitative models for maintenance strategy selection and optimization. Challenges identified by the RCM work group which are considered to impede the achievement of cost-effective operation and maintenance of wind turbines are discussed together with proposed solutions. Standardized and automated collection of in-depth failure and maintenance data, enhanced training of maintenance personnel, and the utilization of quantitative methods for decision support in wind turbine maintenance are identified as important steps to improve the reliability, availability, and profitability of wind turbines.

Among the failure causes identified in this study, vibration has been found to play a central role as a cause for mechanical failure of a variety of wind turbine components. Measures aiming at prevention or early detection of bearing damage are, therefore, concluded to be particularly effective. Another conclusion from this concerning maintenance modeling is that quantitative maintenance–optimization models that are based on the common assumption of independent failure of components run the risk of systematically underestimating the benefit of vibration monitoring.

The merge of practical experience and theoretical expertise realized in the described RCM workgroup is considered vital for the development of decision support tools for wind turbine O&M and an important strength of this study. The RCM analysis presented here follows the methodology of a study described in [154] and the guidelines given in [45].

The implemented RCAM method has covered the following steps:

1. System selection and definition
2. Identification of system functions and functional failures
3. Selection of critical items
4. Data collection and analysis
5. Failure modes, effects, and criticality analysis including failure causes and mechanisms of the dominant failure modes
6. Identification of applicable maintenance actions

The consequences of failure have been assessed for the four criteria:

1. Safety of personnel
2. Environmental impact (in a wind turbine, e.g., discharge of oil or glycol)
3. Production availability (i.e., the impact on electricity generation)
4. Material loss (including primary damage to the component itself, and also secondary damage to other parts of the system)

The determination of maintenance intervals and the comparative analysis of preventive maintenance (PM) measures by means of mathematical models are not included in this case study.

A short description of the methodology of this study together with a tabular compilation of selected results has been published in [155].

7.3.2 Wind Power Turbines for the Case Study

Two wind turbine models have been selected for this case study. In the following section, these models are presented in detail together with their system-level functions and functional failures of interest in this context. The parallel analysis of the two turbines has been chosen to account not only for the differing reliability characteristics of turbines originating from different generations of technology (see, e.g., [156] and [145]) but also for the potentially different applicable PM measures. It is important to note that the turbines have been selected for analysis not due to any abnormal occurrence of failures but due to their relevance for the project partners and to the available experience with O&M of these turbines in the RCM workgroup.

There is usually some variance in design between wind turbines of the same model. A reason for this is the common practice that wind turbine manufacturers have several subsuppliers for major components like, for example, the gearbox. Furthermore, small changes in design have been implemented over the years of production.

7.3.2.1 V44-600 kW Wind Turbine

The Vestas V44-600 kW is an upwind turbine with three blades and an electrically driven yaw system. It was launched in 1996. Its rotor has a diameter of 44 m, a weight of 8.4 t, and a rated rotational speed of 28 r/min. A hydraulically actuated pitch system is used for speed control, optimization of power production, start-up, and aerodynamic braking of the turbine. Additional breaking functionality is provided by a disk brake located on the high-speed side of the gearbox.

The structure of the V44-600 kW system is shown in Figure 7.7. During operation, the main shaft transmits the mechanical power from the rotor

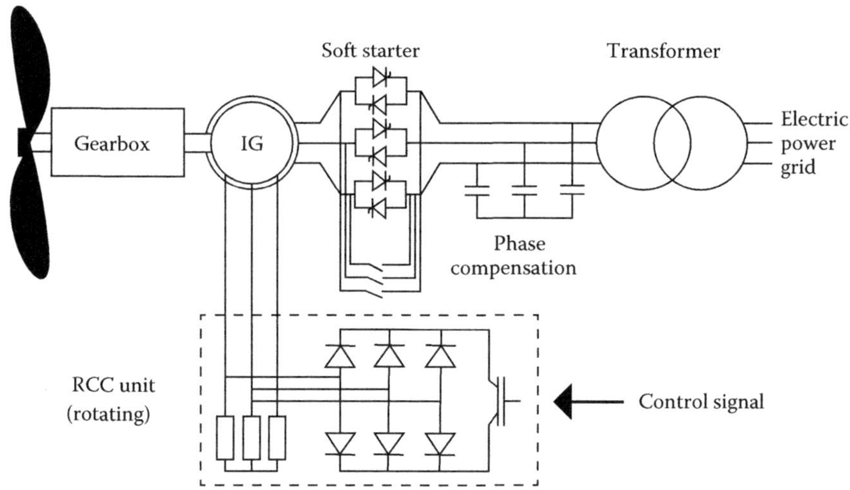

FIGURE 7.7
Structure of the V44-600 kW system with wound rotor asynchronous generator and OptiSlip technology for RCC. (Adapted from Fischer, K. et al., Reliability–centred maintenance for wind turbines based on statistical analysis and practical experience, *IEEE Transactions on Energy Conversion*, **27**(1), 184–195 © 2012 IEEE [15].)

to the gearbox, which has either a combined planetary parallel design or, as in the case of the early V44 turbines analyzed in this study, a parallel shaft design. The gearbox and the generator are connected with a Cardan shaft. The generator is an asynchronous four-pole generator with integrated electronically controllable resistance of the wound rotor (so-called OptiSlip technology, see Figure 7.7), which requires neither brushes nor slip rings. The variability of the rotor resistance is provided by the rotor current control (RCC) unit which is bolted to the nondrive end of the generator rotor and, thus, permanently rotates during wind turbine operation. It consists of a microprocessor unit to which the control signal is optically transmitted, of a power electronics unit and a resistor bundle. As shown in Figure 7.7, the rotor resistance is varied in the way that the resistor bundle is short circuited at varying frequency by means of an insulated gate bipolar transistor (IGBT) in the power electronics unit. This OptiSpeed technology allows the rotational speed of the generator to vary between 1500 (idling) and 1650 r/min.

The generator stator is connected to the electric power grid through a thyristor unit, also called "soft starter." This limits the cut-in current of the asynchronous generator during connection to the grid and smoothly reduces the current to zero during disconnection from the grid. The reactive power required by the generator is partially provided by a capacitor bundle at the bottom of the tower, the power factor correction or phase compensation unit.

The main function of the V44 system is the conversion of kinetic wind energy to electric energy, which is provided to the electric power grid. More

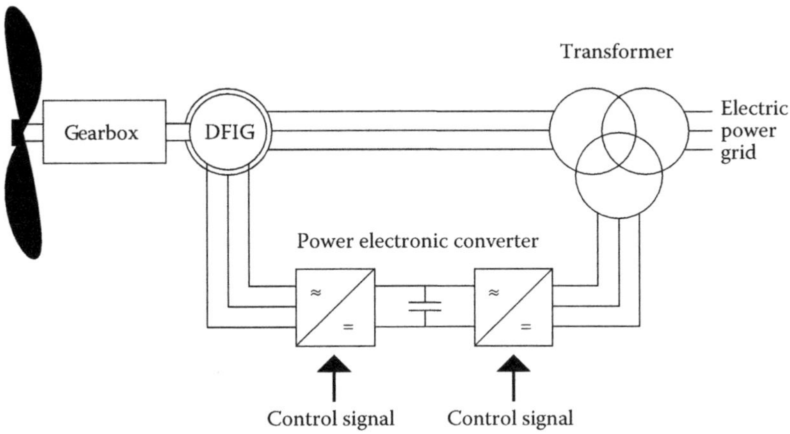

FIGURE 7.8
Structure of the V90-2MW system with DFIG and partially rated converter. (Adapted from Fischer, K. et al., Reliability-centred maintenance for wind turbines based on statistical analysis and practical experience, *IEEE Transactions on Energy Conversion*, **27**(1), 184–195 © 2012 IEEE [15].)

specifically, the system function is to provide up to 600-kW electric power at 690 V and 50 Hz to the electric power grid, at wind speeds of 4–20 m/s and in an operating temperature range of −20 to +40°C.

Failures on the system level that are relevant in this study are both a complete and a partial loss of energy conversion capability of the turbine.

7.3.2.2 V90-2MW Wind Turbine

Figure 7.8 shows the structure of the Vestas V90-2MW system. The first turbines of this type were installed in 2004. Like the V44, the V90-2MW is an upwind turbine with three blades and electrically driven yaw. Its rotor has a diameter of 90 m, a weight of 38 t, and a nominal rotational speed of 14.9 r/min. The pitch control system with individual pitching capability for each blade continuously adapts the blade angle to the wind conditions in order to provide optimum power output and noise levels. In addition, it serves for speed control, turbine start-up and stop by aerodynamic braking. Similarly to the V44, a disk brake is located on the HSS.

In contrast to the V44 turbine, all V90-2MW systems apply hybrid gearboxes with one planetary and two parallel shaft stages, from which the torque is transmitted to the generator through a composite coupling. A major difference from the V44 system is the generator concept: the V90-2MW contains a four-pole doubly fed asynchronous generator (DFIG) with wound rotor. A partially rated converter controls the current in the rotor circuit of the generator, which allows control of the reactive power and serves for smooth connection to the electric power grid.

In particular, the applied DFIG concept (so-called OptiSpeed technology) allows the rotor speed to vary by 30% above and below synchronous speed. The electrical connection between the power converter and generator rotor requires slip rings and carbon brushes. The generator stator is, via a transformer, directly connected to the electric power grid [157,158,231].

The system function of the V90-2MW is to provide up to 2 MW of electric power at 690 V and 50 Hz to the grid, at wind speeds of 4–25 m/s and in a standard operating temperature range of −20 to +30°C. Again in this case, system-level failures of relevance in this study are the complete and the partial loss of energy conversion capability of the turbine.

7.3.3 FMEA Based on Statistics and Practice

For the second stage of the RCAM approach, the critical component should be identified. For this case study, the selection of subsystems for detailed analysis has been based on both turbine model-specific failure statistics and expert judgment. The failure data used for statistical analysis cover the failures of 32 V44-600 kW turbines located in Sweden from the period 1996 to 2005, which have been reported in the database [147]. This database includes failures that led to a wind turbine downtime of at least 1 hour. Statistical data analysis for the V90-2MW system has been carried out based on data from [160].

It includes failures of 57 turbines located in Germany, from the period 2004 to 2008. Both the frequency of failures in terms of the annual failure rate of a subsystem and the mean downtime resulting from a failure are relevant for the criticality assessment of an item. As it is the combination determining the impact on availability and, thus, on the production loss, it is advantageous to combine these two measures by multiplication. This results in the average downtime per wind turbine and year related to failures of a specific subsystem

$$t_{lost} = \frac{\sum_{i=1}^{I} d_i}{\sum_{i=1}^{I} X_i \cdot T_i} \tag{7.1}$$

with d_i being the downtime due to failures of a subsystem in the time interval i, X_i is the number of wind turbines reporting to the database in time interval i, and T_i being the duration of the time interval i.

Figure 7.9 shows the distribution of downtime over the subsystems of wind turbines of interest. Downtime due to external grid failures, inspections, service maintenance, and other external causes has not been taken into account in this analysis. However, in the case of the V90-2MW data, downtime due to retrofits and not failure-related component replacements is included in the data, which to some extent overestimates the relevance of rotor, control system, and hydraulic system failures for the downtime. Note that due to the

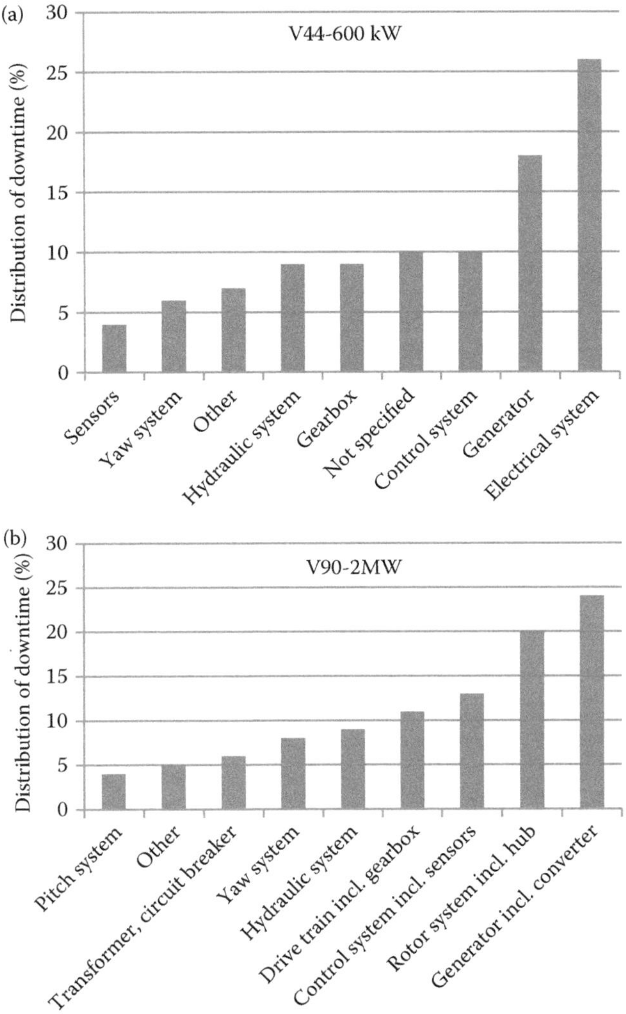

FIGURE 7.9
Distribution of downtime t_{lost} over subsystems in the case of V44-600 kW and V90-2MW systems operating onshore. (Adapted from *Felanalys*, Database of failures for Swedish wind power turbines 1997–2005, used with permission from Vattenfall Power Consultant and the Swedish Energy Authority, 2005. [147]; Balschuweit, R., Stress and damage analysis of wind turbines using the example of Vestas V90-2MW, (German title: *Beanspruchungs-und Schadensanalyse von Windenergieanlagen am Beispiel der Vestas V90-2MW*), MSc thesis, Eng., Sci., Beuth University of Applied Sciences Berlin, Berlin, 2009 [160].)

different failure-reporting schemes of the two data sources and the associated differences in subsystem categories, the results are not suitable for a comparison of the turbines; they do, however, serve the purpose of prioritization intended here.

In addition to failure statistics, expert opinion was taken into account in the identification of the most critical subsystems for in-depth analysis. The judgment of the RCM workgroup members having professional experience with wind turbine O&M was assessed by means of questionnaires, from which a subsystem ranking was obtained. In the case of both wind turbine models, highest priority was given to the gearbox, followed by the generator.

Based on the results of both the failure data analysis and the questionnaire assessment, the subsystems gearbox, generator, electrical system, and hydraulic system were chosen for analysis in the RCM study. In spite of the significant average contribution of the control system to downtime, this system has not been included in the RCM analysis because its failures can hardly be influenced by means of PM.

7.3.4 Resulting Critical Components

The RCAM analysis has identified four major critical components which are: the gearbox, the generator, the electrical system, and the hydraulic system. In the following sections, findings of these components are discussed in greater detail as presented in [15].

7.3.4.1 Gearbox

The function of the gearbox in a wind turbine is the transmission of torque from the rotor to the generator shaft, providing the desired conversion ratio for speed and torque. The failure mode of interest is, thus, the loss of torque transmission capability.

Gearbox failure can have severe consequences: in the case of complete demolition, parts of the gearbox can constitute a risk for personnel. Oil spill of up to 120 l (V44) or 300–400 l (V90) of lubrication oil contained in the respective gearboxes can cause environmental impact. Gearbox failure is among the failures resulting in the longest average downtime and, thus, has a strong impact on production availability, and it can cause severe secondary damage, for example, in the main bearing or the rotor shaft.

According to the experience and knowledge of the RCM workgroup, a failure cause considered valid for all gearbox components during different periods is design deficiencies, that is, that components are under dimensioned with respect to the real loads they are exposed to. This is not without controversy. While Musial et al. [161] state design deficiencies to be a major, mostly manufacturer-independent problem and cause of premature gearbox failure, poor maintenance and a lack of attention to monitoring this component are considered to significantly contribute to gearbox failure by others.

An additional failure cause of general validity for different gearbox components is manufacturing and installation deficiencies, which lead to increased friction or inappropriate high cyclic loading resulting in damage. Training of technicians for improving the quality of installation, service maintenance and repair as well as alignment check and monitoring of the wind turbine drive train is, thus, considered important preventive measures against premature failure of gearboxes.

The bearings, the gearwheels, and the lubrication system have been identified to be the components with highest relevance for gearbox failure and are, thus, discussed in more detail in the following. Failure of shafts in the gearbox is considered to occur as secondary damage only and has, thus, not been included in the RCM analysis.

The majority of gearbox failures, according to the experience of the RCM workgroup often more than 60%, originate from the failure of gearbox bearings. This does not only apply to the V44 and V90 turbine models investigated here. The dominant role of bearing damage for gearbox failure is confirmed by the failure data in [147] to be valid for a variety of geared wind turbines.

The function of the gearbox bearings is to keep the shafts in their positions while allowing rotary motion at minimum friction. The failure mode of interest is the loss of this function, which is often induced by severe mechanical damage in the form of flaking and cracks. Such bearing damages are in turn usually a result of progressed primary damage in form of wear, indentation, smearing, surface distress, corrosion, or electric current damage [162]. In parallel-shaft gearboxes of V44-600 kW turbines as well as in the hybrid planetary-parallel gearboxes of the V90-2MW system, bearing problems have been observed mainly in the HSS and the intermediate-speed shaft bearings. In some cases, up-tower repair of these bearings is possible. However, any failure in a planetary stage requires removal of the gearbox from the nacelle for external repair. The consequences of bearing failure range from partial (due to friction losses) or complete production losses (when severe bearing damage prohibits further operation) to comprehensive secondary damage of other components due to excessive vibration and strong heat generation. The latter implies even a risk of fire.

The main causes of bearing failure identified during the RCM analysis are overloading, inappropriate lubrication, and corrosion. In all these cases, the failure characteristic is an increasing failure rate over time as damage accumulates, a reasoning supported by the quantitatively determined bearing failure characteristics of a 2-MW gearbox published in [163]. Overloading of bearings is often a consequence of the design or installation deficiencies discussed earlier. It can lead to material fatigue and plastic deformation in a bearing. The resulting increased friction and heat evolution cause additional deterioration.

Inappropriate lubrication covers the lack of grease, for example, due to too long lubrication intervals, and also the use of wrong types of lubricant. Both failure causes result in metal-to-metal contact, wear and high contact temperature, which leads to damage of the inner bearing surfaces. This failure mechanism can as well result from the presence of particles in the lubricant to which the bearings are significantly more sensitive than any other part of the gearbox. In addition, inappropriate lubrication also includes overgreasing: an excess of grease between the rolling elements and the races causes the elements to slide instead of rolling, which can result in overheating of the bearing.

Suitable preventive measures to prevent these failure causes are the measurement of the oil pressure (being standard in V90-2MW) and temperature, as well as oil analysis. At present, oil samples are usually taken once a year and analyzed with respect to the content of additives, particles, and moisture as well as signs of corrosion. Online particle counters, which are available as a complement to existing condition monitoring system (CMS), are an option to monitor the oil quality continuously and enable earlier countermeasures in case of observed irregularities.

Another important cause of gearbox bearing failure is the presence of water or moisture in the gearbox oil, a factor that is usually strongly dependent on the location of the wind turbine. Moisture can on the one hand lead to corrosion through oxidation of the metallic components. In addition, hydrogen ingression into the steel reducing the material strength has been put forward as a possible explanation for surface failures [164]. However, there are alternative theories, which include interaction with the additives present in the grease, tribochemical effects having influence on the localized crack behavior [165]. As mentioned previously, oil analysis is a suitable measure to identify moisture in the gearbox oil. An alternative is online moisture detection with an automatic system, but this is not considered to be economic on machines in the kilowatt range. A both cheap and effective option to prevent the ingress of moisture into the gearbox is the installation of a filter dryer in the top casing that dehumidifies the air flowing into the gearbox during thermal cycles.

Vibration-based CMS (vibration-CMS) and complementary measurement of bearing temperatures detect approximately 90% of all impending bearing failures, according to the experience of the RCM workgroup. In this way, secondary damage resulting from excessive vibration and heat generation can be prevented effectively in many cases, however, usually not the bearing damage itself. A major difference between vibration and temperature monitoring is that vibration-CMS typically provides a prewarning time (P–F interval) in the range of several weeks to months while this is only in the range of hours to days in the case of temperature-based detection. The application of vibration-CMS is considered to be state of the art on megawatt turbines. However, vibration monitoring and vibration-based diagnosis of planetary gearbox stages are still challenging at present and an improvement of condition monitoring technology for this purpose is subject to intensive development activities today.

A visual inspection technique increasingly applied to wind turbines to assess the condition of gearbox bearings as well as gearwheels is borescope inspection (also known as endoscopy). Combined with digital imaging technology, this enables even remote condition assessment by gearbox experts.

Borescope inspection is not only applied in case impending failure is detected by vibration or temperature measurements but also without any indication of irregularities. The V90-2MW as well as the majority of other contemporary wind turbines provide an opening in the gearbox casing, which allows easy access for borescope inspection. Due to the compact design of gearboxes, not all components can be inspected with this technique, for example, the HSS bearings of V90-2MW are not accessible for borescope inspection.

The cost effectiveness of measures for prevention and early detection of bearing damages is considered high as it serves to prevent severe secondary damage to the gearbox. In this way, the lifetime of the gearbox can be extended. Up-tower repair may be possible instead of significantly more expensive removal and external repair of the gearbox. In the case of a necessary replacement, a higher residual value of the gearbox can be preserved.

7.3.4.2 Generator

The function of the generator is to convert the mechanical power from the HSS to electrical power fed into the electric power grid. Generator failure is in the large majority of cases caused by winding or bearing failure. This does not only apply for the wind turbines analyzed here, but for a wide variety of wind turbine generators, as shown by Alewine and Chen [166]. According to these authors, rotor failures dominate in smaller machines having mostly a random wound rotor design, while modern generators with form-wound coils are more robust. In 1-MW machines, bearing failures constitute more than half of all generator failures.

Generator Windings: The category of generator windings incorporates both the windings of the stator as well as the wound rotor. The functions of the windings are to lead electric currents while preventing undesired short circuits. The consequences of winding failure range from a limited electric unbalance in the output power to emergency braking followed by standstill of the wind turbine. In addition to the required rewinding of the generator, the consequences can also include secondary damage of the RCC unit (V44) and of the gearbox, the latter resulting from the excessive loading during emergency stops. Failures of the generator windings can be a short circuit due to a failed insulation, or the loss of conduction, that is, the occurrence of an open circuit. Short-circuit failure occurs due to melting of the insulation due to overheating, due to material degradation (aging) of the insulation, or due to mechanical impact, predominantly caused by vibrations (e.g., due to bearing damage). According to [166], a common root cause is thermal cycling damage emerging during cycling between cold ambient temperature and operating temperature due to different thermal expansion behavior of conductor and insulating material. Insulation failure is facilitated in the presence of moisture. The frequency of short-circuit failure increases with time. Measures to prevent failures of the insulation are (1) temperature measurements in the windings, usually being part of the system control and data acquisition system in contemporary turbines, and (2) the early detection of bearing failures by means of vibration monitoring with a CMS.

Open-circuit failure of the generator windings is caused by broken windings or failing contacts at connections. The underlying failure cause can be fatigue as a consequence of excessive vibrations or thermal cycling, or material defects in the copper conductors. To avoid operation at high vibration levels is, thus, also a preventive measure against open-circuit failures. Thermography can be applied to detect initiating damage in the conductor or in contacts since a reduced cross-section usually causes local hot spots. An additional method to identify impending failure of this type is the application of a so-called motor tester during standstill of the generator, which enables diagnosis based on an injected test signal. Among the analyzed wind turbines, only the rotor windings in the V90-2MW generator are accessible for diagnostic measurements. With respect to the failure characteristic, the probability of open-circuit failure of the generator windings is either constant or increasing over time, depending on the underlying failure mechanism.

Generator Bearings: The function of the generator bearings is to keep the generator rotor shaft in its position with a minimum amount of friction for rotary motion. The occurrence of high friction due to cracks or flaking is considered to be the main failure mode. Failed generator bearings can usually be replaced on site.

Bearing damage in the generator can be caused by misalignment or inappropriate lubrication through the mechanisms already discussed in the context of gearbox bearings. Another common cause is electric current flowing through the generator bearings, leading to electrical erosion. Both the mechanically and electrically initiated causes have as well been reported in [166] to be common failure causes in wind turbine generators in general. Bearing damage is, according to the experience of the RCM workgroup, accelerated by insufficient maintenance and accumulates over time. The corresponding failure characteristic is, thus, an increasing failure rate. Preventive measures include a strict compliance with the lubrication scheme, which prescribes lubrication of the generator bearings every 6 months as a part of the regular service or the installation of an automatic lubrication system. According to [167], extending the PM of generator bearings including a change of lubricant effectively reduced the frequency of generator bearing failures in Vestas turbines in the past. Vibration-CMS facilitates an early detection of misalignment, as stated in [166].

Generator Slip Rings and Brushes (V90 Only): Generator components being both maintenance- and failure intensive are the slip rings and carbon brushes providing the electrical contact to the V90-2MW generator rotor. Carbon dust originating from the wear of the brushes favors electric spark over, which damages the slip ring. This issue has been solved successfully by means of a suction system for continuous removal of the carbon dust during operation. In addition, the quality of the brushes has improved significantly, which significantly extended slip ring life to approximately 2 years today. According to [166], carbon brush life is usually affected by the environmental conditions: temperature and humidity. Slip rings and carbon brushes are inspected every 6 months during service and are replaced depending on their condition. Continuous condition monitoring of the brushes, having the benefit of preventing damage from over-worn brushes, can be implemented using electrical contacts in the brushes or visual monitoring of the brush thickness by means of a camera.

7.3.4.3 Electrical System

1. The RCC is the most frequently failing item in the electrical system of the V44-600 kW. This is clearly indicated by the turbine-specific failure statistics and in agreement with both the experience of the RCM workgroup and [167]. In average, the RCC unit has to be replaced at least once during the wind turbine lifetime. In spite of its attachment to the generator, the RCC is discussed as a part of the electrical subsystem here in accordance with the categorization of failure data of the V44.

   The function of the RCC unit is a fast control of the generated electric power by regulating the rotor current. Failure of the RCC unit results in a loss of this current control capability. However, the impact on the system function, that is, if a full or a partial loss of the energy conversion capability is caused, depends on the failed part of the RCC. Due to the fact that the failure of the power electronics unit clearly dominates over the failures of the RCC microprocessor unit and the resistor unit, only failure modes and causes of the first are discussed in the following. Failure of this unit usually permits continued operation of the wind turbines at reduced power of maximum 300 kW. In that case, the generator operates with a fixed rotor resistance. In some cases, failure of the power electronic unit of the RCC is caused by loose contacts or cable twist and, thus, by mechanical impact, for example, due to vibrations. However, more frequent is the occurrence of IGBT failure. According to the RCM workgroup, this has been observed predominantly in the mornings and evenings, that is, during times with large changes in power demand and generation and, therefore, frequent coupling activities in the power grid. This observation suggests that the failure is related to grid disturbances such as voltage dips or frequency variations. The frequency of failure of the RCC power electronics unit tends to increase with component age. Overheating-related failure of this unit occurs also as a secondary failure due to high temperature in the nacelle, that is, after a defect in the cooling system of the gearbox.

The different failure causes explained require different measures for prevention: vibration-related failure can be influenced through the level of vibration, for example, by applying continuous vibration monitoring and preventive bearing replacement before strong vibrations occur.

2. Converter (V90-2MW): In a variable-speed wind turbine with a DFIG like the V90-2MW, the function of the converter is to feed the generator rotor circuit with an electric current of desired amplitude and frequency in order to ensure the desired output of active and reactive power from the stator to the grid. Depending on the operating speed, the converter either transmits electric power from the generator rotor circuit to the grid (super synchronous operation) or vice versa (sub synchronous operation) (see e.g., [168]). As shown in Figure 7.8, the converter system consists of a rotor-side and a grid-side converter in back-to-back configuration. The rotor-side converter controls the current to the rotor whereas the grid-side converter controls the dc-link voltage [169]. In addition to these two power electronic units with IGBT switches, the converter subsystem includes corresponding microprocessor (or control) units. The converter is water cooled by means of a pump-driven cooling circuit. High failure rates of the converter systems found in V80 and early V90-2MW turbines have decreased. However, the converter system is still a source of failure in later turbines of the V90-2MW series. Salty environments, grid disturbances, insufficient cooling, and condensation problems negatively impact the failure frequency of the converter system according to the RCM workgroup. The consequences of converter failure are mainly limited to production loss. In case of failure, the turbine is stopped smoothly by means of the aerodynamic brake so that excessive loading of the drive train components and potential secondary damage of this are avoided.

   The failure modes of the converter identified in this study are (1) open-circuit failure, that is, an interruption of the electrical connection and (2) a malfunction of the rotor current and frequency control. A common cause is the failure of the microprocessor units, a problem that can be addressed, for example, with software updates or by using dust- and moisture-protected circuit boards, but usually not with PM. Another cause is the failure of the power electronic components with potential causes ranging from high ambient temperature, vibration, moisture or condensation problems, and component aging to disturbances from the grid. Based on that, possible preventive measures are those reducing or monitoring the given failure causes, such as reduction of vibration in the power electronics unit by design modifications, or monitoring of vibration and bearing temperatures. Potential design modifications of converters for wind turbines aiming at increased fault tolerance are discussed in [170].

3. Phase Compensation Unit (V44-600 kW only): Besides the RCC unit discussed previously, a second frequently failing component of the V44 electrical system is the capacitor bundle at the bottom of the tower providing reactive power to the generator. Failure of capacitors is a common cause of fire in wind turbines. A lack in manufacturing quality is considered to be the main cause of failure. According to the current maintenance practices, the capacitors are visually inspected during the regular service.

7.3.4.4 Hydraulic System

The functions of the hydraulic system in the considered wind turbines are: to operate the rotor-blade pitch mechanisms for controlling the angle of attack of the wind, which includes aerodynamic braking capability, and to actuate the mechanical disk brake.

The hydraulic system comprises two different pressure levels, but only a single hydraulic pump: the pitch system is operated at a level of 200 bars, while the mechanical brake, connected with a pressure reducing valve, is actuated at a level of 30 bars.

The theoretical worst-case scenario of a full loss of braking capability resulting in overspeeding of the turbine is considered highly improbable as this is prevented by two different safety systems for the pitch system and the mechanical brake. In case of emergency, a safety valve releases the pressure from the hydraulic pitch actuators and in this way allows the blades to be pitched out of the wind solely by the aerodynamic forces; emergency operation of the mechanical brake is ensured by a hydraulic accumulator with stored pressure.

The two failure modes selected by the RCM workgroup for analysis are: the loss of pitching functionality or a too slow pitch, and an unintentional activation of the mechanical brake. The consequences of these include a partial or full loss of power generation capability, but no risk for personnel or the environment. Secondary damage to the drive train can result from pitch errors of single blades, which lead to unbalances and, thus, impose high cyclic load on the drive train components.

The most common cause of too slow pitching or loss of pitching functionality is a failure of a control valve due to wear out. The failure characteristic is, thus, a failure rate increasing with the age of this component. At present, hydraulic valves are maintained with a run-to-failure strategy. A possible preventive measure is a scheduled replacement of the valves at the age of around 10 years. However, a central cause for the wear out of the hydraulic control valves is moisture and particle contamination of the hydraulic oil. A more effective damage-preventing measure in this context is, thus, an additional offline filtration of the hydraulic oil or a measurement of the moisture content for early detection. According to the present common practice, the hydraulic oil of the V44-600 kW is exchanged every 4–5 years, while an oil change in the V90-2MW hydraulic system is usually required after 3 years. It is expected that the implementation of offline filtration significantly extends the oil change intervals for the hydraulic oil.

The other failure mode of interest, the unwanted activation of the mechanical brake, is disadvantageous as it imposes large loads on the wind turbine drive train. It is mostly caused by an error in the control system and can, thus, hardly be influenced by maintenance. However, in some case, it is caused by oil leakage through an upstream valve which causes a pressure buildup at the disk brake, a problem that could be overcome by means of a drainage system [171].

In addition to the scheduled exchange of hydraulic oil explained previously, the present maintenance practices for the hydraulic system cover a visual check of its components to detect oil leakage, the exchange of filters as well as functional tests of the control valves, and the hydraulic pump during the regular service every 6 months.

7.3.4.5 Summary of Results from the RCAM Case Study

This case study presented an RCM analysis of the two wind turbine models V44-600 kW and V90-2MW. The RCM study has been performed in a workgroup involving a wind turbine owner and operator, a maintenance service provider, a provider of condition-monitoring services, and wind turbine component supplier as well as researchers at academia. The analysis forms the basis for the development of quantitative models for maintenance strategy selection and optimization.

Regarding the failure modes, causes, and underlying mechanisms, wide-ranging parallels between the considered wind turbine models are recognized. Compared to the V44-600 kW design, comprehensive additional failure-preventive measures have been found implemented in the V90-2MW series.

Among the failure causes identified in this study, vibration has been found to play a central role as a cause for mechanical failure of a variety of wind turbine components. Measures aiming at prevention or early detection of bearing damage are, therefore, concluded to be particularly effective. A second conclusion from this concerning maintenance modeling is that quantitative maintenance–optimization models that are based on the common assumption of independent failure of components run the risk of systematically underestimating the benefit of vibration monitoring.

Both the frequency and quality of service maintenance have been found to significantly impact the technical condition and failure rate of wind turbines. This underlines the important role of maintenance besides the inherent, design-, and quality-dependent reliability of a wind turbine for its operating availability. A perceived challenge in this context, the lack of experienced technicians resulting from the rapid increase in wind power installations, should be tackled by enhanced training of maintenance personnel. Another step considered important by the RCM workgroup is to overcome the present common practice to aim maintenance decisions at short-time cost savings instead of a long-term minimization of LCC. This can be achieved by developing and introducing data-based decision support tools in maintenance management. The mathematical models for maintenance strategy selection and optimization that are being developed based on the presented RCM study aim at contributing to this process. However, the amount and depth of wind turbine failure and maintenance data available today are insufficient for the application of sophisticated quantitative methods in practice. The standardized and automated collection of in-depth failure and maintenance data is, thus, concluded to be an important step in order to enhance the reliability, availability, and profitability of wind turbines.

7.4 RCAM Stage 3 LCC Analysis for Wind Power Systems

7.4.1 Impact of Using Condition Monitoring for PM

7.4.1.1 Introduction

This case study proposes an LCC model for investigating the benefits of using condition monitoring for PM of wind power systems. Moreover, it compares two different wind turbines one located offshore and one onshore.

The case studies are based on real data from Olsvenne2 at Näsudden (Gotland, Sweden) and Kentish Flats, in the U.K. Interviews have been performed within the case study specifically providing information of the operation data. The RCM project has involved persons from industry and academia and two study visits. From the industry, the following persons have contributed to the study: the Operation Manager at Vattenfall, Gotland, the Project Team Manager of Turbines at Kentish Flats Offshore Wind Farm, Elsam Engineering, Denmark, and the Managing Director on SKF Condition Monitoring Center.

The main conclusion is that CMS benefits maintenance management of wind power systems. Improvements can be especially shown for offshore wind farm maintenance planning. The case study is discussed in detail in [172].

7.4.1.2 The Proposed LCC Model

The LCC calculates the total cost for a technical system during its lifetime. Section 3.7 presents theories for LCC and economic calculations.

The power plant financial concerns could typically be investment, maintenance, downtime, and remainder value. Within this study, it has been assumed that the total LCC (in euros per year) is defined as

$$LCC = C_{inv} + C_{CM} + C_{PM} + C_{PL} + C_{rem} \tag{7.2}$$

where C_{inv} is the cost of the investment, C_{CM} is the cost for CM, C_{PM} is the cost for PM, C_{PL} is the cost for production loss, and C_{rem} is the remainder value.

For this analysis, the present value method is used. The present value of one outlay C to be paid after n years is gained by multiplying it with the present value factor $PV_f(n, d)$ as

$$PV = CPV_f(n, d) \leftrightarrow C(1 + d)^{-n} \tag{7.3}$$

where C is the outlay in (euros), n is the number of years from the present to the date of the outlay, and d is the discount rate, where, for example, a discount rate of 7% gives $d = 0.07$.

7.4.1.3 System Description and Input Data

This case study describes how operation and maintenance are performed at two wind farms owned by two different companies:

1. Näsudden, Gotland, Sweden, owned by Vattenfall
2. Kentish Flats, U.K., owned by Elsam

The following maintenance process description is resulting from discussions within the RCM study reference group including the Operation Manager at Vattenfall at Näsudden and the Project Team Manager at Kentish Flats for Elsam.

Below is a description of the two case studies in terms of system data, maintenance contracts, scheduled maintenance, manuals, and CMS.

7.4.1.3.1 System Data

At Näsudden, more than 100 wind turbines are installed, owned by different companies. Vattenfall owns about 30 of those wind turbines from different manufacturers. Some of the turbines at Näsudden are part of research projects and are prototypes. Here, the turbine Olsvenne, a Vestas V90, has a CMS developed by Vestas called VCMS. This is a prototype, and data from this system cannot be accessed by the personnel of Vattenfall, only by Vestas. Vattenfall has wind turbines in other places in Sweden, for example, on the West Coast and one in Lappland, and there are plans for several new farms to be built mainly offshore. They also have activities in other countries in Europe such as Germany, Poland, and Denmark.

The operation of Kentish Flats, an offshore farm with 30 × 3-MW turbines placed in the North Sea, U.K., has recently been taken over by Elsam, which also operates several turbines and farms in Denmark. Kentish Flats was completed in September 2005. The majority of Elsam's wind turbines in Denmark were taken over by Vattenfall in the summer of 2006. The CMS system VCMS has been installed at Kentish Flats in the summer of 2006. Elsam has access to data from the CMS on this site.

7.4.1.3.2 Maintenance Contracts

Vattenfall has contracts for maintenance with several companies. They have maintenance contracts with Vestas, Siemens, and Enercon. Maintenance contracts are valid for 1 year at a time and the decision has been made not to have any of the turbines insured. For all new farms under Elsam, a 5-year contract with full maintenance is signed with the manufacturer. This is in line with the 5-year warranty period on the wind turbines. For Kentish Flats, a 5-year contract was signed with Vestas.

7.4.1.3.3 Scheduled Maintenance

Today's wind power maintenance is mainly scheduled maintenance. At Vattenfall, scheduled maintenance is carried out approximately twice a year in general. The maintenance is categorized as either minor or major. Minor maintenance takes about 4 hours for two people and major maintenance takes about 7 hours for two people. One hour costs about 54 euros, that is the hourly cost of a maintenance worker from Vestas.

At Elsam, scheduled maintenance is carried out in 3–6 months intervals, for older and smaller turbines. Newer, larger turbines have intervals of 6 months to a year. Older turbines are maintained by Elsam's own maintenance division, Elsam Wind Service, which has 10–15 workers. For larger tasks, Elsam rents external independent services. At Kentish Flats, scheduled maintenance requires two people working 2 days per turbine per year, all at a cost of 750 euros per person per day if maintenance is scheduled, and a cost of 850 euros per person per day if it involves a major overhaul. These costs include all labor-related costs such as training, safety equipment, pensions, offices, transportation, and so on.

7.4.1.3.4 Maintenance Manuals

Vattenfall uses a maintenance manual by Vestas. The manual includes general rules, safety procedures, changing oil filters, and controlling leakages. At Elsam, there is a manual for scheduled maintenance that is followed when maintenance is carried out. For CM, there is no manual in use. Manuals are often of varying quality and as a result interest groups in Germany now want to establish standards for maintenance manuals.

7.4.1.3.5 Similarities and Differences between the Two Wind Farms

Similarities and differences between the two wind farms are listed in Table 7.9. The farms are listed according to number of turbines, CMS installed,

TABLE 7.9

Similarities and Differences between the Two Wind Farms

System	Number of Turbines	CMS	Access to Data	Scheduled Maintenance	Contract on Maintenance
1. Näsudden/ Vattenfall	30 different manufacturers	VCMS[a] on Olsvenne2	No	With half a year interval	1 year
2. Kentish Flats/Elsam	30 Vestas	VCMS[a]	Yes	With half a year interval	5 years with Vestas

Source: Nilsson, J. and Bertling, L., Maintenance management of wind power systems using condition monitoring systems–life cycle cost analysis for two case studies in the nordic system, *IEEE Transactions on Energy Conversion*, **22**(1), 223–229 © 2007 IEEE [172].

[a] Vestas CMS.

whether the company has access to data or not, how often scheduled maintenance is carried out and valid time on maintenance contracts.

7.4.1.4 LCC Analysis

7.4.1.4.1 Overview of the Approach

An LCC analysis has been performed for different scenarios comparing the effect of implementing different maintenance strategies. These strategies are based on the assumption that a CMS is implemented for two different case studies. First, one turbine onshore at Näsudden, and second, one average turbine at the offshore farm Kentish Flats.

At first, three strategies where PM/CM is affected are presented. Then, two strategies where only CM is affected are presented. Finally, one strategy in which only the production loss is affected is presented. The last three strategies are only carried out for the offshore farm while the first three strategies are carried out both for the farm offshore, where an average turbine is observed, and the single turbine onshore. The strategies are based on the following assumptions.

1. The cost of PM is compared to the cost of PM and a CMS. The cost of CM and the cost of production loss are set to zero.
2. The cost of PM/CM is compared to the cost of PM/CM and a CMS. The cost of production loss is set to zero.
3. The cost of PM/CM is compared to the cost of PM, when CM is set to zero, and a CMS. The cost of production loss is also set to zero.
4. Man-days are reduced and CM becomes preventive because of better planning with a CMS. The point at which a certain amount of unscheduled maintenance becomes preventive is determined in order to see when a CMS is profitable.

5. Man-days in the cost of major component replacements are reduced when the components are replaced two at a time instead of one at a time.
6. Availability increases and the cost of production loss is thereby decreased.

The purpose of these strategies is to evaluate the benefit of implementing CMS. In the first three strategies, a percentage for how much the cost must be decreased to make CMS profitable is calculated. The last three strategies give a direct indication of whether the price of a CMS is covered when implementing the strategy.

Possible benefits with CMS are that maintenance could be planned better, the right maintenance can be carried out at the right time, and unnecessary replacements can be minimized [173]. Downtime could be reduced as failures are discovered more easily.

The costs that could be saved with a CMS include costs for transportation to a wind farm offshore. Repairs can be planned in a new way. If one gearbox needs repair, then another gearbox that might go down at a later stage could be repaired at the same time, thanks to the CMS information. This saves money for transportation. On Kentish Flats, it is assumed that the installation of a CMS after a certain period of training time should allow both scheduled maintenance and CM to be optimized. The replacement of major components is of special interest here. The cost of a man-day includes all labor-related cost including training, safety equipment, pensions, offices, and transportation. In the two strategies where man-days are reduced at the farm because of better planning with a CMS, the difference in the total sum of the present value of LCC before and after is calculated.

7.4.1.4.2 Approach for the LCC Analysis

The sum of the costs for each year's LCC is calculated and multiplied with the present value factor $PV_f(n,d)$. This gives the present value for the sum of LCC, PV(LCC) for each year, which is then summarized over the years as the total sum of present value for the LCC, TPV(LCC). This is at first done without a CMS.

The cost of a CMS is then added and the effect of different strategies is calculated. This is first done in three strategies where one wind turbine onshore versus one average wind turbine offshore is observed. Then, one wind farm offshore where maintenance is planned using a CMS is observed. Finally, conclusions are drawn on the possible benefits of using a CMS.

The cost of a CMS is a part of the cost of PM (C_{PM}). The cost of replacing major components is a part of the cost of CM (C_{CM}). C_{PM} could be divided into a cost for a CMS (C_{CMS}) and a cost for scheduled maintenance (C_S).

$$C_{PM} = C_{CMS} + C_S \tag{7.4}$$

The cost of PM and CM ($C_{PM} + C_{CM}$) and a CMS is

$$C_{CMS} + C_S + C_{CM} \tag{7.5}$$

The LCC has been made for different cases with a cost for CMS. The cost for production loss is C_{PL}. The first three strategies are presented below.

Strategy 1: C_{CM} and C_{PL} are set to zero and C_{PM} is decreased so that TPV for the decreased C_{PM} with a CMS equals TPV of C_{PM} without a CMS

$$TPV(C_{CMS} + A_1 C_S) = TPV(C_S) \tag{7.6}$$

where A_1 is a factor that decreases C_S.

Strategy 2: C_{PL} is set to zero and $C_{PM} + C_{CM}$ is decreased, so that TPV for the decreased $C_{PM} + C_{CM}$ with a CMS equals TPV of $C_{PM} + C_{CM}$ without a CMS

$$TPV(C_{CMS} + A_2(C_S + C_{CM})) = TPV(C_S + C_{CM}) \tag{7.7}$$

where A_2 is a factor that decreases $C_S + C_{CM}$.

Strategy 3: C_{PL} is set to zero and C_{CM} is at first set to zero, so that C_S within C_{PM} with a CMS can be increased so that TPV of an increased C_S within C_{PM} with a CMS equals TPV of $C_{PM} + C_{CM}$ without a CMS

$$TPV(C_{CMS} + A_3 C_S) = TPV(C_S + C_{CM}) \tag{7.8}$$

where A_3 is a factor that increases C_S.

C_{CM} can be divided into cost for unscheduled service C_{US} and cost for replacing major components C_{MC}

$$C_{CM} = C_{US} + C_{MC} \tag{7.9}$$

C_{US} can be divided into a cost for man-days (C_M) and a cost for spare parts (C_{SP})

$$C_{US} = kC_M + C_{SP} \tag{7.10}$$

where k is the number of man-days.

C_{MC} can be divided into a cost for man-days (C_M) and a cost for replacing major components (C_R)

$$C_{MC} = kC_M + C_R \tag{7.11}$$

where k is the number of man-days.

C_{CM} has been compared for different scenarios. The next two strategies where the man-days are affected are presented below.

Strategy 4: In C_{US}, kC_M is decreased because the man-days can be reduced in unscheduled service when we plan it better with a CMS. Unscheduled

TABLE 7.10

Costs for Replacements of Major Components

Component	Cost [EUR]
Gearbox	300,000
Generator	150,000
Transformer	100,000
Blade	200,000

Source: Nilsson, J. and Bertling, L., Maintenance management of wind power systems using condition monitoring systems–life cycle cost analysis for two case studies in the nordic system, *IEEE Transactions on Energy Conversion*, **22**(1), 223–229 © 2007 IEEE [172].

service becomes preventive. The point B_1 at which a certain amount of unscheduled service becomes preventive is found to see where a CMS is profitable

$$C_{US} = kC_M, U_S + C_{SP} \text{ becomes } C_{US} = kC_{M,S} + C_{SP} \tag{7.12}$$

Strategy 5: The major components are replaced two at a time instead of one at a time because of a CMS. This means that in C_{MC}, k becomes $k/2$ and kC_M becomes $kC_M/2$. $C_{MC} = kC_M + C_R$ becomes

$$C_{MC} = (kC_M/2) + C_R \tag{7.13}$$

The cost for a CMS affects the production loss. The final strategy in which the availability increases is presented below.

Strategy 6: If availability increases to B_2 from 97.5%, the total sum of the present value of the cost increases by $TPV(LCC)_{Before} - TPV(LCC)_{After}$ because of a CMS.

7.4.1.4.3 Application

The analyses have been made based on the estimates of the costs for the single turbine onshore, Olsvenne2, and the farm offshore, Kentish Flats.

The task of estimating costs is difficult especially as prices can vary a lot. Table 7.10 shows costs of different major components. The assumed investment cost for one CMS is 20,000 euros.

The present value of the total LCC for every year has been calculated, which are then summarized as the total sum of the present value of LCC. The discount rate d is set to 7% and the life of a turbine N is 20 years.

The total sum of the present value of LCC has been calculated in different strategies. In strategy 1, PM must decrease by a certain amount to make a

CMS profitable. In strategy 2, PM and CM must together decrease by a certain amount to make a CMS profitable. In strategy 3, PM with a CMS could increase by a certain amount when CM is zero because of a CMS. In strategy 4, the cost for man-days kC_M is decreased as unscheduled service becomes preventive, that is, the cost for man-days decreases from 1000 euros per man-day into 750 euros per man-day. The point where a certain amount of unscheduled maintenance becomes preventive is found to see where a CMS is profitable. In strategy 5, major components are replaced two at a time and this means that the man-day k is reduced to $k/2$ in the cost for replacements C_{MC}. In the final strategy 6, where production loss is affected, availability increases into a certain percentage. The cost for production loss is thereby decreased.

7.4.1.4.4 Input Data for Olsvenne2

The investment cost is placed in year zero. Connection cost is included in the investment cost. PM and CM for 2 years are included in the investment cost. There is also a cost for basic maintenance materials in both years. These costs are included in C_{PM} and C_{CM} for years 1 and 2. C_{PM} and C_{CM} from year 3 are estimated for each year. A cost for production loss has been calculated as income times unavailability from year 1 and is called CPL. Commercial availability, that is, 97.5%, has been used in this calculation. Unavailability is calculated as 1-availability. A cost for demolition and recycling C_{Rem} has been estimated for the last year. Costs for major component replacements have been calculated. The gearbox is changed twice during a lifetime. The generator is changed once during a lifetime. The transformer and blades are changed 1/10th time during a lifetime. These costs are combined under cost of replacements C_R, according to

$$C_R = C_{R,GB} + C_{R,G} + C_{R,T} + C_{R,B} \tag{7.14}$$

where $C_{R,GB}$, cost of replacing gearboxes for a year; $C_{R,G}$, cost of replacing generators for a year; $C_{R,T}$, cost of replacing transformers for a year; $C_{R,B}$, cost of replacing blades for a year.

The investment cost in one CMS is C_{CMS} in year 0. $C_{CMS} = 20{,}000$ euros. The input data include no CMS and are for one single turbine.

7.4.1.4.5 Input Data for Kentish Flats

The cost of the first year is the investment cost, which includes connection cost. From years 1 to 20, costs for scheduled service are included and are the same every year. Major overhaul is included in scheduled maintenance for most of the years but is entered as a cost in years 4, 6, 9, 13, and 14. This is included in C_{PM}.

Unscheduled maintenance costs decrease from year 1 for 4 years and then increase at the end. They include one visit to each turbine every month in year 1, reduced to one visit every 3 months in year 4. This is included in C_{CM}. A cost for production loss (C_{PL}) has been calculated as income times unavailability from year 1. These costs are calculated together. Here, commercial availability has been used, that is, 97.5%. Unavailability is calculated as 1-availability.

The major component replacements have been estimated per year for 20 years. For example, 11 gearboxes are to be replaced out of 30 during the turbines lifetime, 20 generators are to be replaced during a lifetime, 3 transformers and 3 blades are to be replaced out of 30 during a lifetime, and so on. These costs are included under C_R, accordingto Equation 7.11, and are part of C_{CM}.

The investment cost in 30 CMS is 30 C_{CMS} in year 0. 30 C_{CMS} = 600,000 euros. One scheduled man-day costs 750 euros at Kentish Flats, $C_{M,S} = 750$ euros. One unscheduled man-day costs 1000 euros at Kentish Flats, $C_{M,US} =$ 1000 euros.

The number of man-days k is 43; this means that 43 components are replaced in 43 days including 11 gearboxes, 20 generators, 3 transformers, 3 blades, and so on. Thus 43 components will be replaced during the wind farm's lifetime. The input data include no CMS for the entire farm.

7.4.1.4.6 Results of the LCC Analysis

Results from the calculations of the two cases, Olsvenne2 and Kentish Flats, are presented in Figures 7.10 and 7.11. For Olsvenne2, the results of the total sum of the present value for the LCC and the strategies 1–3 are presented. For Kentish Flats, the results of the total sum of the present value for the LCC and the strategies 1–6 are presented. In strategies 4–6, the man-days at the farm are affected by a CMS and this changes C_{CM}.

1. *Olsvenne2*: The sum of the present value for 20 years, the total sum of the present value for the LCC, was about 3.72 million euros for Olsvenne2 when the discount rate was set to 7%. For present value cost every year as well as C_{PM}, C_{CM}, and C_R, see Figure 7.10.
2. *Kentish Flats*: For Kentish Flats wind farm, the total sum of the present value for the LCC would be about 4.25 million euros per turbine with the discount rate set to 7%. For present value cost every year as well as C_{PM}, C_{CM}, and C_R, see Figure 7.11.
3. *Summary of Results for Implementation of the Strategies*: The implementation of the different strategies shows several ways to cover the cost of a CMS, especially if the calculations are carried out for a large-scale wind farm offshore. For the separate turbine, CM has to be included, but if only the PM is included, it has to be decreased significantly. Table 7.11 shows the changes needed to cover the cost of a CMS in the separate turbine onshore versus an average turbine offshore, and

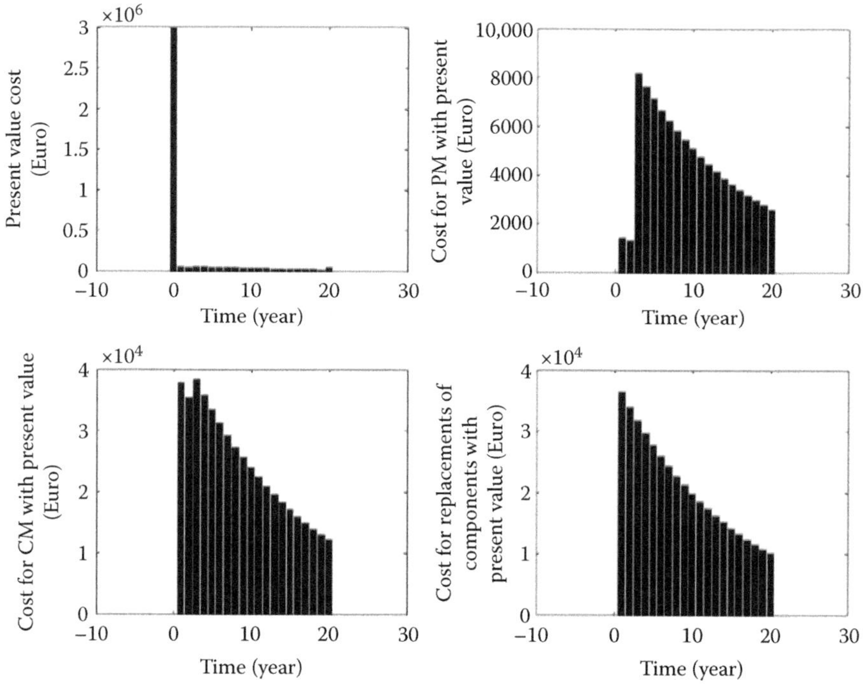

FIGURE 7.10
Results of the LCC analysis for Olsvenne2: present value cost and C_{PM}, C_{CM}, and C_R of major components with replacements every year for Olsvenne2. The cost of replacements of major components is included under C_{CM}. The discount rate was set to 7%. (Nilsson, J. and Bertling, L., Maintenance management of wind power systems using condition monitoring systems–life cycle cost analysis for two case studies in the nordic system, *IEEE Transactions on Energy Conversion*, **22**(1), 223–229 © 2007 IEEE [172].)

how a CMS is justified when maintenance could be better planned for the wind farm offshore.

7.4.1.4.7 *Conclusion*

The LCC analysis has been carried out to gain an understanding of whether a CMS is profitable for the separate wind turbine onshore and the wind farm offshore. The LCC is compared for different alternative maintenance strategies.

The results from the first three strategies show that for a single turbine, PM would have to decrease by 23% if only PM is affected. The result for the farm was 4.5%. If CM is also affected, PM and CM together would not have to decrease by more than 3.5% for the single turbine and 2.5% for the farm. A decrease in CM is needed to justify a CMS, at least for the turbine onshore.

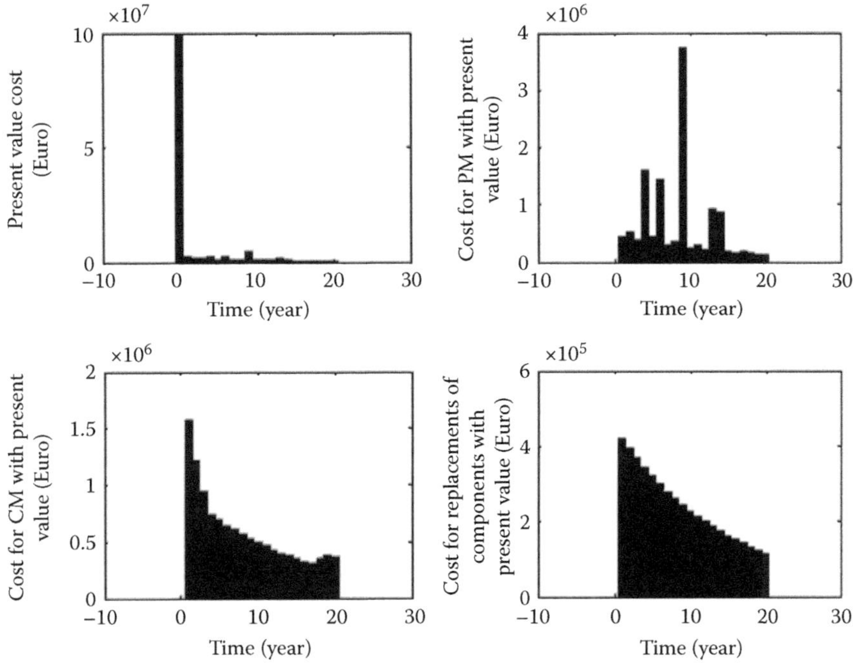

FIGURE 7.11
Results of the LCC analysis for Kentish Flats: present value cost and C_{PM}, C_{CM}, and C_R of major components with replacements every year for Kentish Flats Wind Park. The cost of replacements of major components is included under C_{CM}. The discount rate was set to 7%. (Nilsson, J. and Bertling, L., Maintenance management of wind power systems using condition monitoring systems–life cycle cost analysis for two case studies in the nordic system, *IEEE Transactions on Energy Conversion*, **22**(1), 223–229 © 2007 IEEE [172].)

When an entire farm was observed, the conclusions were that a change from scheduled to unscheduled maintenance of 47% would be enough to make a CMS profitable. Replacing components two at a time instead of one at a time would not be enough to make CMS profitable, but it would have a significant impact. Availability would not have to be increased by more than 0.43% to reduce the cost of production loss to cover the cost of a CMS. There are many ways to cover the cost of a CMS, especially for the wind farm offshore where maintenance could be planned more efficiently with a CMS.

7.4.1.5 Introduction

This case study proposes the LCC analysis approach for maintenance management of wind turbines. The LCC approach for maintenance management presented in this case study is an extension on the case study presented in Section 7.4.2.5. The case study is discussed in detail in [174].

TABLE 7.11

Changes Needed to Cover the Cost of a CMS

Strategy	Onshore, 1 Turbine	Offshore, 1 Average Turbine
1. A change in preventive maintenance.	23%	4.5%
2. A change in preventive- and corrective maintenance.	3.5%	2.5%
3. How much preventive maintenance can increase if corrective maintenance is set to zero.	6.3 times	1.85 times
	Change at a Wind Farm, 30 Turbines	
4. How much of the corrective maintenance needs to be preventive maintenance.	47%	
5. How much can be saved by replacement of two at a time.	EUR 230,000	
6. How much availability must increase.	0.43%	

Source: Nilsson, J. and Bertling, L., Maintenance management of wind power systems using condition monitoring systems–life cycle cost analysis for two case studies in the nordic system, *IEEE Transactions on Energy Conversion*, **22**(1), 223–229 © 2007 IEEE [172].

The case studies are presented based on the data from three different wind turbines rated 3 and 6 MW. Three different scenarios have been studied and the effect of CMS has been analyzed. For any chosen value, the CMS proves to be a profitable option.

An LCC model has been proposed including increasing failure rates of components. Furthermore, two failure rate models have been investigated and the effect of CMS on the maintenance and on the LCC has been studied for both the failure rate models.

7.4.1.6 Theory and Concepts

The proposed LCC model is similar as presented in Section 7.4.2.5 and uses the definitions that are presented in Section 3.7.

The modeling of the maintenance planning has been made using concepts as presented in Chapter 2. The PM by condition monitoring is in focus for the maintenance modeling. Specifically, the use of CMS for the wind turbines is studied. CMS can be defined as follows: "Activity performed either manually or automatically intended to measure at pre-determined intervals the characteristics and parameters of the actual state of an item. CMS is distinguished from inspection in that it is used to evaluate any changes in the parameters of the item with time. CMS may be continuous, over time interval or after a given number of operations" [175]. Techniques that are included in CMS are, for example, vibrations analysis and oil analysis.

7.4.1.7 Proposed Approach

In this case study, an approach for comparing different maintenance strategies using LCC is proposed. Considering the values C_{inv} and C_{rem} in Equation 1 are constant, LCC will directly be affected by the choice of the maintenance strategy. The LCC approach provides a useful tool in making long-term strategic decisions concerning operation and maintenance. The maintenance cost depends on the expected failures in the wind turbine components. Hence, the choice of failure rate model is crucial. Here, two basic failure rate models are used to demonstrate the application of the proposed approach.

The two different models used are a base model where the frequency of failure of each component is constant during the life time of the wind turbine and an aging model where the frequency of failure increases linearly every year because of the aging of components. The aging model is based on the intuitive understanding that the probability of failure for components increases with age due to degrading, stress during abnormal operating conditions, and other factors. Following this argument, it is assumed that the change in failure rate due to aging is seen only in mechanical components and that the failure rate of electrical components is not affected by age. Hence, in both the base and aging models the failure rate for electrical components is assumed constant. The difference between these two models is that in base model the average failure rate is a result of the applied PM while in the aging model the PM is carried out based on the value of the failure rate.

Within both base and aging models, three different scenarios have been carried out.

1. *Scenario 1*: LCC analysis is done considering no CMS. The PM is only the time-based scheduled maintenance applied to the wind turbine.
2. *Scenario 2*: LCC analysis is done considering CMS with an efficiency of 80%. It is considered that for components with CMS, the faults can be predicted in advance in 80% of cases. For these 80% of cases, PM is carried out when the CMS alarm is received. The PM could either be a repair action or replacement of the effected component. Hence, this scenario considers that 80% of CM is replaced by PM.
3. *Scenario 3*: LCC analysis is done considering 100% efficiency for CMS. Hence, in this case 100% CM is replaced by PM. However, failures of components, which are not monitored by CMS, are corrected by CM.

These three scenarios represent three different maintenance strategies. To consider CMS in the LCC analysis, the cost is added to the investment cost of each turbine. Here, the cost of CMS is considered dependent on efficiency, which means the CMS with higher efficiency is more expensive. Application of CMS is modeled as change from CM to PM in this approach. Following assumption has been made in line with [172]; when PM is carried out, maintenance activities are planned in advance, CMS supplies information to predict a failure, ideally, 3 months in advance and during this time the maintenance is planned and the necessary tools can be reserved. Further, PM reduces the longer downtimes, generally, associated with CM, in case of failure of critical components. The reduced downtimes will influence the production losses.

The sum of the costs for each year is calculated taking into account the present value factor and the present value of LCC of each year is summed to calculate the Total Present Value of LCC (TPV(LCC)). This is the final value that will be used to compare the different scenarios for the three types of WT systems.

7.4.1.8 Case Study

7.4.1.8.1 System Description

Case studies are done for three different WTs: the first one is an offshore WT with a rating power of 3 MW; the second is an offshore WT with rating power of 6 MW and the third an onshore WT with rating power of 3 MW. The service contract plays an important role on the evaluation of LCC. Through these contracts, the manufactures and owners decide the time duration during which the costs of maintenance are covered by the manufactures. After the end of the contract period, the maintenance costs are paid by the owner. In the following case studies, a 5-year contract with the manufactures is considered for the offshore wind turbines (WT1 and WT2) and a 2-year service contract is considered for the onshore wind turbine WT3. These assumptions are based on the general practice prevalent currently.

7.4.1.8.2 Maintenance Activities and Turbine Components

In this case study, each WT has been considered with 17 different components, each one with a different failure rate. Different maintenance activities have been taken into account:

1. Inspection
2. Small repair (4 or 8 hours)
3. Major repair (12, 16, or 24 hours)
4. Major replacement (12, 24, 36, 48, or 60 hours)

These maintenance activities present different work times. For instance, a "small repair" can take 4 or 8 hours of work time. A "major repair" can take 12, 16, or 24 hours, and "major replacement" can take 12, 24, 36, 48, or 60 hours of work time depending on the type of failure, component, and turbine. Maintenance activities for offshore wind turbine might include the use of crane ship. The cost of shipping is then added to the basic cost of the maintenance.

Different maintenance activities have been selected for each component of the turbine considering the failure probabilities for different types of failures. It is assumed that different failures will lead to one of the four maintenance actions, listed above. This selection has been made on data regarding each WT. Each turbine has been considered composed of following 17 components:

1. Gearbox
2. Generator
3. Brake system
4. Main shaft/bearings
5. Yaw system
6. Cooling system
7. Composite disc coupling
8. Sensors
9. Crane system
10. Hydraulic and oil systems
11. Rotor blades/hub
12. Pitch system
13. Rotor structure
14. HV component
15. Electrical system
16. Structure parts
17. Accessories

7.4.1.8.3 Application of CMS

In the case studies, second and third scenarios take into account the use of CMS. CMS is generally applied to critical components of wind turbines. Based on the real data, in all the three WTs the CMS is applied on eight specific components:

1. Gearbox (WT1 & WT3)
2. Generator
3. Main shaft
4. Yaw system
5. Blades

6. Pitch system
7. Rotor structure
8. HV component

For the components with CMS, replacement of CM by PM is considered in scenarios 2 and 3. However, for the components without any CMS, the CM is considered unaffected.

The CMS has an initial cost that is added to the investment cost in year 0. The cost of a CMS with an efficiency of 80% is considered 20,000 euros while a CMS with an efficiency of 100% is considered to be 26,000 euros.

7.4.1.8.4 Preventive Maintenance

The PM in the second and third scenarios has been calculated as follows:

$$C_{PM} = \alpha_1 \cdot f_r(t,c) \cdot CMS_{eff} \tag{7.15}$$

$$\alpha_1 = (N_t \cdot N_{hr} \cdot C_{mh,pm} + C_{sp,pm} + C_{cs,pm} \cdot N_{hcp}) \tag{7.16}$$

where N_t, number of technicians [pers]; N_{hr}, number of hours needed to repair a specific component [hr]; $C_{mh,pm}$, the cost of scheduled man per hour [euro/hr · pers]; $C_{sp,pm}$, the cost of spare parts due to a replacement [euros]; $C_{cs,pm}$, the cost of the crane ship per hour [euro/hr]; N_{hcs}, the number of hours of crane ship use [hr]; $f_r(t,c)$, the probability of failure of a specific component during a given time interval in this case a year; CMS_{eff}, efficiency of condition monitoring system.

In the first scenario, where CMS is not used, PM is considered to be time-based scheduled maintenance and the cost is fixed at 6000 euros per year for WT1 and WT3 and it is fixed at 8000 euros per year for WT2.

7.4.1.8.5 Corrective Maintenance

The CM has been calculated as

$$C_{CM} = \alpha_2 \cdot f_r(t,c) \cdot (1 - CMS_{eff}) \tag{7.17}$$

$$\alpha_2 = (N_t \cdot N_{hr} \cdot C_{(mh,cm)} + C_{(sp,cm)} + C_{(cs,cm)} \cdot N_{hcp}) \tag{7.18}$$

where $C_{mh,cm}$, the cost of unscheduled man per hour [euro/hr · pers]; $C_{sp,cm}$, the cost of spare parts due to replacement [euro]; $C_{cs,cm}$, the cost of the crane ship per hour in case of CM [euro/hr]; $(1 - CMS_{eff})$, the inefficiency of the condition monitoring system.

7.4.1.8.6 Production Losses

Due to failures and resulting maintenance actions, the WT unit has to be shut down for a certain time period. This shutdown period causes loss of production, which eventually translates into monetary loss. This economic loss is considered in the LCC analysis as cost of production losses. This is an extension to the approach presented in [172], as in the former a constant value for production loss was assumed for each year. The cost of production losses has been evaluated as follows:

$$C_{PL} = \alpha_3 \cdot N_{hr} \cdot P \cdot C_f \cdot C_{el} \tag{7.19}$$

$$\alpha_3 = f_r(t, c) \cdot (CMS_{eff} \cdot 0.85 + (1 - CMS_{eff})) \tag{7.20}$$

where $\alpha_3 \cdot N_{hr}$, the total downtime [hours]; P, the electric power that could have been produced [kWh]; C_f, the capacity factor; C_{el}, the cost of electricity [euro/kW].

The proposed method to calculate the production losses considers the probability of failure of the components, downtime associated to failure of each component, and models the effect of CMS on the expected downtime. As seen in Equations 7.7 and 7.8, it is assumed that an indication of failure from the CMS will reduce the downtime by 15%.

7.4.1.9 Results from Application Studies

The results from the study show that WT2 (highest rating power and offshore installation) has the highest value of TPV(LCC) and WT3 (lowest rating power and onshore installation) has the lowest for both base and aging models. This is due to maintenance being less expensive for an onshore installation as there are no shipping costs and spare parts transportation costs. However, in all the three WTs, the value of TPV(LCC) decreases when CMS is considered in the analysis. Hence, it can be concluded that CMS is beneficial for the three types of wind turbines analyzed.

Figure 7.12 shows the cost reduction gained from scenario 1 to scenario 3. For instance in the base model, the TPV(LCC) of WT2 decreases by 1.5% in the scenario 2 and by 2.4% in scenario 3 compared to scenario 1, where the CMS is not used. This decrease indicates the total saving that can be gained with the use of CMS during the entire lifetime of the WTs.

The comparison of the three scenarios for WT2 for the aging model is shown in Figure 7.13. The costs for scenario 1 are higher for the entire operating life of the turbine. It can be observed that scenario 2, with CMS efficiency of 80%, gives lower cost and scenario 3, with CMS efficiency of 100%, gives the lowest yearly cost. The comparison in Figure 2 does not include year 0, this is done so that the presented diagram is easy to read.

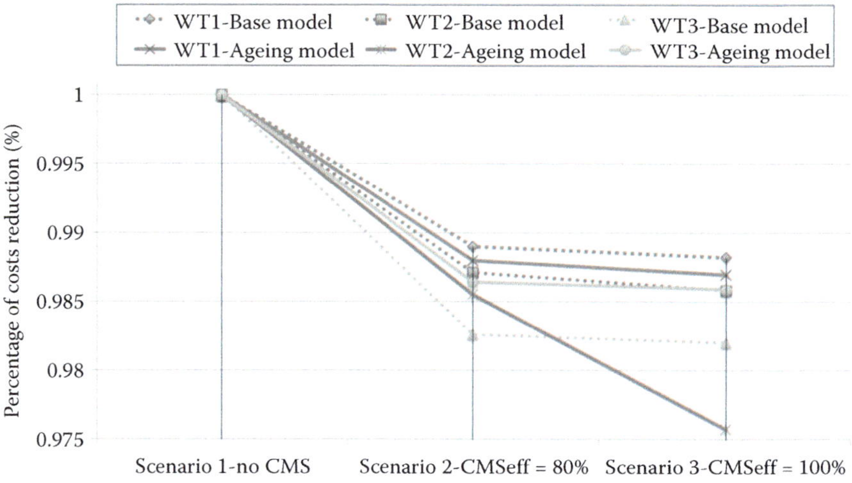

FIGURE 7.12
Cost reduction from strategy 1 to strategy 3. (Puglia, G. et al., Cost efficient maintenance strategies for wind power systems using lcc, in PMAPS 2014, IEEE, Durham © 2014 IEEE [174].)

7.4.1.9.1 Sensitivity Analysis

Discount rate affects the value of TPV(LCC). As the discount rate for future is uncertain, it is of interest to understand the effect of its variation on the final results. Hence, a sensitivity analysis has been carried out for different discount rates. The total sum of TPV(LCC) has been evaluated with three different discount rates: 0%, 7%, 10% for the three scenarios and for each WT considering the aging model. As shown in Figure 7.14, TPV(LCC) decreases when the discount rate increases. It can be observed from the figure that the profitability of CMS decreases with increasing discount rate. Hence, CMS becomes more attractive with lower discount rates.

In Figure 7.14, for the purpose of presentation the values are normalized to the TPV_LCC of scenario 1 at 0% discount rate.

7.4.1.9.2 Comparison with the First Case Study

In Section 7.4.2.5, two WTs were analyzed, a single onshore WT of 3 MW power rating and an average offshore WT of 3 MW power rating. This approach, with the suggested modification, has been used with available data for a single onshore 3-MW WT, a single offshore 3-MW WT. Further, the approach is also applied to a single offshore WT of 6-MW rating power. Hence, a comparison has been done between the results presented in this case study and those obtained from Section 7.4.2.5. Table 7.12 shows the conclusion from the case study in Section 7.4.2.5, for an average offshore WT of 3 MW and in an offshore WT of 3 MW, considered here.

In Table 7.12, the first strategy shows the minimum decrease in cost for maintenance (both PM and CM) to make CMS profitable. Strategy 2 presents

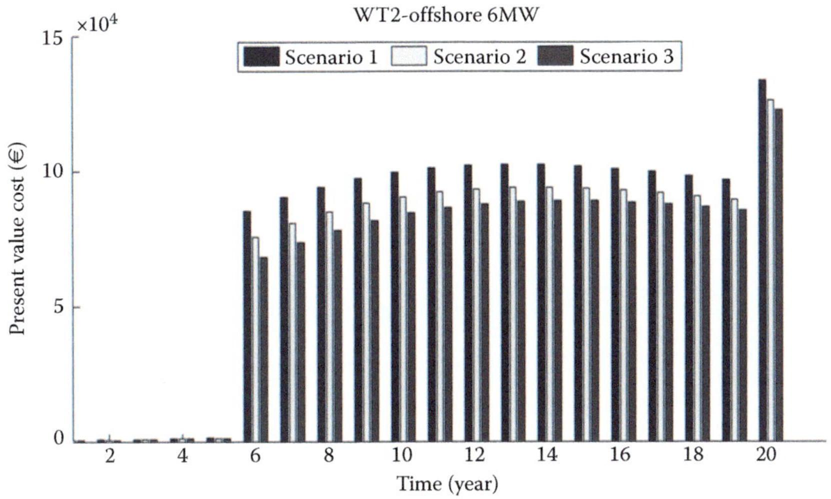

FIGURE 7.13
Comparison between scenarios 1, 2, 3 applied to WT2 in the aging model from year 1. (Puglia, G. et al., Cost efficient maintenance strategies for wind power systems using lcc, in PMAPS 2014, IEEE, Durham © 2014 IEEE [174].)

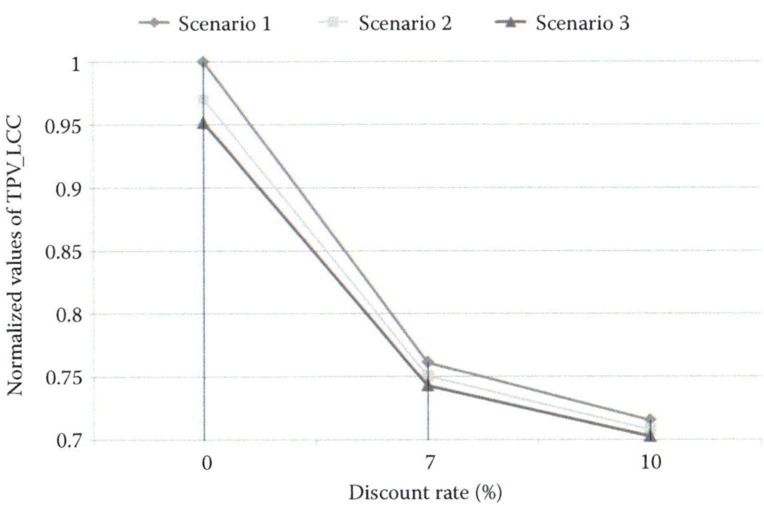

FIGURE 7.14
Profitability of use of CMS for different values of discount rate. (Puglia, G. et al., Cost efficient maintenance strategies for wind power systems using lcc, in PMAPS 2014, IEEE, Durham © 2014 IEEE [174].)

TABLE 7.12

Comparison between the LCC Case Studies

Strategy	LCC 2007	LCC – 2013 Updated
How much the cost of PM+CM has to decrease	2.5%	1%
How much of the CM needs to be PM	47%	27.5%

Source: Puglia, G. et al., Cost efficient maintenance strategies for wind power systems using lcc, in PMAPS 2014, IEEE, Durham © 2014 IEEE [174].

the amount CM that should be replaced by PM in order to make CMS profitable. For these calculations, PL is set to zero [172]. The results from Section 7.4.2.5 show that the sum of CM and PM has to decrease by 2.5% for CMS to be profitable. However, in the current case, the sum of PM and CM has to decrease by 1% to make CMS profitable. The difference in the results is mainly due to two reasons: difference in input data and evaluation of PM and CM based on the failure rate of the most critical components in this current work. When the CM is partially replaced by PM, a change from unscheduled service to schedule service occurs. In the case study in Section 7.4.1, a change from unscheduled to scheduled maintenance of 47% is deemed necessary to make CMS profitable. In this study, the results suggest that CMS will be profitable if the unscheduled service decreases by 27.5% and becomes scheduled.

The comparison between these two studies is considered appropriate, since data used are similar and the LCC analysis takes into account the same type of costs. An implementation of the approach, in this work, includes the evaluation of cost of PM, CM, and PL based on linearly increasing failure rates of components and the effect of CMS. The effect of efficiency of CMS has been considered. Furthermore, within this study a new type of WT has been considered: TPV(LCC) and its changes due to CMS have been calculated for an offshore WT of 6 MW and results show that due to higher power rating, the profitability of CMS increases.

8

RCAM Case Reliability and Maintenance Component Modeling

8.1 Hazard Rate Estimation and Condition-Based Maintenance of Electrical Contacts

8.1.1 Introduction

Infrared thermography is the most common way to assess the condition of high-voltage electrical contacts in electrical distribution and transmission systems. However, thermography has one major drawback as contacts carrying low or no load may not be assessed. This case study proposes a method that solves this issue. A full presentation of the case study can be found in [176].

The proposed method make use of thermography measurements to estimate statistical distribution parameters for the time to failure for a population of electrical contacts. The statistical distributions may then be used to support maintenance decisions for the sometimes large proportion of contacts that may not be assessed directly due to low load. As the measurement results accumulate over the years, more accurate predictions can be made. The proposed method is illustrated using three test cases applied to a population of high-voltage disconnectors with randomly generated thermography measurements.

The main conclusion of this case study is that the proposed method will provide decision support for PM of high-voltage apparatus that may not be assessed directly by thermography, which may constitute a large proportion of the disconnector population. The aim of such a maintenance policy is to reduce the number of hidden contact failures in a disconnector population, hence reducing the risk of a catastrophic power system failure.

The benefit of the proposed method lies in its ability to not only use information on times to failure but also the information on how close to a failure a contact may be. The proposed method is practical because it does not require any new measurements to be made, only to save measurements from those inspections that are being made regularly. A benefit with the proposed method is that as the number of accumulated measurement results will go down. Consequently, any predictions made are expected to improve in time. The bias introduced by only using the measurements from the contacts with the highest load will most likely only have a marginal effect in practice since these contacts are less prone to overheating due to their low loads [176].

8.1.2 Electrical Contact Maintenance

8.1.2.1 Contact Maintenance

PM on electrical contacts has traditionally been carried out as time-scheduled maintenance. As new condition assessment techniques have been developed, more sophisticated maintenance strategies, such as CBM, have been implemented. This change has been caused by increased competition among electricity distribution companies as well as demands from the regulator. As more CBM has been introduced, the importance of infrared thermography has increased. Some companies have abolished scheduled PM altogether and will only take certain equipment offline if a failure or degradation has been indicated by a thermography measurement.

The only two parameters that may be modified in order to increase electrical contact availability are the load current and thermography inspection frequency [3], when basing all contact maintenance on thermography. As a consequence, companies have been increasing the thermography inspection frequency. If the load is too low, however, increasing the inspection frequency will only improve the contact availability marginally and will thus be very inefficient.

8.1.2.2 Thermography

Infrared thermography is the most common method to assess the condition of electrical contacts. It is a nondestructive, noncontact technique, which can be applied on high-voltage electrical equipment online. The thermography camera measures and images the infrared radiation from the object in a specified spectral band. Because of its ease of use and low cost, it has become the preferred method for assessing equipment condition online in many electrical transmission and distribution systems around the world.

A widely used method of using thermography on electrical equipment is by employing the delta-T criteria [177]. This is a qualitative method of estimating the maintenance priorities by using tables of temperature ratings to assess the severity of overheating the equipment [177].

A main drawback of using thermography on electrical contacts is that it is heavily dependent on the load current [177]. A commonly used rule of thumb states that any measurement made at less than 30% of the rated load current is considered too inaccurate [178]. This has also been verified by Monte Carlo simulations [179].

Nevertheless, a survey investigated at which load all thermography measurements were made in the Swedish transmission system (220–400 kV) during 2006. The result of the survey was that more than 50% of all thermography measurements were made at a load current of less than 30% of the rated load [176].

8.1.2.3 Effect of Maintenance

An electrical contact that is found to be failed is assumed to be immediately replaced with an identical one, leaving the contact in an As Good As New (AGAN) state. A common failure mode for electrical contacts is wear of the silver surface, which is there to provide a low electrical resistance. When this surface is worn the contact resistance will increase and consequently the temperature rise. Any standard maintenance tasks such as cleaning and lubricating the contacts are assumed to leave the contact in a state that is conceptually As Bad As Old (ABAO).

8.1.3 Failure Definition

In this study, an electrical contact failure is defined as a contact having a temperature rise greater than the maximum allowed temperature rise, X_{max}, when loaded with its nominal load current. The value for X_{max} can be found using a combination of values found in [180] and [181] as a norm, the maximum allowed temperature rise for silver coated contacts in air is then $X_{max} = 95\,\text{K}$ [179].

8.1.4 Hazard Rate Estimation Using Infrared Thermography

This section presents a method to save and make use of all measurement results to estimate statistical distribution parameters for the time to failure for an equipment population. Briefly, the method involves the following steps:

1. Collect thermography measurement results.
2. Clean data and discard poor quality measurements.
3. Sort measurements by contact age into groups.
4. Fit a probability density function (p.d.f.) to the measurement results for each age group.
5. Calculate the probability of exceeding the maximum temperature rise (failure) for each age group.
6. Estimate the statistical distribution parameters by least squares fitting a reliability function to the probabilities from the previous step.

The following subsections describe the method in more detail.

1. *Collect measurement results*: The first step of the measurement collection procedure is to make sure that all measurements are normalized with respect to the nominal load of the contact, that is, calculate what the temperature rise would be at nominal load. This is done by [179]

$$X = X_m \left(\frac{I_n}{I_{load}}\right)^a \tag{8.1}$$

   where X_m [K] is the measured temperature rise when loaded with I_{load} [A], I_n [A] is the nominal load for the contact and 'a' is a dimensionless object-specific constant.
2. *Data cleaning*: Thermography measurements are heavily dependent on the load current and measurements made at too low load are not useful and must be discarded. A rule of thumb is that any measurement made at less than $0.3I_n$ is considered too inaccurate [178]. This has been verified by Monte Carlo simulations [179].
3. *Sorting by age*: After the data has been cleaned and only accurate measurements remain they must be sorted according to the contact age. This reordering does not lead to any loss of information if the times to failure are assumed to be independent and identically distributed [182].
4. *Fit p.d.f. to measurement results*: The next step is to fit a p.d.f. for all measurements made at a certain contact age. Let the random variable X_j be the calculated temperature rise at nominal current, using Equation 8.1, for contact $j = 1, 2, \ldots, k$, where k is the number of contacts of age t and $t = 1, 2, \ldots$ is describing the contact working age in calendar years. Let $\mathbf{X}_t$ be a vector $(x_1, x_2, \ldots, x_k)$ for contact age t. The p.d.f. describing the contact temperature rise, X, at contact age t is $f_t(x)$. The measurements in $\mathbf{X}_t$ may then be used to fit $f_t(x)$ to a statistical distribution using standard methods such as maximum likelihood estimation (MLE) [183]. The p.d.f. fitting procedure should be followed by a goodness-of-fit test to assess the fit of the model to the measurement set. It is assumed that the thermography measurement errors are symmetrically distributed with an expected value of zero.
5. *Calculation of failure probabilities*: The probability of a contact surviving up to time t is described by the reliability function. Hence, the probability of a contact surviving up to time t given the thermography measurements, $\mathbf{X}_t$, is then given by the conditional reliability function, $R_t(t|X_t)$

$$R_t(t|\mathbf{X}_t) = P(T > t|\mathbf{X}_t) = P(X \leq X_{max}) = \int_0^{X_{max}} f_t(x)\mathrm{d}x \tag{8.2}$$

   Provided that $x > X_{max}$, $t \to \infty$

6. *Estimating statistical distribution parameters*: The final step is to estimate the parameters for the statistical distribution for the conditional reliability function $\hat{R}(t|\mathbf{X}_t)$ by using the vector of probabilities from Equation 8.2 $\mathbf{R}(t|\mathbf{X}_t) = (R1(1|X1), R2(2|X2), \cdots)$. The estimated conditional reliability function $\hat{R}(t|\mathbf{X}_t)$ is found by fitting a reliability function to the probabilities in $\mathbf{R}(t|\mathbf{X}_t)$ by the Method of Least Squares (MLS). As described earlier, there is a need to perform a goodness-of-fit test in order to assess the fit of the model to the calculated values of the reliability function at different contact ages. The p.d.f. $f(t|\mathbf{X}_t)$ describes the time to failure for a disconnector contact, given the measurements in $\mathbf{X}_t$. The estimated p.d.f. $\hat{f}(t|\mathbf{X}_t)$ is found via the relationship [45]:

$$\hat{f}(t|\mathbf{X}_t) = -\frac{\mathrm{d}(\hat{R}(t|\mathbf{X}_t))}{\mathrm{d}t} \tag{8.3}$$

Subsequently, the estimated hazard function is found by [45]

$$\hat{h}(t|\mathbf{X}_t) = \frac{\hat{f}(t|\mathbf{X}_t)}{\hat{R}(t|\mathbf{X}_t)} \tag{8.4}$$

8.1.5 Illustration of Hazard Rate Estimation

The proposed method will be demonstrated by an example estimating Weibull parameters for the time to failure for the electrical contacts of horizontal center break high-voltage disconnector switches. The service role of a disconnector in an electrical network is to provide a visual open circuit in its open position and to carry the load current in its closed position. Since no real data exists to demonstrate the method, it is illustrated by a test case.

1. *Test case*: The test case population comprises 100 horizontal break disconnectors of different ages, see Figure 8.1. Every disconnector comprises six rotating terminal contacts that are measured via thermography once a year; the three center main contacts are not included in the example. In this illustration, this results in 600 measurement points. If a temperature rise above X_{max} is discovered, the contact is considered to be failed. The maintenance action for this failure mode is to replace the entire contact leaving it in an AGAN state. It is assumed that the saving of the measurements starts at an arbitrary point in time with an existing aged contact population, see Figure 8.1.

   In this illustration example, the contact temperature rises are randomly generated from three different statistical distributions in three test cases. The parameter values used are based on laboratory [150]

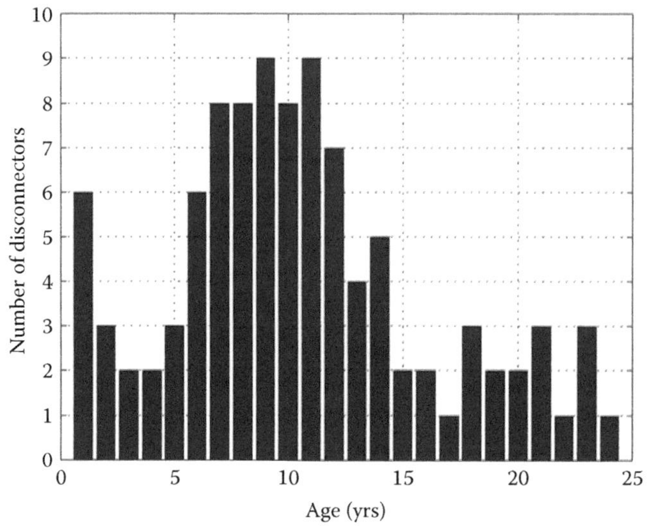

FIGURE 8.1
Test case disconnector population age profile. (Lindquist, T. and Bertling, L., Hazard rate estimation for high-voltage contacts using infrared thermography, in *2008 Annual Reliability and Maintainability Symposium*, pp. 231–237, © 2008 IEEE.) [176]

and field measurements [179] in order to generate realistic measurements. The parameters and distributions are chosen such that the methods sensitivity to the input data may be analyzed. Table 8.1 summarizes the three test cases. In brief, the three distributions are, in Case 1, a Normal distribution with a mean and variance increasing with time. In Case 2, the measurements are generated from a right-skewed Weibull distribution with a large increasing variance and finally, in Case 3 the measurements are generated from a left-skewed Weibull distribution with a small increasing variance. In

TABLE 8.1
Test Case Input Data

	Case 1	Case 2	Case 3
Distribution	Normal	Weibull	Weibull
Parameters	$\mu(t) = 50 + 2.8t$	$\eta(t) = 10 + 0.4t$	$\eta(t) = 20 + 0.4t$
	$\sigma^2(t) = 3.5 + 0.3t$	$\beta = 2$	$\beta = 10$
		$\gamma(t) = 40 + 2.2t$	$\gamma(t) = 40 + 2.2t$

Source: Lindquist, T. and Bertling, L., Hazard rate estimation for high-voltage contacts using infrared thermography, in *2008 Annual Reliability and Maintainability Symposium*, pp. 231–237, © 2008 IEEE. [176]

test cases 2 and 3, the three parameter Weibull, as defined in [178], is used.

For illustration purposes, it is assumed that all measurements were made at $>0.3I_n$ and thus no measurements will be discarded due to low load. X_{max} is set to 95 K (see Section 8.1.3) in all three cases.

Figure 8.2a–c illustrates the randomly generated temperature rise measurements, generated using the parameters in Table 8.1. The figure also shows the maximum temperature rise, X_{max}, as a solid line. The measurements are sorted by age and all points above the maximum temperature rise, X_{max}, are measurements of failed contacts. Figure 8.2a–c represents the outcome after carrying out steps 1, 2, and 3 from Section 8.1.4, that is, collecting, cleaning, and sorting. Figure 8.2a–c represents test cases 1, 2, and 3, respectively.

2. *Results*: This subsection shows the results after carrying out all six steps in Section 8.1.4 for the test case presented earlier. Figure 8.3 shows the resulting reliability function fitted to the survival probabilities for Case 1, calculated for each age. The results in Figure 8.3 follow steps 4, 5, and 6 in Section 8.1.4. In Figure 8.3, the circles represent the probabilities, $R(t|X)$, of exceeding the maximum temperature rise (failure) for each age, using Equation 8.2. The solid line in Figure 8.3 represents the estimated conditional reliability function, $\hat{R}(t|X_t)$.

   The results from the proposed method are compared to the MLE. In Figure 8.4, the estimated p.d.f.'s, $\hat{f}(t|X_t)$, for the time to failure for all three test cases are shown both using the proposed method and the MLE. The time to failure is assumed to be Weibull distributed. When implementing the MLE, only right censored and actual observations of the times to failure are used in the parameter estimation. The MLE was implemented using Matlab's statistics toolbox [184]. Table 8.2 summarizes the comparison of the estimated Weibull parameters for the proposed method and the MLE, including the mean and variance.

   Figure 8.5 shows the hazard rate estimated with the proposed method as well as the MLE. For comparison, the empirical hazard rate is also included. The empirical hazard is simply the quotient of the number of failed contacts of a certain age and the total number of contacts of that age.

3. *Discussion*: From the results in Figure 8.4a–c and Table 8.2, it appears that the proposed method will always provide a more conservative estimate of the time to failure than will the MLE. The MLE on the other hand is very consistent due to the fact that the proportion of right censored observations is very similar in all three cases. The difference in the distribution of the temperature rises in Figure 8.2 does

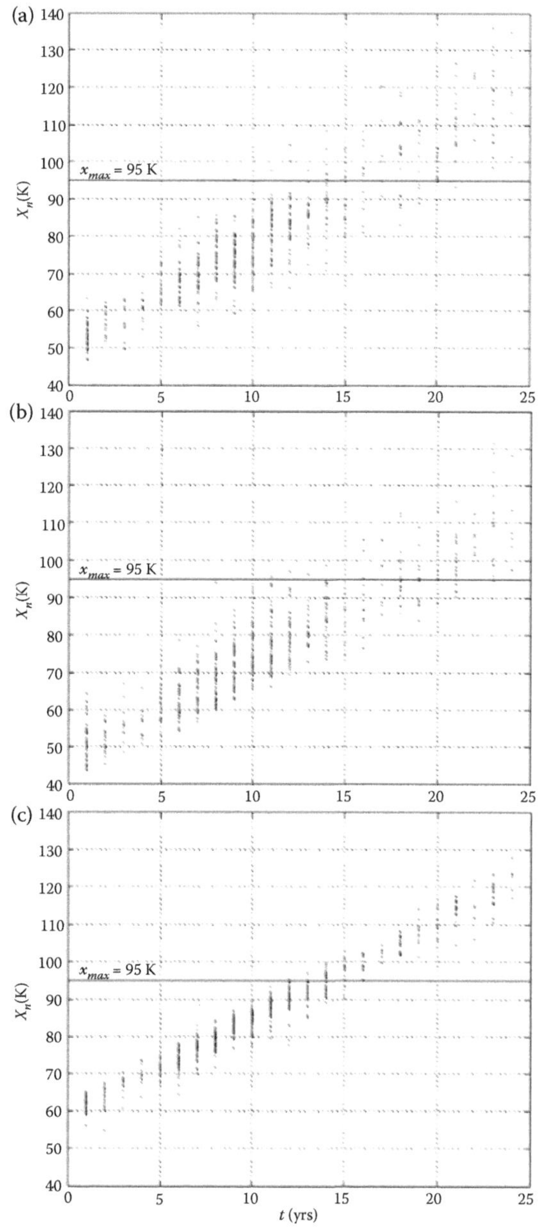

FIGURE 8.2
Randomly generated temperature rise measurements for: (a) Case 1, (b) Case 2, (c) Case 3. (Lindquist, T. and Bertling, L., Hazard rate estimation for high-voltage contacts using infrared thermography, in *2008 Annual Reliability and Maintainability Symposium*, pp. 231–237, © 2008 IEEE.) [176]

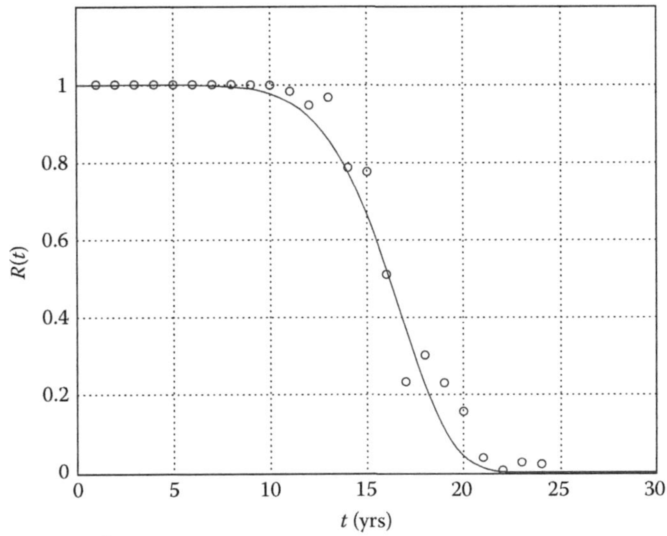

FIGURE 8.3
Reliability function fitted to survival probabilities for each year, Case 1. (Lindquist, T. and Bertling, L., Hazard rate estimation for high-voltage contacts using infrared thermography, in *2008 Annual Reliability and Maintainability Symposium*, pp. 231–237, © 2008 IEEE.) [176]

not, to any large extent, effect the MLE parameter estimation. As can be seen from the results in Figure 8.4a–c, the proposed method is better reflecting the real-world situation in Figure 8.2a–c than the MLE. This is because the proposed method is considering the probability of survival given the contact age whereas the MLE is treating the contact age as the actual time of failure, even though the contact may have been failed for some time. Given this situation, the proposed method is hence a more suitable method for estimating the time to failure compared to the MLE.

8.2 RCAM Stage I and Stage II for Transformers

8.2.1 Survey of Failure Statistics of Transformers and Lifetime Modeling

8.2.1.1 Introduction

The lack of data is a problem in statistical analysis of transformer failures and maintenance. The limited amount of failure statistics is a consequence of two facts. First transformers have relatively long lifetimes and won't

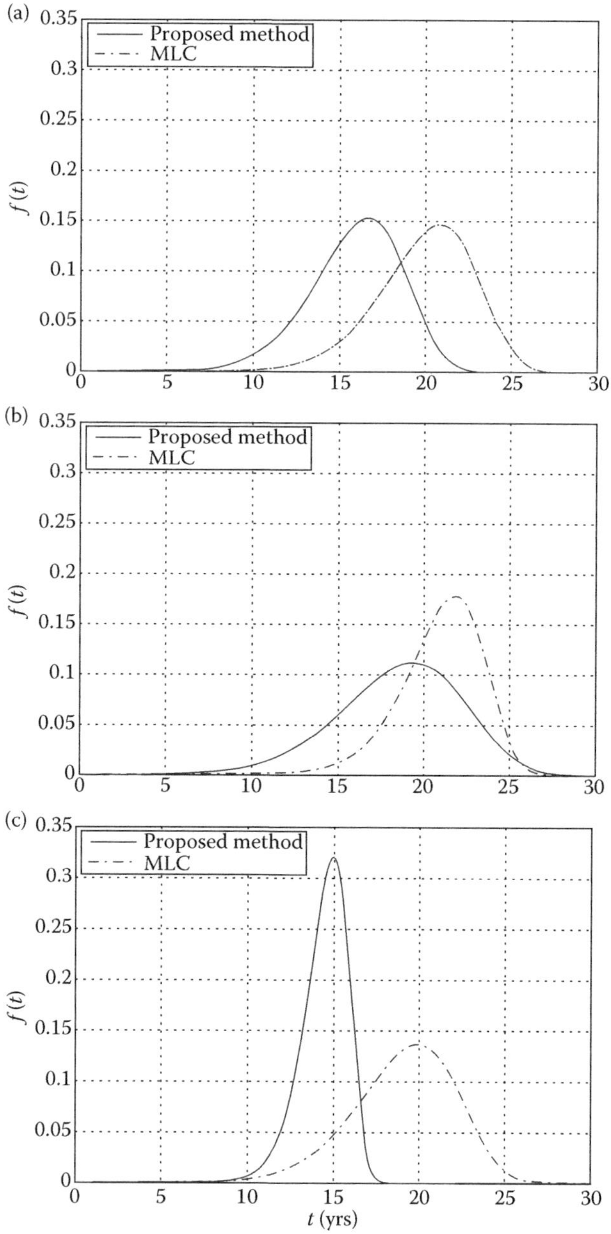

FIGURE 8.4
The resulting Weibull p.d.f. for the time to failure using both the proposed method and MLE, for Cases 1, 2, 3. (Lindquist, T. and Bertling, L., Hazard rate estimation for high-voltage contacts using infrared thermography, in *2008 Annual Reliability and Maintainability Symposium*, pp. 231–237, © 2008 IEEE.) [176]

TABLE 8.2

Summary of Results Comparing the Proposed Method to the MLE (Weibull Parameters)

Parameter	Proposed Method	MLE
CASE 1		
Scale, $\hat{\eta}$	17.0	21.1
Shape, $\hat{\beta}$	7.0	8.3
Mean, μ	15.9	19.9
Variance, σ^2	7.1	8.1
CASE 2		
Scale, $\hat{\eta}$	20.0	22.0
Shape, $\hat{\beta}$	6.0	10.6
Mean, μ	18.6	21.0
Variance, σ^2	12.9	5.7
CASE 3		
Scale, $\hat{\eta}$	15	20.3
Shape, $\hat{\beta}$	13	7.4
Mean, μ	14.4	19.1
Variance, σ^2	1.8	9.2

Source: Lindquist, T. and Bertling, L., Hazard rate estimation for high-voltage contacts using infrared thermography, in *2008 Annual Reliability and Maintainability Symposium*, pp. 231–237, © 2008 IEEE. [176]

fail frequently, second during expansions of the power system transformers are replaced with larger or better suited units before they have reached their end of life. Moreover, transformers are not equal. They have different rated powers, voltages and currents, are different loaded, and are of different designs. Therefore, it is not sure that two transformers are equal in statistical meaning. In order to decide on which part maintenance shall be performed with the aim to extend the lifetime of the transformer, it would be good to have more failure statistics. The statistics should include information about how often different components of the transformer fail and presumed causes for failures of each component. Then the part of the transformer that stands for most failures, which can be avoided by maintenance, could be found.

This section presents the results a review of statistical surveys made on power transformers which is presented in detail in [185]. It shows information about how often different transformer components fail has been found in the literature, however, information about the failure causes on component level has not been found. The case study shows that many transformer fails due to manufacture, design, and incorrect maintenance, which are causes that can't be prevented or limited by maintenance.

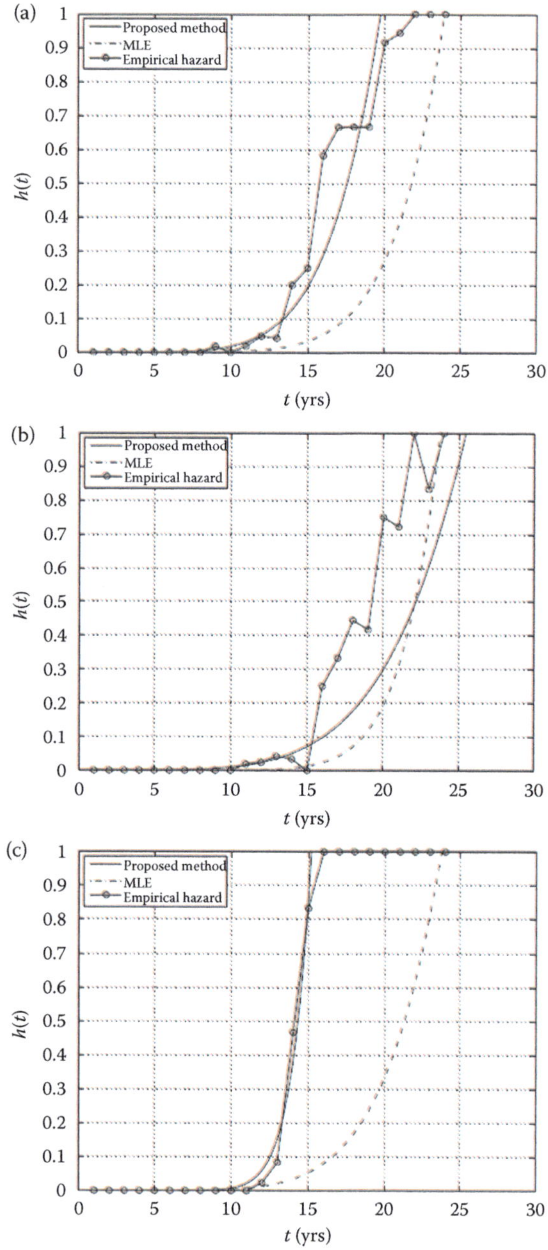

FIGURE 8.5
Estimated hazard rates using the proposed method, the MLE and the empirical hazard for Cases 1, 2, 3. (Lindquist, T. and Bertling, L., Hazard rate estimation for high-voltage contacts using infrared thermography, in *2008 Annual Reliability and Maintainability Symposium*, pp. 231–237, © 2008 IEEE.) [176]

8.2.1.1.1 Transformers

The transformer is used in the electric power system to convert the power between different voltage levels at the transmission or distribution level.

The transformer consists of core, windings, and insulation materials. These parts are enclosed and protected by a tank. An on-load tap changer (OLTC) regulates the voltages. Bushings connect the transformer windings to the net. The core is made of magnetic steel in order to be able to wear a magnetic field. The windings usually consist of copper strands isolated by cellulose. Oil serves as both isolation material and cooling medium in the transformer. Cooling can be either natural or forced. If the cooling is forced, the oil circulation is controlled by pumps. Fans are often used to circulate the air outside the transformer in order to increase the cooling of the oil [186].

8.2.1.1.2 Failures

Different classifications of failures can be seen in the literature. In the CIGRE survey [187], the failures were classified by failed components, origin and presumed cause. When the failures were classified by failed components, the classes windings, magnetic circuit, terminals, tank and dielectric fluid, other accessories and tap changer were used. The different origins were mechanical, dielectric, thermal, chemical, and unknown. The classes for presumed cause were design, manufacture, material, transport or storage, incorrect erection in site, incorrect maintenance, abnormal overload, overfluxing, lightning, external short circuit, loss of cooling, and unknown.

All failures cannot be avoided by preventive maintenance. Among the presumed causes, given above, failures due to design, manufacture, and lightning cannot be avoided by maintenance. However, maintenance can increase the transformers ability to withstand overloads and external short circuits up to some limit. Loss of cooling could be avoided by maintenance of the cooling system. Incorrect erection in site and incorrect maintenance can be avoided by information rather than by maintenance.

8.2.1.2 Failure Statistics

Failures can occur due to different reasons. Some failures can be limited or prevented by maintenance others cannot. Some failures depend on degradation processes, which can be used to describe the lifetime of the equipment, that is, to construct a lifetime model—a model for the degradation process leading to failure. Here a resume of failure statistics is given to see why transformer fails and if they fail due to processes that can be used to formulate a lifetime model.

8.2.1.2.1 CIGRE Statistics

In order to study the reliability of large power transformers in service, determine realistic parameter values, and eventually pinpoint the causes of transformer failures and determine transformer outage times, a group called "Transformer Reliability" WG 12.05 was formed in 1975. In 1978, the group launched a survey that involved transformers and reactors 20 years old or younger, designed for networks with a highest system voltage of at least 72 kV but without limitations on rated power. For practical reasons, the survey was limited to the countries represented in CIGRE SC 12 [187].

The received data included more than 1000 failures that had occurred between 1968 and 1978 in a total population of more than 47,000 unit-years. From this data, failure rates were determined for different populations depending on voltage range, transformer type, age, and the presence of OLTC [187].

Difference was made between failures with forced outage and failures with scheduled outage. A failure with forced outage was defined as a failure that caused an immediate need to disconnect the transformer. That is, when the transformer is disconnected from the net within half an hour after the failure occurred. If the needed maintenance could be planned to a later time, the failure is defined as a failure with scheduled outage [187].

First the failure rates were calculated as a function of winding voltage, types of units, and whether the transformer had an OLTC or not. It was found that the failure rate for a transformer generally is in the order of 2% per unit-year. A comparison of failure rates corresponding to different populations showed that the failure rate increases with voltage if no distinction is made between units with and without OLTC. For power station transformers and auto transformers both with OLTC, the failure rate clearly increased with the winding voltage. For substation transformers, no significant difference was found between different winding voltages [187].

When the failure rate was calculated as a function of the age and the winding voltage, it was only for transformers with winding voltage between 300 kV and 700 kV that the failure rate increases with the age. For smaller units, the failure rates seem to decrease slowly with age [187].

If instead the numbers of failures are studied, it can be seen that 70% of all failures forced an outage. Some diagrams over failed components, origins and presumed causes of the failure, both with forced outage and scheduled outage are also included in the paper [187].

702 failures had occurred in substation transformers. 691 of these failures had occurred in transformers with OLTC and 11 in transformers without.

Figure 8.6 shows that more than 40% of substation transformers with OLTC have failed due to failure in the tap changer, 19% have winding failures, tank and dielectric fluid, terminals and other accessories stand for about 12% of the failures each. Only 2.6% were due to failure in the magnetic circuit. When only failures with forced outage are considered the only differences are those, the windings stand for 26% and tank and dielectric circuit stand for 8% [187].

In Figure 8.7, the main origins for all failures on substation transformers can be seen. The origins were mechanical, 53.1%, and dielectric, 30.8%,

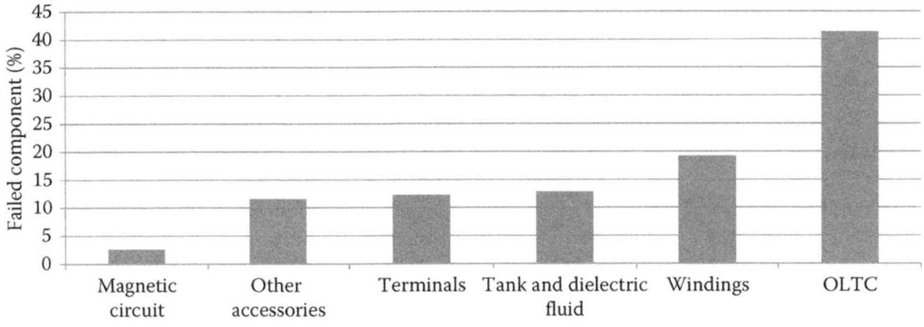

FIGURE 8.6
Substation transformer failures categorized by failed component. Transformers with OLTC. All failures included. (Data from Dietrich, W., *Electra*, (88):21–48, 1983.) [187].

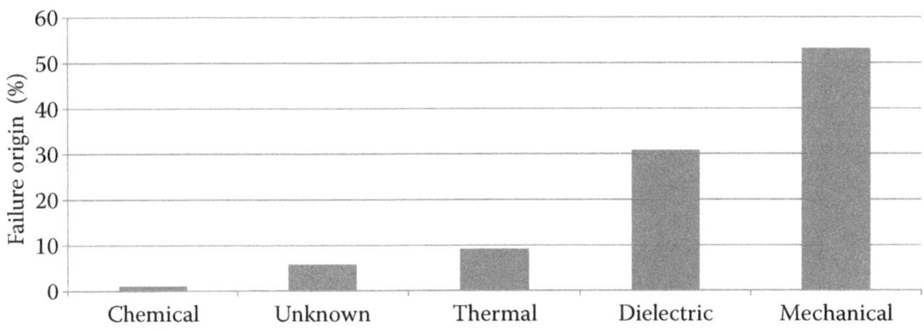

FIGURE 8.7
Substation transformer failures categorized by failure origin. Transformers with OLTC. All failures included. (Data from Dietrich, W., *Electra*, (88):21–48, 1983.) [187]

followed by thermal, 9.2%, unknown, 5.8%, and chemical failures, 1.1%. The origins for failures with forced outage were mechanical, 46.1%, and dielectric, 36.3%, the other origins remained on the same levels as for all failures [187].

Figure 8.8 shows the presumed causes of failures in substation transformers. The presumed cause was unknown in about 30% of all failures, manufacture in 14.7%; design and material stand for 12% each. Incorrect maintenance, lightning, and external short circuits stand for 8%–9% each. The resulting failures were due to transport or storage, incorrect erection in site, abnormal overload, overfluxing, and loss of cooling. The differences in presumed causes between all failures and failures with forced outage were small [187].

They also found that substation transformers more often failed due to the OLTC than power station transformers. The reason was assumed to

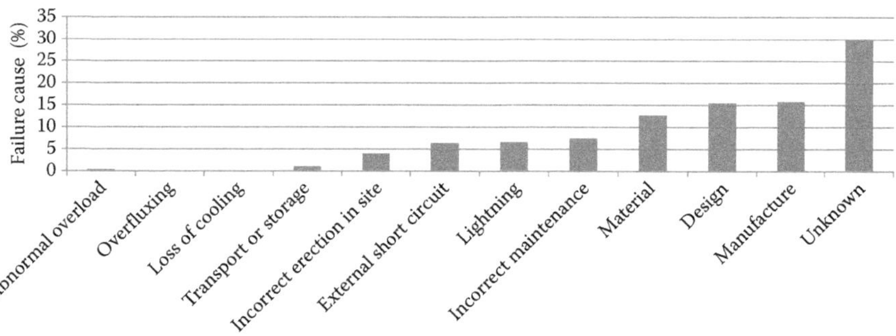

FIGURE 8.8
Substation transformer failures categorized by failure cause. Transformers with OLTC. All failures included. (Data from Dietrich, W., *Electra*, (88):21–48, 1983.) [187]

be less frequent operations for OLTCs and stricter maintenance on station transformers [187].

The total number of failures in substation transformers without OLTC was just 11, the total number of failures in substation transformers was 702. 40% of these, that is, 4.4 transformers, had failed due to the windings. If there had been just one failure more or less, the distribution between the different failed components had changed a lot. This means that due to the limited amount of data, it was impossible to draw any conclusions about the failure causes for transformers without OLTC [187].

According to the later CIGRE study [188], abnormal events like overvoltages and system faults cause more transformer failures than aging of winding isolation, but the aging may make the transformer more vulnerable to these abnormal events.

8.2.1.2.2 Elforsk Survey

In 2006, Elforsk [189] made a survey about the lifetime of transformers. The survey collected data from 8 Swedish transformer owners about their 130 kV transformers. A Bayesian method was used to evaluate lifetime distributions for different groups, depending on manufacture, of transformers. The amount of transformers, which had reached their end of life, was very low only for 3 of 217 transformers. Therefore, it was concluded that the used Bayesian method gave lifetime distributions, which were displaced towards longer lifetimes.

8.2.1.2.3 A Finnish Survey

In Finland, a comprehensive survey of failure statistics was performed, which included investigation of the failure rate of distribution transformers in different electric power systems. The effect of different earthings, network connections, and overvoltage protection were analyzed. The detailed results are unfortunately confidential so the paper [190] only presents general results. Despite collection of data from 15 Finnish utilities, the amount of data was small. If there had been one failure more or one less during a certain year, it had affected the result a lot. When different manufactures were compared, it could be seen that one was better and the others was worse.

8.2.1.2.4 Statistics from the Commonwealth of Independent States

The Commonwealth of Independent States, CIS, is an alliance of 11 former Soviet Republics: Armenia, Azerbaijan, Belarus, Georgia, Kazakhstan, Kyrgyzstan, Moldova, Russia, Tajikistan, Ukraine and Uzbekistan [191].

Sokolov et al. [192] have studied failure statistics from the ZTZ-Service database, which includes over 5000 large transformers, that is, transformers rated 100 MVA and above. They have found three main causes of failure. The first cause is weak construction due to, for example, material defects and underestimation of operational stresses, the second is to high operation stress due to unusual events or operational error, and the last is inadequate maintenance and low quality repair, which can lead to critical deterioration of safety margins. Figure 8.9 shows failure modes for transmission transformers in ZTZ-Service 2000–2005.

According to Sokolov et al. [192], the failure rate of the transformer is high during the first 3 years in operation, a wear in period. After some years with low failure rate, it raises again after 7–15 years due to failures in bushings, load tap changers, and windings. Their conclusion is that the failure profile, in the range 0–35 or 40 years, is a combination of wear in failures, random

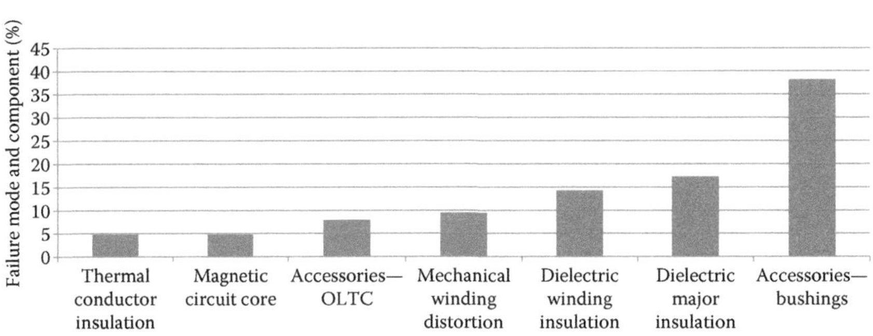

FIGURE 8.9
Failure modes for transmission transformers in ZTZ-Service in 2000–2005. (Data from Drobyshevski, A. et al. Reliability problems with large power transformers and shunt reactors. Typical failure modes and failure causes, in *CIGRE SC A2 Colloquium on "Transformer Reliability and Transients"*, Moscow, Russia, pp. 20–23, June 2005.) [192].

failures, and wear out failures of some weak components such as bushings and tap changers. They did not find any obvious failures due to paper aging.

8.2.1.2.5 Statistics from North America

In a study [193] of the failure rate of transformers in the USA two peaks were seen. The first peak appears during the first year of the transformers operation, due to a wear in period, and the second peak appears after 11–14 years in operation. The second peak in the failure rate was assumed to appear due to lower short circuit withstand capability and/or heating effects caused by the loading of the secondary winding rather than due to thermal aging of the cellulose.

8.2.1.2.6 Statistics from South Africa

Minhas et al. [194] present a failure analysis of power transformers in the Eskom network, South Africa, in the range of 88–765 kV and 20–800 MVA. Six failure modes have been identified in the analysis. These are lightning, core problems, tap changer initiated failures, general aging, short circuit, and others. Others consist of, for example, bushing, unknown, and operational errors.

Figures 8.10 and 8.11 show the failures in transformers on 20–100 MVA and in transformers above 100 MVA, respectively, were distributed between different failure modes.

According to the analysis [194], most failures in small transformers, 20–100 MVA, occur due to aging and tap changers, in medium size transformers, 100–400 MVA, the aging and "other failures" stand for most failures and failures in larger transformers occur due to lightning.

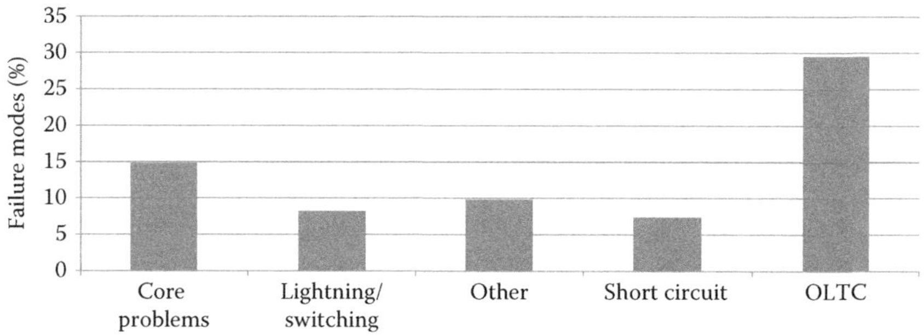

FIGURE 8.10
Failure modes for transformers on 20–100 MVA. (Data from de Klerk, P. et al., Failures in power system transformers and appropriate monitoring techniques, in *High Voltage Engineering Symposium*, in Conference Publication No. 467, pp. 22–27, August 1999.) [194].

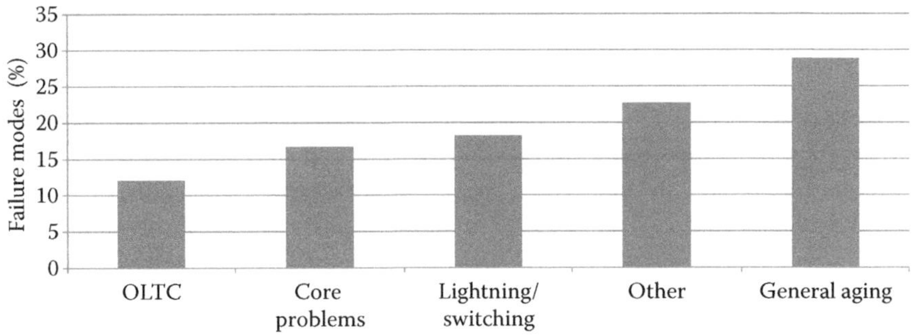

FIGURE 8.11
Failure modes for transformers above 100 MVA. (Data from de Klerk, P. et al., Failures in power system transformers and appropriate monitoring techniques, in *High Voltage Engineering Symposium*, in Conference Publication No. 467, pp. 22–27, August 1999.) [194].

8.2.1.2.7 More about Failure Rates

The National Grid Company, UK, has found that their transmission transformers have a failure rate of about 0.3% per year and no increasing trend in failure rate with age has been seen since 1971. They also conclude that transformer failures are usually triggered by extreme system conditions like lightning strikes and short circuit [195]. According to [196], 75% of the failures occur because of bad health condition, such as dielectric problems.

According to [197], the average life of a transformer is assumed to be somewhere between 25 and 30 years and the life span of OLTCs is confirmed to be in the same range as other transformer equipment. The failure rate of OLTCs varies between 0.1% per year in Germany and other similar nets and 3.7% per year in Thailand. Failure reasons have been found to be poor oil quality, high humidity in the oil and over voltages.

In [198], Lapworth studies different hazard rates and their consequences on mean time to failure. His conclusion is that an average failure rate does not estimate a reasonable probability of failure of transformers during their whole lives. A more realistic likelihood of failure would have a low failure rate until some time when the failure rate suddenly increases. For transformers, he suggests that the failure rate starts to increase due to damage caused by unusual system events, such as short circuits and lightning strikes, not due to aging of paper or oil.

Lapworth [199] also writes that one of the main uses with reliability data is the estimation of end of life for the transformer. About end of life assessment and risk management, he also writes "The interest here is not so much in the likelihood of experiencing a problem with a typical transformer, but in assessing the likely consequences and risks involved in continuing to operate a transformer which is suspected of being faulty." Then the aim would be to derive expected values for time to failure and failure costs.

8.2.1.2.8 Conclusion

The statistics from different surveys are hard to compare. Some surveys have classified the failures by failed part, others by cause or origin. Some surveys even use a mixture of these classes. Despite these difficulties, some conclusions can be drawn. The references agree that most failures in transformers with OLTC occur due to the OLTC. However, large transformers fail more seldom due to the OLTC than smaller units. The windings seem to be the part that suffers from most failures in transformers without OLTC, and the second in transformers with OLTC.

The main origins for failures are mechanical and dielectric followed by thermal.

The main cause is in many cases unknown. When the failure cause is known, manufacture, design, and material stand for most failures, followed by operational stresses due to for instance lightning and short circuits in the net and incorrect maintenance. Even though some surveys also mention aging as a cause of many failures others say that no failures due to aging have been seen.

According to [194], around 30% of all transformer failures occur due to general aging.

8.2.1.3 Aging

Even if the statistics do not show that aging is a main cause of failure in the transformer, aging can be an underlying cause. For instance, aging can affect the transformers ability to withstand short circuits and overloads. In addition, one way to construct a life time model is to model the physical or chemical aging processes. Therefore, the aging processes for the transformer's parts are described.

According to the Loading Guide [200], there is no simple and unique end-of-life criterion for the transformer. According to [201], the aging of transformers is related to the aging of windings, tank, bushings, and OLTCs. Aubin et al. [188] agree and state that the transformers lifetime is closely related to its aging processes.

8.2.1.3.1 Aging of the Windings and Its Isolation

According to [195], the most life limiting component in the transformer is the paper winding isolation. Operating temperature, moisture, oxygen, and acidity levels in the oil affect the aging of the paper. By using standard techniques to evaluate the hot spot temperature in the windings and relate this temperature to laboratory aging rates for paper gives that paper lives in about 55 years if there are no thermal design weaknesses.

Even though the aging of the cellulose isolation is a time function depending on temperature, moisture content, oxygen content, and acid content, the model given in IEC Loading guide [200] only uses the temperature as a free parameter. The relative aging rate V for the isolation exposed to the hot spot temperature Θ_h in °C is

$$V = 2^{(\Theta - 98)/6} \quad (8.5)$$

for nonthermally upgraded paper and

$$v = e^{\frac{15{,}000}{110+273} - \frac{15{,}000}{\Theta_h + 273}} \quad (8.6)$$

for thermally upgraded paper.

The relative aging rate is very sensitive to changes in the hot spot temperature. For nonthermally upgraded paper, the relative aging rate equals 1 when the hot spot temperature equals 98°C if the temperature is reduced to 92°C the aging rate is halved. For upgrade paper 110°C corresponds to an aging rate of 1 [200].

The loading guide also gives equations for the evaluation of the loss of life for the isolation. The loss of life L during the time period $[t_1, t_2]$ is

$$L = \int_{t_1}^{t_2} V \mathrm{d}t \quad (8.7)$$

As a reference the loading guide [200] says that if the transformer is operated at the hot spot temperature of 110°C and the isolation is thermally upgraded, well-dried, and oxygen free, 25% of the isolations tensile strength will remain after 15.41 years.

According to the CIGRE working group 12.09 [188], it has been overemphasized to base the expected life of the transformer solely on the aging of the insulation due to temperature. The reason for the big interest of the isolation lifetime is assumed to be that its material properties are easy to measure in a laboratory and they believe that one or more of the isolation properties can be used to estimate the lifetime of the transformer. The dielectric properties of the paper insulation aspect the lifetime of the transformer, but the dielectric strength of the paper is not significantly influenced when the paper is aging. It has also been found that different properties are affected with different rates, when the isolation ages [188].

The CIGRE working group also criticizes that the loading guide assumes that the paper insulation will break first at the hot spot, due to high temperature failures, and not at the most vulnerable point, due to mechanical stresses [188]. It has been observed that different transformers with about the same age have DP values that differ a lot from each other, therefore, Jarman et al. [195] have found that the most critical factor in determining the aging rate of the isolation appears to be the thermal design of the transformer.

If the paper and press board shrinks due to aging, it will lead to slack windings with increased risk for short-circuit damage as a consequence [195].

The aging of windings depends on thermal stresses due to overloads; therefore, it depends on the loading history [201].

8.2.1.3.2 Aging of the Tap Changer

When the OLTC is operating, there are sparks between the contacts. The operational stresses lead to erosion of the contacts and the oil is carbonized due to the sparks. The amount of produced particles depends on the load current and the temperature [201].

8.2.1.3.3 Aging of Bushings

According to [201], bushing age due to thermal stresses depends on the operating load. A bushing failure leads often to heavy transformer failures, it can even lead to explosion and fire in the transformer [202].

A study in Russia showed that the main causes of bushing failures are old transformer oil and high water content in the oil. The transformer oil ages and breaks down faster with higher oil temperature. An important factor for the formation of water in the oil is the amount of aromatic rings in the oil [202].

8.2.1.3.4 Aging of the Tank

The metal in the tank could be affected by corrosion. The amount of corrosion depends on the age of the tank and the maintenance history [201].

8.2.1.4 Maintenance

All failures cannot be avoided or limited by preventive maintenance. However, some failures due to aging could be limited or avoided by preventive maintenance. To get a clue of which failures that can be prevented or limited by maintenance today, a description of different maintenance tasks performed to transformers today or under investigation follows.

There is a distinction between preventive and corrective maintenance. Corrective maintenance is performed in order to set a failed component back in operation. Preventive maintenance is planned and performed before the component has failed with the aim to reduce the probability of future failure [203].

According to the maintenance instruction [204], most regular maintenance performed on transformers is inspections, tests, and cleaning. A common test is dissolved gas analysis (DGA) of the insulation oil. If there are any leakages, gaskets are tightened, if there is rust on the tank it is treated and painted and the silica gel in the dehydrating breather is changed if it is needed. Further maintenance tasks are performed if tests and inspections show that it is needed. Even if the control equipment is alarming for a failure in the transformer, the performed maintenance is based on inspections and tests [205].

8.2.1.4.1 Maintenance for Isolation Oil

Gas analysis and condition tests of the isolation oil shall be made every year or once in 4 years, depending on the size of the transformer [206]. Depending on the results from these tests further maintenance are made. Gas analysis tell something about the condition of the transformer, while condition tests of the oil only tell the condition of the oil isolation [207].

If the condition of the oil is under the acceptable limit, it can be maintained. The oil can be switched, degased, dried, or reclaimed. If the amount of water in the oil is too high, the transformer can be vacuum dried, since most water are bounded to the cellulose it is no idea to only maintain the oil [207].

In [208], a new drying method based on the mineral Zeolite's capacity to adsorb water molecules is investigated. The method is said to be cheap, environmental friendly, and able to use in-service. So far as no disadvantages have been proved with this drying method. The transformer can be reclaimed during its normal operation. During the reclamation, the oil passes through podsol filter, where the podsol adsorbs pollutants.

Afterwards the oil is degased. Normally the oil passes through the filters 8–12 times before the reclamation is completed. The reclamation removes in this way acids, which speeds up the natural aging of the oil, thereby can the reclamation extend the lifetime of the oil and the transformer [209].

Elforsk [210] made a study of the effects of different maintenance performed on four transformers standing in the same surrounding. The four transformers were subject to different maintenance and their results from different tests before and after the maintenance were compared. The different maintenance tasks were reconditioning, reclaiming, overhaul at the fabric and no maintenance. The first study was made during 2000 and a second was made 3 years later in order to investigate the long-time effect of the performed maintenance [210,211]. One conclusion was that the reclaimed oil still showed "new oil properties" after 3 years. Another conclusion was that 3 years was a too short time to be able to see changes in the isolation.

8.2.1.4.2 Maintenance for OLTC

According to the maintenance instructions for OLTCs [212,213], the maintenance consists mostly of inspections, measurements, and lubrication. The contacts are switched if they are old or in bad condition. The OLTC shall be maintained either after a certain number of connections, specified by the manufacture, or after 3, 6, or 8 years depending on the size of the transformer [206].

To ensure a high oil quality in the OLTC, it is important that the oil is dried and filtered. Usually the oil is continually dried by the dehydrating breather, therefore, it is important to check and change the silica gel in the breather.

The oil can either be filtered continually or at special maintenance occasions. It is also important to check the condition of the contacts and switch them if necessary. Maintenance on OLTCs is made at certain intervals either after a certain number of operations or after a certain number of years [197].

Even if the OLTC stands for a major part of all transformer failures, some people say that it do not fail if it is maintained correctly according to the manufacturer instructions [214].

Hällgren and Öfverbeck [215] derived optimal maintenance intervals for OLTCs from statistical data. According to their statistics, preventive maintenance for OLTCs consists of cleaning and lubricating the motor drive mechanism, cleaning and replacing contacts, replacing leaking gaskets, and oil filtering. The studied failures have mostly small consequences, such as failing voltage regulation, abnormal network voltages, and in some cases disconnection of costumers. In many failure cases, cleaning and lubrication of the motor drive were enough as repair. The expected total maintenance cost was minimized and an optimal maintenance interval was found to be about 30,000 operation cycles. Due to the increased risk of failures with larger consequences, when the maintenance interval is enlarged a new interval was derived with a higher expected consequence cost. The recommended maintenance interval was set to 15–20,000 operation cycles.

According to [216], Hydro-Québec developed a maintenance strategy for their OLTCs. The reason to develop the strategy was that two-thirds of their transformers with OLTC are 25 years old or more. Moreover, they have found that about 50% of the failures, requiring replacement or transportation to repair shop, occurred due to failure in the OLTC. The maintenance strategy consists of two procedures. In the first procedure, an OLTC family is analyzed based on all failure and maintenance data available for that family. Typical failures and appropriate maintenance tasks are identified and the underlying reasons for each failure are determined. The maintenance tasks are also divided into minor and major work solutions, and a cost analysis is performed for the major work solutions. Major work solutions consist of overhaul, replacement, and retrofit. Then the second procedure called OLTC Life Decision Making Process is used on an individual OLTC. The age, number of operations, failure, reliability, and preventive maintenance are under consideration. The question is if major work is required on the OLTC or not. If not, minor work is performed and the OLTC is returned to service. To choose the best major work solution on this OLTC, the condition of the whole transformer and network planning is investigated. If any major solution extends the life of the transformer to a reasonable cost it is performed, otherwise the transformer is replaced [216].

8.2.1.5 Lifetime Models

Many transformers are replaced when they reach their economical or strategic lifetimes. Economical lifetime is a theoretic concept and refers to the time up to the replacement time which gives the optimal profit from an alternative investment [217]. When the strategic lifetime of the transformer is out, it is replaced because of changes in the power system. The old transformer is no longer suited to transform between the wanted voltage levels either because it is not designed for these voltage levels and currents or the power supplier wants a more reliable transformer in that place [214]. In this chapter, the focus will be on lifetime models, trying to describe the transformers lifetime in years depending on different measures or a known history of the transformer. Some of the models are indeed models describing the transformers age.

8.2.1.5.1 Lifetime Models Not Including Maintenance

8.2.1.5.1.1 A Bayesian Method using Perk's Distribution Chen and Egan [218] present a Bayesian method for transformer lifetime estimation, which uses a simplified Perk's hazard function and so-called Iowa curves. The method can be used on a transformer population with known ages and retire ages on the transformers. Iowa curves or Iowa Survivor curves are a set of 18 curves that can be used to predict retirement patterns of assets. The Perk's hazard function extends the Iowa curves to a continuous spectrum of distribution functions. In their model, Chen and Egan use a Bayesian method to find the parameters that gives the distribution function, which describes the data best. Perk's distribution is said to be better suited to describe data than the Weibull function since it can assume a combination of hazard functions in a single distribution. The Perk's distribution function is:

$$f(t) = \frac{e^{\beta \cdot t + \alpha}}{1 + e^{\beta \cdot t + \mu}} \cdot \left(\frac{1 + e^{\beta \cdot t + \mu}}{1 + e^{\mu}} \right)^{-exp(\alpha - \mu)/\beta} \tag{8.8}$$

and the cdf is

$$F(t) = 1 - \left(\frac{1 + e^{\beta \cdot t + \mu}}{1 - e^{\mu}} \right)^{-exp(\alpha - \mu)/\beta} \tag{8.9}$$

where α, β, and μ are three parameters and t is the time.

8.2.1.5.2 Lifetime Models Based on the Aging of the Cellulose Isolation

8.2.1.5.2.1 Loss of Life Based on Loading Guide Weekes et al. [219] start with the aging model for the cellulose given in the loading guide [20] and evaluate the insulation loss of life for a converter transformer during the spring and summer months. With coolers, the loss of life is found to be 1.2 days during 1 day in the summer and 0.3 days/day in the spring. These figures can be compared to 3.8 days/day during the summer if no restrictions on the load are used. If the load is restricted to 90% when the temperature rise above 15°C the loss of life of insulation remains on 1.25 days/day during the summer.

Weekes et al. [219] suggest that these loss of life calculations can be used, once the owner has decided which risk he wish to take, to decide about seasonal operating restrictions.

8.2.1.5.2.2 Elapsed Life Based on CO, CO_2, and Furfurals Pradhan and Ramu [220] state that the remaining life of a transformer can be determined by the theory of stochastic processes and that the theory is beyond the scope of their investigation. Instead they have used monitoring data from experiment transformers. From these data, a fair correlation between DP value and amount dissolved CO, CO_2, and furfurals in the oil was found. An expression for the elapsed life of the transformer, which depends on an evaluated DP value, is given. The DP value is calculated as a function of the amounts of CO, CO_2, and furfurals in the oil. According to the paper, the elapsed life in years is

$$\text{Elapsed Life} = 20.5 \cdot \ln\left(\frac{1100}{DP_{value}}\right)$$

where it has been assumed that unaged transformer cellulose insulation has a DP value of about 1100. A natural question is if anyone else has used the theory of stochastic processes in order to predict the remaining life of the transformer.

8.2.1.5.2.3 Probability of Failure Based on the Inverse-Power Law and Arrhenius Model In [201,221 and 222], Zhang et al. present a lifetime model for electrical systems and use transformer in a secondary substation as a component example. The inverse power law describes the life of electrical equipment in the presence of the electrical or mechanical stress S. The equation is

$$L = L_0(S/S_0)^{-n} \tag{8.10}$$

where S_0 is a scale parameter for the lower limit of stress, L_0 is the corresponding lifetime, and n is the stress endurance coefficient.

The Arrhenius model

$$L = L_0(S/S_0)^{-n} \cdot e^{-BT}, \quad T = 1/\Phi_0 - 1/\Phi \tag{8.11}$$

is a life model for an electrical equipment exposed to thermal stresses due to overheating of insulation materials. B is proportional to the activation energy for the main thermal degradation reaction in the insulation, T is the thermal stress, Φ is the absolute temperature, and Φ_0 is a reference temperature.

If the inverse power law is multiplied with the Arrhenius model, an overestimation of the total stresses is determined. By introducing a correction factor $(S/S_0)^{bT}$, where b is an coefficient depending on the reaction of the material due to combined stresses, a better estimation is found.

The Weibull function

$$P(t > L) = 1 - \exp\left(-\left(\frac{t}{L_{63\%}}\right)^{\alpha}\right) \tag{8.12}$$

describes the likelihood of failure at given stresses. Determining $L_{63\%}$ from the multiplication of 4, 5 and the correction factor gives the expression for the probability of failure in 7

$$P(t > L) = 1 - \exp\left(-\left(\frac{S}{S_0}\right)^{\alpha(n-bT)} \cdot \left(\frac{t}{L_0}\right)^{\alpha} \cdot e^{\alpha bT}\right) \tag{8.13}$$

The assumption is that a failure will occur immediately if t is larger than the lifetime L. While using this model, both n and α can be found from two aging tests.

With m different failure causes, each with probability P_i for $i = 1, \ldots, m$ and causing a_i percentage of all failures, the total failure probability P_m is

$$P_m = \sum_{i=1}^{m} a_i \cdot P_i \tag{8.14}$$

According to the results given in [222], 80% of the transformer failures occur in the age interval 10–38 years. Performed regular maintenance on transformers makes it possible to avoid an extreme increase of the failure rate. The main cause for the aging of transformers seems to be wear-out problems [222].

8.2.1.5.3 Models Which Include Maintenance

8.2.1.5.3.1 A Bayesian Method and Equivalent Age In the report [189], Elforsk presents an equivalent age for the transformer as

$$L_e = \circ A + \sum LK + \sum LM$$

where $\circ A$ is the transformers age, LK are the stresses from lifetime consuming factors, and LM are the stresses from lifetime intensifying factors. Lifetime consuming factors are, for example, short circuits in the transformer and lifetime intensifying factors are, for example, different maintenance tasks. The LK's are positive but the LM's are negative numbers. According to the report, this is an example of how the lifetime can be modeled in a Bayesian lifetime analysis. However, no values are proposed on LK or LM.

For six different groups, depending on the manufacture of transformers, Elforsk [189] evaluated the density function of the lifetime. A Weibull function with parameters based on an expert's assumptions of the transformer lifetime was used as priori which was updated by lifetime observations according to Bayesian theory.

8.2.1.5.3.2 Failure Rate Based on DGA Results Zhang et al. [223] present a model which also uses Bayesian theory, but with the purpose to derive an optimal maintenance strategy. Before they reach at the optimal maintenance strategy, they have to derive the failure rate of transformer based on DGA results. The model presented in [48] includes the probability, $p(k)$, for an event k in the transformer and the reduction of the probability of event k due to maintenance m, $\Delta p(k,m)$. The risk reduction due to maintenance m performed at time t is $\Delta CR(k,m,t) = \frac{\Delta p(k,m)}{p(k)} \cdot CR(k)$, where $CR(k)$ is the risk for event k. The probability reduction due to maintenance is found from $\Delta p(k, m) = p_{before}(k) - p_{after}(k)$, where $p_{before}(k)$ is the failure probability before maintenance and $p_{after}(k)$ is the failure probability after maintenance.

If the number of components, that can be maintained, in the transformer is N and $k=1,...,N$, the number of different maintenance on component k is L_k and $m=1,...,L_k$, then $I(k,m,t)=1$ if maintenance task m performed on component k begin at time t and zero otherwise [48].

Since the aim is to reduce the risk of failure as much as possible by performing maintenance, the objective function is

$$max\sum_{k=1}^{N}\sum_{m=1}^{L_k}\sum_{t=1}^{T}\Delta CR(k,m,t)\cdot I(k,m,t)$$

where $t=1, \ldots, T$ is the time interval under consideration. In order to find the failure probabilities needed to perform the maintenance optimization, Zhang et al. [223] estimate the failure rate of the transformer. It is assumed that the failure probability follows a Weibull distribution with known scale parameter α and an unknown shape parameter β. Moreover, it is assumed that the shape parameter β follows a normal distribution $\beta \sim N(\mu, \sigma^2)$, where μ is the mean and σ is the variance.

The shape parameter β is determined with the use of Bayes'rule and DGA results.

Bayes' rule is

$$P(\beta|DGAresults) = \left(\frac{P(DGAresults|\beta)P(\beta)}{P(DGAresults)}\right)$$

For simplicity, Zhang et al. [223] assume that the total amount of combustible gases, G, also follows a normal distribution, that is, $G \sim N(\omega\beta + k_i, \omega^2\gamma^2)$, where $i=1,...,4$ corresponds to different conditions, k_i is the average amount of combustible gases in condition i and ω is known.

Then the posterior of β can be found by using the linearity of the normal distribution and the Bayesian rule. Zhang et al. found that the mean and the variance of β are

$$\mu = E(\beta|G) = \frac{\sigma^2}{\sigma^2+\gamma^2} \cdot \frac{G-k_i}{\omega} + \frac{\gamma^2\mu}{\sigma^2+\gamma^2}$$

$$\sigma = \mathrm{Var}(\beta|G) = \frac{\sigma^2\gamma^2}{\sigma^2+\gamma^2}$$

At the end of their paper, Zhang et al. [223] present an example of their method where DGA results from a transformer over an 8-year period have been used. In the example, they use $\omega = 15$, $k = 60$, $\sigma^2 = 2$, and $\gamma^2 = 20$. The shape parameter β is found to equal 1. Since the empirical failure rate of transformers is 2% and $\beta = 1$ the scale parameter $\alpha = 50$.

8.2.1.5.3.3 A Markov Model Gives "Mean Time to First Failure" In [224], a model is presented, which can be used to evaluate "mean time to first failure." A maintenance model for the transformer is simplified to a Markov chain model, in which the transformer can be in four different stages, including the stage at which the transformer has failed. The other three stages represent different transformer conditions. In this model, it is assumed that the transformer successively degrades until it fails, that is, if the transformer is at one of the two stages corresponding to the best conditions it cannot fail. From stage 1, good condition, it can only reach stage 2, ok condition, and from stage 2 it can only degrades to stage 3, not so good condition, and eventually from stage 3 the transformer can reach stage 4, fail.

The transformer can also move from any stage to a stage where the condition is better, if maintenance is performed. In the paper, two maintenance models are mentioned: the perfect and imperfect maintenance models. In the perfect model performed maintenance represents by a transition to the next better stage, that is, from stage 3 to 2 or from 2 to 1. In the imperfect maintenance model it is further assumed that an inspection at stage 1 can cause a move to stage 3.

The probabilities with which the transformer moves from one stage to another are evaluated from the maintenance model for the transformer. At the first three stages, inspections and tests are performed with different probabilities. The test results categorize the oil condition as good, requiring reconditioning and poor or adverse. From good oil condition, the transformer moves to one of the three transformer conditions directly. In the other oil conditions, maintenance must be performed. The maintenance included in the model is oil filtering and oil replacement. Depending on the transformer condition and the oil condition, these maintenance tasks are given different probabilities. After the maintenance, the transformer is at one of the three stages with different probabilities. The actual probabilities are not given in the paper [224], but can be found from real transformer data. The inspection rate can be varied in order to receive high reliability with minimum costs.

8.2.2 Failure Modes and Effects Analysis of Transformers

8.2.2.1 Introduction

This case study shows the results from an RCM analysis of the transformer components. The results have been gained from a reference group involving both academics and the industry partners. Specifically the study has involved access to data from ABB as well as access to their approach for the analysis of transformers. The case study is discussed in detail in [225]. The study includes an FMEA of the transformer and a review of lifetime models, finally it propose a method for ranking of the transformers.

8.2.2.2 FMEA for the Transformer

The fault trees for the transformer presented in this section are a result of RCM analysis with a team of experts on transformers, academics, and a literature study. In order to analyze the transformer and compose a fault tree, the transformer has been divided into different subcomponents: windings, tank, bushings, cellulose isolation and cooler and oil insulation, see Figure 8.12. Fault trees and a short explanation for them are presented below. A fault tree for the tap changer is also given although this part is seen as steering equipment rather than a part of the transformer.

Table 8.3 shows a summary of the RCM analysis including the results from the FMEA from this case study of transformers. In the following sections more details will be presented from the FMEA analysis for each sub component of the transformer.

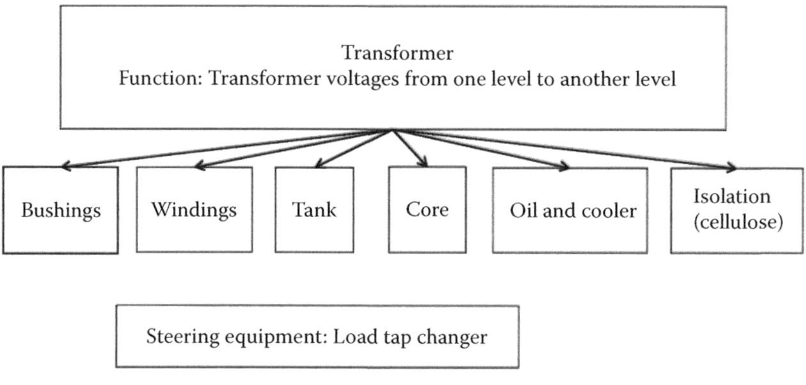

FIGURE 8.12
Fault tree for a transformer, the transformer's parts.

TABLE 8.3

Resulting FMEA for Transformers

Component	Function	Failure Mode	Failure Causes
Bushings	Connect windings to the work	Short circuit	External contamination
			Water ingress
Core	Wear magnetic field	Loss of efficiency	DC magnetization
Solid insulation	Insulation of windings	Cannot supply windings	Overload
			Oil deterioration
Transformer oil	Isolate and cool the active part	Overheated oil	Oil deterioration
		Contamination of oil	Aging, Short circuits
Tank	Enclose TRAFO oil	Oil leakage	Corrosion
	Protect the active part		Careless handling
Tap selector	Regulate the voltage level	Cannot change voltage level	Wear
Diverter switch	Maintain a coherent current	Contact failure	Contamination of oil
			Lack of maintenance
Windings	Conduct current	Short circuit	Transient overvoltage
			Hot spot
			Movement of TRAFO
Cooling system	Cool the active part	Temperature rise	Wear

8.2.2.2.1 Core

The core's function is to carry magnetic flux. The failure mode of this function is a reduction of the transformer's efficiency. The cause can be a mechanical fault in the core, due to DC magnetism or displacement of the core steel during the construction, that is, a construction fault [226]. Figure 8.13 shows a fault tree for the core.

8.2.2.2.2 Windings

The windings belong to the active part of a transformer, and their function is to carry current. The windings are arranged as cylindrical shells around the core limb, where each strand is wrapped with insulation paper. Copper is today the primary choice as winding material. In addition to dielectric stresses and thermal requirements, the windings have to withstand mechanical forces that may cause windings replacement. Such forces can appear during short circuits, lightnings, short circuits in the net, or during a movement of the transformer [227].

A fault in the windings can occur due to material faults in the cellulose isolation, see Section 8.2.2.2.4 for further information. In Figure 8.14, a fault tree for the windings is given.

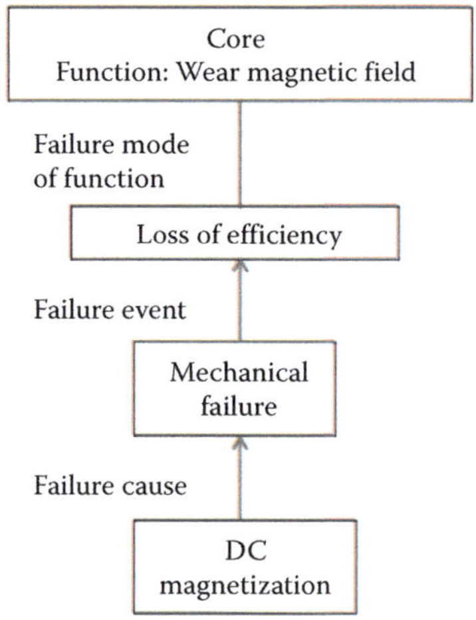

FIGURE 8.13
Fault tree for the core.

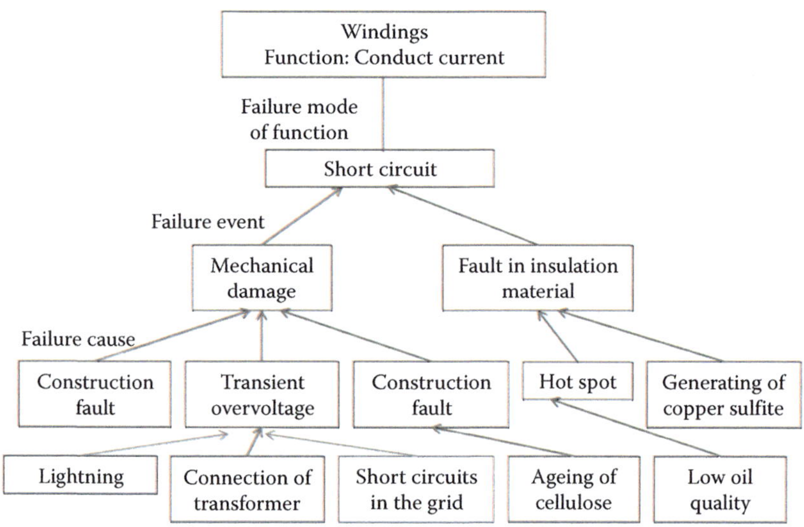

FIGURE 8.14
Fault tree for the windings.

8.2.2.2.3 Tank

The tank is primarily the container of the oil and a physical protection for the active part of the transformer. It also serves as support structure for accessories and control equipment. The tank has to withstand environmental stresses, such as corrosive atmosphere, high humidity, and sun radiation [227]. The tank should be inspected for oil leaks, excessive corrosion, dents,

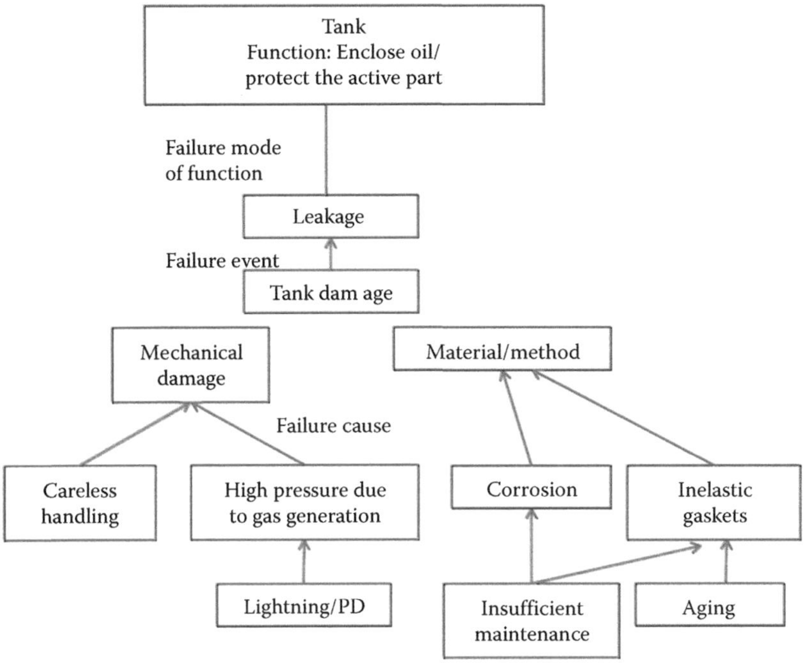

FIGURE 8.15
Fault tree for the tank.

and other signs of rough handling. Internal arcing in an oil-filled transformer can instantly vaporize surrounding oil which can lead to a high gas pressures that may rupture the tank [228]. In Figure 8.15, a fault tree for the tank is given.

8.2.2.2.4 Solid Insulation

The solid insulation in a transformer is cellulose-based products such as press board and paper. Its function is to provide dielectric and mechanical isolation to the windings. Cellulose consists of long chains of glucose rings. When degradation of the cellulose occurs, these chains get shorter. Degree of polymerization (DP) is the average number of these rings in the chain and indicates the condition of the paper. New paper has an average DP number of 1200–1400. A DP less than 200 means that the paper has a poor mechanical strength and may no longer withstand short circuit and other mechanical forces that appear. The solid insulation is the weakest link in the transformer insulation system, due to the degradation of the cellulose is irreversible and it is often not economically defensible to replace it. The aging of cellulose is accelerated by water, oxygen, and heat [229].

An analysis of the paper can tell the moisture content and the DP number of the paper insulation. There are few available methods that maintain

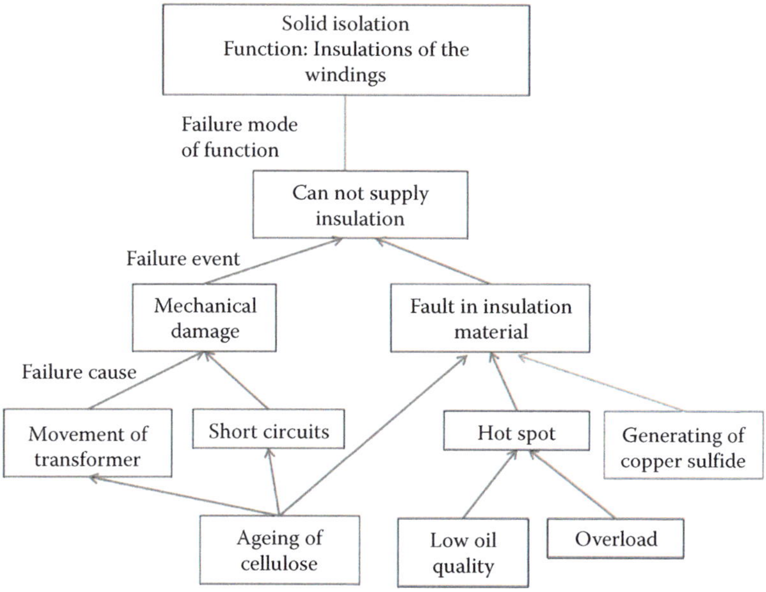

FIGURE 8.16
Fault tree for the cellulose insulation.

the paper insulation directly. One method is to dry the transformer, which reduces the water content in the insulation. Other maintenance like degassing of the oil also reduces the gas content in the paper as a new equilibrium between the gas in the oil and in the paper is reached [230].

Figure 8.16 shows a fault tree of the cellulose.

8.2.2.2.5 Cooler and Oil Insulation

The functions of the cooler and oil insulation are two: cool the active part of the transformer and electrical insulation between the different parts [227]. In Figure 8.17, a fault tree of the oil is given.

8.2.2.2.5.1 Cooler The cooling system can be based on either natural or forced circulation of oil and air or water. If the cooling system is based on natural circulation of oil and air, there are no pumps or fans. In the following presentation, it is assumed that the oil and air or water circulation are forced by pumps.

The function of the cooler is to cool the active part of the transformer. The cooler can be unable to cool due to absented cooling of the oil. This can happen due to two fault causes: either because of malfunction of the oil circulation or because of too high temperature in the second cooling medium, air or water.

The first can be caused by the failure of the oil pump or by dirt and particles in the oil, that increases the viscosity of the oil. Particles can be formed in the

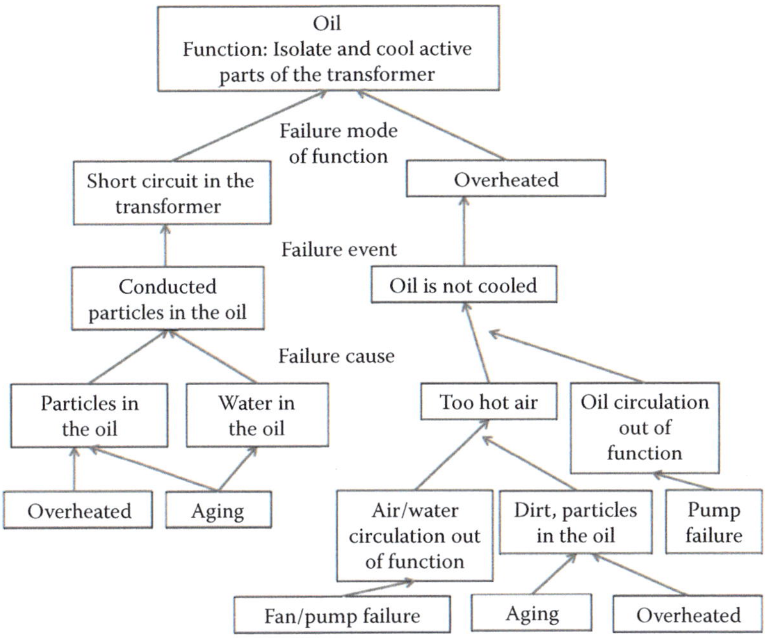

FIGURE 8.17
Fault tree for the oil insulation.

oil due to aging. In transformers with forced oil circulation, there is a risk that the pump breaks.

If the pump is unable to circulate the oil, the oil temperature inside the transformer will rise and there is a risk that the transformer will be overheated [214]. Too high temperature in the second cooling medium is caused by bad air/water circulation due to the failure of the fan or pump. High temperature outside the transformer can be the reason to high temperature of the cooling air.

8.2.2.2.5.2 Oil Insulation The transformer oil is a highly refined product from mineral crude oil and consists of hydrocarbon composition of which the most common are paraffin, naphthene, and aromatic oils. The oil serves as both cooling medium and part of the insulation system. The quality of the oil greatly affects the insulation and cooling properties of the transformer. The major causes of oil deterioration are due to moisture and oxygen coupled with heat.

Another function of the oil is to impregnate the cellulose and isolate between the different parts in the transformer. If the isolation fails, there is a short circuit. A short circuit can appear if there are conducted particles present in the oil. Conducted particles are, for example, water that appears in the oil as a result of the aging process of cellulose, and other particles, for example, metal, these particles are also a result of aging.

8.2.2.2.6 Bushings

There are many sorts of bushings, and they are classified either according to isolation material on ends, or inside the bushings or to the construction. The bushings on an oil filled transformer are air-to-oil bushings, which means that the isolation material on the ends are air and on the other its oil. The isolation material inside the bushing can be oil, oil-impregnated paper, resin-impregnated paper or gas. Oil-insulated bushings can either have an own closed oil system or share oil with the transformer. The dielectric strength of the isolation material decides the length of the bushing. There are two main constructions of bushings: solid and capacitance-graded bushings. The solid bushing has a central conductor and porcelain or epoxy insulators at either end. They are used in small distribution transformers as well as in large generator step-up transformers. Capacitance-graded bushings are used for all voltage ratings above 25 kV. At predetermined radial intervals, between the conductor and the insulator, conducting layers of oil-impregnated paper or other insulation material are located [232].

The function of the bushings is to isolate electrical between tank and windings and to connect the windings to the power system outside the transformer [227]. Figure 8.18 shows a fault tree for the bushing.

The main failure mode of the bushing is short circuit. A short circuit in the bushing can either happen due to material faults in the isolation or due to damage. A damage on bushings of porcelain can occur due to earthquakes or sabotage like stone throwing [214]. A damage can also occur during handling, shipping, or due to flying parts from other failed equipment [232]. A material fault in the isolation can occur due to water or dirt. Damages, cracks in the porcelain, and bad gaskets can give water an entrance to the oil isolation in the bushing [214].

It is important that the gasket between the transformer tank and the bushing is absolutely tight so that no air or water are allowed to enter the transformer. As the gasket ages, it becomes inelastic and allows water and air to enter [214].

Both water and dirt in the bushings is a consequence of insufficient maintenance.

Periodic inspections and maintenance are made on bushings in order to find potential problems. It is important that the oil level remains on a normal level. If the oil level decreases, it is a sign of a leak in the bushing. If the oil level sinks resulting in exposure of the paper isolated parts, it can lead to a dielectric failure in the bushing. If the leak is on the air side, it can give water an entrance to the bushing [232].

If the transformer is situated in a highly polluted environment, the bushings shall be washed regularly [232].

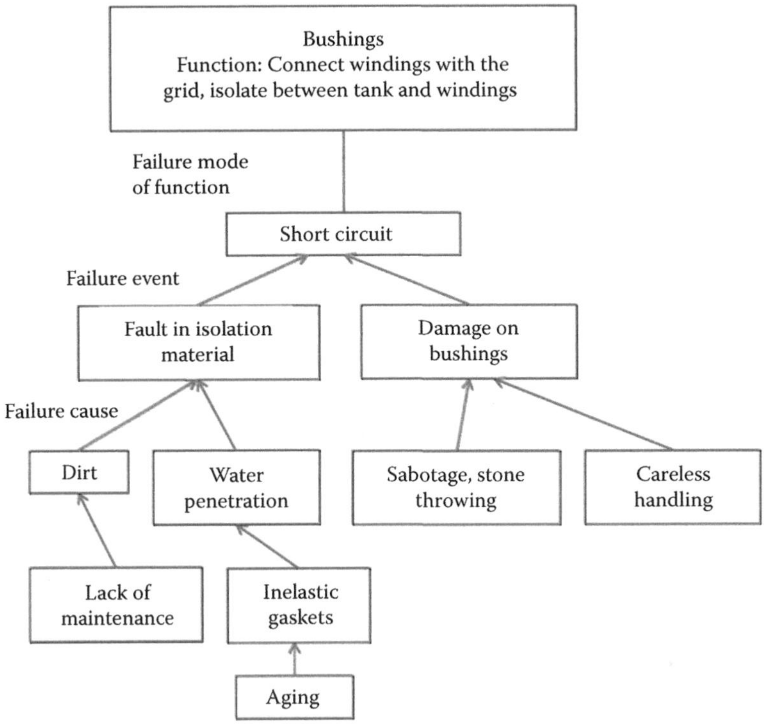

FIGURE 8.18
Fault tree for the bushings.

8.2.2.2.7 *Tap Changer*

The function of an OLTC is to regulate the voltage level by adding or subtracting turns from the transformer windings. The OLTC is built in two separate sections: the diverter switch and the tap selector. Due to the fact an interrupting of the supply is unacceptable for a power transformer, these are fitted with a complex mechanism that changes turns ratio without interrupting the load current. To obtain a noninterrupted flow current, the tap change procedure is performed in two steps [232].

1. The next tap is preselected by the tap selector at no load.
2. The diverter switch transfers the load current from preselected tap.

The tap selector makes the new tap connection before releasing the old and avoids the high current from the short-circuited turns by temporarily placing a large diverter resistor in series with the short-circuited turns before breaking the original tap connection. To avoid contamination of the transformer oil, the diverter switch has its own oil filled housing separate from the rest of the transformer.

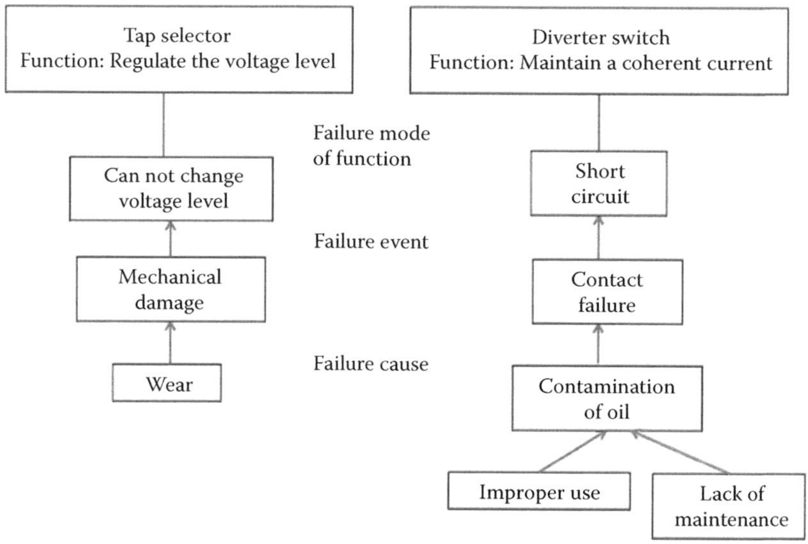

FIGURE 8.19
Fault tree for the tap selector and the diverter switch.

An annual inspection should be carried out to read the counting device that counts number of operations performed by the diverter switch. These readings are used to determine when an overhaul should be performed. The oil in the diverter switch housing has to be cleaned with certain intervals. The only parts that should require maintenance during the service life according to [233] are the diverter switch contacts that may need to be replaced, and the Motor Drive Mechanism which provides the drive to allow the OLTC to operate.

8.2.2.3 Different Lifetime Definitions

This section presents an overview on lifetime definitions from literature survey and results from discussions within the RCM analysis group.

First the definition of relative aging rate and loss of life in the cellulose is described according to the IEC standard. Then some lifetime definitions used in other literature are given and at the end a proposed life time model is described.

8.2.2.3.1 Relative Aging Rate

In the IEC standard [234], a method to calculate the relative aging rate of the cellulose isolation, depending on the hot spot temperature, is given. This aging rate is then used to evaluate the isolations loss of life.

The model presented in [234] is only based on the insulation temperature as the controlling parameter. The relative aging rate $V = 1,0$ corresponds to a temperature of 98°C for nonthermally upgraded paper and to 110°C for thermally upgraded paper. Since the temperature distribution is not uniform, the part that is operating at the highest temperature will undergo the greatest deterioration. Therefore, the rate of aging is referred to the winding hot spot temperature. The relative aging rate for nonthermally upgraded paper is defined according to the following equation:

$$V_n = 2^{(\Theta_h - 98)/6} \tag{8.15}$$

and for thermally upgraded paper

$$V_n = e^{\left(\frac{15,000}{110+273} - \frac{15,000}{\Theta_h + 273}\right)} \tag{8.16}$$

where Θ_h is the hot spot temperature in°C.

8.2.2.3.1.1 Loss of Life Calculation The model for the loss of life can also be found in [234]. The loss of life L over a certain period of time is defined as follows:

$$L = \int_{t_1}^{t_2} V_n \mathrm{d}t \tag{8.17}$$

where

- V_n is the relative aging during interval n according to 1 and 2
- t_n is the nth time interval
- n is the number of each time interval
- N is the total number of intervals during the period considered

8.2.2.3.2 Some Lifetime Definitions

8.2.2.3.2.1 Economical Lifetime Economical lifetime is a theoretic concept and refers to the time up to the replacement time which gives the optimal profit from an alternative investment [217].

8.2.2.3.2.2 Technical Lifetime The technical lifetime for a component is defined as the time until the component is unable to perform the desired function and it is impossible to repair the component [236].

8.2.2.3.2.3 Strategic Lifetime When the strategic lifetime of the transformer is at the end, it is replaced because of changes in the power system. The old transformer is not able to transform between the wanted voltage levels, either because it is not designed for these voltage levels and currents or the power supplier wants a safer transformer in that place [214].

8.2.2.3.2.4 Equivalent Lifetime Different transformers are situated on different places in the power system and naturally they are exposed to different stresses like, for example, short circuits. Therefore, in order to compare different transformers with each other, it can be useful to define an equivalent lifetime.

In the report "Livslängdsdefinitioner för krafttransformatorer" Elforsk defined the equivalent lifetime as

$$L_e = \circ A + \sum LK + \sum LM$$

where $\circ A$ is the transformers age, *LK* are the stresses from lifetime consuming factors, and *LM* are the stresses from lifetime intensifying factors. LK's are positive but LM's are negative numbers [236].

8.2.2.3.3 A Stochastic Transformer Lifetime

The lifetime of an equipment is stochastic. When working with maintenance, there is a need to connect the equipment's lifetime with the maintenance actions performed on it. Among the lifetime definitions above, there is only one, the definition of equivalent lifetime that takes maintenance and failures into account. However, that definition uses the age of the transformer to day and values on stresses from lifetime consuming and intensifying factors, that is, it is not stochastic. Here, a definition of a stochastic lifetime that takes the stochastic influences of maintenance and failures into consideration is proposed.

8.2.2.3.3.1 Proposed Definition The lifetime, T, of the transformer is a stochastic variable, that is, a stochastic function that depends on a couple of stochastic variables with different distributions

$$T(X, Y, Z) = X - \sum_{i=1}^{n} Y_i + \sum_{j=1}^{m} Z_j$$

where X is the lifetime according to the design of the transformer measured from the moment when the transformer is put into operation until it is shutdown. Y is a stochastic vector with n elements, each one corresponding to the impact of one "fault" type on the lifetime. Z is a stochastic vector with m elements, each one corresponding to the impact of a maintenance action on the lifetime.

It is possible that some of the Y's depend on time or, for example, the magnitude of a short circuit current. Y can, for example, depend on the time until the fault is discovered or the time until maintenance is done to correct the fault.

8.2.2.4 Ranking of Transformers

In [237,238], a condition-based evaluation method to assess the lifetime status of power equipment is represented. The method is developed by ABB and is a unit-oriented approach that concentrates on the serviceability for use of an individual unit. The objective of this method is to identify the most vulnerable units in a population by ranking them on the basis of certain aspects. This ranking is then basic data for decision making regarding to take appropriate actions on the highest priority units, that is, the most critical ones in the population. These activities could be either suitable maintenance activities or other adequate measures such as replacement or refurbishment. Some results from a case study done on eight GSU transformers applying this method are also represented in [237]. The following description of the method is considering transformers.

8.2.2.4.1 The Method Procedure

The method procedure could be divided in following five steps.

1. Identify a set of evaluation criteria
2. Collect data
3. Design inference net by formulate rules and relations
4. Determine the evaluation criterion value
5. Rank the units

1. *Identify a set of evaluation criteria*: The first most important question to ask is on what the evaluation shall be based on. To answer this question, one has to identify the different aspects (or risks) that could jeopardize the functionality or suitability for use of a unit in the population under consideration. Exactly which aspects that should be involved depend on the type of transformer, location, load, and so on. From the case study on a population of transformers in [237], following aspects were further investigated:
   a. Thermal aspects
      - Brittleness of aged paper
      - General aging of insulation (including oil)
      - Possible heating of core bandages

   b. Electrical aspects
      - General view based on influential
      - Factors including design, aging, and sign

   c. Mechanical aspects
      - Tilting (axial forces)
      - Buckling (radial forces)

   d. Loadability
      - Short-time emergency loading
      - Long-time emergency loading

The aspects that one wants to assess are denoted as evaluation criteria. The most important aspects for the evaluation have to be identified, and these will then constitute the set of evaluation criteria.

2. *Collect data*: Enough and relevant in data for each evaluation criterion is necessary. The input data could come from different sources:
   - Design and manufacturing
   - Normal operation and environmental conditions
   - Exceptional lifetime events in the power system
   - Unit maintenance, relocation, repairs, and so on
   - Monitoring and diagnostics
   - Oral information from experts in the field
3. *Designing the inference net by formulate rules and relations*: This step aims to develop evaluation procedures that shall determine a value for each evaluation criterion. These procedures can be seen as knowledge-based inference nets that describe causalities and influence factors. The input is unit data related to a certain aspect, and the output is an evaluation criterion value for the unit. The inference net consists of rules and relations that describe the dependencies between different input data for each evaluation criteria. The net could be a combination of different formulated rules, such as conventional formulas, weighting procedures, or logical connections. The inference should be designed so that the resulting evaluation criteria value is a relative measure of the technical risk. To determine which rules each evaluation criteria are required, it is necessary to first describe the model dependencies and relationships between the known information. The rules in the inference net for the case study in [238] are defined by transformer experts and the evaluation procedure is data driven.
4. *Determine the evaluation criterion value*: The evaluation criterion value for each unit is determined by using the inference nets developed in step 3.
5. *Rank the units*: Perform partial ranking by ranking the units with the respect to one evaluation criterion at time outgoing from the values from step 4. Table 8.1 is the resulting partial ranking list from the case study in [238]. The lower ranking number the worse case the unit represents within the population. In cases there it was impossible to significantly discriminate the units they were given the same ranking number.

As can be seen in Table 8.4, various evaluations criteria can rank the transformer in different order. Hence, a ranking number for one aspect is not directly comparable to ranking number for another aspect. To perform an

TABLE 8.4

Ranking Numbers for Eight Transformers in the Case Study

Units	aF	rF	bp	co	gia	el
TRAFO 1	6	6	1	1	1	3
TRAFO 2	7	6	3	4	2	4
TRAFO 3	7	6	3	4	2	4
TRAFO 4	1	1	2	6	5	7
TRAFO 5	1	1	7	6	6	6
TRAFO 6	1	1	3	2	2	5
TRAFO 7	4	1	6	2	6	8
TRAFO 8	4	1	8	6	2	2

Source: Sundermann, U. and Petersson, A. N. F. L., Life assessment: Ranking of power transformer using condition based evaluation, a new approach, in CIGRE Paris Conference Paper, pp. 12–204, 1998. [237]; Franzén, A. and Karlsson, S., *Failure Modes and Effects Analysis of Transformers*: Royal Institute of Technology KTH, Department of Electrical Engineering, TRITA-EE 2007:040, January 2007. [225]

Note: The lower ranking number the worse case within the population.

Abbreviations: aF, axial force; bp, brittlepaper; co, thermal heating of core; el, electric aspects; gia, general insulation aging; rF, radial force.

overall ranking, that is, a ranking list that takes more than one aspect in consideration, the user of the equipment must decide which type of aspect to emphasize and if some weighting procedure should be used. Sometimes it may be more appropriate to perform a ranking which just involves some of the determined aspects. Table 8.4 is used to illustrate a constructed example how an overall ranking list could be performed. Each aspect i is assigned a weight w_i based on the relative importance to the overall transformer condition. We denote the calculated partial ranking number for aspect i and unit j as a_{ij}. Here, just for illustration purpose, the thermally aspects in a weighting procedure are emphasized by assigning $w_i = 10$ to the aspects bp, co, and gia ($i = 3, 4$, and 5) and $w_i = 1$ to the other aspects ($i = 1, 2$, and 6). In other words, this overall ranking is derived under the assumptions that the thermally aspects are 10 times more critical than the other three aspects.

Then $a_{orank,j}$ is calculated for the eight units according to Equation 8.18.

$$a_{orank,j} = \frac{\sum_{i=1}^{6} w_i a_{ij}}{\sum_{i=1}^{6} w_i} \quad \text{for } j = 1, \ldots, 8 \tag{8.18}$$

The result of the overall ranking calculations is represented in Table 8.5. The $a_{orank,j}$ column is the rounded values from the calculations and can be seen as the relative technical risk measure under the assumptions that the ranking number in Table 8.4 is scaled in that way. The Ranking column shows in which order actions should be taken within the population.

TABLE 8.5

Results of Overall Ranking Example of the Eight Transformers

Unit	$a_{orank,j}$	Ranking
TRAFO 1	1	1
TRAFO 2	3	3
TRAFO 3	3	3
TRAFO 4	4	4
TRAFO 5	6	7
TRAFO 6	2	2
TRAFO 7	5	5
TRAFO 8	5	6

Source: Franzén, A. and Karlsson, S., *Failure Modes and Effects Analysis of Transformers*: Royal Institute of Technology KTH, Department of Electrical Engineering, TRITA-EE 2007:040, January 2007. [225]

8.2.2.4.2 A Method Extension

Often not all of the transformers in a population are equally critical. A failure by some transformer can cause more consequential damage than others. An extension of the above-described method is represented in [239], there also the strategic or economic importance of the units to the user is considered.

8.2.2.4.2.1 Model Parameters In the extended model, there are two parameters:

1. Technical risk
2. Importance

The technical risk is the result from the steps 1–5, that is, the ranking list. Importance is cost consequences in a broad sense. Apart from economical losses, the importance parameter could, for example, also include potential risk consequences a unit failure brings. The goal is to scale the units in the population in terms of ascending importance. To derive this importance scale, the population must be investigated and it is up to the transformer owner to decide what the importance parameter should include.

8.2.2.4.2.2 Risk Index and Interpretation of Technical Risk and Importance Graph To obtain the risk index value, one has to plot each unit in a Technical Risk and Importance diagram. Scale the technical risk of each unit on a low-to-high scale, for example, 0–10. Then also order the strategic or economic importance of each unit on a scale, for example, 1–10. The closer an unit falls the "high priority area" the higher risk index value it will be assigned. Hence, units at the same distance from the "high priority area" have the same risk index value. The higher risk index value a unit has, the higher priority regarding to investments for maintenance, replacements, and so on. To illustrate an example, we use the results from the $a_{ovrank,j}$ column in Table 8.5.

Assuming that the results are appropriately scaled in a technical risk sense the technical risk is scaled for these eight units on a low-to-high 1–8 scale. Thus the numbers are "switched" in the $a_{ovrank,j}$ column to obtain this. After investigation of the population, it is concluded that it is appropriate to use a 1–10 scale for the importance and further that the units 1–2 have a low importance (importance = 1), 3–6 medium importance (importance = 5) and 9–10 have a high importance (importance = 9). Figure 8.20 shows the results in a Technical Risk and Importance graph.

According to the risk index method, units 6, 7, and 8 have the highest priority within the population. Note, for example, that the unit 1, that has the highest priority according to the overall ranking list (Table 8.5), is here assigned a lower priority due to the low importance within the population.

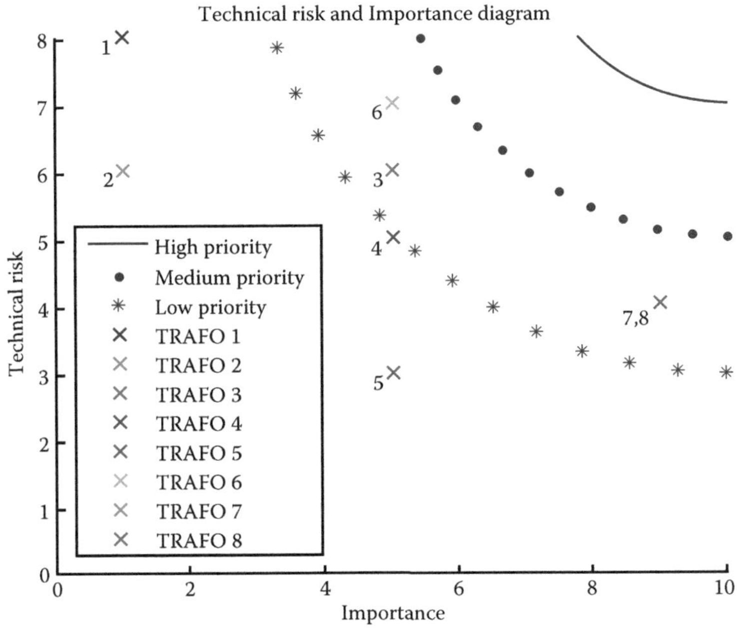

FIGURE 8.20
Combined equipment assessment considering total technical risk and importance aspects. The shorter distance the unit falls to the high priority area, the higher risk index it will be assigned. (Adapted from Franzén, A. and Karlsson, S., *Failure Modes and Effects Analysis of Transformers*: Royal Institute of Technology KTH, Department of Electrical Engineering, TRITA-EE 2007:040, January 2007.) [225]

8.3 Condition-Based Maintenance Management Framework

This section presents an approach for early fault detection which is then shown for a case study of gearbox bearings. The approach uses novel technologies from Artificial Neural Network Approach and input data from SCADA alarms [240]. The analysis shows example on RCAM Stage II where the gearbox has been previously defined as a critical component (Stage I). This was a result from system analysis which is further presented in Chapter 7.

8.3.1 Introduction

Gearbox has proven to be a major contributor toward downtime in wind turbines. The majority of failures in the gearbox originate from the gearbox bearings. An early indication of possible wear and tear in the gearbox bearings may be used for effective predictive maintenance, thereby reducing the overall cost of maintenance.

This section introduces a self-evolving maintenance scheduler framework for maintenance management of wind turbines. Furthermore, an artificial neural network (ANN)-based condition monitoring approach using data from supervisory control and data acquisition system is proposed. The ANN-based condition monitoring approach is applied to gearbox bearings with real data from onshore wind turbines, rated 2 MW, and located in the south of Sweden. The results demonstrate that the proposed ANN-based condition monitoring approach is capable of indicating severe damage in the components being monitored in advance.

The ANN model in the approach is capable of estimating the operating parameter being modeled with high accuracy. The presented statistical method for anomaly detection can detect an anomaly even when the error between the estimated and measured parameter is small. The ANN model uses the information rich data stored in the SCADA system, which otherwise often goes unnoticed. Furthermore, it also integrates the use of SCADA alarms and warning, with the condition monitoring. The ANN-based condition monitoring approach is applied to gearbox bearings with real data from onshore wind turbines, rated 2 MW and located in the south of Sweden. The results show that the ANN approach is able to indicate damage in the gearbox bearings almost a week before the vibration-based CMS. The gearbox was replaced after 3 months of detection of damage in the bearings. During this time, there exists an opportunity to make an optimal maintenance decision with a possibility to reduce the overall replacement cost.

8.3.2 Condition-Based Maintenance Management Framework

This section presents a framework for maintenance management of wind turbines, the self-evolving maintenance scheduler (SEMS). The SEMS framework considers a short window of time, which exists between an indication of impeding failure from CMS and the eventual failure of the component. The overall logic of the SEMS framework is presented in Figure 8.21.

According to the SEMS framework, any alarm from the vibration-based CMS or the ANN-based condition monitoring approach will give intimation to the maintenance personnel to perform an onsite inspection of the specific component. The main outcome of this inspection is to judge the extent of damage to the component. The maintenance planning is initiated after the information from the inspection is available, considering remaining useful life of the damaged component, forecast of power from the wind turbine and the forecasted weather windows suitable for maintenance. The maintenance decision can be optimized by considering various factors like opportunistic maintenance and minimization of the loss of production due to downtime. The SEMS framework relates the indication of impending failure from CMS to the maintenance activity, which could be scheduled replacement of the damaged component.

A feedback loop is shown in Figure 8.21. Through this feedback, information is given to the ANN model about a maintenance activity done on the component being monitored. This feedback enables the system to update the ANN model to keep in tune with changing operating conditions in the wind turbine, hence, giving a self-evolving feature to the framework. The approach for the updating the ANN model is presented in [241].

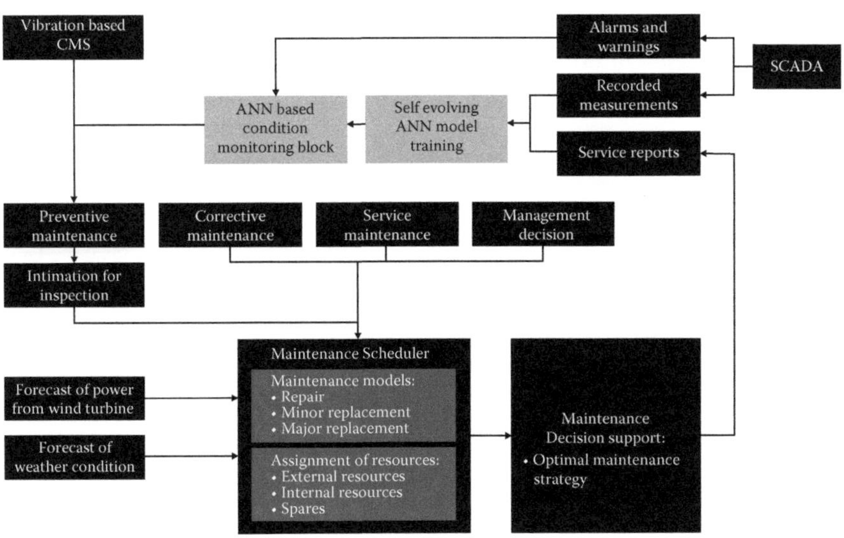

FIGURE 8.21
The SEMS Framework. (Banglore, P. and Bertling, L., An artificial neural network approach for early fault detection of gearbox bearings, *IEEE Transactions on Smart Grid*, **6**(2) © 2015 IEEE.) [240]

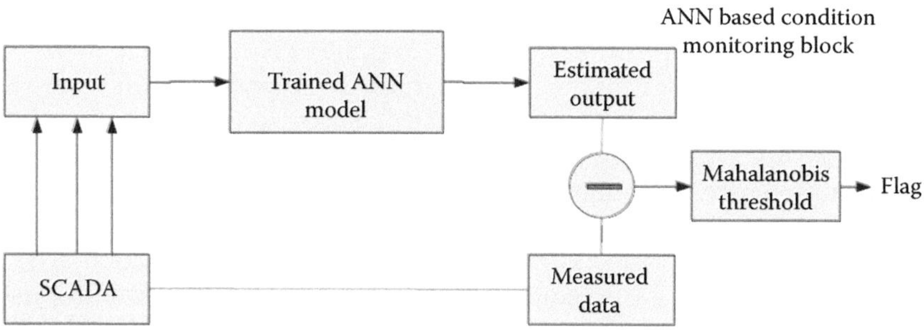

FIGURE 8.22
ANN-based condition monitoring block. (Banglore, P. and Bertling, L., An artificial neural network approach for early fault detection of gearbox bearings, *IEEE Transactions on Smart Grid*, **6**(2) © 2015 IEEE.) [240]

The ANN-based condition monitoring block in Figure 8.21 is expanded and presented in Figure 8.22.

The proposed ANN-based condition monitoring approach is based on a trained ANN model, which is used to estimate an operating parameter for the component being monitored. The estimated operating parameter is then compared to the actual parameter stored in SCADA. The MD is evaluated, which can then indicate to an abnormal behavior of the monitored component. Such an anomaly will raise a flag and initiate an onsite inspection of the component, in line with the SEMS framework.

8.3.3 ANN-Based Condition Monitoring Approach

8.3.3.1 ANN Model

ANN is based on the biological structure of the neural networks in living organisms. Connection of simple processing units together enables delivery of a complex process. Figure 8.23 presents the NARX structure, used in this study. In Figure 8.23, $u(t)$ represents the input vector at time instant "t" and $y(t)$ is the corresponding output. "w" and "W" are the weight vectors in the hidden layer and the output layer, respectively.

The dynamic behavior of an NARX model can be described as

$$y_t = F(y_{t-1}, ..., y_{t-q+1}; u_t, ..., u_{t-q+1}) \tag{8.19}$$

In Equation 8.19, F is the nonlinear function, "q" is the delay line memory vector [242]. In this study, the delay vector q has a value 1. By doing so, the output at time instant $t-1$ is used as an input to estimate the output at time instant t thereby simulating the thermal inertia in the gearbox system.

Levenberg–Marquardt backpropagation (LM) training algorithm is used for training the neural network. LM method has the combined advantage of Newton's method, which converges rapidly and gradient descent method,

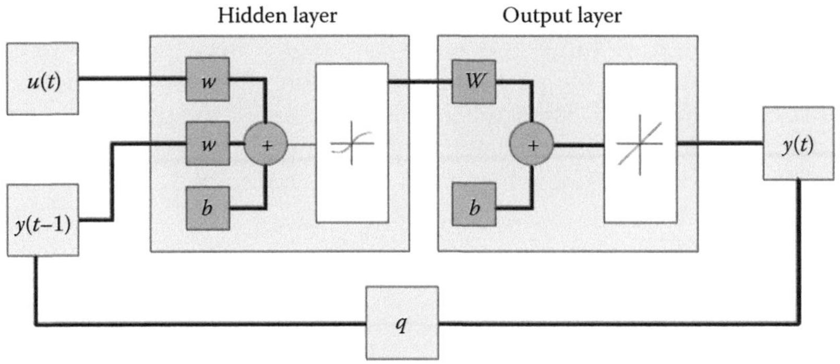

FIGURE 8.23
NARX ANN model structure. (Banglore, P. and Bertling, L., An artificial neural network approach for early fault detection of gearbox bearings, *IEEE Transactions on Smart Grid*, **6**(2) © 2015 IEEE.) [240]

which is assured to converge. The LM method is an efficient method to train neural networks with high precision [243].

The network is trained by minimizing the cost function represented in

$$\zeta(w) = \frac{1}{2N} \cdot \sum_{i=1}^{N} [d(i) - F(x(i); w)]^2 \tag{8.20}$$

where $\{x(i), d(i)\}_{i=1}^{N}$ is the training sample and $F(x(i);w)$ is the nonlinear approximation function for the network models, w is the weight vector and N is the total number of sample points.

According to the LM method, the weight vector is updated as

$$\Delta w = [H + \lambda I]^{-1} g \tag{8.21}$$

where H is the Hessian matrix approximated as per Equation 8.22 and g is the gradient vector defined as per Equation 8.23. I is an identity matrix with dimensions same as H and λ is a scalar parameter used to switch between Newton's method and gradient decent method.

$$H = \frac{1}{N} \sum_{i=1}^{N} \left[\frac{\delta F(x(i); w)}{\delta w} \right] \left[\frac{\delta F(x(i); w)}{\delta w} \right]^T \tag{8.22}$$

$$g = \frac{\delta \zeta(w)}{\delta w} \tag{8.23}$$

If the value of λ is zero, Equation 8.21 reduces to Newton's method and if λ is large enough to overpower H the method is similar to gradient descent method. The aim of the method is to move toward Newton's method, which is fast near minimum value and hence value of λ is reduced at each consecutive step as long as the performance function of the network defined by Equation 8.20 is reduced. The value of λ is increased, if the performance function increases for a consecutive step. The value is increased or decreased by a factor of 10 and the initial value of λ is considered to be 1.

8.3.3.1.1 Anomaly Detection Using MD

An error in the ANN model is defined as the difference between estimated 10-min average temperature value, obtained from the ANN model and the actual temperature value recorded in SCADA. The error is calculated for each 10-min time step. Based on the analysis of the output of the ANN model, it was realized that using a simple root mean squared (rms) value of the error is not sufficient to detect an anomaly in the monitored component. Hence, there was a requirement to use advanced statistical method for the detection of outliers.

MD is a unit less distance measurement, which has the ability to capture correlation of variables in a process or a system. MD gives a univariate distance value for a multivariate data. MD has been applied successfully to capture outliers in different fields of application (see [244,245]). Hence, the MD measure was selected to detect outliers in the given data set.

MD can be calculated as depicted

$$\mathrm{MD}_i = \sqrt{(X_i - \mu)C^{-1}(X_i - \mu)^T} \tag{8.24}$$

$X_i = [X_1, X_2, ..., X_m]$ is the ith observation vector where m is the total number of parameters. $\mu = [\mu_1, \mu_2, ..., \mu_m]$ is the vector of mean values obtained from the healthy/reference data set and C is the covariance matrix.

The threshold for anomaly detection is calculated based on the MD values for errors obtained during the training stage. As the training is done on the data from a period of operation with no recorded abnormal operation of the component, it is considered that the MD values obtained during training are healthy values and hence are used as a reference for anomaly detection.

The Mahalanobis for the training data set is calculated considering the following parameters:

$$X_{ref} = [\text{Error, Target value}] \tag{8.25}$$

$$(MD_{ref})_i = \sqrt{((X_{ref})_i - \mu_{ref})C_{ref}^{-1}((X_{ref})_i - \mu_{ref})^T} \tag{8.26}$$

The reference vector is used to calculate the covariance matrix C_{ref} and the mean value vector μ_{ref}. $(MD_{ref})_i$ represents the MD for the ith row of reference vector. The "error" in Equation 8.25 is the difference between the temperature estimated by the ANN model and the measurement recorded in SCADA and "target value" is the recorded SCADA measurements used during training.

The MD values obtained during the training stage can be represented accurately by a two-parameter Weibull probability distribution function defined

$$f(x) = \beta\eta^{-\beta}(x)^{\beta-1}e^{-(\frac{x}{\eta})^{\beta}} \tag{8.27}$$

A new value of MD, obtained during the condition monitoring stage is calculated using

$$X_{CMstage} = [\text{Error, SCADA measurement}] \tag{8.28}$$

$$(MD_{CMstage})_i = \sqrt{((X_{CMstage})_i - \mu_{ref})C_{ref}^{-1}((X_{CMstage})_i - \mu_{ref})^T} \tag{8.29}$$

The "SCADA measurement" in Equation 8.28 is the actual temperature recorded in the SCADA during the condition monitoring and "Error" is the difference between the ANN estimated and actual measured quantities. The MD value is considered as an anomaly if $f(MD_{CMstage})_i < 0.01$. Hence, any MD value obtained during monitoring stage, which has a probability of occurrence less than 0.01 is considered to indicate an anomaly in the component.

8.3.4 Analysis of SCADA Alarms and Warnings

SCADA alarms and warnings are important indicators toward the operating state of the wind turbine components. However, if there are a large number of wind turbines in the wind farm, the number of SCADA alarms and warning occurring may become overwhelming. It might become difficult to keep track of each kind of alarm and warning occurring in wind turbines. To overcome this difficulty, an approach is proposed to club the analysis of SCADA alarms and warning with the above-mentioned condition monitoring approach. To achieve this combined analysis, a classification procedure is suggested in this case study. The alarms and warnings occurring in the SCADA system are classified using a three character classification system. The first and second characters are classified as per Table 8.6. The alarms and warnings are divided in nine major categories depending on the component or system which is responsible for the alarm or warning. The second character of the classification is decided based on the subcomponent or subsystem, which is responsible for the SCADA alarm or warning. The third character is based on the way the alarm or warning is acknowledged and is done as per Table 8.7.

The classification can be visualized through an example; the classification code 712 represents a remote acknowledged alarm in the transmission system in gearbox lubrication system.

TABLE 8.6

First and Second Character for SCADA Alarms and Warnings Classification

1st Character Which component/system has caused the alarm/warning		2nd Character Which subcomponent/system has caused alarm/warning	
Environment	1	Wind speed	1
		Ambient temperature	2
		Shadow detection	3
		Ice detection	4
		Environment other	5
Generator	2	Generator control system	1
		Electrical machine	2
		Mechanical components +	3
		Rotor and control	4
		Invertor/convertor	5
		Generator other	6
Grid	3	Transformer	1
		Grid control system	2
		Grid other	3
Nacelle	4	Temperature control	1
		Nacelle other	2
Rotor	5	Blades and pitch system	1
		Hydraulic system	2
		Rotor control system	3
		Rotor other	4
Tower	6	Tower	1
Transmission	7	Gearbox lubrication system	1
		Mechanical brake	2
		Gearbox	3
		Gearbox lubrication control	4
		Main shaft	5
Turbine	8	Turbine control system	1
		Manual stop	2
		Turbine other	3
Yaw	9	Yaw control system	1
		Mechanical components	2

Source: Banglore, P. and Bertling, L., An artificial neural network approach for early fault detection of gearbox bearings, *IEEE Transactions on Smart Grid*, **6**(2) © 2015 IEEE. [240]

TABLE 8.7

Third Character for SCADA Alarms and Warnings Classification

3rd Character How is	
Auto	1
Remote	2
Local	3

Source: Banglore, P. and Bertling, L., An artificial neural network approach for early fault detection of gearbox bearings, *IEEE Transactions on Smart Grid*, **6**(2) © 2015 IEEE.) [240]

8.3.5 Case Study

This section illustrates for a real practice application: the SEMS framework and the proposed ANN-based condition monitoring approach.

Continuous SCADA data and maintenance records are used for two different wind turbines, of the same manufacturer, rated 2 MW and located onshore in the southern part of Sweden.

8.3.5.1 Gearbox

The gearbox used in wind turbines under consideration is a planetary gearbox combined with two-stage parallel shaft gearbox. This is a common configuration used in the wind industry due to its large ratio and power capacity. The gearbox has a flexible mounting and is connected to the generator shaft using composite coupling. The brake disc is mounted on the high speed shaft (HSS) of the gearbox coupled to the composite coupling. Several parameters of the gearbox such as bearing temperature, lubrication oil temperature, and lubrication oil pressure are monitored and recorded in SCADA system.

Figure 8.24 shows a schematic of a three-stage planetary gearbox with different bearings. Five different bearings are labeled: (1) PCB-A (planet carrier bearing rotor end); (2) PCB-B (planet carrier bearing nonrotor end); and (3) HSS-A/B/C (high-speed shaft bearings A, B, and C). Temperature measurement for these five bearings is available in SCADA system. A model for the normal behavior for each of the five bearings is created using ANN. These five ANN models will be used collectively to monitor the condition of the gearbox.

8.3.5.2 ANN Models

The ANN models for the five different bearings are trained on the SCADA data from 2011, where there have been no recorded damages in the gearbox. The approach presented [241] is used to decide the training dataset to train the ANN models. The five ANN models have a set of inputs and output

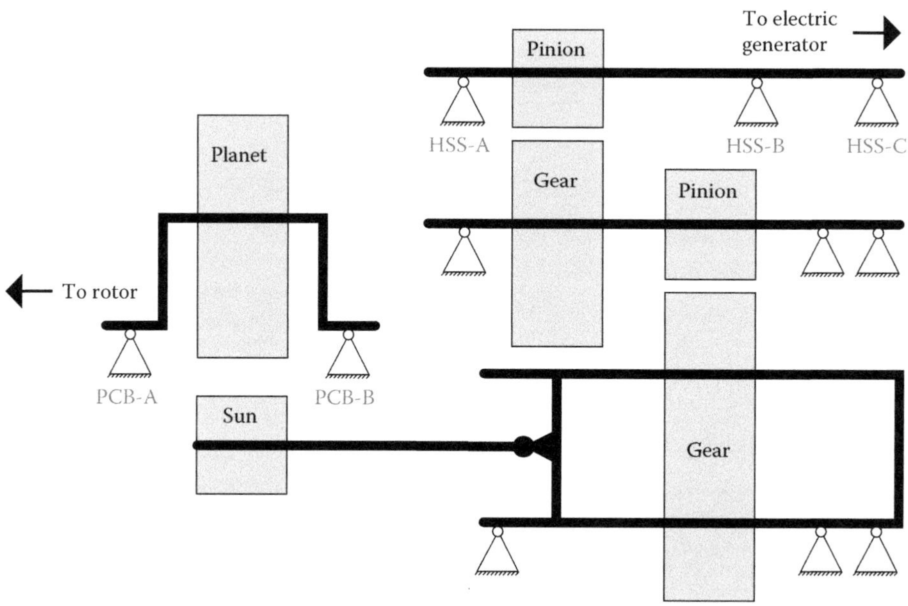

FIGURE 8.24
Schematic of gearbox with different bearing locations. (Adapted from Oyague, F., Gearbox reliability collaborative (GRC) description and loading, in *Nat. Renew. Energy Lab., Washington, DC, USA, Tech. Rep. NREL/TP-5000-47773*, November 2011.) [246]; (Banglore, P. and Bertling, L., An artificial neural network approach for early fault detection of gearbox bearings, *IEEE Transactions on Smart Grid*, **6**(2) © 2015 IEEE.) [240]

parameters. Table 8.8 presents the inputs and outputs for different bearing models. All the values presented in Table 8.8 are 10-min average values that are extracted from the SCADA database. It is assumed that the maintenance decisions are taken at discrete intervals of time. In line with this assumption a period of 3 days is selected, over which the output MD is averaged. The averaging of the distance measure over a period of time also reduces the possibility of false alarms due to erroneous signals from SCADA or shortcomings of the ANN model. Total number of SCADA alarms and warnings that have appeared in the time span of 3 days are considered and presented along with the MD on the same x-axis. Hence, all the results presented in this section are average MD values, and total number of SCADA alarms and warning over 3 days.

CASE I: TURBINE A1 WITH RECORDED GEARBOX BEARING FAILURE

The results for five bearing models for the time period from January 2012 to December 2012 are presented in Figures 8.25–8.29.

TABLE 8.8

Inputs and Output for ANN Models

Output/Input	PCB-A	PCB-B	HSS-A	HSS-B	HSS-C
Power generated	x	x	x	x	x
Gearbox oil temperature °C	x	x	x	x	x
Nacelle temperature °C	x	x	x	x	x
Rotor rotations per minute	x	x	x	x	x
PCB-A temperature °C	–	x	–	–	–
PCB-B temperature °C	x	–	–	–	–
HSS-A temperature °C	–	–	–	x	x
HSS-B temperature °C	–	–	x	–	x
HSS-C temperature °C	–	–	x	x	–

Source: Banglore, P. and Bertling, L., An artificial neural network approach for early fault detection of gearbox bearings, *IEEE Transactions on Smart Grid*, **6**(2) © 2015 IEEE. [240]

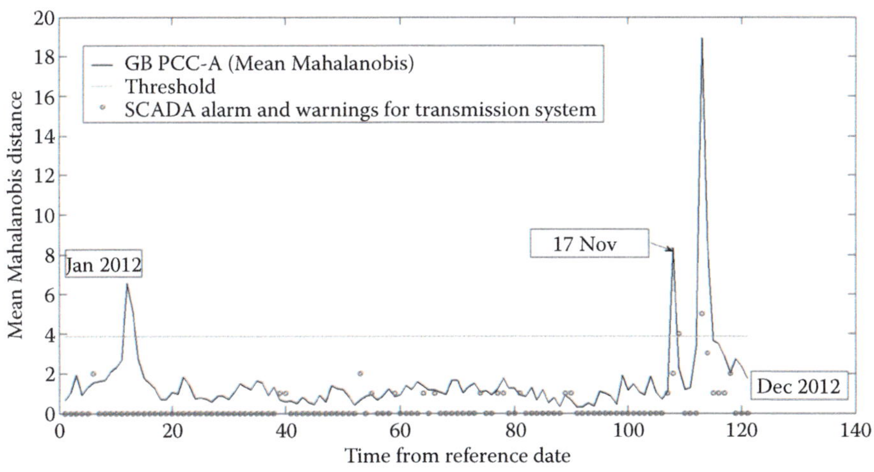

FIGURE 8.25
Average MD calculated for condition monitoring period of 1 year for PCB-A bearing for Turbine A1. (Banglore, P. and Bertling, L., An artificial neural network approach for early fault detection of gearbox bearings, *IEEE Transactions on Smart Grid*, **6**(2) © 2015 IEEE.) [240]

From Figure 8.25, it can be observed that on November 17, 2012, the MD has crossed the threshold. For turbine A1, the vibration-based CMS raised an alarm on November 23, 2012. After an inspection of the gearbox on November 28, 2012, it was found that the gearbox PCB-A was damaged due to spalling in the bearing as can be seen in Figure 8.30. The gearbox was replaced in February 2013.

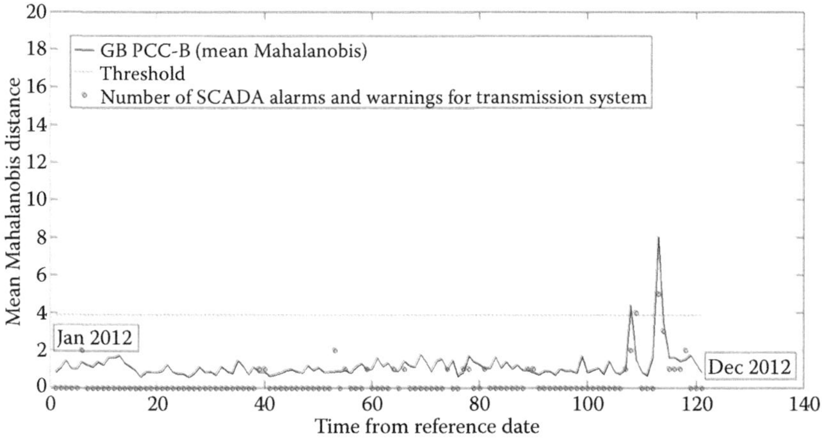

FIGURE 8.26
Average MD calculated for condition monitoring period of 1 year for PCB-B bearing for Turbine A1. (Banglore, P. and Bertling, L., An artificial neural network approach for early fault detection of gearbox bearings, *IEEE Transactions on Smart Grid*, **6**(2) © 2015 IEEE.) [240]

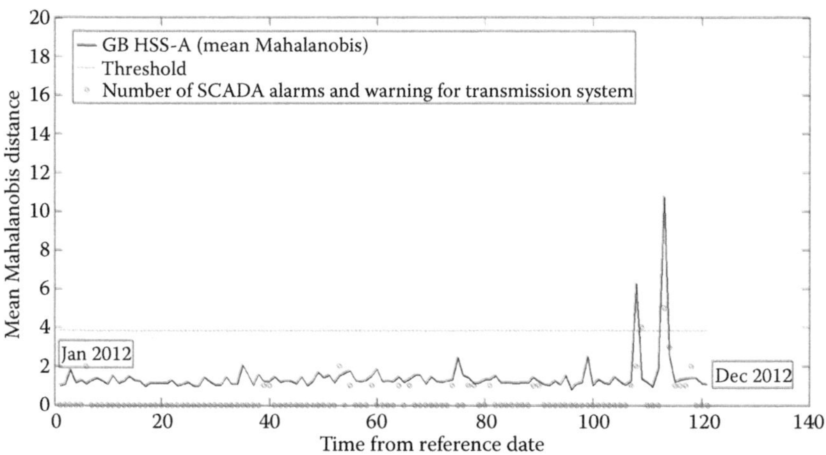

FIGURE 8.27
Average MD calculated for condition monitoring period of 1 year for HSS-A bearing for Turbine A1. (Banglore, P. and Bertling, L., An artificial neural network approach for early fault detection of gearbox bearings, *IEEE Transactions on Smart Grid*, **6**(2) © 2015 IEEE.) [240]

The proposed ANN-based condition monitoring approach was able to detect the deterioration in the bearing almost a week before the vibration-based CMS raised an alarm. From Figures 8.25 through 8.29, it can also be observed that the number of SCADA alarms and warnings for the transmission system were higher than normal during the period when the MD has crossed the threshold. Furthermore, it can be observed that the MD is higher for PCB-A bearing compared to the other four bearings. Hence, the amplitude of the MD can give

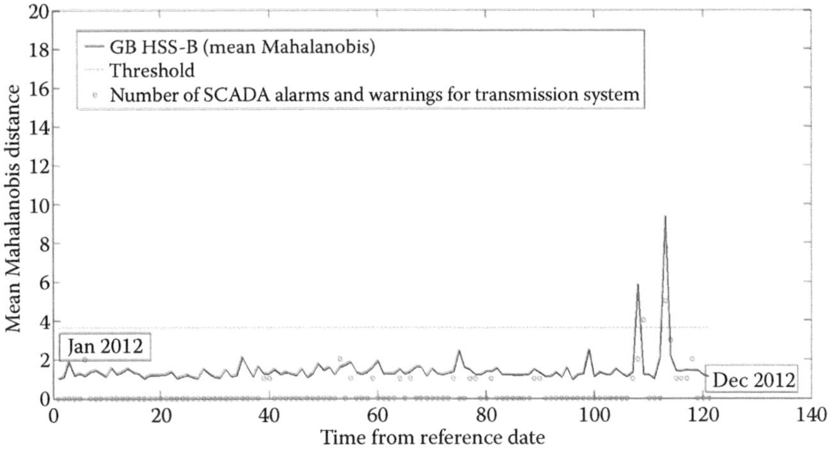

FIGURE 8.28
Average MD calculated for condition monitoring period of 1 year for HSS-B bearing for Turbine A1. (Banglore, P. and Bertling, L., An artificial neural network approach for early fault detection of gearbox bearings, *IEEE Transactions on Smart Grid*, **6**(2) © 2015 IEEE.) [240]

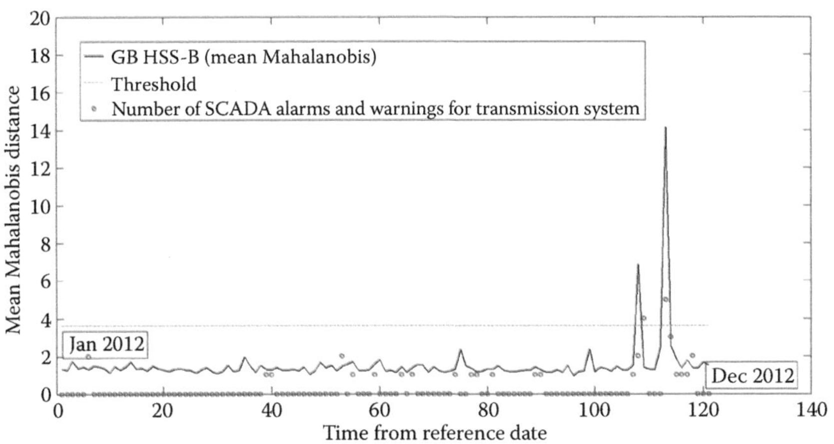

FIGURE 8.29
Average MD calculated for condition monitoring period of 1 year for HSS-C bearing for Turbine A1. (Banglore, P. and Bertling, L., An artificial neural network approach for early fault detection of gearbox bearings, *IEEE Transactions on Smart Grid*, **6**(2) © 2015 IEEE.) [240]

an indication to the bearing which might be the reason for the abnormal operation. Such an indication could aid focused inspection on a particular bearing thereby providing a possibility to reduce inspection time.

FIGURE 8.30
Spalled PCB-A bearing in turbine A1, realized after inspection following indication from vibration-based CMS. (Banglore, P. and Bertling, L., An artificial neural network approach for early fault detection of gearbox bearings, *IEEE Transactions on Smart Grid*, **6**(2) © 2015 IEEE.) [240]

CASE II: TURBINE A2 WITHOUT RECORDED GEARBOX BEARING FAILURE

In order to validate the ANN-based condition monitoring approach, an application to second wind turbine, rated 2 MW is presented. There has not been any recorded gearbox bearing failures in the wind turbine for the monitoring period of January 2012–December 2012. The output from the ANN models should not show any abnormal operation for the monitoring period. The results are presented in Figures 8.31 through 8.35. From the results, it can be seen that the gearbox bearings do not show any abnormal operation. It can be observed that the number of SCADA alarms and warnings are high during the beginning of the year. However, the MD has not crossed the threshold value and hence, these alarms do not present any immediate harm to the component under consideration. Hence, the proposed ANN-based condition monitoring approach has been validated for a case with recorded gearbox failure (turbine A1) and for a case without any recorded failure (turbine A2).

8.3.5.2.1 *Comparison with Previously Proposed Approaches*

The proposed ANN-based approach is compared with the previously published similar approaches in [247,248].

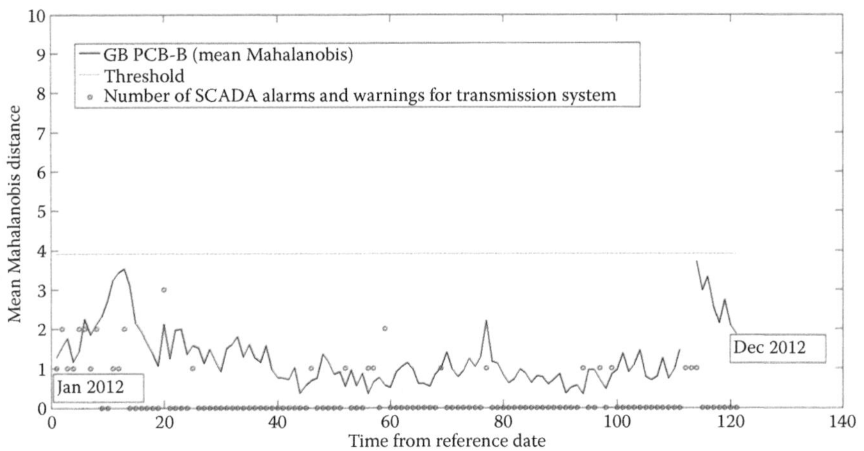

FIGURE 8.31
Average MD calculated for condition monitoring period of 1 year for PCB-A bearing for Turbine A2. (Banglore, P. and Bertling, L., An artificial neural network approach for early fault detection of gearbox bearings, *IEEE Transactions on Smart Grid*, **6**(2) © 2015 IEEE.) [240]

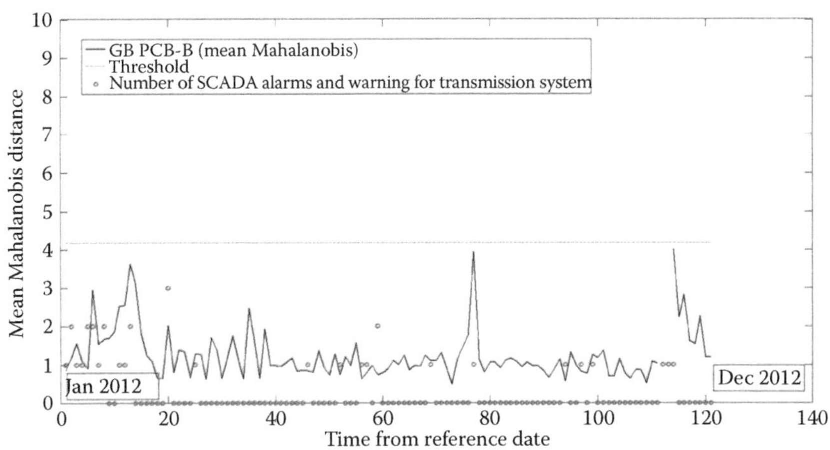

FIGURE 8.32
Average MD calculated for condition monitoring period of 1 year for PCB-B bearing for Turbine A2. (Banglore, P. and Bertling, L., An artificial neural network approach for early fault detection of gearbox bearings, *IEEE Transactions on Smart Grid*, **6**(2) © 2015 IEEE.) [240]

In [247], intelligent system for predictive maintenance has been introduced. The proposed approach uses ANN to model the normal behavior of the gearbox bearing temperature. A similar model is presented in [248], where the 10-min average temperature data is used to model the normal behavior of the temperature characteristics of the gearbox bearing using neural networks. In both the proposed approaches, anomaly detection is done based on the

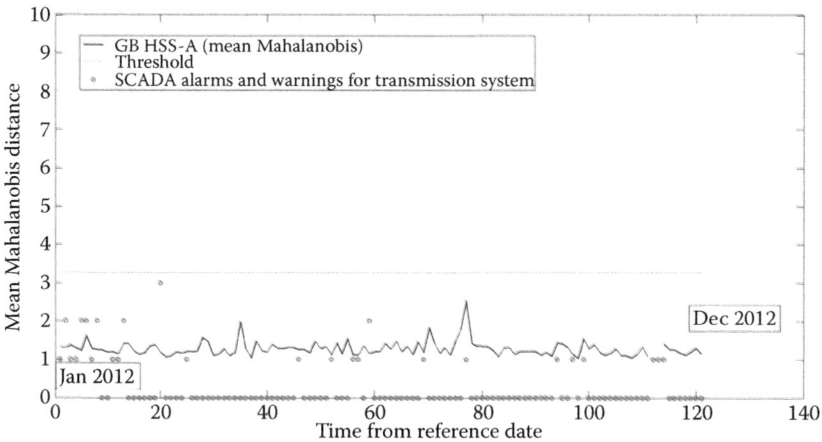

FIGURE 8.33
Average MD calculated for condition monitoring period of 1 year for HSS-A bearing for Turbine A2. (Banglore, P. and Bertling, L., An artificial neural network approach for early fault detection of gearbox bearings, *IEEE Transactions on Smart Grid*, **6**(2) © 2015 IEEE.) [240]

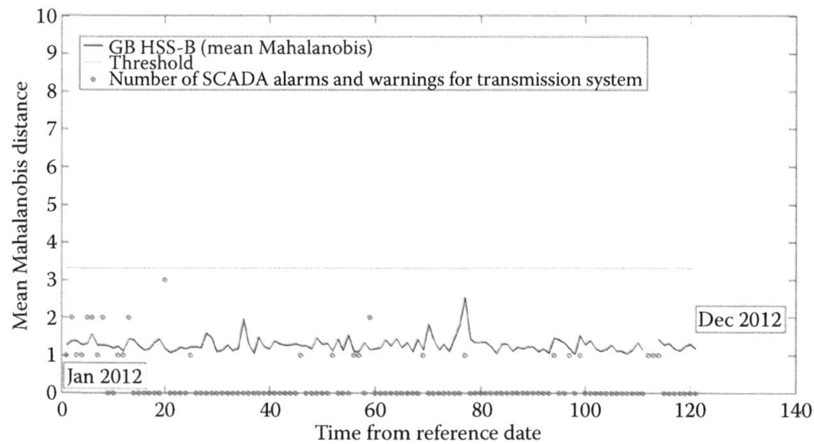

FIGURE 8.34
Average MD calculated for condition monitoring period of 1 year for HSS-B bearing for Turbine A2. (Banglore, P. and Bertling, L., An artificial neural network approach for early fault detection of gearbox bearings, *IEEE Transactions on Smart Grid*, **6**(2) © 2015 IEEE.) [240]

error between the ANN estimated temperature and the actual measured temperature. In [247], the anomaly is detected based on increasing error value in relation to a predefined confidence interval and in [248] the anomaly is detected based on increase in the duration and frequency of errors. However, with regards to the case studies presented it was found that the temperature might not deviate significantly, from the normal operating range, even when

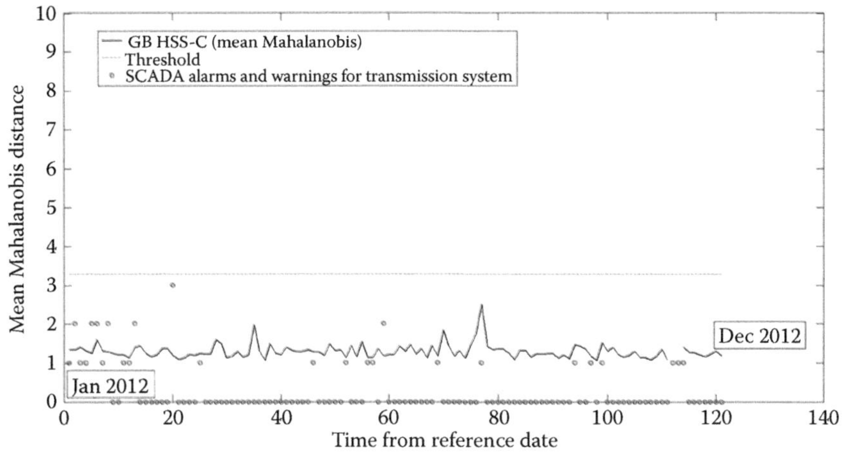

FIGURE 8.35
Average MD calculated for condition monitoring period of 1 year for HSS-C bearing for Turbine A2. (Banglore, P. and Bertling, L., An artificial neural network approach for early fault detection of gearbox bearings, *IEEE Transactions on Smart Grid*, **6**(2) © 2015 IEEE.) [240]

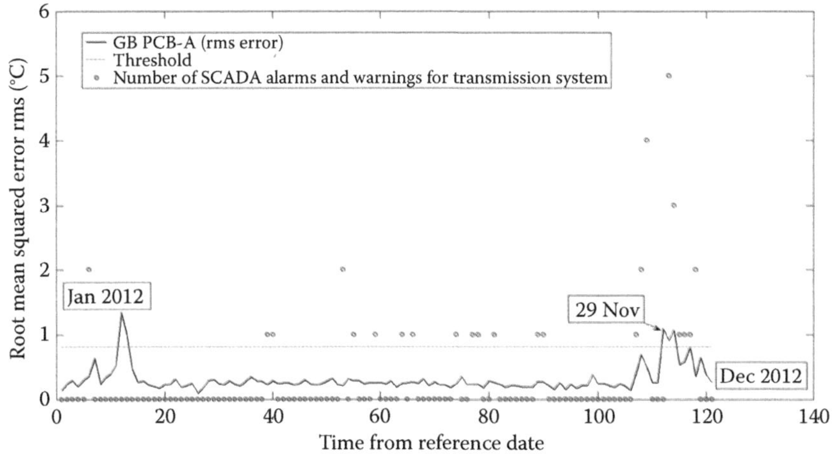

FIGURE 8.36
Average rms error calculated for condition monitoring period of 1 year for PCB-A bearing for Turbine A1. (Banglore, P. and Bertling, L., An artificial neural network approach for early fault detection of gearbox bearings, *IEEE Transactions on Smart Grid*, **6**(2) © 2015 IEEE.) [240]

an anomaly is present. The rms value of the error for turbine A1, which has a recorded gearbox failure, is shown in Figure 8.36.

Figure 8.36 shows the rms value for error averaged over a time interval of 3 days and presented for a period of 1 year for turbine A1. It can be observed from Figure 8.16 that the threshold value is set at 0.8°C. Even with a low threshold for difference in ANN estimated temperature and actual temperature, the system is not able to capture the fault in advance. Hence, to overcome this shortcoming of using solely the error value, an advance statistical outlier detection technique of MD is proposed in this study. Furthermore, the approach proposed in [248] does not define a simple threshold value for anomaly detection. The presented approach provides an easy understanding of the output of the ANN model by providing a threshold value.

The ANN models presented in [247,248] are trained on dataset created manually. However, the ANN model presented in this section is using an automatic approach to decide the training dataset. The approach makes it simpler to use the presented ANN-based approach, especially when the number of wind turbines is large.

As opposed to [247,248], the approach presented here also considers the SCADA generated alarms and warnings and aids the maintenance/operation personnel to distinguish the important alarms and warnings.

9

RCAM Case Maintenance Optimization

9.1 Introduction

This chapter provides a selection of different case studies. All the studies proposed different types of optimization models for reliability and/or maintenance evaluations. The case studies include electrical distribution systems, wind farms, and components in nuclear power plants.

9.2 Multiobjective Optimization for Electrical Distribution Systems

9.2.1 Introduction

One of the major goals for managers of electric power distribution systems is maximum asset performance. To reach this goal, both minimal life-cycle cost and maintenance optimization become crucial, while meeting demands from customers and regulators. This creates the necessity for determining optimal balance between PM and CM in order to obtain the lowest total cost.

This case study investigates the problem of balancing between PM and CM as a multiobjective optimization problem, with customer interruptions on one hand and the maintenance budget of the network operator on the other. Moreover, the proposed optimization model is solved with meta-heuristics in conjunction with an evolutionary particle swarm optimization algorithm. The case study is discussed in detail in [249]. An application study is presented which is performed on the urban electrical distribution system, that is, the Birka system, earlier presented in Chapter 6.

9.2.2 Proposed Optimization Model

This case study proposes an approach resulting in maintenance policy optimization. The solutions to the maintenance policy optimization are developed by utilizing component reliability importance indices. These are in turn derived from Monte Carlo reliability simulations.

The method is based on five essential concepts as follows:

1. Reliability Monte Carlo simulations
2. Component reliability importance indices
3. Multiobjective approach
4. Optimization heuristics
5. Selection of optimum (results)

9.2.2.1 Event-Driven Monte Carlo Simulation

The Monte Carlo simulation method is based on an event-driven approach, that is, a simulation with variable time steps, where the time between significant events is identified. Here significant events are state changes of components and minimal cut-sets are used to calculate the system behavior as a result of component status. The simulation runs for a specified number of cycles in the current set-up. It is assumed that all components are functional at the start of simulation. The next step is to generate the time-to-failure for every component. Time-to-failure is based on the component's failure rate, which in turn is based on component parameters, for example, level of maintenance and inherent failure rate. Then move forward in time to the first event, which here is a failed component. The changed status of the failed component is then recorded along with the effects of the failure. The list of coming events is consequently updated with the generated repair time for the failed component. This is followed by the identification of the next event in the simulation, probably the repair of the failed component but not impossibly an occurrence of another failed component and its consequences (e.g., a second-order minimal cut-set failure). Reliability data is obtained by recording the effects of the different system states and their durations, such as the utilized component importance index, and more traditional reliability data, such as customer interruption costs and SAIFI and SAIDI [79]. The simulation approach and its implementation into the reliability analysis tool RADPOW are presented in Chapter 3.

9.2.2.2 Component Reliability Importance Index

Component reliability importance indices provide the connection between component and system reliability performance [45]. The component reliability importance index used in this research paper, [250], corresponds to the expected total customer interruption cost caused by the studied component over a specific time interval (usually 1 year). The index is calculated by accumulating the total interruption cost caused by every interruption to the finally causing component over all simulated years, and this index was developed for calculation by simulation [250]. The accumulated cost for every component is then divided by the total simulation time in order to obtain an expected interruption cost per time unit. The index is defined as follows:

$$I_i^M = \frac{K_i}{T} \quad \text{[euros/yr]} \tag{9.1}$$

where K_i is the total accumulated customer interruption cost over the total simulated time T for component i. The index gives an indication of which components should be prioritized for PM actions (or in some cases redesigns of the structures that result in the high value of I^M). Moreover, I^M gives information on components that do not incur much interruption cost for the network. It might, for example, be beneficial to reduce PM for these components. This information needs to be adequately combined with knowledge about available maintenance actions, their costs, and estimated effects for reaching the best possible solution.

9.2.2.3 Multiobjective Approach to the Optimization Problem

The task of finding the optimal balance of preventive maintenance (PM) and corrective maintenance (CM) is approached as a multicriteria/multiobjective optimization problem. There are customer's demands for electrical power delivery on one hand, and the maintenance cost for the DSO on the other. In this case study, the total customer interruption cost is used as the measure of the performance of the electric distribution system from the customer perspective. The maintenance costs are closely related to the analyzed electrical distribution system including its components, structure, and available resources.

The multiobjective approach can be extended by studying every load point's availability as an individual objective. Some cases might, for example, call for pure Pareto improvements [251], where all customers are viewed separately, that is, improvements that reduce costs or at least keep costs at current levels for all parties involved. To study all customers independently while requiring Pareto improvements narrows down the feasible solution space. Furthermore, with more objectives, the solution space quickly becomes difficult to grasp with the increasing number of load points.

The two objectives (customer interruption cost and cost of maintenance) do not entirely point the solution in two different directions since the cost of CM to a certain degree correlates with the customers' inconvenience.

9.2.2.4 Heuristic Optimization Approach: Approximate Gradient Evaluation Bicriteria Optimization Method

The proposed optimization model in this case study is based on an aggregated auxiliary objective function that incorporates the two objectives of the customer interruption cost and the maintenance cost of the electrical distribution system (including both CM and PM). The model is referred as Approximate Gradient Evaluation Bicriteria Optimization Method (AGE-BOM) [249].

A scaling between the customer interruption cost and the maintenance cost of the electrical distribution system is introduced. This scaling is varied in order to obtain a number of nondominated solutions with specific tradeoff between customers and DSO. The resulting objective function of the optimization is

$$mins \cdot C^{IC} + C^{CM} + C^{PM} \quad [\text{euros/yr}] \tag{9.2}$$

where C^{IC}[euros/yr] is the expected yearly system customer interruption cost, C^{CM}[euros/yr] the cost of CM, C^{PM}[euros/yr] the cost of PM, and s is the scale factor (tradeoff). The unit of the scale factor, s, becomes DSO money per unit of customer money. The scale factor constitutes a translation of the expected customer interruption costs into terms of DSO costs. C^{IC} is obtained from simulations [250] and depends on the maintenance strategy. The values of C^{CM} and C^{PM} are based on the maintenance strategy; see the case study for an example. The three different costs in the objective function depend on the network and its components. They constitute component failure rates, repair times, and network structure and operation. In addition, the simulation delivers the component reliability importance index, I^M [euros/yr], for every component [250]. The index I^M corresponds to the average yearly customer interruption cost caused by the specific component. How C^{CM} and C^{PM} are calculated depends on the specific electric distribution system under study. For this case study these costs are presented as follows:

$$C^{IC} = \sum_{L} \lambda_L (k_L \cdot P_L + c_L \cdot P_L \cdot r_L) \quad [\text{euros/yr}] \tag{9.3}$$

where C^{IC} is the total expected yearly customer interruption cost for the system, P_L [kW] average power, λ_L [f/yr], and r_L [h/f] are reliability indices for every load point L, and k_L [euros/f, kW] and c_L [euros/kWh] are cost constants representing the customer types and composition for every load point L.

It is assumed that the caused interruption cost is linearly dependent on the failure rate of the component, when no other data are changed, that is, assuming that a relative change in failure rate results in the same relative change in customer interruption cost. Given maintenance actions and estimates of failure rate changes and maintenance cost/savings caused by these, a cost–benefit ratio can be developed. This is the ratio between the change in interruption cost and the cost/savings of the investigated action. By doing this for all available actions for all components, the available actions can be ranked.

The optimization, which is a steepest descent method, commences with a leap. The leap introduces the best cost–benefit ratio actions for each component into the solution. This is done despite violating the assumption above.

The leap is followed by a stepwise approach that does not violate the above assumption. In other words, all available maintenance actions are evaluated, but only the most profitable one is selected, given that it is expected to result in a better objective function. Figure 9.1 illustrates the optimization process for this case study.

The start condition for the optimization is that all components are at their initial (current) state. Then a reliability calculation (simulation) is performed. The index, I^M, which is an output from the simulation, is used in order to estimate the impacts on the objective of all maintenance actions available (see Equation 9.4). The estimates are then used to select all seemingly beneficial maintenance actions (this is the leap). This is done despite the fact that every maintenance action is evaluated individually, neglecting the consequences of all the other actions. This approach does not warrant a local optimum being reached, and therefore, proceed with more cautious "steps," that is, a new simulation based on the maintenance actions chosen from the previous step/leap is continued with. All available maintenance actions as before are evaluated, but here only select the most beneficial maintenance action (hence, this is called a step). The steps and reliability calculations are then performed until no more improvements are found. The achieved optimal point is stored. One optimization cycle is then accomplished and the scale, s, is incremented. The calculation continues with a leap starting from the previous optimum. This is continued until there are no more scales to optimize for.

In both leap and step, the following approximate evaluation of each individual maintenance alternative is made

$$P_{ij} \approx \Delta\lambda_{ij} C_i^{CM} + \Delta C_{ij}^{PM} + \Delta\lambda_{ij} \frac{I_i^M}{\lambda_i} \quad \text{(euros/yr, action)} \tag{9.4}$$

where P_{ij} is the expected change of the objective function j if maintenance alternative is implemented for component i, s is the scale factor of customer interruption costs, λ is the component failure rate, and Δ denotes the change from the current maintenance policy in the optimization process. C_i^{CM} corresponds to the expected CM cost that a failure of component i incurs for the DSO. C_{ij}^{PM} depends on the cost difference of introducing the studied maintenance action compared to the current policy. Equation 9.4 is approximate since the last term is an approximation of how the customer interruption cost will change. For each component in the leap, the lowest value of P is chosen for the next simulation. In the steps, only one action is chosen, that is, based on the lowest value, given that this value is below zero.

9.2.2.5 Results and Selection of Optimum

Given that the method has not found local optima, the suggested approach will deliver a number of optimal points, exemplified in the case study below. These points will be located on the Pareto border and are all optimal from a

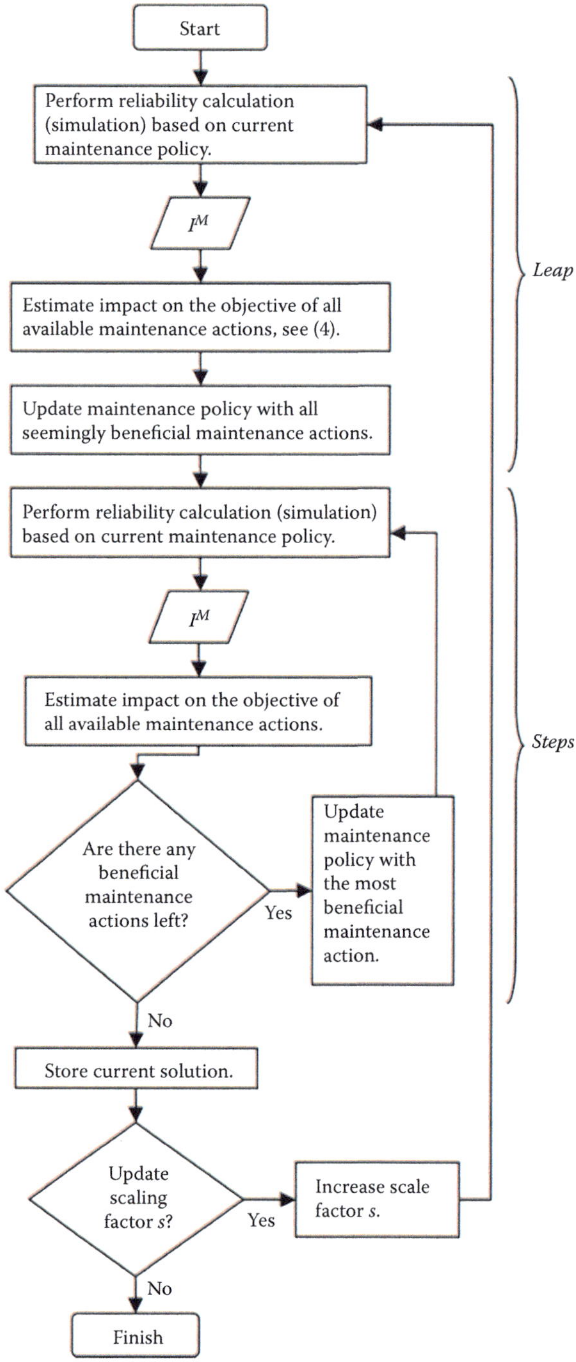

FIGURE 9.1
Flowchart for the optimization process. (Adapted from Hilber, P. et al., *IEEE Transactions on Power Systems*, **22**(4), 1675–1682, 2007 [249].)

specific point of view. The solution that is selected by the decision maker depends on many factors such as the current status and behavior of the network. The different solutions provide the decision maker with important information on the expected consequences of the different maintenance policies.

9.2.3 Application

The proposed optimization method has been demonstrated within an application study. The application study has been performed for the urban electrical distribution system, the Birka system, which was previously presented in Chapter 6 and Section 6.3.3.1.

9.2.3.1 The Birka System

The Birka system is located in the southern parts of Stockholm. The system (see Figure 6.3) includes a 220/110 kV station (Bredäng) and one 110/33 kV, 33/11 kV station (Liljeholmen). These two stations are connected with two parallel 110 kV cables. From the Liljeholmen station, there are two outgoing 33 kV feeders, Högalid (HD) and Railway (SJ); there are also 32 outgoing 11 kV feeders (LH11), here represented by one average set of components (28–35). The model includes 178 components, numbered 1–58, with 16 copies of components 28–35. The components are divided into five types: circuit breakers (e.g., c2), cables (e.g., c5), transformers (e.g., c3), bus bars (e.g., c1), and fuses (e.g., c34). In the network, every component has a specific failure rate and repair rate. In total, this electrical distribution system serves approximately 38,000 customers where the load point SJ consists of one customer, that is, the railway. The load point LH11 represents one average load point of 32 actual outgoing feeders, which in total serve 14,300 customers. The load point HD feeds approximately 23,400 customers. In the model, piecewise constant failure and repair rates for every component are assumed based on investigations of the network [252]. Note that the constant failure rates should be interpreted as population average values for the specific type of component population, averages that can be modified by selected maintenance policies for the individual. Furthermore, the components are assumed to be independent.

9.2.3.2 Maintenance

The maintenance actions are modeled on an aggregated level in this application study, that is, if the components should be maintained as of or if the PM should be increased or decreased. This is modeled as if there were three different PM alternatives for each component in the electrical distribution system.

1. Keeping the current PM level, average failure rate is assumed to remain unchanged, no change in cost for PM and CM.
2. Improving the PM, the average failure rate is assumed to be halved for the studied component, the additional cost of this is one cost unit.
3. Decreasing the PM, the average failure rate is assumed to be doubled for the studied component, cost savings: one cost unit.

It is assumed, furthermore, that the cost of one CM action is β cost units. In the basic case, β is set at 10 cost units. Here, the cost of CM is chosen in relation to the maintenance effects in order to introduce incentives into the operator's maintenance budget to perform PM. The relationship between changes in failure rate and cost of CM is expressed as

$$\Delta C^{CM} = \beta \sum_{i} (\lambda_i^{new} - \lambda_i) \tag{9.5}$$

where i denotes component number. The change in PM is calculated according to the formula

$$\Delta C^{PM} = \text{sum(incr maint)} - \text{sum(decr maint)} \tag{9.6}$$

where sum(incr maint) corresponds to the number of components with increased PM actions and sum(decr maint) is the number of components with decreased maintenance.

It is worth noting that the assumed maintenance alternatives in general "punish" relocation of maintenance resources in terms of total number of component failures. Consider the example of two components, both with the same initial failure rate, λ, and both being at alternative 1. By moving maintenance resources, that is, moving one component to alternative 2 and the other to alternative 3, this results in the sum 2 1/2λ (compared to 2λ, before maintenance reallocation).

9.2.3.3 Results

The optimization routine results in a number of optimal points (solutions) which are all optimal from a specific point of scale. In Figure 9.2, a number of optimal points are displayed, as well as the starting point (which is the present situation). Note that since every optimization is built on the results from a separate simulation, some of the optimal points are dominated by other optimal points. A point is dominated when another point exists that is better in respect of at least one criterion without being worse in any other criteria. The fact that every optimization is based on one or more (individual) simulations explains the existence of these points.

In Table 9.1, more details are found for the solutions presented. Solutions 7–13 all dominate the "initial point." Even when considering SAIDI

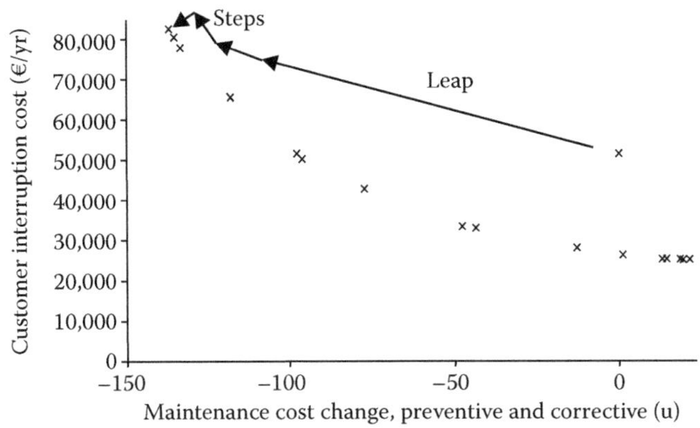

FIGURE 9.2
Optimal calculated solutions for the Birka system. The x-axis corresponds to changes in maintenance budget in comparison to today's budget. Note the starting point for the optimization (not an optimum), located at (0, 51,912). The arrows illustrate the optimization process to one optimum. (Adapted from Hilber, P. et al., *IEEE Transactions on Power Systems*, **22**(4), 1675–1682, 2007 [249].)

and SAIFI, solutions 8–13 dominate the initial point, despite SAIDI and SAIFI not being directly included in the optimizations. Solutions 7–13 are probably more interesting than the others, since they do not aggravate the situation for any of the two parties involved. This is, however, only true after looking at the total customer interruption cost. After studying every load point separately, it can be seen in Table 9.1 that the interruption cost for node SJ is higher for solutions 1–13 than for the starting point. One approach to this somewhat problematic situation might be to state that solutions 7–13 constitute Pareto improvements from a system perspective, which implies that resources are being utilized for the common good of the customers. Another approach might be to put constraints on the optimization, ensuring that the reliability offered to customers does not fall below current levels or to penalize customer node interruption costs that are above today's level. To investigate this issue further, the utilized customer objective is split up into three new objectives, that is, one measure for every load point.

9.2.3.4 Results from the DSO Perspective

When the optimization problem from the perspective of the DSO is studied, one approach is to see how much maintenance budget is decreased, without decreasing the service to the customers. This is achieved by identifying the solution with the nearest lower customer interruption cost compared to the solution of today (0, 51,912). The maintenance cost difference between the

TABLE 9.1

Optimization Results from the Birka System Case Study

	Meas.							
Solution	C_{LH11}^{IC} [euros/yr]	C_{HD}^{IC} [euros/yr]	C_{SJ}^{IC} [euros/yr]	ΣC^{IC} [euros/yr]	C^{PM} [units]	Σ comp. failures [f/yr]	SAIFI [int/yr]	SAIDI [h/yr]
Org.	47,009	4283	620	51,912	n/a	4.01	0.144	0.271
1, 2, 3	73,734	8049	1160	82,943	−144	4.66	0.211	0.451
4	72,103	7204	1046	80,353	−142	4.64	0.192	0.434
5	70,612	6302	909	77,823	−140	4.63	0.175	0.416
6	58,492	6361	917	65,770	−124	4.57	0.174	0.356
7	46,181	4726	788	51,695	−87	2.91	0.121	0.280
8	45,589	4225	725	50,539	−85	2.90	0.113	0.271
9	39,038	3398	727	43,163	−66	2.86	0.104	0.227
10, 11	29,930	3147	694	33,771	−33	2.57	0.098	0.178
12	29,449	2934	698	33,081	−29	2.56	0.092	0.174
13	24,767	2929	701	28,397	3	2.43	0.089	0.151
14, 15	23,763	2495	515	26,773	17	2.42	0.079	0.141
16	23,242	2232	409	25,883	30	2.35	0.074	0.135
17	23,052	2213	344	25,609	32	2.33	0.074	0.134
18	23,116	2091	347	25,554	36	2.31	0.073	0.133
19	23,042	2095	307	25,444	37	2.30	0.073	0.133
20	23,170	2068	306	25,544	39	2.28	0.073	0.133

Source: Hilber, P. et al., *IEEE Transactions on Power Systems*, **22**(4), 1675–1682, 2007. [249]
Note: Org.; Represents the non optimized original solutions.

starting point and the optimal point now reached gives us an estimate of the efficiency of today's maintenance policy. That is, how much it is possible to save on today' s maintenance policy without reducing average customer service. This approach suggests solution number 7 for the case study, which would significantly reduce the cost of PM. Likewise, this operation in reverse is performed by going down from today's (0, 51,192) solution to the Pareto border in order to localize the point that, given today's budget, will give us the lowest customer interruption cost. This approach suggests solution 13. According to Table 9.1, this solution, with the utilized assumptions, would result in an almost halved customer interruption cost. While the PM is increased for this solution, the cost of CM is lowered, resulting in a slightly lower maintenance cost than that of today.

9.2.3.5 Results for One Optimum (Number 10)

To exemplify a specific solution one of the reached optima has been studied in more detail. It is assumed that DSO chooses solution number 10. This might

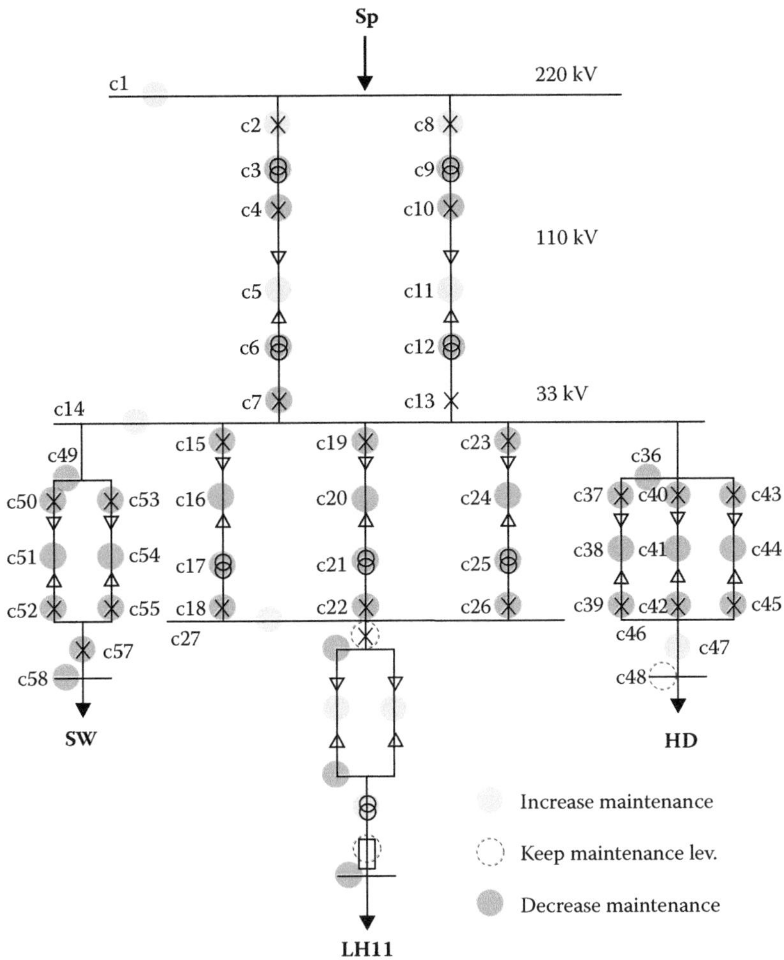

FIGURE 9.3
Illustration of optimum number 10. (Adapted from Hilber, P. et al., *IEEE Transactions on Power Systems*, **22**(4), 1675–1682, 2007 [249].)

be motivated by the fact that this point has a suitable combination of lowered customer interruption costs and lowered maintenance cost. One additional advantage of this point is that the interruption cost for node 3 is relatively close to the starting value.

The resulting maintenance plan stipulates that the PM level should be increased for 56 components while being decreased for 89 components and kept the same for 33 components. Figure 9.3 presents an illustration of the suggested actions for optimum number 10 applied to the electrical distribution system.

9.3 Maintenance Support Optimization for Offshore Wind Farms

9.3.1 Introduction

This section presents a case study capturing the stage III of an RCAM analysis which is to define optimal maintenance plans.

This case study proposes an analytical model for the cost-based optimization and selection of a maintenance support organization for an offshore wind farm. The model considers decisions regarding the location of the maintenance accommodation, the number of technicians, the choice of transfer vessels, and the use of a helicopter. The model includes an analysis of the transportation strategy using alternative transportation means, a queuing model of maintenance activities, and an economic model of the maintenance support organization. A detailed presentation of the case study is provided in [253].

The model includes an analysis of a transportation strategy using alternative transportation means, a queuing model of maintenance activities, and an economic model of the maintenance support organization. An example based on a generic 100 wind turbine 5-MW wind farm is used to demonstrate the application of the model. The results show the benefit of the production losses of the different options, which enables the identification of an optimal maintenance support organization based on the reliability, logistic costs, and electricity price. The most cost-efficient maintenance support organization in the case study consists of an offshore accommodation with technicians on service 24 hours a day, 7 days a week. The solution suggests transportation by use of a crew transfer vessel (CTV) equipped with a motion compensated access system.

9.3.2 Model Description

An offshore wind farm consists of N_{WT} wind turbines with an installed capacity of P MW each and a season-dependent capacity factor of C_f^s.

9.3.2.1 Time, Season, and Environmental Conditions

Historical data regarding environmental conditions at offshore wind farms are generally collected in time series of 1- or 3-h steps. In the proposed model, it is assumed that the time series are indexed by $t \in T = \{0, 1, ..., N_{T-1}\}$, where T is a time set with a number of N_T steps Δt. The time series for the wind speed and significant wave height are denoted w_t and h_t, respectively.

The year is divided into the four seasons of winter, spring, summer, and autumn indexed by $s \in S = \{win, spr, sum, aut\}$. $T^s \subset T$ denotes the subset of the time set belonging to each season indicated by N_T^s time steps.

9.3.2.2 CM and PM of Wind Turbines

The wind turbines are subject to CM and PM activities as earlier defined in Chapter 2.

The CM is carried out after a failure occurs. If a failure occurs, the wind turbine would stop operating until the required repair has been performed. Because of this possibility and the limited accessibility of offshore wind turbines, the occurrence of unexpected failures can have a severe impact on turbine availability and, therefore, on the revenue from production.

Failures of wind turbines are typically classified into minor and major failures according to the logistic needs for repair. To repair a minor failure, the wind turbine can be accessed by both workboat and helicopter, while a major failure always requires a workboat. This classification is necessary due to the significant differences in dimension and weight of spare parts as well as the equipment required to perform the maintenance. The failure rate for minor and major failures per season is denoted by λ_m^s and λ_M^s, respectively, and the durations of the related CM activities in the wind turbine are denoted by r_m and r_M. Each repair action is assumed to require one maintenance team.

The PM of wind turbines typically covers the annual service maintenance and condition-based maintenance. The PM activities require r_{PM}^s hours and N_{PM}^s maintenance teams per wind turbine. It is assumed that the PM activities consist of short duration activities that may be performed on several discontinuous work shifts. In case of PM, the wind turbine is stopped for the duration of the maintenance work only.

9.3.2.3 Maintenance Technicians and Work Shift

The maintenance technicians at the wind farm are assumed to be organized into teams of N_{team} technicians working in shifts of d_{shift}^s [h] duration. A maintenance team works for d_{team}^s [h] per season. It is distinguished between permanently employed and supplementary maintenance technicians. Permanently employed maintenance technicians can perform both CM and PM activities, while CM activities are always prioritized due to their crucial impact on turbine availability and production as explained above. The minimum duration of a maintenance activity in the wind turbine is r_{min} [h]. If the number of permanent maintenance technicians would not be sufficient to perform all PM activities scheduled for a season, supplementary maintenance technicians working mainly on PM would be hired for the season in agreement with common practice at existing offshore wind farms.

A variety of work shift arrangements are possible, for example, working 7 days a week for 12 hours a day. A subset $T_m \subset T$ represents the working hours when maintenance technicians are available to perform maintenance.

For a given work shift arrangement, the total number of maintenance teams to cover the work through the year equals the number of maintenance teams per shift multiplied by a work shift multiplier denoted by M_{shift}.

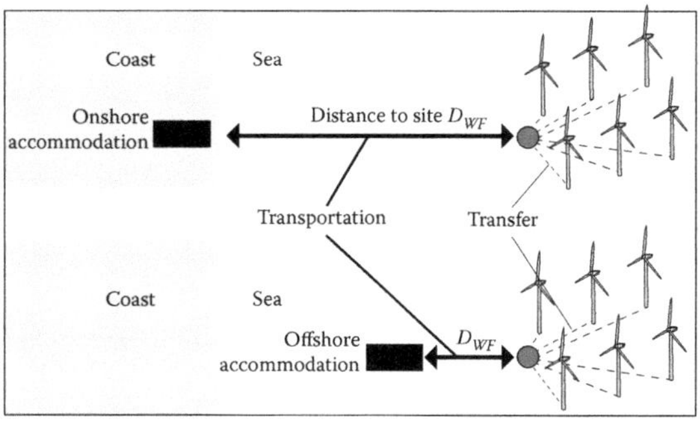

FIGURE 9.4
Illustration of the logistics of maintenance technicians and transportation. (Adapted from Besnard, F. et al., A model for the optimization of the maintenance support organization for offshore wind farms, *IEEE Transactions on Sustainable Energy*, **4**(2), 443–450 © 2013 IEEE [253].)

9.3.2.4 Accommodation and Transportation

The accommodation for the maintenance technicians is located at a distance of D_{WF} [km] from the wind farm either onshore or offshore (see Figure 9.4).

The main transportation mean to the wind turbine is a workboat referred to as crew transfer vessel. The access to the wind turbine using the CTV is constrained by a maximum significant wave height H_b [m] and maximum wind speed W_b [m/s]. The speed of the vessel during transport to the site is v_b [km/h], and the average time needed for the transfer of a maintenance team to a wind turbine once it has been transported to the site is denoted as τ [h] (see also Figure 9.4). A CTV can carry a N_b^{max} maximum of technicians.

An alternative to the transfer of technicians using CTVs is the hoisting of the technicians by means of a helicopter. A helicopter is not constrained by the significant wave height. The maximum wind speed for access by helicopter is W_h [m/s]. It is assumed that only one maintenance team can be transported by helicopter at a time. The speed of the helicopter is v_h [m/s] and the hoisting time per team of technicians is denoted by τ [h].

9.3.2.5 Economical Parameters

The electricity price C_{el} [euros/MWh] includes both incomes from potential incentives and the electricity market price. The yearly cost for each permanent technician is c_{tech} [euros/yr]. The cost for supplementary technicians is assumed to be the same as for permanent technicians, yet proportional to the duration of their employment.

The seasonal chartering cost and daily operating costs for one CTV are denoted by C_b^{yr} [euros/yr] and C_b^{day}, respectively. The yearly cost and cost per failure for the helicopter are C_h^{yr} and C_h^f. The fixed overhead costs for the accommodation, maintenance coordination, and support are denoted by C_{over}.

9.3.3 Proposed Optimization Model

Figure 9.5 summarizes the steps in the proposed optimization model for the evaluation of a maintenance support organization including a helicopter. Steps A–C focus on determining the total downtime per failure as a function of the number of maintenance teams. The number of supplementary technicians $n^s_{PM}(n)$ for performing the PM and the number of CTVs $b^s(n)$ are calculated in Step D. The wind farm availability, the cost of production losses, and the cost of the support organization are assessed in Step E. The final step is to determine the optimal number of maintenance teams n^*.

Each step in the model is discussed separately below for the case of a support organization including a helicopter. The model can be simplified if only CTVs are used.

9.3.3.1 Weather and Work Shift Delays, and Use of the Helicopter

The expected maintenance delay due to the weather and work shift constraints can be statistically determined for a given duration of maintenance activity based on environmental time series for the site. The approach used is described in the flowchart in Figure 9.6. It is similar to the approach presented in [254]. The novelty of this algorithm is its capability to include an alternative transportation means, that is, the use of a helicopter in the present work to improve accessibility during harsh weather. The algorithm presented in Figure 9.5 is described in more detail below.

Step 1: For each time step in the time series, the algorithm evaluates the maintenance delay, due to weather and work shift restriction, to perform a repair activity of duration r [h]. This results in $WT_b(t,r)$ [h] in the case of the CTV and $WT_h(t,r)$ [h] in the case of the helicopter.

It is assumed that h is a multiple of the time step Δt, that is,

$$r = r_{\Delta t} \cdot \Delta t, r_{\Delta t} \in Z^+,$$

where $r_{\Delta t}$ represents the number of time steps. If r is not a multiple of Δt, a linear interpolation of the results is used.

If a failure occurs at time $t \in T$, the earliest time for performing the maintenance is $t + k \in T$ when the weather and working hour constraints are fulfilled each instance for the duration of the maintenance activity $r_{\Delta t}$. Therefore, using the CTV, the maintenance delay time is

$$\begin{aligned} &WT_b(t,r) = min\{k \cdot \Delta t\} \ \ s.t \\ &k \in Z^+, \forall l \in \{t+k; \ldots ; t+k+r_{\Delta t}-1\} \\ &w_l \le W_b, h_l \le H_b, l \in T_m \end{aligned} \tag{9.7}$$

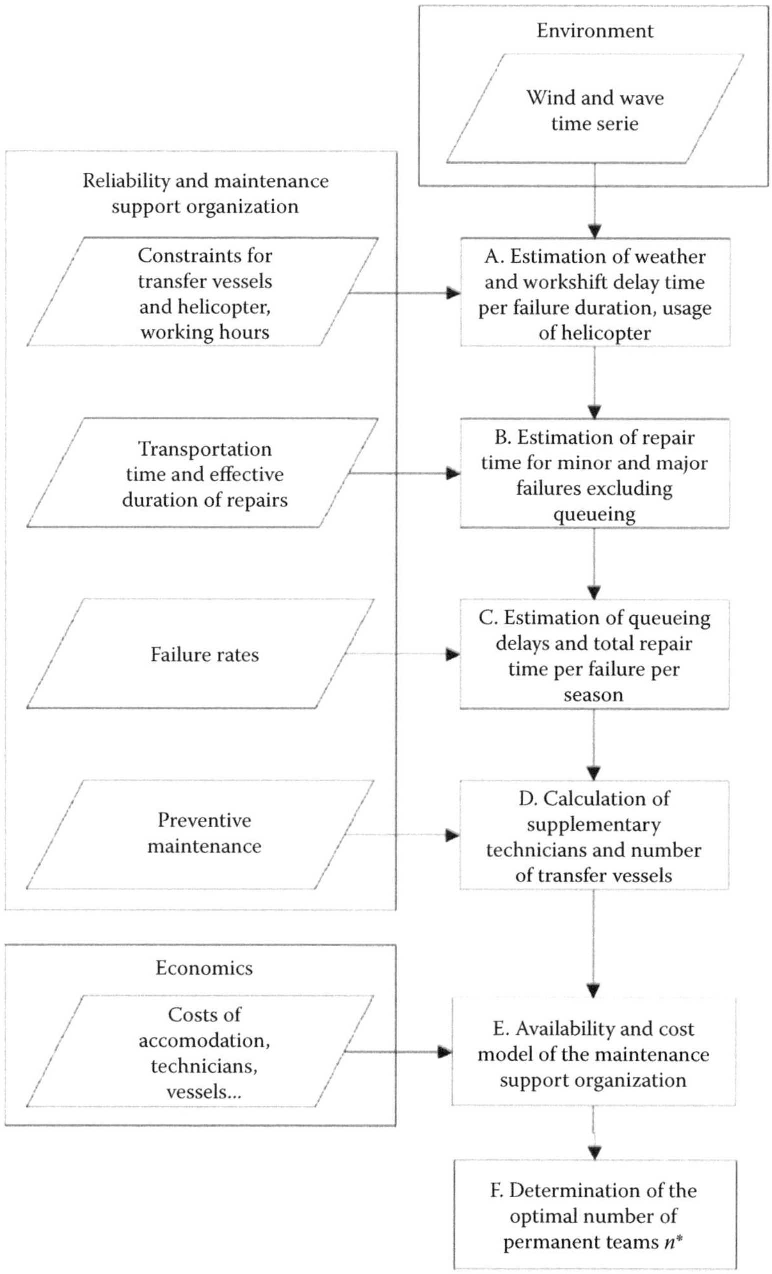

FIGURE 9.5
Work flow of the different steps of the wind optimization model. (Besnard, F. et al., A model for the optimization of the maintenance support organization for offshore wind farms, *IEEE Transactions on Sustainable Energy*, **4**(2), 443–450 © 2013 IEEE [253].)

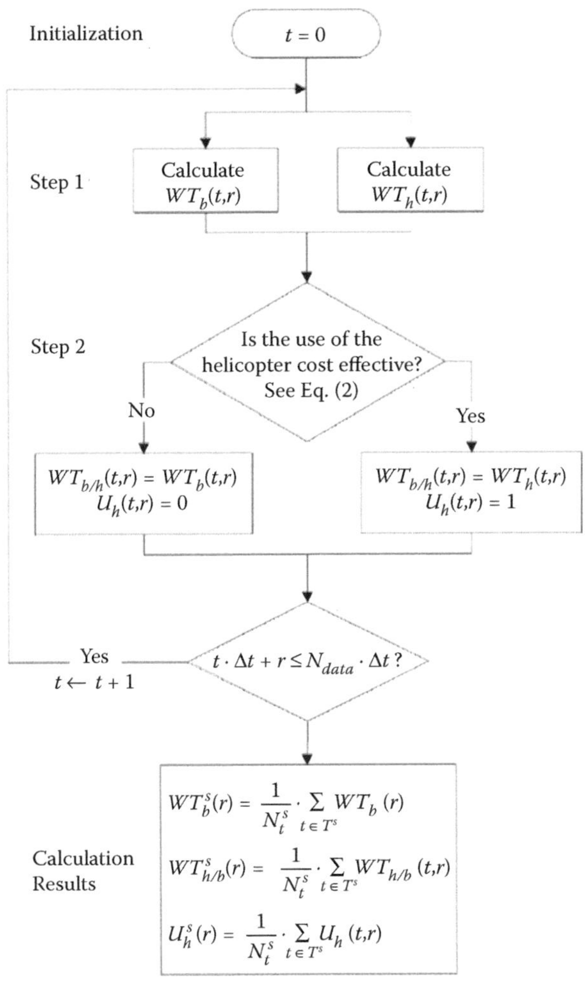

FIGURE 9.6
Flowchart of the algorithm for estimating the weather and work shift delay and usage of helicopter for a maintenance activity of r [h] for the wind optimization model. (Besnard, F. et al., A model for the optimization of the maintenance support organization for offshore wind farms, *IEEE Transactions on Sustainable Energy*, **4**(2), 443–450 © 2013 IEEE [253].)

The same equation may be used in the case of the helicopter, with weather constraints modified accordingly.

Step 2: The second part of the algorithm is an evaluation of the usage of the helicopter. The helicopter is used if the cost benefit for the reduction of production losses is higher than the helicopter cost per event

$$(WT_b(t,r) - WT_h(t,r)) \cdot P \cdot C_f^s \cdot C_{el} > C_{heli}^f, t \in T^s \tag{9.8}$$

Calculation results: The calculation results based on the algorithm are season-specific estimates of delays due to weather and work shift restrictions as a function of the duration of the repair activity, $WT_b^s(r)$ [h] and $WT_{b/h}^s(r)$ [h], for a support organization with and without the use of the helicopter, respectively. Moreover, the probability of usage of the helicopter per minor failure and season, $U_h^s(r)$ [%], is determined. The accessibility using the CTV is defined as the percentage of working hours when wind turbines can be accessed using the CTV without weather delay for the minimum duration of maintenance activity hours.

9.3.3.2 Repair Times Excluding Queuing of Maintenance Activities

The repair time for the minor and major failures excluding queuing delays, denoted by d_m^s and d_M^s, are the sum of (i) the delay due to weather and work shift constraints, (ii) transportation and transfer time, and (iii) duration of the effective maintenance work r_m or r_M. In the case of major failures, in which a CTV would be required to access the wind turbine, the repair time would be

$$d_M^s = WT_b^s(r_M) + \left(\frac{D_{WF}}{v_b} + \tau_b\right) + r_M \tag{9.9}$$

while in the case of minor failures and partial use of a helicopter, the following results:

$$d_m^s = WT_{b/h}^s(r_m) + (1 - U_h^s(r_m)) \cdot \left(\frac{D_{WF}}{v_b} + \tau_b\right) + U_h^s(r_m)) \cdot \left(\frac{D_{WF}}{v_h} + \tau_h\right) + r_m \tag{9.10}$$

Note that if a maintenance activity requires more time than a single work shift, it should be divided into subactivities which can be performed within one shift.

The average failure rate and repair time per failure and per season can be calculated as

$$\lambda^s = \lambda_m^s + \lambda_M^s \tag{9.11}$$

$$d_{CM}^S = \frac{\lambda_m^s \cdot d_m^s + \lambda_M^s \cdot d_M^s}{\lambda^s} \tag{9.12}$$

The average repair rate for a failure is given by

$$\mu^s = \frac{1}{d_{CM}^s} \tag{9.13}$$

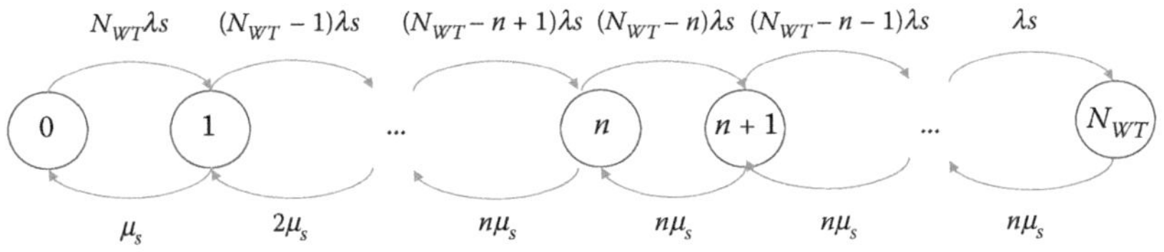

FIGURE 9.7
Markov diagram for the queuing model of maintenance activities. The state numbers represent the number of wind turbines in the wind farm that are in a failed state. (Besnard, F. et al., A model for the optimization of the maintenance support organization for offshore wind farms, *IEEE Transactions on Sustainable Energy*, **4**(2), 443–450 © 2013 IEEE [253].)

9.3.3.3 Queuing of Maintenance Activities

A backlog of maintenance activities may occur when there are not enough maintenance teams to simultaneously perform maintenance on all failed wind turbines. This is especially prevalent during harsh weather conditions, during which failures may accumulate and the existing maintenance workforce may not be sufficiently large to perform work on each failure when wind turbines are accessible.

The backlog of maintenance activities can be represented by a Markov model as depicted in Figure 9.7. The Markov model consists of the states i, in which i represents the number of failed turbines. The failure transitions shown in the upper part of the diagram occur at a rate which is proportional to the number of wind turbines in operation. The repair transitions given in the lower part of the diagram occur at a rate which is proportional to the minimum number of wind turbines in repair and the number of permanent maintenance teams. For an introduction to the Markov modeling, see Chapter 3.

The Markov model is solved in steady state, which is assumed to be realistic due to the long lifetime of the wind turbines, high transition rates, and expected similar state probability distributions throughout the different seasons. In steady state, a recursive relationship exists between the probability of being in each state i denoted by $P(i)$

$$\begin{aligned} &P(i)\cdot(N_{WT}-i)\cdot\lambda^s = P(i+1)\cdot(i+1)\cdot\mu^s, \forall i < n \\ &P(i)\cdot(N_{WT}-i)\cdot\lambda^s = P(i+1)\cdot n\cdot\mu^s, \forall i \geq n \\ &\sum_i P(i) = 1 \end{aligned} \tag{9.14}$$

The system of equation can be solved to calculate. The average queuing time per failure can be calculated by using Little's law. It is the ratio of the average length of queuing (average number of failed wind turbines waiting for a maintenance team to perform the repair work) and the average number of failures per time unit [255]:

$$Q^s(n) = \frac{1}{\lambda^s} \cdot \frac{\sum_{i>n} P(i) \cdot (i-n)}{\sum_i P(i) \cdot (N_{WT} - 1)} \tag{9.15}$$

The average total downtime per failure and season including queuing is

$$d^s_{CM} + Q^s(n) \tag{9.16}$$

9.3.3.4 Supplementary Maintenance Teams and CTVs

For a support organization to be feasible, it is necessary that there are a sufficient number of maintenance teams to perform all the CM and PM activities. To evaluate the need for supplementary maintenance technicians for PM, the effective working time of the technicians needs to be calculated. Since the helicopter is used only for minor failures, this is done in the case of the CTV. When wind turbines are accessible using the CTV, a maintenance technician can perform the work for an effective time of

$$\epsilon^s_{shift} = \frac{d^s_{shift} - 2 \cdot \left(\frac{D_{WF}}{v_b} + \tau_b\right)}{d^s_{shift}} \tag{9.17}$$

The total number of maintenance team hours to be performed per season using the CTV, therefore excluding the maintenance performed by helicopter is

$$r^s_{tot} = r^s_{PM} \cdot N^s_{PM} + (1 - U^s_h(r_m)) \cdot \lambda^s_m \cdot r_m + \lambda^s_M \cdot r_M \tag{9.18}$$

Due to the fact that a small portion of repair work using the helicopter can be carried out during times when the turbines are not accessible by CTV, there is no additional need for technicians for the work performed using the helicopter. The number of supplementary maintenance teams for PM for the season can be estimated as

$$n^s_{PM}(n) = max\left(0; int\left(\frac{4 \cdot r^s_{tot}}{\epsilon^s_{shift} \cdot AC^s_b \cdot d_{team}} - n \cdot M_{shift}\right) + 1\right) \tag{9.19}$$

where n is the number of seasons per year, AC^s_b the probability of accessibility using the CTV, and d_{team} the working time for each team per year.

It is assumed that a CTV may be chartered each season. The number of CTV per season can then be calculated as

$$b^s(n) = int\left(\left(n + \frac{n^s_{PM}(n)}{M_{shift}}\right) \cdot \frac{N_{team}}{N^{max}_b}\right) + 1 \tag{9.20}$$

9.3.3.5 Availability and Economic Model

The availability per season can be calculated as follows:

$$A^s(n) = 1 - \frac{1}{2190} \cdot (\lambda^s \cdot (d^s + Q^s(n)) + r^s_{PM}) \tag{9.21}$$

where the dominator 2190 represents the total number of hours during a season.

The yearly availability is calculated as

$$A(n) = \frac{1}{4} \cdot \sum_{s \in S} A^s(n) \tag{9.22}$$

For estimating the total yearly cost of using the CTV, it is assumed that the CTV travels to the wind farm every day that the wind farm is accessible. The total yearly cost is then expressed as follows:

$$C_b^{tot}(n) = \frac{1}{4} \cdot \sum_{s \in S} b^s(n) \cdot (C_b^{yr} + 365 \cdot AC_b^s \cdot C_b^{day}) \tag{9.23}$$

where 365 refers to the number of days per year and 4 to the number of seasons.

The total yearly cost of the helicopter is the sum of the fixed cost and cost of each event when the helicopter is used:

$$C_h^{tot} = C_h^{yr} + N_{WT} \cdot C_h^f \cdot \sum_{s \in S} \lambda_m^s \cdot U_h^s(r_m) \tag{9.24}$$

The total cost for the technicians is calculated as

$$C_{tech}^{tot}(n) = C_{tech} \cdot \left(n \cdot M_{shift} + \frac{\sum_{s \in S} n_{PM}^s(n)}{4} \right) \tag{9.25}$$

The cost for the maintenance support organization is the sum of the cost of the overhead, technicians, and transportation:

$$C_{org}^{tot}(n) = C_{over} + C_{tech}^{tot}(n) + C_b^{tot}(n) + C_h^{tot} \tag{9.26}$$

9.3.3.6 Optimal Number of Permanent Maintenance Teams

The optimal number of permanent teams for the support organization is determined by minimizing the sum of the cost for the support organization and the production losses:

$$\hat{n} = arg_{n \in Z^+} min\{C_{losses}^{tot}(n) + C_{org}^{tot}(n) \tag{9.27}$$

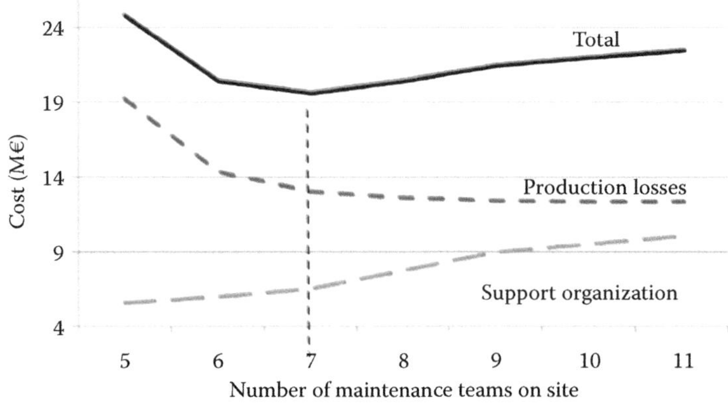

FIGURE 9.8
Cost based on the organization scenario 1 (see Table 9.5) as a function of the number of maintenance teams. The optimal number of maintenance teams ($\hat{n}$) is 7. (Besnard, F. et al., A model for the optimization of the maintenance support organization for offshore wind farms, *IEEE Transactions on Sustainable Energy*, **4**(2), 443–450 © 2013 IEEE [253].)

For each support organization scenario investigated, the optimal number of permanent technicians is determined numerically by calculating the cost of the support organization and the cost of the production losses for a range of possible numbers of maintenance teams, and by selecting the solution based on the lowest total cost, as illustrated in Figure 9.8. Note that due to the effect of the number of boats, the cost for the support organization is not linear.

9.3.4 Application

The proposed optimization model has been demonstrated by means of a case study of a fictitious offshore wind farm consisting of $N_{WT} = 100$ wind turbines located 60 km from a harbor. Each turbine has a rated capacity of $P = 5$ MW. The season-dependent capacity factors are provided in Table 9.2. Wind and wave data are based on real data from the 160 MW Horns Rev offshore wind farm located 15 km off the coast of Esbjerg in Denmark. The time series covers 5 years and, therefore, provides a sound basis for statistical analysis. The time step is set to $\Delta_t = 3$ h.

The generic set of reliability and service maintenance data used for the case study is summarized in Table 9.2 and corresponds to an average of four minor failures and one major failure per turbine and year. The durations of the minor and major repairs are $r_m = 8$ h and $r_M = 16$ h, respectively.

Each maintenance team is assumed to consist of $N_{team} = 3$ technicians, which is recommended for safety reasons. The technicians work in shifts of a duration of $d^s_{shift} = 12$ h.

TABLE 9.2

Reliability Data and Capacity Factor

Season	λ_m^s [1/yr]	λ_M^s [1/yr]	r_{PM}^s[h]	C_f [%]
Winter	5	1.2	6	53
Spring	3	0.8	30	41
Summer	3	0.8	30	38
Autumn	5	1.2	6	48

Source: Besnard, F. et al., *IEEE Transactions on Sustainable Energy*, **4**(2), 443–450, 2013. [253]

Offshore wind turbines are generally serviced once a year during spring or summer. The scheduled maintenance includes condition-based maintenance activities assumed to be 6 h per season, with the yearly service maintenance employing 48 h with two maintenance teams. It includes such activities as follows [256,257]:

1. Change of lubrication systems and oil filters
2. Checkup of brushes and slip ring for DFIG machines
3. Test of safety systems, strength testing of bolts
4. Oil sampling and analysis of the gearbox lubricant
5. Visual inspection of the blades

This case study compares several maintenance support organizations with respect to different locations of the maintenance accommodation, work shift arrangements, and transportation means for the purpose of identifying the most cost-effective solution. The accommodation locations and work shift arrangements investigated are summarized in Table 9.3. The difference in the work shift arrangements between onshore- and offshore-based organizations is related to working hour regulations in Europe, which differ for onshore and offshore accommodations. The overhead costs only include the relative difference in the cost of accommodation and work shift supervision.

Two different types of CTVs are investigated, CTV1 and CTV2. The main difference between the vessels is that the CTV2 is equipped with an access

TABLE 9.3

Accommodation and Work Shift Arrangements

Location	D_{WF} [km]	Work Shift	d_{team} [h]	M_{shift}	C_{tech} [euros/yr]	C_{over} [keuros/yr]
Onshore	60	12/7	1450	3	60,000	–
Offshore	10	12/7	2100	2	80,000	3400
Offshore	10	24/7	2100	4	80,000	4000

Source: Besnard, F. et al., *IEEE Transactions on Sustainable Energy*, **4**(2), 443–450, 2013. [253]

TABLE 9.4

Vessels and Helicopter Characteristics

	$H_{b/h}$ [m]	$W_{b/h}$ [m/s]	$v_{b/h}$ [km/h]	$\tau_{b/h}$ [min]	$N_{b/h}^{max}$	$C_{b/h}^{yr}$ [keuros/yr]	C_b^{day}
CTV1	1.5	15	40	30	12 pers.	900	1.2 [keuros/day]
CTV2	2	15	40	30	12 pers.	1200	1.6 [keuros/day]
Heli.	n.a.	17	220	5	1 team	1200	1 [keuros/h]

Source: Besnard, F. et al., *IEEE Transactions on Sustainable Energy*, **4**(2), 443–450, 2013. [253]

system, in the form of a gangway or stabilizing platform to enable access to the wind turbine at higher significant wave height. Table 9.4 summarizes the features of the transportation means investigated in the present study.

The electricity price is assumed to be 150 euros/MWh guaranteed price in Germany [258] for offshore wind power. The costs of the CTVs and helicopter were kindly provided by Vattenfall and are summarized in Table 9.4.

9.3.5 Results and Discussion

The main results from application study of the maintenance support organization for a wind farm is presented in Table 9.5. The results are in the range of the observed availability of 95%–97% at the Horns Rev wind farm [259], whose support organization is similar to that in scenario 3. Moreover, the availability results seem consistent with the advantages of the logistic solution investigated.

It can be observed that the organization scenario 10, which consists of an offshore accommodation with 24/7 work shifts and the use of CTV2, offers the most cost-efficient solution closely followed by the options 12, 2, and 4. All options include the use of CTV2, clearly more cost beneficial than using CTV1.

The cost benefit of using the helicopter differs with the type of CTV as well as the work shift arrangement. The yearly availability increase is in the range of 0.2%–0.7% for an organization with 24/7 work shifts and type CTV2 vessels as well as for an organization with 12/7 work shifts and CTV1, respectively. The use of the helicopter is cost beneficial in all cases except for the case of an offshore accommodations with 24/7 work shifts and CTV2.

It can also be observed that an offshore accommodation is cost beneficial only in the case of the 24/7 work shift. This can be attributable to the relatively low increase in the availability and work efficiency due to the location of the accommodation alone. The benefit would increase with a longer distance from the shore and harsher weather conditions. It can be noted that the availability increases by almost 1% for each logistic solution by using 24/7 work shifts instead of 12/7 work shifts.

TABLE 9.5

Summary of Results of Analyzing Maintenance Support Organization Scenarios for the Wind Application

Scenario Description					Results								
Org.	**Location**	**Workshift**	**CTV**	**Heli.**	n^*	$n_{PM}^{spr}(n^*)$	b^{win}/b^{spr}	U_h^{yr}	$A(n^*)$	$C_{org}^{tot}(n^*)$	$C_{org}^{losses}(n^*)$	**Total cost**
1	Onshore	12/7	CTV1	No	7	0	2-2	n.a.	95.7%	6.6	12.9	19.5
2	Onshore	12/7	CTV2	No	6	0	2-2	n.a.	96.8%	6.9	9.6	16.5
3	Onshore	12/7	CTV1	Yes	6	0	2-2	25%	96.3%	7.3	11.2	18.5
4	Onshore	12/7	CTV2	Yes	4	5	1-2	14%	96.6%	6.6	10.0	16.7
5	Offshore	12/7	CTV1	No	6	0	2-2	n.a.	96.0%	9.0	11.9	21.0
6	Offshore	12/7	CTV2	No	4	4	1-2	n.a.	96.8%	8.5	9.7	18.2
7	Offshore	12/7	CTV1	Yes	6	0	2-2	25%	96.7%	10.3	9.7	19.9
8	Offshore	12/7	CTV2	Yes	4	4	1-2	14%	97.2%	9.8	8.2	18.0
9	Offshore	24/7	CTV1	No	4	0	1-2	n.a.	97.0%	10.9	9.2	20.0
10	Offshore	24/7	CTV2	No	3	0	1-1	n.a.	97.9%	9.6	6.2	15.8
11	Offshore	24/7	CTV1	Yes	3	0	1-1	12%	97.5%	10.2	7.5	17.7
12	Offshore	24/7	CTV2	Yes	3	0	1-1	7%	98.1%	10.8	5.6	16.4

Source: Besnard, F. et al., *IEEE Transactions on Sustainable Energy*, **4**(2), 443–450, 2013. [253]

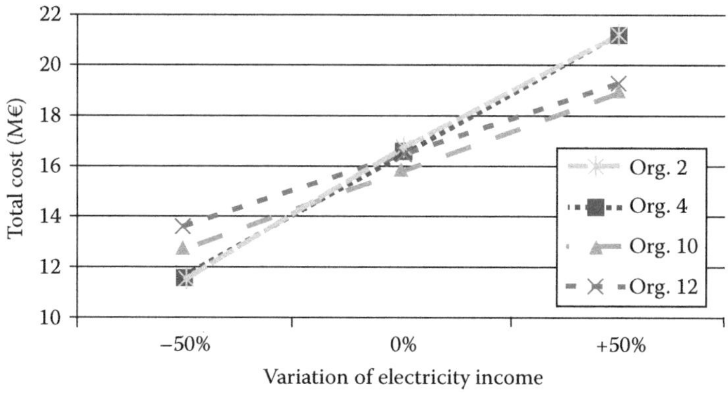

FIGURE 9.9
Sensitivity of the total cost of variations in the electricity price for different organization scenarios (see Table 9.5). (Besnard, F. et al., A model for the optimization of the maintenance support organization for offshore wind farms, *IEEE Transactions on Sustainable Energy*, **4**(2), 443–450 © 2013 IEEE [253].)

The electricity price is a major parameter influencing the results. This parameter also varies considerably between different countries and with time, for example, depending on the local incentive system. The effect of the electricity income on the results is depicted in Figure 9.9. Although the optimal number of maintenance teams was recalculated as a function of the electricity income, the total cost is almost linearly dependent on the electricity income. As expected, due to a higher resulting availability of the wind turbines, the solutions with offshore accommodation and 24/7 work shift are more cost beneficial as the electricity income increases.

Finally, a sensitivity analysis was performed on the input parameter of the failure rate which afflicted with the highest uncertainty. The analysis was performed with a variation of ±50% for the organization scenarios 2, 4, 10, and 12; the results are depicted in Figure 9.10. As expected, the benefits of the offshore accommodation and the benefits of using the helicopter increase with increasing failure rates. It can also be observed that the results are more sensitive to variations in turbine reliability than variations in electricity income.

9.4 Opportunistic Maintenance Optimization for Water Pump System

9.4.1 Introduction

This case study proposes a model for opportunistic maintenance optimization where replacement schedules for shaft seals in feed-water pump

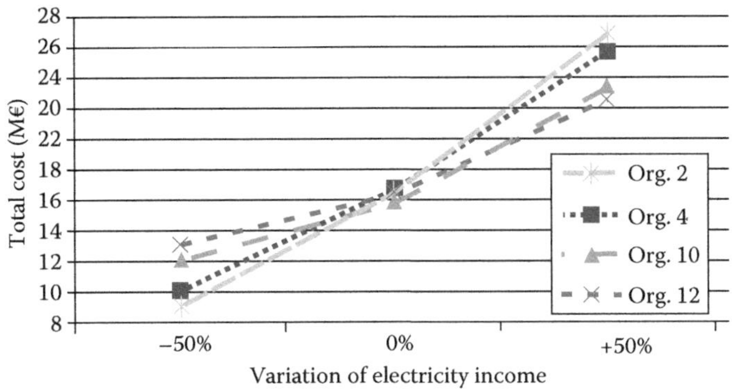

FIGURE 9.10
Sensitivity of the total cost to variations in the wind turbine failure rate for different organization scenarios (see Table 9.5). (Besnard, F. et al., A model for the optimization of the maintenance support organization for offshore wind farms, *IEEE Transactions on Sustainable Energy*, **4**(2), 443–450 © 2013 IEEE [253].)

systems in nuclear power plants are constructed. The feed-water pump system is important for the availability of the entire nuclear power plant. The case study proposes an LCC model for the maintenance optimization and provides an example of RCAM analysis stage III.

Results show that the optimization model is dependent on, for example, the discount interest and a limit for when the optimal solution goes from nonopportunistic to opportunistic is calculated. The circumstances for which opportunistic maintenance could be used have been investigated given different values of discount rates and remaining life at start of the planning period. The case study is discussed in detail in [260].

9.4.2 Overview of Main Concepts and Approach

This section briefly present the main concepts and the overall approach of the case study. The underlying theory and definitions have been presented earlier in Chapter 3 and the definitions of LCC are presented in Section 3.6.

For this case study, availability is the main parameter used for reliability performance, that is, the probability that the component or system is capable of functioning at a time *t*.

The model proposed for the modeling of maintenance in this case study is an opportunistic technique. Opportunistic maintenance refers to the situation in which PM is carried out at opportunities. A typical example is when one component is out for maintenance and it is decided to take out another component for maintenance ahead of the maintenance plan, since it is considered to be rational. Opportunistic maintenance is typically maintenance carried out in a way that is cost saving as at least two maintenance activities are performed at the same time. For minimizing the total cost for maintenance and production loss, opportunistic maintenance is an interesting approach. The opportunistic maintenance optimization model can be described as a deterministic application of a general optimization model [261].

The case study can be summarized with the overall approach as follows:

1. The total cost of maintenance is calculated.
2. This cost is minimized according to some constraints and discounted to model the value of money in time.
3. A sensitivity analysis is made where the different parameters are varied in relation to the discount rate.

9.4.3 Application

9.4.3.1 System Description

The case study shows on application study of the feed-water system at Forsmark 1. Forsmark 1 is a Boiling Water Reactor (BWR), and its function is explained here to provide an understanding of the importance of the feed-water system and its function. In the reactor tank, there is fuel from Uranium that, when nuclear atoms are split, generates large amounts of heating energy [262].

Nuclear fission is started by pulled out control rods from the core. Heat that is generated at nuclear fission is transferred to the water, which is boiled into steam. The steam produced in the reactor is led in large steam pipes to a turbine facility. The difference in pressure between the reactor and the condenser gives the steam force on its way to the turbine, where the steam's heat energy is transformed into kinetic energy. A generator is connected to the shaft of the turbines. The generator's rotor is rotating at the same speed as the turbines. In the generator, kinetic energy is transformed into electrical energy. The electrical energy leaves the plant from a switchyard that divides the electrical power into different lines that connect to the Swedish national grid.

The steam still has a large energy when it leaves the low-pressure turbines. This energy is cooled off by large amounts of cooling water. The water is brought into the cooling water channel and is pumped into a condenser, which is a large heat exchanger placed under the low-pressure turbines. The cooling water is led into water chambers and passes through the condenser through a large number of pipes, where it gathers the heat of the steam. From the outlet chambers, the cooling water is led in a tunnel. The steam that has turned into water again is collected at the bottom of the condenser. The water collected at the bottom of the condenser is called condensate. It is to be returned to the reactor and is therefore passed through a heat- and pressure-increasing process. After a condensate cleaning process, there are feed-water pumps that can give the water the pressure necessary for it to pass into the reactor. The condensate is at this moment changing name to feed water, which will be heated even more in high-pressure preheaters. These gain steam from the high-pressure turbine to warm up the feed water.

The feed water is pumped into the reactor again and the primary circle is closed. The feed water will replace the amount of water that has boiled to steam and left the reactor. The flow is adjusted continuously in relation to the steam output so that the water level in the reactor is preserved. In all situations, there must be total control of the neutron flux in the core and the pressure and the water level in the reactor tank. These parameters affect each other mutually, and the feedwater pump system is important for keeping the process stationary.

9.4.3.2 System Modeling

The feed-water pump system works with a special type of redundancy so that two out of three pumps always must be in operation and one is redundant. Reactors 1 and 2 at Forsmark nuclear power plant are constructed so that they have two turbines with three feed-water pumps on each turbine. If two pumps out of three on one turbine were to go down for some reason, there would be a loss in power of 25% on the actual reactor. If three pumps were to go down, the loss in power would be 50%. As a first step, production loss is not considered, since one pump at a time is observed. To gain a model that can consider production loss, the entire system with three pumps must be observed. This is planned to be done in future work.

The mechanical failures that dominate in feed-water pumps are failures on shaft seals in the pump. They are today replaced when they fail as they are expensive components, that is, no PM is carried out. Indications of failures are that hot water is leaking from the shaft. When this phenomenon appears, or when temperature sensors show that the temperature is over a certain limit, inspections are made. Then a decision on whether the shaft seal should be replaced or not is made. Each pump has two shaft seals, and an interesting question is if it is beneficial to replace both of them when one is failing. In this modeling stage, one pump with two shaft seals is observed.

9.4.3.3 Model

9.4.3.3.1 Total Cost Model

In [13], a total cost model for distribution systems is constructed. Costs included are costs of failure, or cost of CM, cost for PM, and cost of interruption. For production systems, the interruption cost is the cost for production loss (PL). This gives the following model.

The cost for PL can be modeled as a cost per energy unit times Energy Not Supplied (ENS) which, with the typical type of redundancy, would be fourths of the total production loss, depending on how many pumps are down. When observing one pump only, the PL cost is not included in the total cost. The pump is always running and no cost for PL appears. When observing three pumps and when two pumps must be running the cost for PL is taken care of by always running two pumps.

$$TC = C_{PL} + C_{CM} + C_{PM} \tag{9.28}$$

The cost for CM is the cost for maintenance that is done after the failure is observed, that is, the cost of failure. At Forsmark today, only CM of shaft seals is performed. Shaft seals used in the pump in the feed-water pump system are expensive components, and no trials have been made yet to do PM replacements of the shaft seals.

The cost for PM is the cost for maintenance done before the failure has been observed, and opportunistic maintenance is one type of PM. In this case study, an attempt is made to see if it is beneficial to make replacements of two shaft seals at one time, that is, to perform opportunistic replacements.

9.4.3.3.2 *Opportunistic Deterministic Optimization Model*

There are two shaft seals on the pump, which are replaced either one or two at a time. The deterministic model to be given below decides that the replacement of one or two shaft seals should be carried out when, or before, the life length of the shaft seal ends. This decision depends on the constant cost for performing the maintenance and the spare part cost for the shaft seal. The total cost for maintenance is summed over time and minimized subject to some constraints. Two binary variables indicate if maintenance is performed at all at a certain time, and if a certain shaft seal is replaced at a certain time. The first constraint says that if a replacement is carried out, then maintenance is performed. The first binary variable that indicates maintenance is forced to be one if the other binary variable, which indicates replacement, is 1. The next constraint says that the shaft seal must be replaced within the interval of the life length and the final constraint says that the first replacement must be performed before the initialized life length, which is given by the age of the shaft seal at the beginning of the time period.

1. *Planning period:* The time period used in this approach is 6 years. Replacements of shafts are carried out after 11,000 h on average. This is the assumed life length of the shaft seals in the model. The system is an aging system, but is treated with a deterministic approach. The life lengths for shaft seals that were observed were all exceeding 11,000 h and therefore the choice was made to use a life limit of 11,000 h. The initialized life length, which is the life length of the shaft seal at the beginning of the time period, needs to be set at the beginning. The remaining life at the beginning is 1 time step for shaft seal 1 and 10 time steps for shaft seal 2, in the basic case.
2. *Time steps:* The 6 years are divided into hours and the time step has then been chosen to 1000 h. This gives the contract time 55 time steps and the life length 11 time steps in the basic case.

3. *Input data:* Working orders from Forsmark 1 have been studied and an average time for exchanging one or two shaft seals has been calculated. An average time for the life length of the shaft seals has also been calculated, with data extracted from working orders. The time for exchanging shaft seals times the working cost is not the only cost included in the constant cost. Other costs should be included to get a more realistic model. This is, however, difficult to estimate for a single isolated component.
4. *Costs for maintenance:* The cost associated with performing maintenance is a constant cost d. The cost associated with the replacement, a cost per shaft seal called c, is the spare part cost. An estimation of data that depends on whether both shaft seals are replaced at the same time, or if only one is replaced, was made. The average times for exchanging one or two shaft seals are different. This together with the spare part cost for k shaft seals ($k=1,2$) gives the total cost for replacing k shaft seals:

$$n_P \cdot t_{RPT}(k) \cdot C_{WT} + k \cdot C_{SP} \tag{9.29}$$

   where

   - k is the number of shaft seals being replaced (one or two)
   - n_P is the number of people required to perform the maintenance
   - $t_{RWT}(k)$ is the total working time for the maintenance in hours for k shaft seals
   - c_{WT} is the labor cost in SEK for the working time per hour
   - c_{SP} is the cost per spare shaft seal in SEK

   The data is extracted from working orders at Forsmark nuclear power plant. The data given yields the spare part cost $c = 8626$ euros, and the constant cost $d = 482$ euros used in the optimization model.
5. Mathematical model description

   Sets and indices
   $i \in I = \{1,2\}$ (number of shaft seals)
   $t \in T = \{1, \ldots, 55\}$ (time steps)
   T_L Life length of shaft seal (TL = 11)
   τ_i Remaining life at the beginning of the planning period for shaft seal i.
   Decision variables

$$x_{it} = \begin{cases} 1 & \text{if } \textit{ shaft seal } i \textit{ is replaced at time } t, \ i \in I, \ t \in T \\ 0 & \textit{Otherwise} \end{cases}$$

$$z_t = \begin{cases} 1 & \text{if } \textit{ maintenance is carried out at time } t \\ 0 & \textit{Otherwise} \end{cases}$$

Objective function

$$min\sum_{t=1}^{T}(dz_t + c\sum_{i=1}^{2}x_{it})$$

Constraints: A shaft seal can be replaced only when maintenance is performed:

$$x_{it} \leq z_t, 1 = 1, 2, \quad t \in T$$

Each shaft seal must be replaced within its life T_L :

$$\sum_{t=l}^{1+T_L-1} x_{it} \geq 1, i = 1, 2, l = 1, ..., T - T_L + 1$$

The first replacement must be carried out before the remaining life τ_i has expired:

$$\sum_{t=l}^{1+\tau_i} x_{it} \geq 1, i = 1, 2$$

Finally the variables must take binary values:

$$x_{it}, z_t \in \{0, 1\}$$

Cost discount: The total cost is discounted, that is, recalculated into a present value using a so-called discount factor:

$$disc(t) = \frac{1}{(1+r)^t}$$

Using $d_t = disc(t)d$ and $c_t disc(t)c$ for $t\{1, \ldots, 55\}$ yields the adjusted objective function:

$$min\sum_{t=1}^{T}(d_t z_t + c_t\sum_{i=1}^{2}x_{it})$$

The nominal discount rate r_n depends on the real interest r_1, and the inflation r_2, according to:

$$1 + r_n = (1 + r_1)(1 + r_2).$$

The mathematical model can then be summarized as

$$min\sum_{t=1}^{T}\left(d_t z_t + c_t\sum_{i=1}^{2}x_{it}\right) \tag{9.30}$$

subject to

$$x_{it} \leq z_t, 1 = 1, 2, \quad t \in T \tag{9.31}$$

$$\sum_{t=l}^{1+T_L-1} x_{it} \geq 1, i = 1, 2, l = 1, ..., T - T_L + 1 \tag{9.32}$$

$$\sum_{t=l}^{1+\tau_i} x_{it} \geq 1, i = 1, 2 \tag{9.33}$$

$$x_{it}, z_t \in \{0, 1\}, i = 1, 2, t \in T \tag{9.34}$$

9.4.3.4 Analysis

The opportunistic deterministic optimization model has been implemented in AMPL and GAMS, and MATLAB has been used to analyze and describe the results graphically. All programs used for the analyses are commercially available tools. A sensitivity analysis where the different parameters were varied has been made.

The first conclusion of the sensitivity analysis was that the result depends on the discount rate. Varying the other parameters ($T, \tau_i, T_L, c,$ and d) shows that they also are important for the decision of maintenance. It is necessary to study which of these parameters affect the schedule of maintenance the most. Patterns could be found for when opportunistic maintenance is the best solution and when it is not the best solution. Conclusions that can be drawn from the varying of the different parameters are presented in the text below.

Critical nominal discount interest, r_n^*, is defined as the highest nominal discount interest where the opportunistic solution is optimal. Over this interest, the optimal solution is to replace shaft seals when they fail. That is, the nominal discount interest, r_n, gives the optimal solution as follows:

$$r_n = \begin{cases} \leq r_n^* & \text{opportunistic} \\ > r_n^* & \text{nonopportunistic} \end{cases} \tag{9.35}$$

The other parameters have been varied and solutions observed in relation to this discount interest. Observing r_n^* for different choices of remaining life, τ_i, shows that larger difference between τ_1 andτ_2 gives a lower interest r_n^*.

Varying T_L indicates that if the shaft seals' life lengths are one time step longer than in the basic case, the optimal solution, when one shaft seals' remaining life length is zero, is to replace the first one at once.

If the constant cost d is large enough in relation to the spare part cost c, opportunistic replacements would be optimal. The constant cost d is increased until the opportunistic solution is reached. When the solution is the nonopportunistic and d is increased until the opportunistic solution is reached, d does not have to be increased much until this happens.

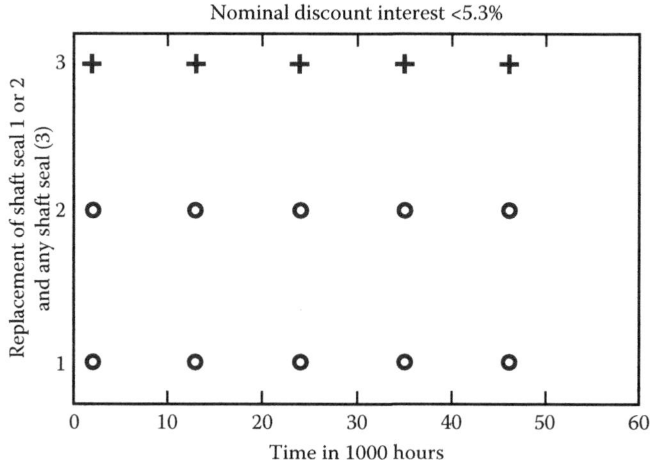

FIGURE 9.11
Results of the deterministic optimization model with nominal discount interest <5.3%. The optimal solution is to replace opportunistic as soon as the first shaft seal breaks. (Adapted from Nilsson, J. et al., An opportunistic maintenance optimization model for shaft seals in feed-water pump systems in nuclear power plants, in 2009 IEEE Bucharest PowerTech, pp. 1–8, Bucharest, © 2009 IEEE [260].)

9.4.3.5 Results

The results show that the discount rate has a large effect on how maintenance should be performed on shaft seals in the pump in the feed-water pump system. The optimal solution for the basic case is such that opportunistic replacements are used if the nominal discount rate is lower than 5.3%. Then the optimal solution is to replace both seals as soon as the first one breaks. The next replacement takes place when the life has ended for the shaft seals, see Figure 9.11. When the nominal discount interest is higher than 5.3%, the optimal replacement schedule is corrective, that is, the shaft seals should be replaced when they fail only, see Figure 9.12. A nominal discount interest of 5.3% corresponds to, for example, a real interest of 2.6% and an inflation of 2.6%. These are low figures of real interest and inflation and a reasonable result is therefore that the result with nominal interest higher than 5.3% should be the most interesting. This finding had not been possible if opportunistic maintenance had been ignored in the problem formulation. However, in the basic case, the remaining life at start of the components differs a lot, (τ (1,10)), which gives a low critical nominal discount interest. A more realistic case would be that τ_i does not differ this much, as the shaft seals often are put in at the same time from start. Less difference between remaining life at start for the components gives a higher critical nominal discount interest and thereby it is more likely that the optimal solution is the opportunistic in most cases.

Table 9.6 shows r_n^* for some points of remaining life at start for shaft seals 1 and 2. At the same remaining life at start for both components, the optimal

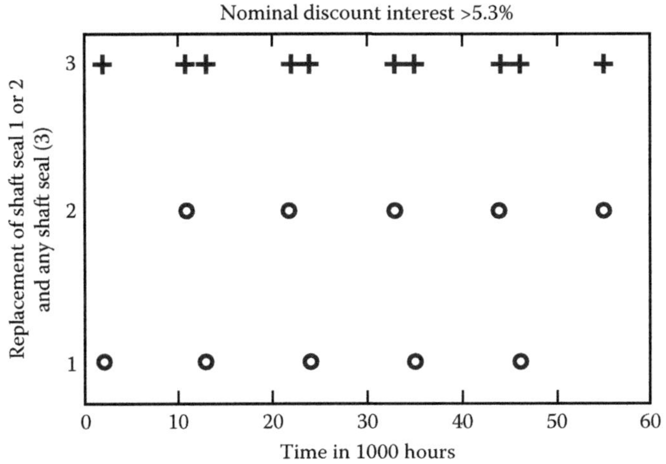

FIGURE 9.12
Results of the deterministic optimization model with nominal discount interest >5.3%. The optimal solution is to use only corrective replacements, with no opportunistic approach, as the shaft seals fail. (Nilsson, J. et al., An opportunistic maintenance optimization model for shaft seals in feed-water pump systems in nuclear power plants, in 2009 IEEE Bucharest PowerTech, pp. 1–8, Bucharest, © 2009 IEEE [260].)

TABLE 9.6

Remaining Life at Start and Critical Nominal Discount Interest

τ_1	τ_2	r_n^*	τ_1	τ_2	r_n^*	τ_1	τ_2	r_n^*
0	0	–	2	0	24.0%	4	0	11.5%
0	2	24.0%	2	2	–	4	2	24.0%
0	4	11.5%	2	4	24.0%	4	4	–
0	6	7.5%	2	6	11.5%	4	6	24.0%
0	8	5.5%	2	8	7.5%	4	8	11.5%
0	10	4.5%	2	10	5.5%	4	10	7.5%
6	0	7.5%	8	0	5.5%	10	0	4.5%
6	2	11.5%	8	2	7.5%	10	2	5.5%
6	4	24.0%	8	4	11.5%	10	4	7.5%
6	6	–	8	6	24.0%	10	6	11.5%
6	8	24.0%	8	8	–	10	8	24.0%
6	10	11.5%	8	10	24.0%	10	10	–

Source: Nilsson, J. et al., An opportunistic maintenance optimization model for shaft seals in feed-water pump systems in nuclear power plants, in 2009 IEEE Bucharest PowerTech, pp. 1–8, Bucharest © 2009 IEEE. [260]

solution is to replace both shaft seals at the same time, and this is independent of the interest rate.

For example, the point $\tau = (48)$ in the table shows the critical nominal interest $r_n^* = 11.5\%$ when the remaining life at start is 4 time units for shaft seal 1 and 8 time units for shaft seal 2. This means that if the nominal interest $r_n = 12\%$, the shaft seals should be replaced when they fail. The total maintenance cost over 6 years is about 63,200 euros at these discount rates. Discounting by 12%, instead of 11.5%, gives a cost that is about 1000 euros less.

9.4.3.6 Application of Model

The proposed maintenance model is generic and could be used for any production system, such as a subsystem in the nuclear power plant or a system of wind turbines. Different input data, with small corrections of the model, should give a maintenance schedule for the required time period and time step. Maintenance of the shaft seals used in the feed-water pumps connected to reactors 1 and 2 at Forsmark nuclear power plant is today only corrective. A question is if maintenance of the shaft seals can be more efficient with a PM program? The opportunistic maintenance optimization model shown here is a general model that first has been applied for the aircraft industry, and now with some corrections used for nuclear power. Input data differs for the different applications even different systems within nuclear power could differ when it comes to input data.

10

RCAM Case Hydropower Systems

10.1 Introduction

Operators of hydropower plants were early adopters of RCM [263]. This chapter shows on results from RCM introduced for hydropower plants in Sweden. Experience gained from Norway is presented in [264]. Main findings in this case study results from a joint project supervised by Bertling and Backlund [265] and also previous studies [70].

The implementation of RCM at Vattenfall is referred to as the VVM RCM method. The commercial software BI-Cycle tool has been used for the VVK (Vattenfall Vattenkraft) RCM analysis. The case study shows how the maintenance plans have changed after the introduction of RCM, for example, more PM activities have been introduced due to high costs if a failure occurs. This chapter closes with a comparison of the three RCM approaches: the Moubray RCM, the RCAM (reliability-centered asset maintenance), and the VVK RCM.

In hydroelectric power plants, the level of maintenance is often reasonably well adapted to operations due to many years of operational experience. Reliability is well controlled and the operational costs are so low that decrease in production leads to only minor cost reductions. An analysis of maintenance must therefore be relatively simple to be profitable [264]. A conclusion from the study is that it is more important to achieve dynamic maintenance planning than to design the perfect maintenance programme.

When designing a maintenance plan, many factors need to be taken into consideration. Some of the questions that arise are:

- What kind of failures can occur?
- How severe are the consequences of the failures?
- What kind of maintenance tasks should be performed?
- If scheduled maintenance, how often should the maintenance tasks be conducted?

Large costs are associated with maintenance and loss of production. Maintenance tasks cost money but can reduce the chance of failures. Some failures lead to costly repairs and expensive downtime for production, while others result in costs that can be tolerated. To find an optimal maintenance strategy costs for conducting various maintenance tasks must be compared to the costs and consequences resulting from a failure. RCM at Vattenfall Vattenkraft has resulted in different ways to decide what maintenance to perform than earlier. To be able to further improve maintenance and see what the effects of RCM are so far, there is a need to evaluate what has been done.

10.1.1 Approach

Vattenfall Vattenkraft will start maintaining some of their plants according to the results from RCM. No statistical material depicting the resulting reliability or costs of RCM was therefore available for the case study. A purely quantitative approach is consequently not possible. So far the results from RCM are suggestions of maintenance tasks to be performed and intervals for these tasks. To see and evaluate the effects of RCM, a qualitative study of the RCM structure and process is conducted. The maintenance strategy resulting from RCM will also be compared to how maintenance was performed earlier. The information used to analyze the process and results from RCM is obtained through printouts from the software BI-Cycle used in VVK RCM. BI-Cycle is used in the RCM analysis for calculations and documentation. These printouts show how the RCM analysis has been performed and what the resulting maintenance strategies are. Information about the situation prior to RCM is obtained through old maintenance plans. Furthermore, an interview has been held with a member of the maintenance personnel who helped to conduct the analysis included in the RCM process. To evaluate the effect of a change in inspections, quantitative models have been applied. These models use data obtained through the printouts from BI-Cycle.

10.2 Terminology

10.2.1 Maintenance and Reliability

The terminology used for maintenance planning in this case study follows the presented concepts in Chapter 2. The fundamental measure of reliability is the availability which has been introduced in Chapter 3.

10.2.2 Short Description of a Generator

Generators are used to convert kinetic energy into electric energy [266]. In a hydropower plant, water causes the turbine to rotate. An axis connected between the turbine and the generator provides the necessary kinetic energy to the generator. The generator is built up of mainly two parts: rotor and stator. The generators used in Vattenfall's hydropower plant can vary in design and performance significantly. Generators can have long lifetimes, and developments in design and performance are made continuously. Therefore, characteristics can vary between generators, for example, whether they are old or new or large or small.

10.2.2.1 Rotor

The rotating part of a generator is called the rotor and is connected to the turbine that delivers the mechanical torque. The rotor carries direct current and delivers a dynamic magnetic field through its rotation. The magnetic field is produced by the field poles located on the rotor. The rotor uses direct current from a magnetizing system.

The rotor is composed of the rotor ring, rotor poles, and the core [266].

The rotor ring mechanically holds the poles in place and is fitted around the core. The size of the rotor ring is constructed so that the angular velocity will correspond to an AC frequency of 50 Hz. The ring is fastened to the core with bolts [266].

Rotor poles are attached to the rotor ring and give rise to the magnetic field. The poles are made up of windings that carry direct current. The pole core is isolated from the windings. The winding is connected to the exciter through slip rings and coal brushes [266].

10.2.2.2 Stator

The stator is stationary and encompasses the rotor. The purpose of the stator is to convert the energy in the dynamic magnetic field into electricity. This is accomplished with stator windings that deliver electricity when magnetic flux produced by the rotor passes through them and cause induction [266].

The stator frame is fastened to a concrete foundation. The stator is often flexibly attached to the foundation to allow for expansion. This is desired because the stator may expand when in operation [266].

The laminated core carries the magnetic flux. The stator winding is wound around the core. In the core, there are air channels used for cooling. The core is securely fastened so no vibrations will arise [267].

The stator winding is an isolated copper conductor. It should be able to handle the induced voltage and current. The coils can be isolated with different materials. These different isolators age differently and have different characteristics [267].

10.3 Comparison of Three RCM Methods

10.3.1 Three RCM Methods

Two RCM methods were introduced in Chapter 4 that are the RCM Moubray and RCAM. In this chapter, an RCM method developed and applied by

Vattenfall Vattenkraft will be presented, from now referred to as VVK RCM. This chapter will briefly describe, discuss, and compare these methods for RCM.

10.3.2 RCM Questions According to Moubray

As previously presented in Chapter 4, the RCM method involves asking seven questions about the studied system:

1. What are the functions and associated performance standards of the asset in its present operating context?
2. In what ways does it fail to fulfill its functions?
3. What causes each functional failure?
4. What happens when each failure occurs?
5. In what way does each failure matter?
6. What can be done to predict or prevent each failure?
7. What should be done if a suitable preventive task cannot be found?

These steps will be described one by one and then some additional features of this method are pointed out.

What are the functions of the asset? To answer this question, the asset's functions are divided into primary and secondary functions. The primary functions are the main purposes of the asset while secondary functions are additional properties that the asset is expected to meet. Functions should be described by a verb, an object, and a standard of performance.

In what ways does it fail to fulfill its functions? The next step is to identify in what way the asset can fail to perform its functions established in step one. There could be several ways the asset fails to fulfill its desired functions.

What causes each functional failure? Each functional failure may have several causes, failure modes. It is at this level that the maintenance of the system is to be done. It is stressed that the analysis must be applied at an appropriate detail level otherwise the work may become very extensive or in the other case, become meaningless.

What happens when each failure occurs? The effects of the failure should be recorded. This includes evidence that a failure has occurred, environmental or safety threats, effects on production, physical damage, and how to restore the system after the failure.

In what way does each failure matter? This step analyzes what consequences each failure leads to. First the failures are classified as apparent or hidden. If occurring on their own, hidden failures will not be noticed. Evident failures are failures which will become evident if occurring on their own. Evident functional failures are classified according to three groups that describe what the consequences of a failure are. The three groups listed below are ordered according to importance:

1. Safety and environmental consequences
2. Operational consequences
3. Nonoperational consequences

Operational failures affect costs in connection with production and operation. Nonoperational failures only effect the cost of repairing.

What can be done to predict or prevent each failure? Examine if there is any maintenance which can be done to prevent or predict the failure. These tasks are called preventive tasks. Predetermined tasks which may be used are scheduled restoration and scheduled discard. These are often appropriate when dealing with age-related failures. To use these strategies, there must be a point in time when there is an increase in the probability of failure. Condition-based tasks are used to identify potential failures. If condition-based tasks are feasible, the problem of how frequently to perform these tasks must be answered. This can be a difficult problem if reliable information about the failure probabilities and P–F intervals is hard to acquire. Condition-based tasks are feasible if a potential failure condition is possible to identify, the P–F interval is reasonably constant and not too short, and that monitoring the item at intervals shorter than the P–F interval is possible.

What should be done if a suitable preventive task cannot be found? If no appropriate preventive task is feasible or worth doing, there are three choices: redesign, no scheduled maintenance, or to conduct failure finding tasks. Failure finding tasks are intended for hidden failures. When deciding which option to choose the consequences of failures must be considered. If the consequence is nonoperational, economy can rule the choice but when there are safety or environmental consequences redesign might be the only option.

In the RCM Moubray method, the RCM process is summarized as a decision diagram. When a maintenance strategy is found that is feasible and worth doing it is chosen and further analysis of other maintenance tasks is not required. Whenever possible, scheduled on-condition tasks should be chosen. Otherwise scheduled restoration tasks and then scheduled discard tasks are selected. The last choice when dealing with less severe consequences (operational and nonoperational) is no scheduled maintenance or redesign. If the consequence involves environmental or safety hazards, the problem must be addressed and no scheduled maintenance is not an option.

Characteristics of RCM Moubray: The Moubray's RCM method has a predetermined preference of maintenance strategies. The method steers towards performing preventive tasks rather than corrective tasks after a failure. Of the preventive strategies, condition-based maintenance is preferred to predetermined maintenance. Environmental and safety consequences have a high priority in the analysis. Since the process stops when an acceptable maintenance strategy is found, it is possible that another type of strategy would be more efficient if it was also evaluated. On the other hand, some work can be saved and the process is made faster this way.

Characteristics of VVK RCM: VVK RCM is in many aspects very similar to RCM Moubray but also contains some major differences. It approximately follows the same steps but differs when the maintenance strategy is chosen. Since Vattenfall Vattenkraft has a large number of plants with similar systems and components they have chosen to make a standard RCM analysis for different generator systems. Vattenfall Vattenkraft has grouped similar generators together in 10 different categories. For each generator group, an RCM standard analysis has been done. This has been performed for a specific unit which then has served as a standard for this generator group. These standard analyses are then used as a base when making an RCM analysis for each individual generator unit at each plant. Approximately 10% of the intervals are changed when applying a standard to a specific unit [268].

A computer software program BI-Cycle is used in the VVK RCM analysis as an aid to perform and document the analysis. A flowchart of each asset is also used. The VVK RCM analysis requires six steps.

1. Define functions
2. Establish functional failures
3. Establish failure modes
4. Risk analysis (this step is called consequence analysis in the VVK documentation [46])
5. Establish possible maintenance tasks
6. Analysis and decision on maintenance strategies

Define functions. Functions are divided into primary and secondary functions. Primary functions describe the reasons why the asset is acquired. Secondary functions are other functions the asset is expected to perform. Secondary functions could, for example, be related to safety or appearance. Each function in the standard analysis is studied, if necessary functions are added or removed. Since the same standard is applied to all generators of the same kind, there is no exact level of performance included in the description of the functions.

Establish functional failures. For each function, all relevant functional failures are considered. The standard analysis is used and functional failures are added or removed if needed.

Establish failure modes. All reasonable causes for each functional failure are now listed. These are called failure modes. Again the standard analysis is used as a reference.

Risk analysis. The risk analysis is performed for each failure mode and contains several steps. First, the cause of the specific failure mode is described. To describe the causes for a failure mode, there is a choice to select between several predefined causes or the possibility to define one. Next, a description of the effects of this failure mode is documented. This text should include the chain of events leading to a functional failure if no preventive maintenance is applied. All evidence that a functional failure will occur and any alarms which are activated up until the time of a functional failure should be included. Safety and environmental issues as well as physical damage on equipment must also be included. Finally, a description of what restoration that must be performed to get the equipment running again and in what way the failure effects production must be added. If an alarm is triggered and reaches the operating personnel it is specified. The consequences of a failure are assessed. This is done by considering eight different categories and choosing one of three levels which describe the severity of the consequences for each category. The different categories are:

1. Breakdown costs
2. Costs of preventive maintenance compared to run-to-failure maintenance
3. Length of interruption of production
4. Working environment hazards
5. Flooding due to water released from the dam
6. Environmental hazards
7. Safety hazards
8. Efficiency loss in the plant.

The consequences of the first three categories are assessed after the analysis of the maintenance methods is performed. The three levels deciding how severe the failure consequence is for each category listed above have previously been defined by the plant owner depending on the location and importance of the plant. To evaluate the risk defined as

$$\text{(Probability of failure)} \cdot \text{(consequence of failure)} = \text{risk} \tag{10.1}$$

and to determine what kind of maintenance to use, VVK RCM uses a risk matrix (Figure10.1) with four probability intervals and three degrees of severity of the consequences. The risk matrix yields a number between 1 and 5 which corresponds to the assessed risk. If the number yielded is 3 or above, a preventive strategy is recommended. If the risk is one or two, corrective maintenance may be used. In the risk matrix, some cells contain two numbers. If, for example, a failure occurs in intervals between 21 and 50 years and the consequence is classified as serious the risk number is 3 if the consequence is categorized as a safety consequence, otherwise, the risk number is 2.

VVK RCM requires three approximations of failure intervals. Minimum, best, and maximum approximations have been defined in the following way.

Minimum approximation is the time until the failure mode will occur. This can sometimes correspond to a potential failure which is point P in Figure 2.2.

Probability \ Consequence	A Minor	B Serious	C Severe, catastrophic
Failure occurs more often than once a year	2 / 3 Safety, efficiency loss, environmental	4	5
Failure occurs in intervals of between 1 and 20 years	2	3	4
Failure occurs in intervals of between 21 and 50 years	1	2 / 3 Safety	3
Failure occurs in intervals of over 50 years	0	1	2 / 3 Interruption of production, flooding

FIGURE 10.1
Risk matrix for VVK RCM. (Adapted from E. Persson, *Metoder för kostnadsoptimering av underhållsåtgärder och riskvärdering av felsätt*. Master thesis KTH, 2004 [70].)

Best approximation is the time until the failure mode will occur and this leads to a functional failure (point F in Figure 2.2). This approximation is also used to determine which probability interval to select in the risk matrix (Figure 10.1).

Maximum approximation is the maximum time it will take until the failure mode and the functional failure will occur. This approximation is only used to specify the uncertainty in the approximations.

Consequently in cases where the failure mode leads without a delay to a functional failure, the minimum and best approximations should be the same.

Possible maintenance tasks. The different maintenance possibilities are studied in this step. A decision tree is used for help. First, corrective maintenance tasks are documented. Second, condition-based maintenance which is divided into continuous and scheduled maintenance is specified if it is possible to conduct. If the probability of failure increases with time and there is a possibility to use predetermined tasks, these are recorded. For a hidden failure, it is investigated whether there are any feasible failure finding tasks. Finally, possible redesign or education of personnel to avoid the failure is documented.

Analysis and decision on maintenance strategies. To economically decide what kind of maintenance to use, some additional input data is needed. This input data includes stop time and costs due to loss of production, the cost of personnel to perform maintenance, cost of new parts for damaged equipment and other costs in connection with the maintenance tasks and breakdowns.

There is also the possibility to specify if the maintenance can be planned ahead of time and how much this could reduce costs for the loss of production. VVK assumes that all spare parts needed are available at the plant in question. If this is not the case, it should be noted but it will not affect the analysis and costs calculated by the software. Based on this information and the consequences of a failure, a decision on maintenance strategy and intervals for maintenance is taken.

10.3.2.1 Characteristics of VVK RCM

Breakdowns and failures are fairly uncommon when dealing with hydropower plants. For generators, most of the failure modes analyzed have never even occurred at the plant studied. Since few failures occur it is often hard to assess the failure rate of components. This makes it hard to decide on optimal maintenance strategies.

The use of a standard analysis has the advantage that it is more cost efficient. Each component gets thoroughly analyzed for the standard template and few aspects are overlooked. When fitting the standard analysis to individual components, there is a danger of simply assuming that the standard analysis fits and not taking into consideration the individual features of the specific component analyzed. Another aspect of the standard analysis is that they are constructed for a specific unit and not a functional unit. This might lead to certain details that are specific for the generator analyzed to be included in the standard analysis. When applying the standard analysis to a generator, this issue should be kept in mind.

The VVK RCM has an extensive approach where all feasible maintenance tasks are compared. This requires a lot of work but yields a resulting maintenance strategy that is optimal regarding the tasks studied.

A shortcoming that VVK is aware of is the treatment of spare parts in the analysis where it is assumed that spare parts are always available at the plant in question. In some cases where the delivery time of spare parts is long, this could lead to a miscalculated maintenance strategy.

A question which is hard to handle is the possibility to co-maintain certain assets so that the total production downtime decreases. This has been handled by the possibility to specify how many percent of downtime costs can be saved for each component. This is very hard to approximate since knowledge of other parts and how often they need to be maintained has to be taken into consideration simultaneously. The RCM analysis should focus on one failure mode at a time and yield a best possible maintenance strategy for it. When the task planning is performed, a system view will be used and suitable intervals for maintenance in order to be able to co-maintain assets will be chosen.

10.3.2.2 Reliability-Centred Asset Management

The aim of RCAM is to relate preventive maintenance to the total maintenance cost and system reliability. The method was first proposed by L. Bertling and applied to power distribution systems [13,14]. The method start with the total system is modeled and reliability of the system is studied. To perform this step, component reliability information is needed. Critical components for the systems reliability are determined by a sensitivity analysis. The critical components are studied further and failure modes are determined. Impacts of preventive maintenance tasks for the failure modes are analyzed and modeled. One of the following models for the reliability of the components due to maintenance is used.

1. Component reliability is assumed to be affected only by preventive maintenance. A constant ratio between failure rate and the effect of preventive maintenance is proposed.
2. The second model expands the first model to also be able to be dependent of time.

The information deduced in the critical component analysis is used when comparing different preventive maintenance strategies with respect to reliability and cost.

When all the critical components have been studied focus again switches towards system analysis. The preventive maintenance for the individual components is linked together, and reliability and cost of maintenance for the entire system are studied. Through this system analysis, the final maintenance strategy is determined.

10.3.2.2.1 Characteristics

To apply the RCAM approach, one needs to be able to deduce reliability information for components and understand the impact of maintenance on reliability and cost. The information needed for RCAM is hard to obtain for many components due to the high level of research needed.

The RCAM model was developed for power distribution systems. These are typically systems with a high modeling complexity. For networks, there is a lot to gain from analyzing system performance and reliability. Failures in electrical distribution systems can have many effects. Either the entire electrical distribution system can fail, just a part of it or even just a performance reduction in the network may be the consequence.

An advantage of the RCAM method is its modeling and mathematical approach. This leads to a more standardized approach with fewer opportunities for personal decisions than other RCM methods. The result obtained specifies the specific amount of improvement in the reliability of the system and total cost resulting from the new maintenance strategy.

TABLE 10.1

Comparison of RCM Moubray, RCAM, and VVK RCM.

	RCM Moubray	RCAM	VVK RCM
Preferred strategy	Yes	No	No
Multiple tasks allowed	Sometimes	Yes	Yes
Input	Qualitative judgments, costs	System structure, statistics, effects of maintenance on reliability	Detailed analysis of system, costs and lifetimes of components, risks
Output	Maintenance strategies and tasks intervals and costs	Quantitative measures for reliability, comparisons of different maintenance plans	Maintenance strategies and tasks intervals, expected costs

10.3.2.2.2 Similarities between the RCM Methods

VVK RCM and RCM Moubray have a great deal in common. The main difference between the methods is that RCM Moubray has a decision diagram where certain maintenance strategies are preferred and if feasibly picked. VVK RCM has a more complex way of deciding maintenance strategy. All tasks are considered and with the help of a risk and cost analysis, the best maintenance strategy is picked. RCAM differs mainly since its approach is purely mathematical. Therefore, RCAM needs more data input and requires a higher level of research than the other methods to be applied. The method was developed for a structurally complex system. For generators which the VVK RCM focuses on, the structure would be very simple since nearly all failures lead to a functional failure and little would be gained from the system analysis which plays a major part in the RCAM method. The method has the advantage of being able to view the system as a unit when deciding maintenance strategies while RCM Moubray and VVK RCM view the system at component and failure mode level when maintenance strategies are determined. Some of the differences between the methods are shown in Table 10.1.

Preferred strategy in Table 10.1 means that the method has a predetermined order in which maintenance strategies should be chosen. For example, RCM Moubray has a decision tree which steers towards scheduled condition-based tasks. Multiple tasks allowed mean that more than one maintenance task can be used to prevent a failure mode.

10.4 VVK RCM Generator Analysis

In this section the VVK RCM method will be studied in more detail. The study will only treat the RCM analysis which has been performed for one particular plant component, the generator. The material studied includes reports generated from the software BI-Cycle which is used when performing VVK RCM. BI-Cycle is a specialized software used for maintenance analysis and decisions [91]. VVK has analyzed in total 25 different failure modes for the generator. These failure modes give rise to 8 failures affecting secondary functions and 17 failures affecting the primary function "deliver electric power."

10.4.1 What Affects VVK RCM Results?

10.4.1.1 Equipment Lifetimes

Figure 10.2 illustrates the approximations of min, best, and max. Min being an approximation for the time until a failure mode occurs and best being the approximated time until a functional failure occurs. Max describes the uncertainty in the analysis and is the maximum time until a functional failure occurs. All failures have the same timescale, years, and none are expected to occur at a time interval of less than 10 years.

According to Figure 10.2, failures are often expected to occur in intervals greater than 20 years. Age-related failures should be analyzed carefully when

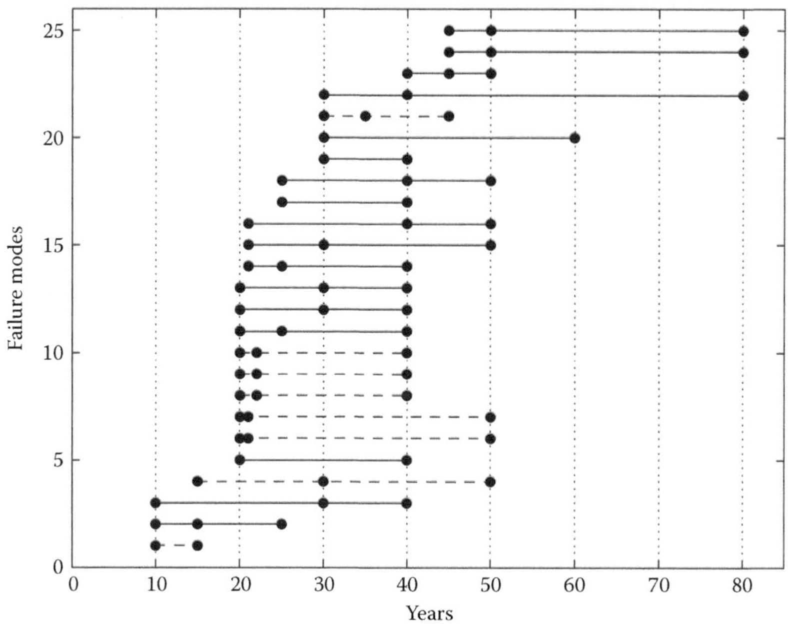

FIGURE 10.2
Min, best, and max approximations of failure times. Dashed lines correspond to secondary functions. (Adapted from O. Wilhelmsson, *Evaluation of the introduction of RCM for hydro power generators at Vattenfall Vattenkraft*. Master Thesis written at the Royal Institute of Technology KTH, Department of Electrical Engineering, 2004/2005 [265].)

the time between failures is so long. If a component is expected to last for 30 years, it is important to establish where in the life cycle the unit is at the moment to be able to find an efficient maintenance strategy. VVK RCM uses a standard analysis and it could be easy to simply apply this to the generator without adapting it for the generator in question. If a failure starts to develop at random as is the case for many of the studied failures, the problem mentioned above will not apply.

In cases when the probability of failure changes with component age, it could be effective to change the frequency of maintenance tasks in time but VVK RCM only allows the user to define one interval for each task. This issue is not of great importance for VVK RCM since the uncertainty of the time until a failure occurs is often large and therefore it would be hard to determine at which point in time to start performing inspections with a shorter interval. More realistic is to revise the inspection interval after each check performed reflecting the present condition of the asset.

10.4.1.2 Risk Analysis

The risks calculated using Figure 10.1 from the consequences and probability of each failure mode are very similar for the generator. There are 5 failure modes that have been approximated to occur at intervals within 1–20 years. The rest are approximated to occur at intervals between 21and 50 years. All the failure modes have resulted in a risk number of 3 or 4. With risk numbers 3 and above, preventive maintenance is recommended. Therefore, the RCM analysis has in all cases resulted in a preventive strategy. Scheduled condition-based maintenance tasks have been chosen for all but two failure modes. These two are hidden failures and are maintained by failure finding tasks. In total, two continuous condition monitoring tasks and three predetermined maintenance tasks have been selected.

10.4.1.3 VVK RCM Detail Level

VVK RCM has been performed at the generator level. One primary function has been identified, deliver electric power. For the corresponding functional failure, no electric power, 17 different failure modes have been determined. For these failure modes, appropriate maintenance tasks have been chosen. Another possibility could have been to go into more depth and analyze each generator component alone, for example, a rotor and stator analysis or even more in depth. This would most likely increase the cost for the procedure but would in turn result in a clearer comprehension of the causes of failures. A reason for not going into this depth is the high reliability of the system which does not justify such an analysis.

An appropriate detail level is to take failure modes that have occurred in the past and failure modes that are currently being prevented through maintenance into consideration [92].

Another advantage of going into more depth in the analysis is to link each failure mode uniquely to a maintenance task. The present level of failure modes results in several causes for each failure and consequently several maintenance tasks for each failure. This causes some problem when thinking in terms of P–F curves when deciding maintenance strategy.

Some failure modes are vaguely formulated. Two examples of vaguely formulated failures are "oval rotor" and "loose slot wedge." No definitions are given to when a rotor is assumed to be oval or a slot wedge is loose. It is possible to understand how the analysis team has chosen to define loosely by reading their failure effect text and checking their intervals for min and best approximation of failure. In the case with an oval rotor, min and best have been set to the same. Consequently, the analysis group's definition of oval rotor is when the rotor is sufficiently oval to cause a generator stop. Another group might have defined min for oval rotor to be when the first sign of deviation from roundness can be detected.

10.4.2 Many Maintenance Tasks for Each Failure Mode

A complication when choosing suitable maintenance intervals is that for each failure mode many different maintenance tasks have been chosen. Most failure modes result in severe consequences and there are often several causes. For each cause, there can be several different maintenance tasks. Only 2 of the 25 failure modes have resulted in only 1 maintenance task, for some as many as 6 different tasks have been chosen. This makes it hard to analyze P–F and maintenance intervals since different tasks can detect potential failures at different stages. To get better intervals for maintenance, it might in some cases be advantageous to divide failure modes into subfailure modes and handle them separately. If there are several inspection tasks, they must be analyzed together. For example, a failure has three different maintenance tasks. If the P–F interval is 5 years it is safe to conduct each individual maintenance task every 12th year. As long as they are spaced out evenly this would mean that a task is undertaken every fourth year which is less than the P–F interval of 5 years.

10.4.3 Task Intervals

Figure 10.3 shows intervals for maintenance tasks that have been proposed by VVK RCM. The large number of tasks to be performed at 10-year intervals is due to that all tasks to maintain the current and potential transformers are conducted at 10-year intervals. Most tasks are to be performed with a frequency of up to 10 years. Three tasks have longer intervals. It is questionable if it is appropriate to schedule task intervals of magnitudes greater than 10 years. A lot can happen in 10 years and the need for these tasks could change. The RCM list should not be a static

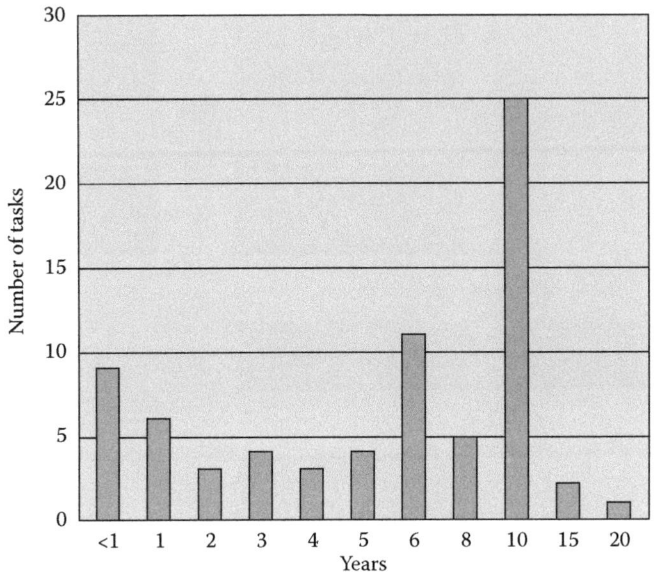

FIGURE 10.3
VVK RCM task intervals for the generator. (Adapted from O. Wilhelmsson, *Evaluation of the introduction of RCM for hydro power generators at Vattenfall Vattenkraft*. Master Thesis written at the Royal Institute of Technology KTH, Department of Electrical Engineering, 2004/2005 [265].)

list but needs to be updated continuously after rounds and maintenance work [268].

10.5 Changes in Maintenance Due to VVK RCM

To see what the VVK RCM process has yielded, old maintenance plans [129] for a generator are analyzed and compared with the plans that VVK RCM have resulted in [46]. First old maintenance plans will be described and their characteristics will be analyzed. Then maintenance suggested by VVK RCM will be compared to the old maintenance plans.

In the VVK RCM analysis, the exciter, the brake system, and the axes are not part of the generator system. In old maintenance plans, these parts were included. For the comparative study, equipment not part of the VVK RCM definition of the generator system will be ignored.

10.5.1 Evolution of Generator Maintenance Plans

The plans for generator maintenance have evolved in two steps illustrated in Figure 10.4 during the last 8 years. In 1998, there was a plan with general

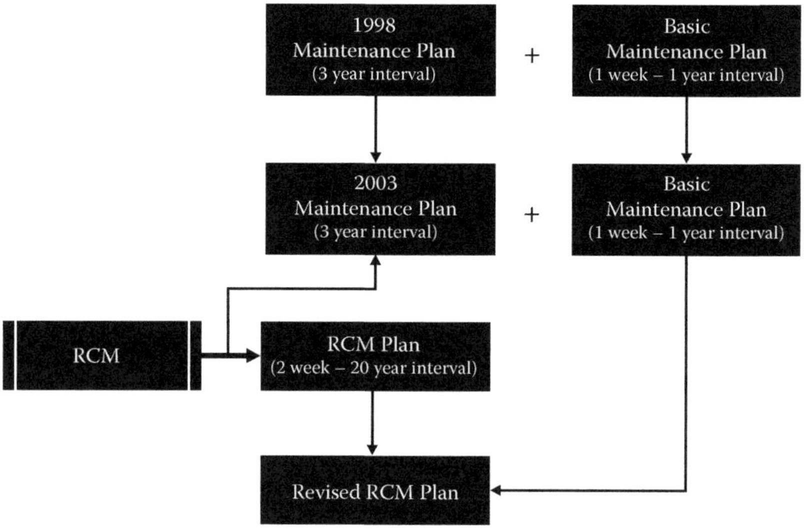

FIGURE 10.4
Evolution of maintenance plans for hydro power plans. (Adapted from O. Wilhelmsson, *Evaluation of the introduction of RCM for hydro power generators at Vattenfall Vattenkraft*. Master Thesis written at the Royal Institute of Technology KTH, Department of Electrical Engineering, 2004/2005 [265].)

maintenance that was conducted every third year. There was also a plan with basic maintenance that should be performed at intervals between 1 week and 1 year. In 2003, new improved plans were used. Since the VVK RCM process had already started, the 2003 plans have most likely been influenced by it. The plans used in 2003 have their structure in common with the 1998 plans. VVK RCM has yielded plans that are developed in a fundamentally different way which will be explained in the following sections. The RCM plan is not yet in use and will go through a final check and revision before being used. The revision will plan the tasks from a system view in a cost efficient way. In this work, the 1998, 2003, RCM, and the current basic maintenance plans have been used.

10.5.1.1 1998 Maintenance Plan

Before the introduction of RCM at VVK, generator checks were carried out every third year. The maintenance plan used was not customized for individual generators. The same plan was used each time the generator check was conducted. In the plan from 1998 that has been studied, few maintenance tasks were performed since a larger renovation of the generator was planned for 2001. When a 3-year check was performed, any observations of interest were recorded, if urgent they were dealt with. If anything alarming was found there was a choice to perform a major generator overhaul [130].

The 1998 maintenance plan contains many features that should be checked. The features listed are mostly parts of the generator such as stator winding or slot wedges. There are also failure modes such as rust due to vibration or unusual heating of the stator windings. In the maintenance plan, there was no information in what way the different functions should be checked or recorded. An example of what the 1998 plan looked like is given in Table 10.2. Since the objects in the 1998 plan were so general the way to inspect pieces of equipment was decided through experience or old protocols [130]. For each object in the plan, there may have been several tasks performed. This depended on what was found during the generator check. If an inspection resulted in a question whether the piece of equipment was in good shape, more thorough tests could be performed.

Tasks that should be performed more often were listed in another list with basic maintenance. Most of these tasks are meter readouts or cleaning tasks.

In total for the generator there were 40 maintenance tasks and components in the 1998 maintenance plan. As mentioned earlier, each of these could in turn contain several tasks. Eight additional tasks were performed according to the basic maintenance plans. The tasks were divided according to generator components as presented in Table 10.3.

10.5.1.1.1 Analysis

Some characteristics of the 1998 maintenance plan are listed below.

- Nearly all parts of the generator are mentioned in the maintenance plan. If maintenance is done properly and all items in the plan are checked this would lead to that the entire generator is inspected rigorously every third year. This strategy for maintenance is a comprehensive approach.
- The fact that it is not stated in what way maintenance is to be performed and why can lead to varying tasks when maintaining. Too much or too little effort might be put into the maintenance. There is

TABLE 10.2

Example of How Some Items in the 1998 Maintenance Plan could look like after Maintenance Tasks were Completed

Component/Task	Task Planned	Task Completed	Observations
Stator			
Ventilation channels	X	X	
Rust due to vibration	X	X	

Source: O. Wilhelmsson, *Evaluation of the introduction of RCM for hydro power generators at Vattenfall Vattenkraft*. Master Thesis written at the Royal Institute of Technology KTH, Department of Electrical Engineering, 2004/2005 [265].

TABLE 10.3

Comparison of Maintenance Plans, 1998, 2003, and RCM. Objects (components and Tasks) in 1998 and 2003 Maintenance Plans and Number of tasks and Failure Modes in Parenthesis for the RCM Plan

Maintenance Plan/Component	Stator	Rotor	Other	Total
1998 Maintenance Plan	12	11	17	40
2003 Maintenance Plan	22	17	13	52
RCM Maintenance Plan	25(9)	17(9)	17(5)	59(23)
RCM Maintenance Plan averaged over 3 years	19	11	19	49

Source: O. Wilhelmsson, *Evaluation of the introduction of RCM for hydro power generators at Vattenfall Vattenkraft*. Master Thesis written at the Royal Institute of Technology KTH, Department of Electrical Engineering, 2004/2005 [265].

also a problem to follow up the results of the last inspection and compare results and trends if no corresponding tests had been performed last time. Difficulties would also arise if a long time employee with knowledge of what maintenance needs to be done is replaced. The new employee would surely bring new tasks and also some old methods would be forgotten.

- The 1998 plan gives the maintenance personnel freedom to conduct the maintenance methods they see appropriate. Depending on the current status of the equipment, varying methods can be used.
- The choice to maintain all equipments every third year is probably a compromise. Some tasks might rather be performed more often and some more seldom. Having a period as short as 3 years could result in that more costly maintenance tasks are not chosen. If intervals would be dynamic more tasks for maintaining could come into question.
- The maintenance process is static and no new ideas evolve such as redesign.
- There are scheduled meter readouts but no documented scheduled analysis of this data. The data should be trended and analyzed but it is not documented how and at what time intervals this should be done.

10.5.1.2 2003 Maintenance Plan

In 2003, a new maintenance plan was introduced. This plan can be customized for the individual generators. Since the RCM process had started, this plan is most likely influenced by it. The plan is clearly divided into different generator parts and subparts. For each subpart, it is stated what should be checked, for example, rust, cracks, or dirt. However, no method to conduct

TABLE 10.4

Example of How Items from the 2003 Maintenance Plan can look like after Maintenance has been Performed

Stator Core					
Piece of Equipment	**Focus of Maintenance**	**Observations**	**Condition (0–3)**	**Date**	**Activity Performed**
Ventilation channels	Deformation	Small bumps	1		
	Dirt	Particles	3		Vacuumed

Source: O. Wilhelmsson, *Evaluation of the introduction of RCM for hydro power generators at Vattenfall Vattenkraft*. Master Thesis written at the Royal Institute of Technology KTH, Department of Electrical Engineering, 2004/2005 [265].

these inspections is proposed. An illustration of a part of the plan is presented in Table 10.4.

A scale from 0 to 3 is used to state the condition of the asset and which measures to be taken.

0. Good condition
1. Keep under supervision
2. Attend to and restore at next stop
3. Attend to and restore immediately

The interval of these checks is still 3 years like the preceding 1998 maintenance plan.

10.5.1.2.1 Analysis

This method is similar to the 1998 plan in the way it lists all possible generator parts, but it also goes further than the proceeding plan by defining what to look for on each piece of equipment. The focus still lies on listing all possible equipments and not in analyzing what prevents failures like the RCM method does. The 2003 plan is more detailed and in depth when listing generator parts than the preceding 1998 plan. When compared in Table 10.3, it is clear that more items have been added to the stator and rotor inspections. This method will achieve a very thorough inspection if conducted correctly since all pieces of equipment will be checked. The time between checks is constant, 3 years and is not adjusted to fit the individual maintenance tasks.

10.5.1.3 RCM

The RCM analysis has been described in earlier chapters. The method studies failures and finds ways to prevent them. Economically and safety efficient maintenance strategies are then chosen and performed. The major difference from earlier strategies is the way it focuses on possible failures and then determines preventive maintenance strategies whereas old plans listed all the different parts of the generator.

10.5.2 Changes Resulting from RCM

10.5.2.1 Plan Structure Difference

As mentioned earlier, the RCM plan is developed in a fundamentally different way than the plans before. This makes it harder to make certain that all assets will be maintained. It is easier to list all pieces of equipment as the earlier maintenance plans did. A piece of equipment can now easily be overlooked if it is not determined to have a significant effect on a failure that has been analyzed. On the other hand, tasks can be cut down if they are unnecessary and do not have any impacts on failure probabilities. This problem is easily overcome with a quick comparison of pieces of equipment maintained in the RCM plans and equipment inventory or old maintenance plans.

When focus is directed towards pieces of equipment it might be hard to find new ways of predicting failures that are not directly connected to the equipment. For example in the RCM analysis, there is a task to monitor ozone in the generator chamber in order to check for damaged insulation. This task is not found in the old maintenance plans which are equipment orientated. The checks of ozone levels might still have been done but are not mentioned in the old maintenance plans since these are equipment orientated.

10.5.2.2 Plan Detail

The method to check equipment is specified in the RCM plans. This makes certain that reoccurring checks are conducted in the same manner. This helps when comparing trends and differences between inspections. Furthermore, intervals for analyzing trends of data are included in the RCM plan which is not the case in the old plans. Since the RCM plans go into such detail compared with the old plans, it is possible that maintenance is performed too strictly according to the plan. Common tasks that ought to be done but are not contained in the RCM plans might be overlooked. Old plans gave the maintenance personnel freedom to conduct tasks they felt suitable. Since the RCM plan goes into such detail, it might appear to contain more tasks than the old plans. This is only a visual perception and has not been established in this analysis.

10.5.2.3 Task Intervals

Another major difference is that RCM allows different maintenance intervals for different pieces of equipment. The spread in time of planned tasks for RCM is presented in Figures 10.5 and 10.6 for stator and rotor tasks. The

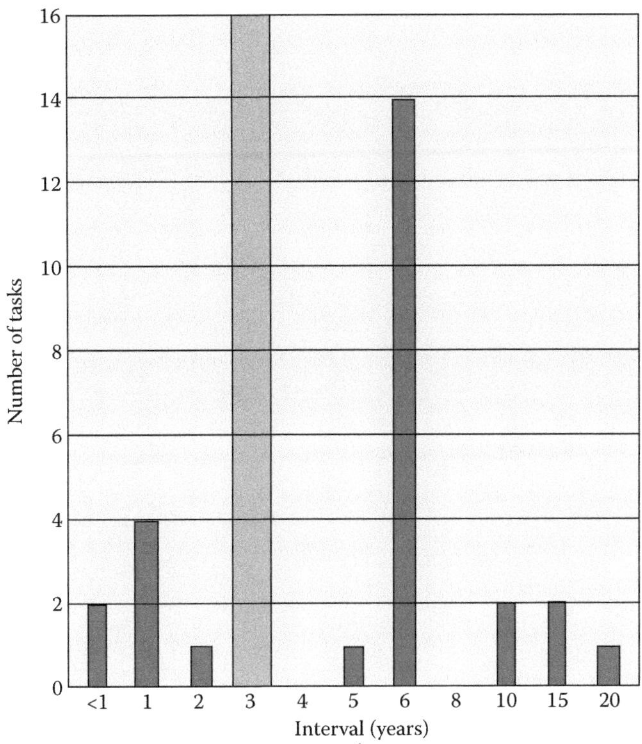

FIGURE 10.5
RCM maintenance intervals for the stator. The column at 3-year intervals corresponds to how tasks were performed before. (Adapted from O. Wilhelmsson, *Evaluation of the introduction of RCM for hydro power generators at Vattenfall Vattenkraft*. Master Thesis written at the Royal Institute of Technology KTH, Department of Electrical Engineering, 2004/2005 [265].)

intervals chosen for maintenance tasks vary greatly in length. Intervals range from less than 1 year up to 20 years. In the old plans, maintenance is performed every 3 years. A graph for the old plans would be a single pillar at 3 years. For expensive tasks that are performed seldom it is a difficult task to determine appropriate intervals. In the old plans, the larger tasks were performed when indications led personnel to believe there might be something wrong with the equipment. RCM has determined when these checks should be performed beforehand. This might lead to expensive tasks being performed even if other indications show there is nothing wrong with the equipment.

When constructing maintenance plans from the suggested maintenance task intervals, it might not be profitable to conduct tasks at many different intervals. The maintenance tasks conducted for the rotor shown in Figure 10.6 could, for example, be grouped so that all tasks are performed

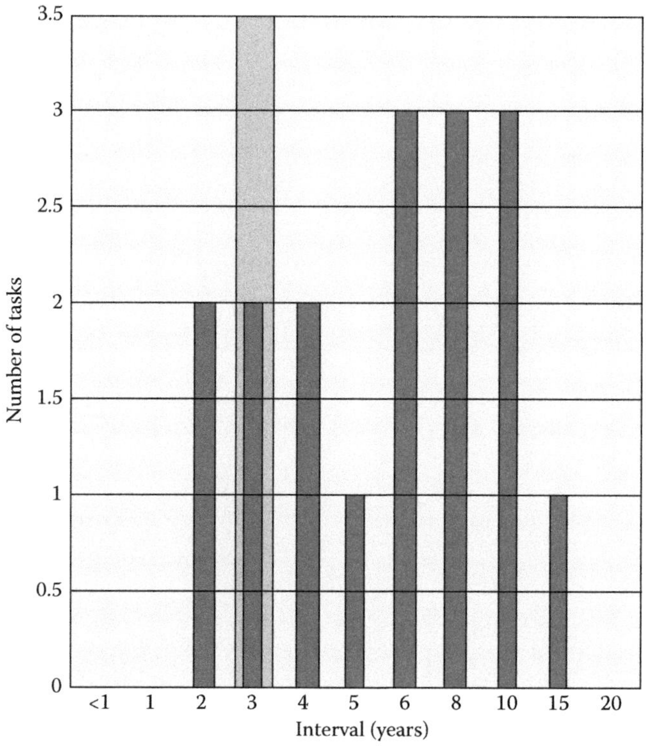

FIGURE 10.6
RCM maintenance intervals for the rotor. The column at 3-year intervals corresponds to how tasks were performed before. (Adapted from O. Wilhelmsson, *Evaluation of the introduction of RCM for hydro power generators at Vattenfall Vattenkraft*. Master Thesis written at the Royal Institute of Technology KTH, Department of Electrical Engineering, 2004/2005 [265].)

at 2- or 6-year intervals. The tasks that are suggested to be performed at 3- and 4-year intervals could be performed every second year and the rest every sixth year. When making these decisions, the individual tasks and how they are affected if the intervals are changed must be taken into consideration. The possibility to use other maintenance tasks that better fit the adjusted intervals could also be investigated. The LCC approach presented in later section could be of help in assessing what the effects are of changing inspection interval.

10.5.2.3.1 Number of Tasks

A list of the number of objects included in the three different maintenance plans is presented in Table 10.3. To be able to compare, RCM intervals are given averaged over 3 years since these tasks are performed with different intervals. The objective has been to compare the number of tasks before RCM to the number of tasks to be performed after the RCM analysis. When comparing the number of tasks performed, one must bear in mind that the RCM analysis focuses on possible failures and then decides what to do to prevent them. Therefore, many RCM tasks apply to the same piece of equipment and failure. For example, three tasks are performed to check the laminated core of the stator whereas it could be mentioned only as laminated core in the 1998 maintenance plan or listed as three parts of the laminated core in the 2003 plan. The frequency of the tasks performed is also different. The suggested intervals for maintenance after RCM are different for most tasks, while intervals for tasks before RCM were performed every third year. In Table 10.3, tasks that are simple and are conducted very often like meter readouts have not been included as these would completely dominate the amount of tasks performed. One must also be aware that these numbers do not reflect the complexity or costs of the tasks undertaken. Because of these differences, the number of objects in each plan presented in Table 10.3 should not be compared directly with one another.

10.5.2.3.2 Stator Analysis

To get a closer look at the differences of the methods and to be able to see if something has been overlooked in the RCM method, it was decided to further investigate the planned activities proposed for the stator.

In the 1998 plan, there are 11 items linked to the stator. Some are very general entries such as winding or pressure fingers and plates. Two of the 11 items are rust due to vibration and abnormal heating of the stator winding, that is, states of the equipment. The other nine items are listed as pieces of equipment.

The 2003 plan contains all the items from the 1998 plan and some additional. Some items belonging to the stator from the 1998 plan are also divided into smaller pieces of equipment. In total, there are 22 pieces of equipment belonging to the stator included in the 2003 plan.

Due to the fact that the maintenance plans are constructed in such different ways and they list procedures differently, it has not been possible to perform a fair comparison between the 1998 plan and the RCM plan for the stator. With the limited information given, it seems as all the stator parts checked in the 1998 plan are also checked in some or many ways in the RCM plan.

Figure 10.5 shows that the tasks for the stator after RCM are performed at many different intervals. The tasks performed at short intervals are mostly cheap. These tasks include meter readouts and trend analysis of data. Also general cleaning of the generator should be performed once a year. The rest of the tasks are performed more seldom, a majority at 6-year intervals. These are tasks that before might have been carried out at 3-year intervals. The tasks performed at intervals of 15 and 20 years are expensive and were not planned at regular intervals before. These major tasks were performed when maintenance personnel found reason to conduct such a check. The main difference

Damaged stator winding isolation

Before RCM

	Interval	Cost/task	Cost for 60 Years
Visual inspection of the winding	3	6860	137200
DSD measurement (uncertain interval)	6	89184	891840
Measure ozone (uncertain interval)	2	1200	36000
Measure and analyse trends in strand currents	0,019	4	12480
Change the stator winding	60	24000000	24000000
Coil repair	20	2000000	6000000
			31077520

After RCM

	Interval	Cost/task	Cost for 60 years
Visual inspection of the winding	6	6860	68600
DSD measurement	6	89184	891840
Measure ozone	2	1200	36000
Measure and analyse trends in strand currents	0,038	4	6240
Change the stator winding	60	24000000	24000000
Coil repair	20	2000000	6000000
			31002680

Difference in cost
74840

FIGURE 10.7
Changes in cost due to VVK RCM for the damaged stator winding insulation failure mode. (Adapted from O. Wilhelmsson, *Evaluation of the introduction of RCM for hydro power generators at Vattenfall Vattenkraft*. Master Thesis written at the Royal Institute of Technology KTH, Department of Electrical Engineering, 2004/2005 [265].)

from earlier is that major inspections now are scheduled but were before conducted on a basis of necessity. Due to lack of information, it is difficult to determine if all the tasks that will be performed after RCM were done before. One specific failure mode affecting the stator is "Damaged isolation stator winding." For this specific failure mode, the changes due to RCM seem to be that certain tasks are performed more seldom. Visual inspections and scheduled meter readouts will not be performed as often as before. A comparison of the costs shows that over a 60-year period which is the expected time until the stator winding is changed the savings amount to 0.2% of the expected total maintenance costs (Figure 10.7).

10.5.3 Conclusion

The effects of the introduction of RCM are difficult to determine. An attempt to make an initial comparison can be made by comparing the amount of tasks performed according to the plans. The quantity of tasks resulting from RCM is difficult to compare to before RCM due to the general nature of the plans prior to RCM. For the stator which has been studied, the amount of different tasks seems to be quite similar. The total amount of tasks performed also depends on the time intervals between tasks. Most tasks determined by RCM are performed at intervals greater than 3 years and hence are performed more seldom than before when most tasks were performed every 3 years.

Of the tasks proposed by VVK RCM, the majority are scheduled condition monitoring tasks. Three tasks are failure finding tasks, three are predetermined maintenance tasks, and two are continuous condition monitoring tasks. It seems as the tasks performed before and after RCM are quite similar. Since maintenance personnel has carried out the RCM analysis, there is a chance that similar maintenance as before has been selected. "In general generator service personnel are trained in repair and not necessarily in diagnostics. As a consequence repair and maintenance programs may suffer" [131]. Some new tasks have been added to the plans. Especially tasks that involve measurements and analysis of data are now stated specifically. These tasks were not in the plans prior to RCM.

Another result from RCM is that all intervals for tasks are now determined. Before RCM, larger tasks were performed when operating personnel felt they were needed. This change should lead to a better control of which tasks that are performed and easier budgeting of costs. The downside is that larger tasks are performed simply because they are stated in the plans and not always because they are needed.

The greatest change is the way in which the plans are designed. After RCM, the tasks in the plans are sorted according to failures and how to detect them early and prevent them, whereas before the generator parts were simply listed. RCM has thus made it clearer why and what the purpose is for each task. The tasks are also more thoroughly described and therefore give better opportunities to compare results over time.

10.6 Proposed Inspection Interval Models

When planning maintenance, a crucial decision to make is the intervals between preventive maintenance tasks. The choice of these intervals has direct influence of the cost of system maintenance. Reliability is most likely affected by the length of time between maintenance tasks are performed. Ultimately, the purpose of maintenance tasks is to avoid or postpone failures so increasing the intensity of the tasks performed should also change the reliability of the system. The main changes resulting from VVK RCM are different intervals between maintenance tasks. Most tasks to be performed according to VVK RCM are condition monitoring tasks. To evaluate if changes of maintenance intervals proposed through VVK RCM are cost efficient, a life-cycle cost (LCC) approach has been used. A model for optimizing intervals for inspections is proposed and then results from VVK RCM are analyzed using this model. The model proposed is valid when maintenance tasks are performed by inspections. The possibility to use a Markov model has also been investigated. Finally, the two proposed methods are compared.

10.6.1 LCC Approach

The idea behind the model is that more frequent inspections will result in a greater chance to detect a potential failure. If a potential failure is detected, a functional failure and corrective repair can be avoided which saves money since functional failures often lead to longer downtime of production and more expensive repairs than if the asset was repaired before a functional failure occurred.

The LCC approach will take costs for inspections, repairs, and loss of production into account. The costs for loss of production are here assumed to be embedded in the costs of repairs and the costs of inspections. The total cost will therefore be

$$C_{LCC} = C_R(p) + C_{Itot}(tI) \tag{10.2}$$

where

$C_R(p)$ is the expected cost of repairing the asset to a good as new condition including loss of production. Depending on the probability of detecting potential failures p, this cost takes on values in the range between CPR and CCR.

$C_{Itot}(tI)$ is the expected total cost for performing all inspections until t_{CR}, given the inspection interval t_I including loss of production. No revenues or operation costs are used.

10.6.1.1 Definitions and Available Data

The model proposed will use approximations that have been performed in VVK RCM. The data which is taken from VVK RCM and used in the model is presented below. For each of the failures analyzed in VVK RCM, the costs and times below have been approximated. These values will be treated as fixed.

C_{CR} is the expected cost of restoring the asset to a good as new condition after a breakdown including loss of production (corrective repair).

C_{PR} is the expected cost of restoring the asset to a good as new condition before a breakdown has occurred including loss of production (preventive repair).

C_I is the expected cost of performing an inspection including loss of production.

t_{CR} is the expected time until failure if no restoration tasks are performed.

t_{PR} is the expected time until preventive repair that will set the asset in a good as new condition if inspections detect a potential failure.

The two variables below will be used in the LCC approach where:

p is the probability of discovering that an asset is in the potential failure condition and repairing it to a good as new condition before a breakdown occurs given an inspection interval t_I. In other words, the probability that the asset is repaired for the cost CPR at time $t_P R$ before a failure has occurred instead of a failure occurring resulting in the cost for repairs CCR at time t_{PR} given the inspection interval t_I.

t_I is the intervals between inspections.

10.6.1.2 Model

The proposed LCC approach is illustrated using a simple example. The example used will consist of the values in Table 10.5 and will be referred to as case 1. When actual failures are analyzed, these values are taken from VVK RCM.

10.6.1.2.1 Expected Cost of Renewal

In reality the time between repair and the time between failures are probabilistic. In the LCC approach, there are only two possible times when repairs are performed. Either before a failure at the time t_{PR} or after a failure at the time t_{CR}. These two ways of repairing will restore the asset to a good as new condition. The cost to repair after a failure C_{CR} is often larger than

TABLE 10.5

Case 1 Data

C_{CR}	10,000,000 SEK
C_{PR}	3,000,000 SEK
C_I	10,000 SEK
t_{CR}	30 years
t_{PR}	25 years

Source: O. Wilhelmsson, *Evaluation of the introduction of RCM for hydro power generators at Vattenfall Vattenkraft.* Master Thesis written at the Royal Institute of Technology KTH, Department of Electrical Engineering, 2004/2005 [265].

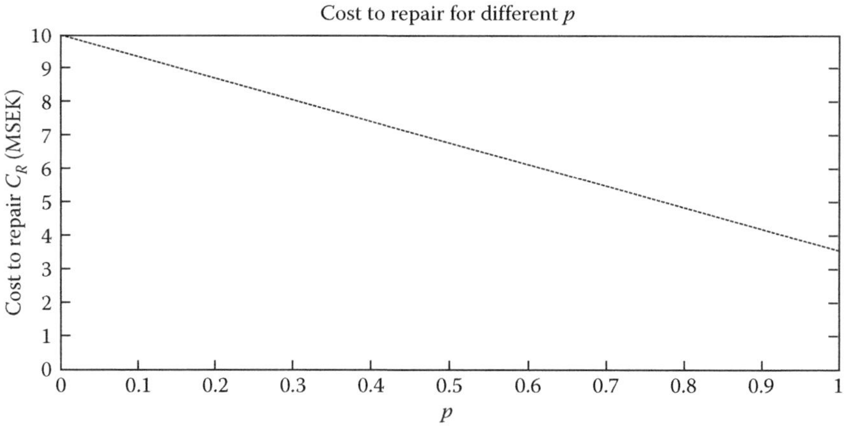

FIGURE 10.8
Repair cost CR for case 1 versus the probability of detecting and repairing a potential failure before a breakdown occurs. (Adapted from O. Wilhelmsson, *Evaluation of the introduction of RCM for hydro power generators at Vattenfall Vattenkraft*. Master Thesis written at the Royal Institute of Technology KTH, Department of Electrical Engineering, 2004/2005 [265].)

costs of preventive repairs C_{PR} undertaken before the failure occurs. The difference in costs is due to that when a failure occurs additional equipment other than the actual failed component may be harmed and repairs cannot be planned ahead of time. Furthermore, the downtime is often longer for corrective repairs which leads to greater costs because of loss of production. The expected cost of restoring the asset to a good as new condition can be calculated through Equation 10.3

$$C_R(p) = p \cdot C_{PR}\frac{t_{CR}}{t_{PR}} + (1-p) \cdot C_{CR} \qquad (10.3)$$

where p is the chance of discovering and repairing a potential failure before the failure occurs. The lifetime is chosen to be t_{CR} so C_{PR} is weighted to accommodate for the different lengths between preventive repairs and corrective repairs. The cost $C_R(p)$ is linear in p and displayed in Figure 10.8 for case 1.

10.6.1.2.2 Cost of Inspections

Inspections are performed in order to determine the state of an asset. After an inspection, a decision is made if the asset should be repaired. In the proposed LCC method, it is assumed that inspections do not affect the state of an asset. In reality, inspections could also increase the probability of failure. Errors in dissembling and assembling during inspections could, for example, affect the state of the asset.

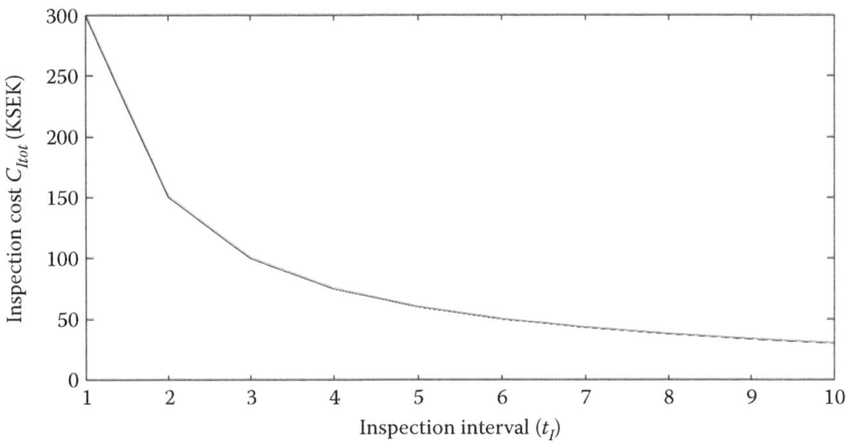

FIGURE 10.9
Total inspection cost C_{Itot} for different inspection intervals for case 1. (Adapted from O. Wilhelmsson, *Evaluation of the introduction of RCM for hydro power generators at Vattenfall Vattenkraft*. Master Thesis written at the Royal Institute of Technology KTH, Department of Electrical Engineering, 2004/2005 [265].)

Performing inspections more frequently improves the probability to detect a potential failure before a breakdown occurs. The benefit of determining the state of an asset and the cost of inspections must be weighted against each other. The cost of an inspection is C_I and the time between inspections is t_I. The total cost of inspections C_{Itot} is calculated for the same time interval t_{CR} as was done for repairs

$$C_{Itot}(t_1) = C_I \frac{t_{CR}}{t_I} \tag{10.4}$$

Figure 10.9 shows how the interval between inspections affects the total cost of inspections (see Equation 10.4). The figure shows that the total cost is sensitive to changes in inspection intervals when intervals are short. When inspection intervals are long, only small savings are made when increasing the inspection interval by 1 year.

10.6.1.2.3 Choosing Inspection Interval

Inspection intervals should be shortened if it is profitable. If inspections are performed at a certain interval, what happens if the interval is decreased by 1 year? The right graph of Figure 10.10 shows how the inspection cost increases when the inspection interval decreases by 1 year. This is calculated by subtracting the cost for inspecting at intervals of $n+1$ from the cost of performing inspections at intervals of n, where $1 < n < 9$ in case 1. To find out if this is beneficial, this cost must be compared with how much is saved

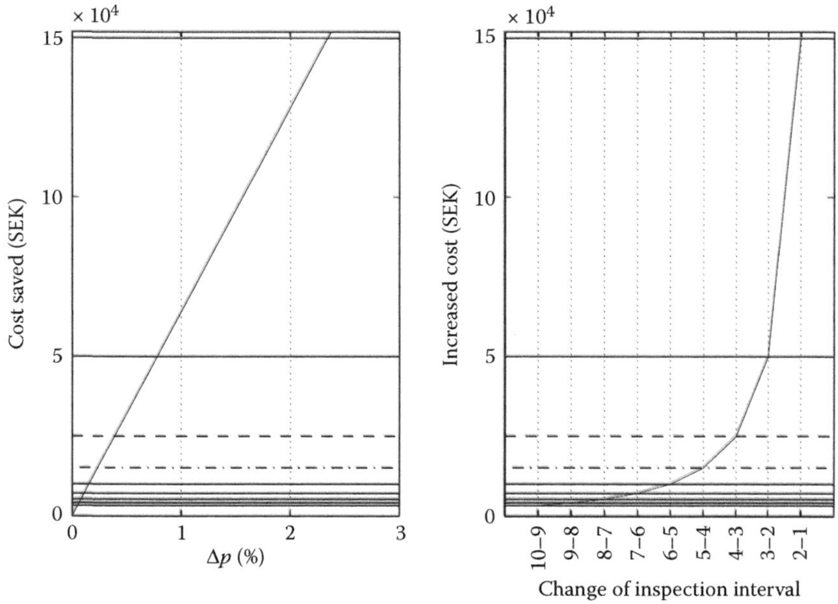

FIGURE 10.10
(Left) Expected cost saved due to an increase in p. (Right) Extra cost for performing inspections at a 1-year shorter interval. (Adapted from O. Wilhelmsson, *Evaluation of the introduction of RCM for hydro power generators at Vattenfall Vattenkraft*. Master Thesis written at the Royal Institute of Technology KTH, Department of Electrical Engineering, 2004/2005 [265].)

because of a greater chance to discover a potential failure before a breakdown occurs, that is, an increase in p shown in the left graph of Figure 10.10. This graph simply shows the negative slope of Equation 10.3.

In case 1, say inspections are performed every fifth year. Should the inspection interval be decreased to 4 years? In the right graph of Figure 10.10, the 5–4 cost is checked and then the dotted/dashed line is followed to the left graph. This graph reveals that the decrease in inspection interval is economically profitable if this leads to a 0.2% or greater increase in the chance to detect a potential failure and repair it before a breakdown occurs. If the interval is decreased to 4, is it also worth decreasing to 3? The same procedure is performed, this time starting at 4–3 and following the dashed line arriving at 0.4% or more increase in p. The two graphs can also be summarized in one (see Figure 10.11) which shows a direct relationship between changes of inspection interval length and the least needed increase in p for the change to be economically justified.

The increase in p needed to justify an inspection interval change of more than 1 year is calculated by adding each years incremental change in p together. Using the example above, the change of the inspection interval from

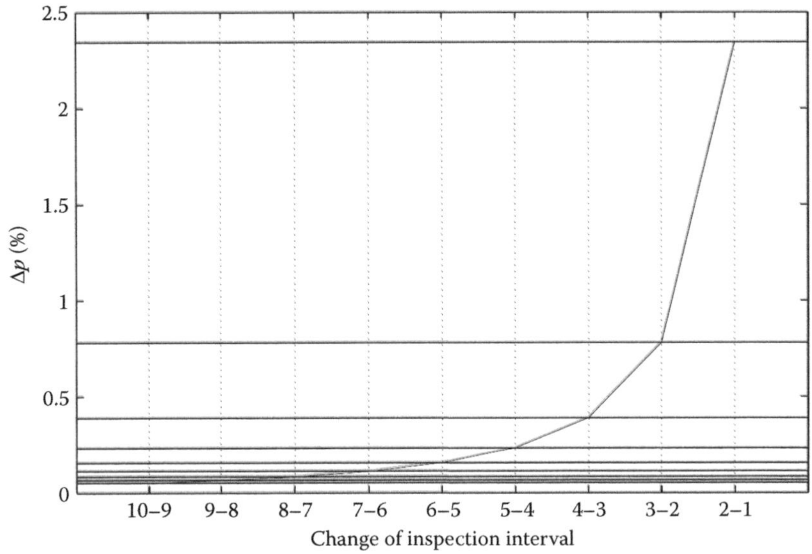

FIGURE 10.11
Increase in p needed to justify a 1-year shorter inspection period. (Adapted from O. Wilhelmsson, *Evaluation of the introduction of RCM for hydro power generators at Vattenfall Vattenkraft*. Master Thesis written at the Royal Institute of Technology KTH, Department of Electrical Engineering, 2004/2005 [265].)

every fifth year to every third year needs to result in a 0.2% + 0.4% = 0.6% increase in p to be economically beneficial.

10.6.1.2.4 *Inspection Intervals versus Failure Detection*

So far only information from VVK RCM has been used. To determine the optimal inspection intervals, it is now necessary to make some assumptions.

If the probability to detect failures and prevent breakdowns is known for all inspection intervals a simple calculation can give the optimal inspection interval. Often this relationship is not known, which is the case for the inspections and failures investigated in this case study. Some characteristics of this relationship can be deduced by reasoning.

- If inspections are performed continuously, the chance to detect a potential failure should be close to 100% and if inspections are performed very seldom the chance to detect a failure approaches 0%.
- The probability to detect and prevent breakdowns is greater the shorter the inspection period is.
- The probability of detecting a potential failure will decrease rapidly once the inspection interval is longer than the normal time from evidence of an upcoming failure becomes evident until the time of failure, the P–F interval.

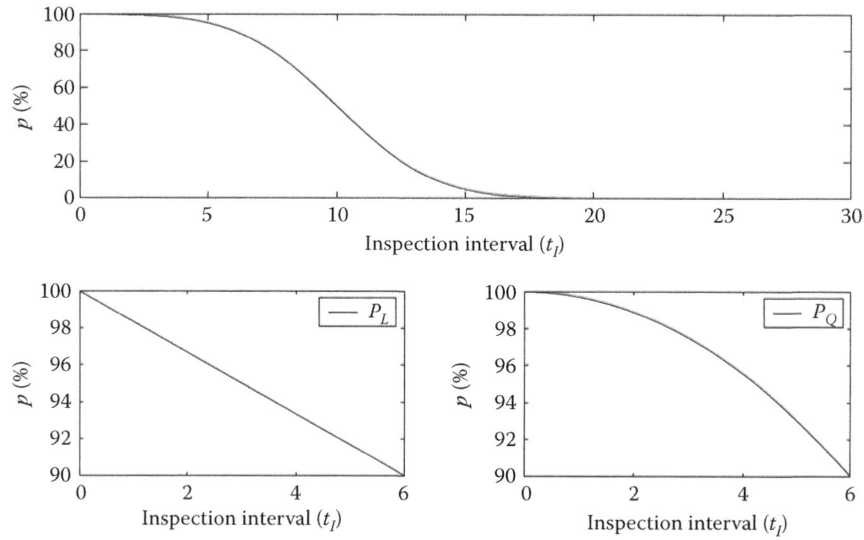

FIGURE 10.12
(Top) An approximated model of the relationship between inspection intervals and probability to detect a potential failure before a breakdown occurs. (Bottom) Two ways of approximating the top 10% of the top graph. To the left a linear function and to the right a quadratic function. (Adapted from O. Wilhelmsson, *Evaluation of the introduction of RCM for hydro power generators at Vattenfall Vattenkraft*. Master Thesis written at the Royal Institute of Technology KTH, Department of Electrical Engineering, 2004/2005 [265].)

These assumptions will result in a relationship between inspection intervals and probability of detecting a potential failure resembling the top graph in Figure 10.12. In the top graph of Figure 10.12, the survival function of the normal distribution is used to depict this relationship. With the type of failures analyzed in VVK RCM with expensive corrective maintenance compared to the cost of preventive maintenance, only the top part of this graph where the chance of detecting a potential failure is high is of interest. This top part has been approximated with two relationships

1. A linear curve PL
2. A quadratic curve with an increasing negative slope PQ

The bottom graphs of Figure 10.12 show one example of each approximation.

Using the relationships in the bottom graphs of Figure 10.12, the total costs of inspections and repairs for different inspection intervals can be calculated. This is simply done by adding Equations 10.3 and 10.4, that is, the cost of repairs and the cost of inspections. The bottom graph of Figure 10.13 shows what the resulting cost for case 1 is for the two different approximations of p.

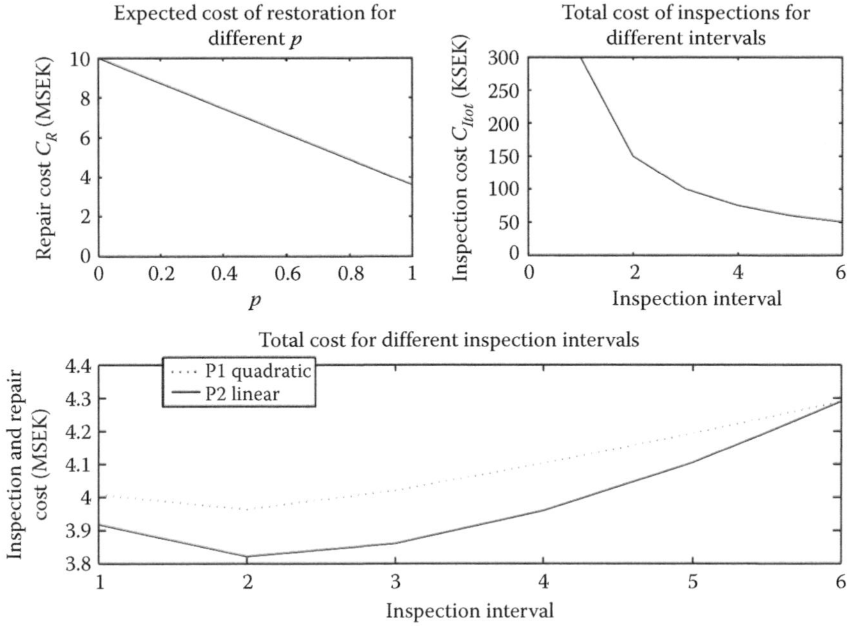

FIGURE 10.13
(Top left) Expected cost of renewal for different p. (Top right) Cost of inspections until the expected time till renewal. (Bottom) Total expected cost of maintenance until renewal for p values shown in Figure 10.12. (Adapted from O. Wilhelmsson, *Evaluation of the introduction of RCM for hydro power generators at Vattenfall Vattenkraft*. Master Thesis written at the Royal Institute of Technology KTH, Department of Electrical Engineering, 2004/2005 [265].)

In both cases, an optimal strategy is to conduct inspections every second year. By choosing an inspection interval of 2 years rather than 6 an expected saving of about half a million SEK is made during one lifetime. The result is also consistent with Figure 10.10. According to the linear relationship PL in Figure 10.12, 1 year shorter inspection interval increases the possibility to detect a potential failure by 1.67%. Using this information in Figure 10.11, it is clear that an increase from 3 to 2 years between inspections is economically beneficial but not from 2 to 1.

10.6.1.2.5 Validity of the Model

This model cannot solely be the base for deciding inspection intervals but can enlighten certain aspects when making the decision of interval length. Some aspects that need to be taken into consideration when using the model are:

- The model does not take into account other than economic factors such as risk of personal injuries or environmental hazards unless they are included in the costs for a breakdown.

- Approximation of how the probabilities p change for different intervals is hard to obtain. By making guesses, some information can still be obtained from the model.
- Using only the information in VVK RCM, the LCC approach yields the needed improvement in the probability to detect failures for an increased intensity of inspections to be justified. These results are built on data from VVK RCM only, no additional assumptions are made.
- No costs are discounted.
- The total cost as a function of inspection interval might not have an unique minimum. This can occur, for example, when the chance of detecting a failure and preventing a breakdown is constant over several inspection intervals. In most normal cases, this will not be a problem.
- It is hard to apply the model to relatively large inspection intervals because changes of these large inspection intervals generate a small change in costs. The resulting needed improvement in the probability of detecting and preventing breakdowns will be very small. For shorter intervals, the associated costs are higher which leads to larger increases in p.
- The model only handles inspections and is not capable of analyzing tasks that affect the state of the equipment such as minor repairs.

To optimize inspection intervals, it is not necessary to know the exact quantity of p for all intervals. Only the improvement in percent from changing from one interval to an other needs to be known.

10.6.2 Application of the Model

The LCC model described in the previous sections is now used to evaluate some of the changes made due to VVK RCM. Costs and time intervals will be taken from the VVK RCM analysis. These values are enough to yield the necessary change in p for a change of inspection interval to be profitable. To assess the optimal inspection interval guesses will be made for values of p corresponding to different inspection intervals. The proposed model is applied to a period of t_{CR} year.

10.6.2.1 Stator Winding Insulation

The stator winding insulation is one of the most crucial components of the generator. It is a deteriorating component and a failure can have one of many causes. In VVK RCM, two types of major inspections are proposed. The two are visual inspection and diagnostic measurements. A failure leads to either replacement of the entire stator winding or replacement of a coil.

TABLE 10.6

RCM Data Stator Winding

C_{CR}	3,200,576 SEK
C_{PR}	2,047,808 SEK
C_I	6860 SEK
t_{CR}	20 years
t_{PR}	20 years

Source: O. Wilhelmsson, *Evaluation of the introduction of RCM for hydro power generators at Vattenfall Vattenkraft.* Master Thesis written at the Royal Institute of Technology KTH, Department of Electrical Engineering, 2004/2005 [265].

The proposed LCC model will be applied to the coil failure and the visual inspections. When visual inspections are performed signs of deterioration are looked for. The deterioration can be due to partial discharges or corona activity for example. The aging process is slow so the chance of detecting a potential failure is high [46].

The data in Table 10.6 is taken from VVK RCM. t_{PR} and t_{CR} are in this case the same. This means that it is possible to remove the equipment from operation just before a functional failure will occur if monitored correctly.

Figure 10.14 shows the percent increase in p that is needed to justify a change in inspection intervals. VVK RCM has proposed an inspection interval for visual inspections of 6 years. Earlier a visual inspection was most likely performed every third year during a generator check. According to the right graph of Figure 10.15, the money saved by decreasing inspection intervals from 3 to 6 years is around 23,000 SEK over a 20-year period. The left graph of Figure 10.15 shows that improving p with 1% is worth 11,500 SEK. The conclusion that can be drawn is that according to VVK RCM a change of intervals from 3 to 6 years does not result in a decrease of p that exceeds 2%. In other words, the cost saved in changing inspection intervals from 3 to 6 years needs to be at least as large as the extra costs corresponding to the extra risk taken due to a larger probability of not detecting and being able to prevent a functional failure. In this case, the change in probability of detecting and preventing a functional failure should not increase more than 2% for the change of interval to be profitable.

It is also clear that the cost for repair is far greater than the cost of inspections. The cost of inspections increases rapidly when inspection intervals become short. Note that so far the analysis has been based.

Figure 10.16 shows the total lifetime costs for some relationships between p and inspection intervals. The optimal inspection interval varies between 3 and 8 years depending on which relationship that is used. Although the

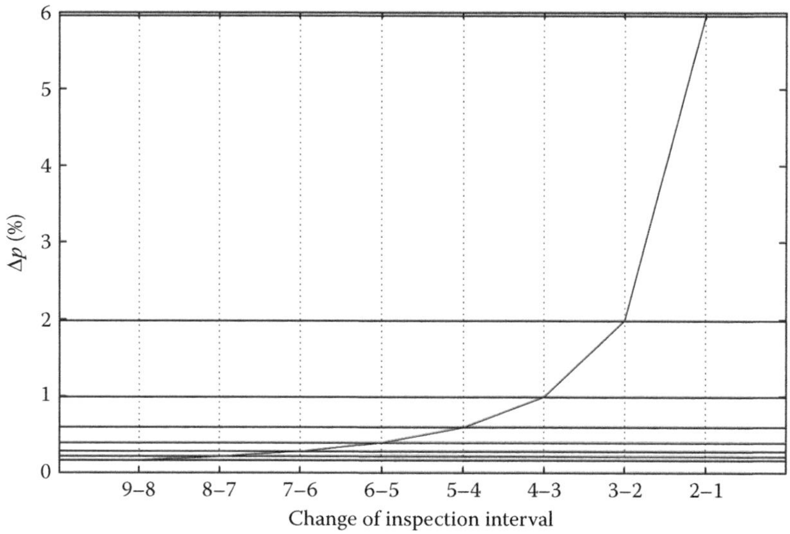

FIGURE 10.14
Results of the application to the stator winding. Needed improvement in the probability to detect a potential failure to justify a change of inspection interval. (Adapted from O. Wilhelmsson, *Evaluation of the introduction of RCM for hydro power generators at Vattenfall Vattenkraft*. Master Thesis written at the Royal Institute of Technology KTH, Department of Electrical Engineering, 2004/2005 [265].)

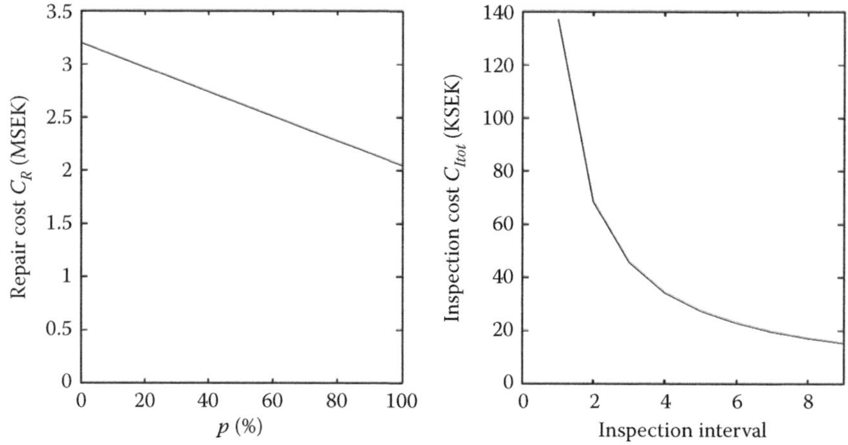

FIGURE 10.15
(Left) Restoration cost when the chance to detect and prevent a breakdown is known. (Right) Cost of inspections during a life time for the stator winding. (Adapted from O. Wilhelmsson, *Evaluation of the introduction of RCM for hydro power generators at Vattenfall Vattenkraft*. Master Thesis written at the Royal Institute of Technology KTH, Department of Electrical Engineering, 2004/2005 [265].)

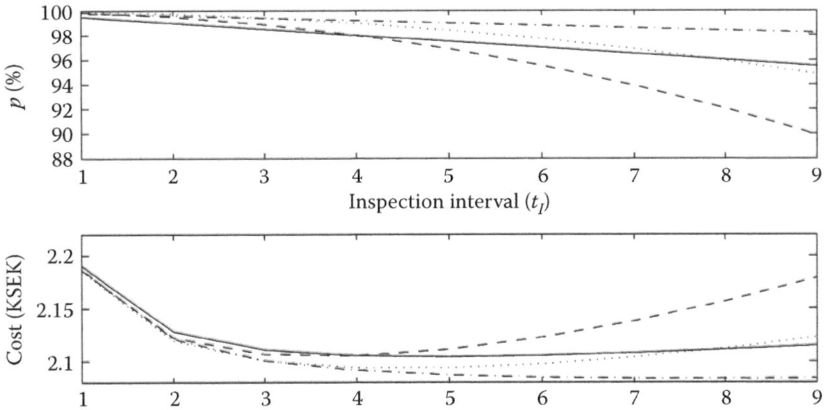

FIGURE 10.16
(Top) Two linear and two quadratic approximations of p. (Bottom) Corresponding expected cost as a function of inspection interval for the stator winding. (Adapted from O. Wilhelmsson, *Evaluation of the introduction of RCM for hydro power generators at Vattenfall Vattenkraft*. Master Thesis written at the Royal Institute of Technology KTH, Department of Electrical Engineering, 2004/2005 [265].)

optimal inspection interval varies, the curves are relatively flat and small changes in the inspection intervals do not make a large impact on the total cost.

10.6.2.2 Stator Cover Plate Bolts

Bolts can come undone due to a locking error or due to aging and wear from assembling and dissembling. A loose bolt can cause major damage to the stator winding. Inspections are performed by checking that all stator cover plate bolts are tight. If the bolts are found to be in an unacceptable state the locking devices for the bolts are changed and all bolts are tightened. According to VVK RCM, the inspections shall be performed every fifth year [46].

Before VVK RCM, the generator checks were performed every third year. The exact way to check the bolts was not specified. It will be assumed they were checked in the same way before and after VVK RCM. Intervals of inspections before RCM were likely to have been performed during every generator inspection, that is, every third year. The data entered into the model is presented in Table 10.7.

The resulting costs and probabilities needed to economically justify changes of intervals are shown in Figure 10.17. VVK RCM proposes an inspection interval of 5 years. The bottom graph of Figure 10.17 indicates that increasing the inspection interval from 3 to 5 years would require that the chance to discover and prevent the failure from occurring should not decrease by more than 1%. The top two graphs indicate that the majority of the cost generated during the lifetime is cost due to repairs. Some relationships between inspection intervals and p and their impact on the total

TABLE 10.7
RCM Data Bolts

C_I	6060 SEK
C_{PR}	601,448 SEK
C_{CR}	2,184,516 SEK
t_{PR}	30 years
t_{CR}	40 years

Source: O. Wilhelmsson, *Evaluation of the introduction of RCM for hydro power generators at Vattenfall Vattenkraft*. Master Thesis written at the Royal Institute of Technology KTH, Department of Electrical Engineering, 2004/2005. [265]

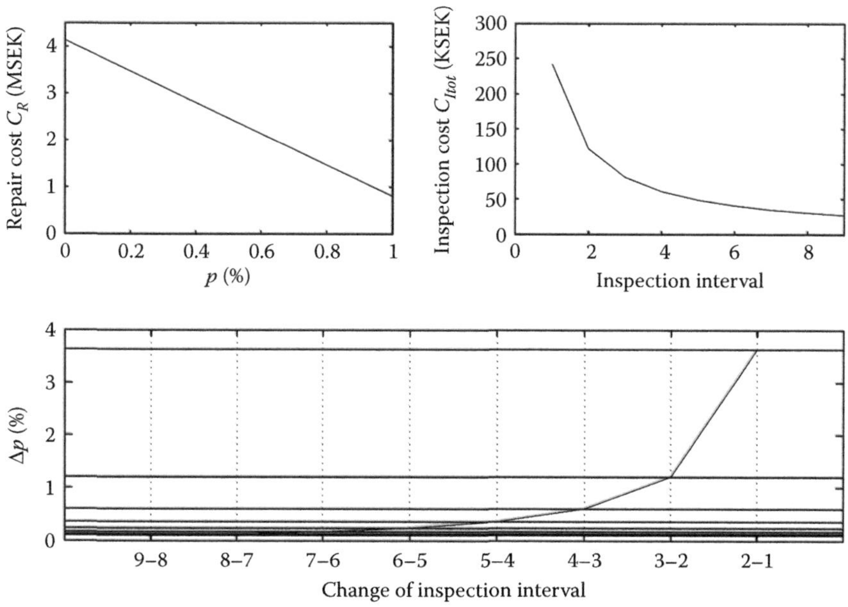

FIGURE 10.17
(Top left) Expected cost of restoration. (Top right) Cost of inspections. (Bottom) Results of the application to the cover plate bolts. Needed improvement in the probability of detecting a potential failure to justify a change of inspection interval, a function of inspection interval for the stator winding. (Adapted from O. Wilhelmsson, *Evaluation of the introduction of RCM for hydro power generators at Vattenfall Vattenkraft*. Master Thesis written at the Royal Institute of Technology KTH, Department of Electrical Engineering, 2004/2005 [265].)

cost are shown in Figure 10.18. Clearly the shapes of the assumed probabilities have a great impact on the optimal inspection interval. The optimal interval varies between 3 and 6 years. The total lifetime cost can vary with about 200,000 SEK depending on inspection interval.

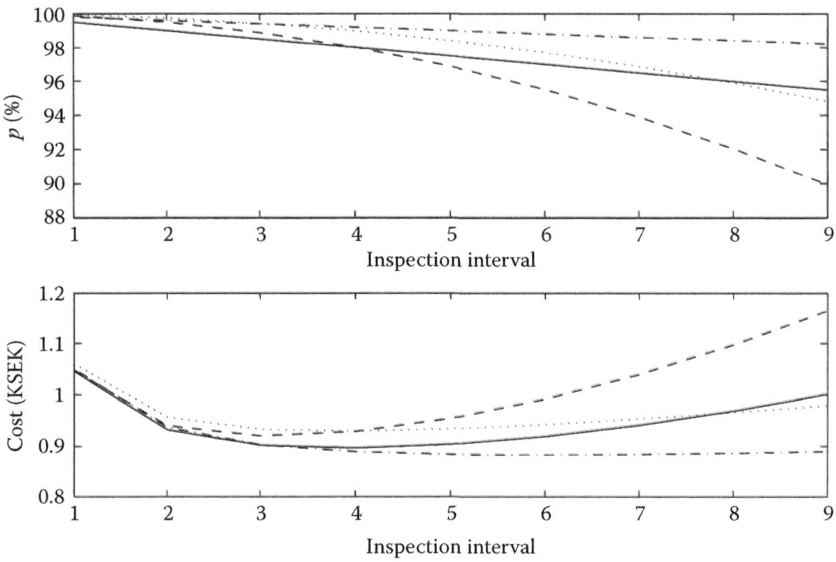

FIGURE 10.18
(Top) Two linear and two quadratic approximations of p. (Bottom) Corresponding expected cost as a function of inspection interval for the bolts. (Adapted from O. Wilhelmsson, *Evaluation of the introduction of RCM for hydro power generators at Vattenfall Vattenkraft*. Master Thesis written at the Royal Institute of Technology KTH, Department of Electrical Engineering, 2004/2005 [265].)

10.6.2.3 Clogged Cooling Channel in the Cooler

When the cooler is clogged or there is air in the system the winding temperature will rise. A sensor will note when the temperature has risen to an unacceptable level and will relieve the generator from service [46]. Data from VVK RCM for this failure is given in Table 10.8. The proposed method for maintaining this failure mode is to check if the inside of the cooler is clean enough. The temperature of the stator should also be checked every second week and analyzed. After a failure occurs, a cleaning task of the cooler is performed. The visual cleanliness inspections will be studied.

According to VVK RCM, the inspections should be performed every fourth year. Figures 10.19 and 10.20 show the results yielded from the model. For this failure mode, it has been chosen to assume values for p for inspection intervals nearly up to t_{CR}. Figure 10.19 shows that if inspections are performed with 9-year intervals the resulting values for p are assumed to be around 20%. It seems quite reasonable that if checks are only performed once in the assumed time until failure the chance to discover a potential failure is small.

It can be noted that relatively large improvements in p are needed to reduce inspection intervals compared to the other failures which have been studied.

TABLE 10.8

RCM Data Cooler

C_I	8460 SEK
C_{PR}	377,328 SEK
C_{CR}	881,992 SEK
t_{PR}	10 years
t_{CR}	10 years

Source: O. Wilhelmsson, *Evaluation of the introduction of RCM for hydro power generators at Vattenfall Vattenkraft*. Master Thesis written at the Royal Institute of Technology KTH, Department of Electrical Engineering, 2004/2005 [265].

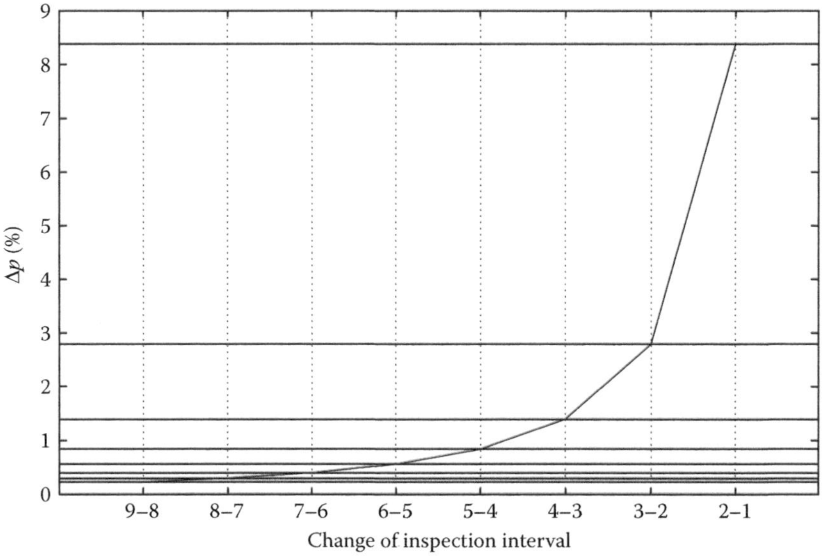

FIGURE 10.19
Results of the application to the cooler. Needed improvement in the probability to detect a potential failure to justify a change of inspection interval. (Adapted from O. Wilhelmsson, *Evaluation of the introduction of RCM for hydro power generators at Vattenfall Vattenkraft*. Master Thesis written at the Royal Institute of Technology KTH, Department of Electrical Engineering, 2004/2005 [265].)

10.6.3 Conclusion

Table 10.9 summarizes the application of the LCC approach proposed. The costs presented in the table cannot be compared to each other since they are saved over different periods of time. In all three cases studied, the intervals have been lengthened. Of the failures analyzed, the largest tolerated decrease

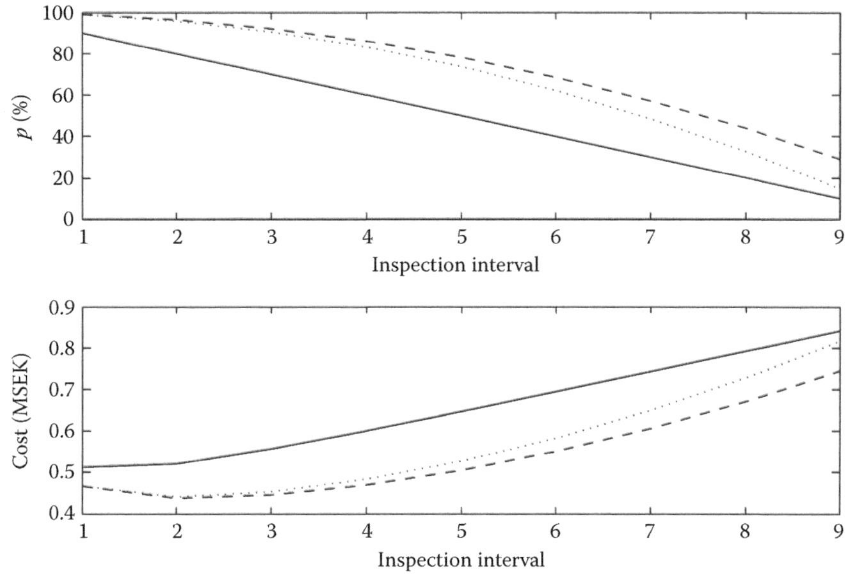

FIGURE 10.20
(Top) A linear and two quadratic approximations of p. (Bottom) Corresponding expected cost for the cooler as a function of inspection interval. (Adapted from O. Wilhelmsson, *Evaluation of the introduction of RCM for hydro power generators at Vattenfall Vattenkraft*. Master Thesis written at the Royal Institute of Technology KTH, Department of Electrical Engineering, 2004/2005 [265].)

TABLE 10.9
Results from the VVK Model

Failure Mode	Inspection Interval Change (Years)	Cost Saved Due to Interval Change (SEK)	Largest Change in p for the Change to be Profitable (%)
Stator winding	3–6	23,000	−2
Stator plate bolts	3–5	32,000	−1
Cooler	3–4	7000	−1.4

Source: O. Wilhelmsson, *Evaluation of the introduction of RCM for hydro power generators at Vattenfall Vattenkraft*. Master Thesis written at the Royal Institute of Technology KTH, Department of Electrical Engineering, 2004–2005 [265].

in p for the change of inspection interval to be profitable is for the visual inspection for the stator winding. This change of inspection interval should only be done if it does not decrease p the chance to detect and prevent a functional failure with more than 2%.

Since all the necessary data is not available it is crucial to use caution when analyzing the results above. The proposed model cannot handle all failure modes addressed in VVK RCM so a general judgment of VVK RCM cannot be drawn. When choosing between two inspection intervals that are fairly long and the effect of the inspections is not certain, it is reasonable to perform the inspections at the shorter interval since the decrease in inspection intervals will not induce at large extra cost. More caution should be used when the intervals are shorter since a decrease then could result in a significantly higher inspection cost. The resulting required changes in p are small for the three failures studied. This is due to high costs for corrective repairs compared to costs for inspections. This has the effect that inspections should be performed relatively often according to this model.

This model could be used as an aid when constructing maintenance plans resulting from RCM as discussed earlier. If RCM has suggested maintenance to be performed with many different intervals it might still be profitable to perform some tasks together. When assessing the effects of a change of inspection interval for these tasks, this model could be used as an aid.

10.6.3.1 Markov Model

10.6.3.1.1 Markov Processes

The state a component is in at time t is denoted $X(t)$. If a component is in the state i at time t this is denoted by $X(t) = i$. The model is assumed to have $(r+1)$ states, $0, 1, \ldots, r$.

The Markov property is defined. Given that a system is in state i at time t, the future states $X(t+v)$ do not depend on the previous states $X(u), u < t$ [40].

A Markov process not dependent of time t but only the time interval v for the transition is said to have stationary transition probabilities [40].

$$P(X(t+v) = j|X(t) = i) = P(X(v) = j|X(t) = i) = P_{ij}(v) \text{ for } t, v > 0 \quad (10.5)$$

This leads to a process with no memory. A Markov process with stationary transition probabilities is often convenient and realistic for modeling technical systems [40].

Now the transition rate matrix A is defined.

$$Q = \begin{bmatrix} -a_{00} & a_{10} & a_{20} & \cdots & a_{r0} \\ a_{01} & -a_{11} & a_{21} & \cdots & a_{r1} \\ \vdots & \vdots & \vdots & \ddots & \vdots \\ a_{0r} & a_{1r} & a_{2r} & \cdots & -a_{rr} \end{bmatrix} \quad (10.6)$$

a_{ij} is the intensity of transferring from state i to state j. The diagonal elements are

$$a_{ij} = \sum_{k=0, k \neq j}^{r} a_{jk} \quad (10.7)$$

The time spent in a state j is exponentially distributed with parameter a_{jj}. That is, the transition rate is constant in time. The state equations may now be written as $A \cdot P(t) = \dot{P}(t)$. The model proposed uses the asymptotic (steady-state) solutions. These results are obtained when the time goes to infinity. The state equations can then be simplified to Equation 6.6 [40].

$$A \cdot P = 0 \tag{10.8}$$

To solve the state equation, the fact that the sum of all state probability must be one is used. The probability to find the system in a certain state at an asymptotic point in time is then obtained. The visit frequency to state j, that is, time between multiple entries into a state can then be calculated by $1/P_j a_{jj}$ [40].

10.6.3.2 Model

A Markov model is used to evaluate the change of reliability of the system due to a change in inspection intervals. The idea of the model is in part influenced by a more complex model presented by J. Endrenyi, G. J. Anders, and A. M. Leite da Silva [159]. The component to be maintained is assumed to be deteriorating. This is modeled by states the component can move through from a good as new state to finally a state where the component is not functioning due to a breakdown, failed state. The states in between represent increasing deterioration (see Figure 10.21).

A continuous time Markov model with identical transition rates λ_d between the deterioration states is used. The transition rate is simply the number of deterioration states n divided by the expected time to failure if no maintenance is performed t_{CR}.

10.6.3.2.1 Inspections and Repairs

Inspections are in reality performed at deterministic intervals. In the Markov model, these will be assumed to be performed probabilistically with an intensity of $\mu_I = 1/t_I$ where t_I is the deterministic time between intervals. When an inspection is performed, a decision is taken to either repair the asset to a good as new state or to return the asset to the same condition it was prior to the inspection. If the choice to repair the asset is made, the asset is put back into service for some time before the repair takes place. The probabilities of the outcomes of inspections can be dependent on from which deterioration

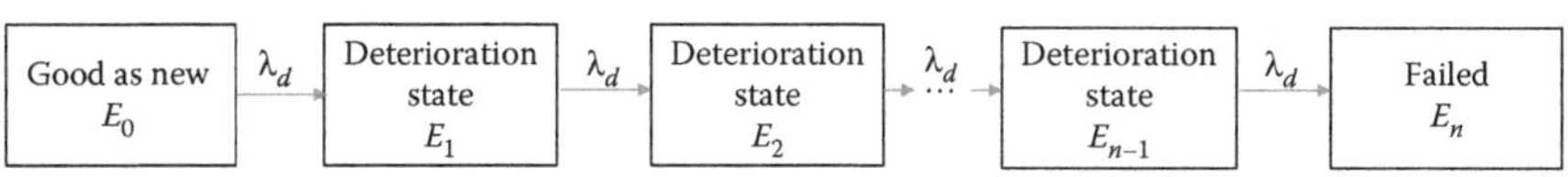

FIGURE 10.21
Deterioration process described by states. (Adapted from O. Wilhelmsson, *Evaluation of the introduction of RCM for hydro power generators at Vattenfall Vattenkraft*. Master Thesis written at the Royal Institute of Technology KTH, Department of Electrical Engineering, 2004/2005 [265].)

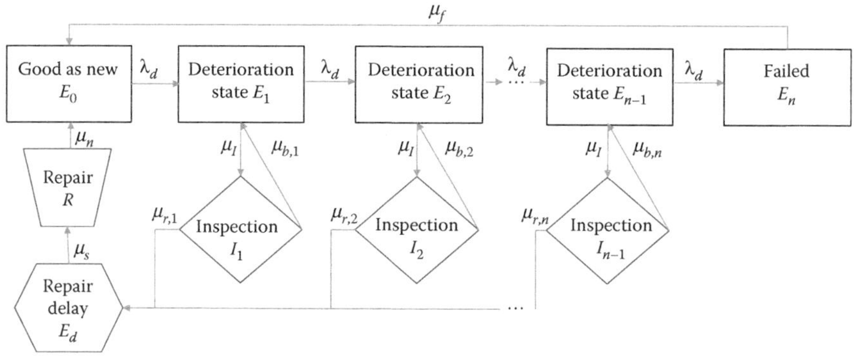

FIGURE 10.22
Deterioration process, inspections, and repairs. (Adapted from O. Wilhelmsson, *Evaluation of the introduction of RCM for hydro power generators at Vattenfall Vattenkraft*. Master Thesis written at the Royal Institute of Technology KTH, Department of Electrical Engineering, 2004/2005 [265].)

state the asset is taken into inspection. In the proposed model, these probabilities are represented by intensities with large values. μ_r represents the intensity that the asset is scheduled for repairs and μ_b is the intensity with which the asset is returned to the deterioration state from which the inspection was performed. When the decision of an inspection is to repair an asset some time elapses before the actual repair takes place. This is represented by μ_s which is the intensity to go into the repair state when a decision has been made to repair. μ_n is the inverse of the time it takes to repair the asset. Finally μ_f is the inverse of the time it takes to repair the asset after failure. Figure 10.22 shows the deterioration states, inspection states, repair and time delay state and transition rates of the system. The intensities and states are listed in Table 10.10.

The mean time between failures, stationary distribution, and other relations can be obtained by using standard calculation procedures for Markov chains. The mean time between failure (MTBF) is the mean time between consecutive transitions from a functioning state into a failed state E_n [40].

10.6.3.2.1.1 Application of Two Extreme Cases The shortest MTBF is achieved if the intensities of inspections μ_I are zero or the intensity to repair when inspected μ_r is set to zero. The MTBF will simply be t_{CM} plus the duration of any inspections performed. The longest MTBF occurs if the intensity to return to a deteriorating state from an inspection state is set to zero. The system then represents a process where the equipment is repaired to a good as new state after every inspection. The number of intervals n can greatly affect the MTBF. In an extreme case, a system is repaired to a good as new state after every inspection. An inspection takes 1 day to conduct and occurs at 5-year intervals. The expected time until failure is 20 years if no repairs are performed. The repair time is 36.5 days and the time between an inspection

TABLE 10.10

Intensities and States

Intensities		
λ_d	Deterioration	(Known)
μ_I	Inspection	(Variable)
$\mu_{r,1} - \mu_{r,n-1}$	Repair after inspection	(Unknown)
$\mu_{b,1} - \mu_{b,n-1}$	Return to deterioration state after inspection	(Unknown)
μ_n	Duration of repair after inspection	(Known)
μ_f	Duration of repair after failure	(Known)
μ_s	Delay until repaired after inspection	(Unknown)
States		
E_0	Equipment is good as new	
$E_1 - E_{n-1}$	Equipment states with increasing deterioration	
E_n	Failed state	
E_d	Equipment in service waiting to be repaired	
$I_1 - I_{n-1}$	Inspection states, corresponding to $E_1 - E_{n-1}$	
R	Repair state, due to a decision after inspection	

Source: O. Wilhelmsson, *Evaluation of the introduction of RCM for hydro power generators at Vattenfall Vattenkraft*. Master Thesis written at the Royal Institute of Technology KTH, Department of Electrical Engineering, 2004/2005 [265].

that results in a decision to repair and the time the actual repair takes place is 2 years. These figures are summarized below.

$$T_{CM} = 20\,\text{Years}$$

$$\mu_I = \frac{1}{5}\,\text{Years}^{-1}$$

$$\mu_b = 0\,\text{Years}^{-1}$$

$$\mu_r = 365\,\text{Years}^{-1}$$

$$\mu_s = \frac{1}{2}\,\text{Years}^{-1}$$

$$\mu_n = 10\,\text{Years}^{-1}$$

For this system, choices of n between 3 and 100 can result in an MTBF ranging from 67 years to 350 years (Figure 10.23). The time between repairs approaches 7 when the number of intervals increases. Seven years between repairs is reasonable since after the inspection after 5 years there is a 2-year delay until repairs are performed. To use the model in practice, it must be

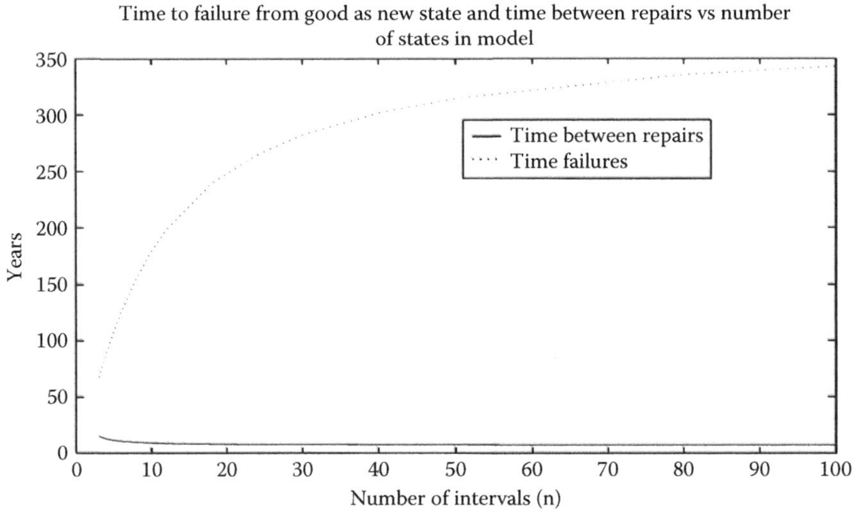

FIGURE 10.23
Mean time between failures and mean time between repair when assets are repaired after inspections and returned to a good as new state for different number of intervals n. (Adapted from O. Wilhelmsson, *Evaluation of the introduction of RCM for hydro power generators at Vattenfall Vattenkraft*. Master Thesis written at the Royal Institute of Technology KTH, Department of Electrical Engineering, 2004/2005 [265].)

possible to distinguish the different states through inspections. Thus there exists an upper limit of n, the number of states.

10.6.3.2.1.2 Influences of the Decision after Inspections Now the possibility to return the asset to the same state as before the inspection is studied. Several different examples of intensities for μ_b and μ_r are of interest.

Two examples of these relationships are given in Figure 10.24 where 10,000 deterioration states are used. The relationship in the left graph is linear and for the first state an inspection with certainty results in returning the asset without repairs and for the last state inspection will lead to repairs for certain. For the right graph, inspections that are performed when being in the first quarter of the states will not result in any repairs being performed while inspections performed from any of the last quarter states will lead to repairs with certainty.

By altering the inspection interval, it is now possible to see how the MTBF is affected. Table 10.11 shows the results from different number of states, different repair intensities and different inspection intervals. The expected result that more frequent inspections lead to a longer MTBF can be seen. Repair intensities μ_2 according to the right graph of Figure 10.24 leads to longer

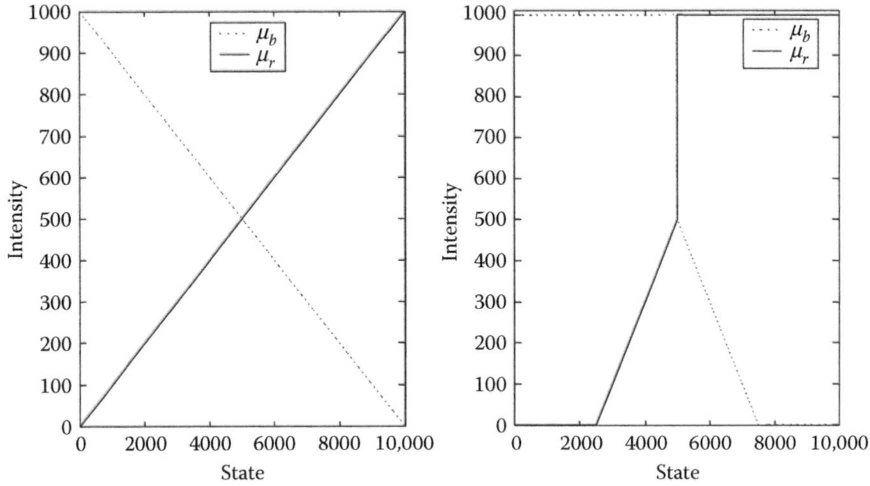

FIGURE 10.24
Intensities for repairing the asset μ_r and intensities for returning the asset to the same deterioration state as before inspection μ_b versus deterioration state. This example contains 10,000 states. Left graph intensities will be referred to as μ_1, right graph intensities μ_2. (Adapted from O. Wilhelmsson, *Evaluation of the introduction of RCM for hydro power generators at Vattenfall Vattenkraft*. Master Thesis written at the Royal Institute of Technology KTH, Department of Electrical Engineering, 2004/2005 [265].)

TABLE 10.11
MTBF

	$n=5$		$n=10$		$n=16$		$n=50$	
Inspection Intervals	μ_1	μ_2	μ_1	μ_2	μ_1	μ_2	μ_1	μ_2
4 Years	68	68	97	101	114	122	140	158
5 Years	55	55	73	75	82	87	94	104
6 Years	48	48	59	61	65	69	73	79

Source: O. Wilhelmsson, *Evaluation of the introduction of RCM for hydro power generators at Vattenfall Vattenkraft*. Master Thesis written at the Royal Institute of Technology KTH, Department of Electrical Engineering, 2004/2005 [265].

Note: n, number of states; μ_1, repair intensities shown in left graph of Figure 10.24; μ_2, right graph intensities.

MTBF than μ_1. Furthermore, the more states in the model the longer time until failure.

10.6.3.2.1.3 Expected Cost To economically optimize inspections performed to prevent a failure, costs of inspections and repairs must be accounted for. Using the same notations as in Section 10.6.1, entry into state E_n induces a

TABLE 10.12

Costs and Intervals

Cost	Time Interval
C_{CR}	MTBF
C_{PR}	MTBR
C_I	μ_I^{-1}

Source: O. Wilhelmsson, *Evaluation of the introduction of RCM for hydro power generators at Vattenfall Vattenkraft*. Master Thesis written at the Royal Institute of Technology KTH, Department of Electrical Engineering, 2004/2005 [265].

cost of C_{CR}. The time between this cost has to be paid is MTBF. The time between entrance into state R, MTBR defines how often the repair cost CPR has to be paid. Finally the cost of inspections, C_I, is paid with the intensity C_I. Table 10.12 summarizes the costs and intervals at which they are induced.

The expected average cost per year C_{tot} can now be obtained by summing up the costs according to Equation 10.9

$$C_{tot} = \frac{C_{CR}}{MTBF} + \frac{C_{PR}}{MTBR} + C_I\mu_I \tag{10.9}$$

The variables MTBF and MTBR are both dependent of μ_I. When the interval between inspections grows the MTBF decreases while the MTBR increases.

This might be somewhat contradictory since with increased inspection intervals there is a greater possibility to leave the asset in service and still be confident that the next inspection will reveal and prevent the failure.

10.6.3.3 Application of Markov Model

The results obtained from the Markov model greatly vary depending on the choice of how many states that are used in the model. Furthermore, the unknown intensities reflecting the outcomes of inspections have a large impact on the results. Since the uncertainty about these quantities is large the purpose of this section is merely to illustrate some possibilities the Markov model has to offer. The Markov model will be applied to the data from the loose bolts and some problems will be discussed.

10.6.3.3.1 Bolts

The data for the failure due to loose bolts (from LCC approach) is entered into the Markov model (Table 10.13). Assuming that inspections performed

TABLE 10.13
Data Bolts

C_I	6060 kr.
C_{BF}	601,448 kr.
C_{AF}	2,184,516 kr.
t_{CM}	30 years

Source: O. Wilhelmsson, *Evaluation of the introduction of RCM for hydro power generators at Vattenfall Vattenkraft*. Master Thesis written at the Royal Institute of Technology KTH, Department of Electrical Engineering, 2004/2005 [265].

during the last quarter of the states will lead to repairs and if inspections are made in the first three quarters of the states, the asset will be returned to the deterioration state it was taken from and the number of states are 10, the resulting cost C_{tot}, MTBF and MTBR are displayed in Figure 10.25.

Figure 10.25 shows that the cost-efficient inspection interval is 2 years according to the top graph. The bottom graph shows that the mean time between failures is extremely high, nearly 4000 years when the inspections are performed every year. When the inspection interval is increased the MTBF quickly decreases to around 120 years for inspection intervals of 9 years. The mean time between repairs MTBR varies less. MTBR starts at 30 years and increases to slightly more than 50 years when the inspection interval is 9 years. This means that according to this model, if inspections are performed at nine year intervals failures will occur every 120 years and repairs that are performed to prevent a failure are performed every 50 years in average. Put in another way, in 120 years one failure will occur and repairs preventing failures will be performed 2.4 times.

10.6.4 Model Comparison

With perfect information, the two models proposed should yield the same optimal inspection interval. Since information is expensive and hard to obtain, the models can complement each other and shed light from different angels.

The main differences between the models are:

- The approach of the methods is different. While the LCC uses the costs of different activities to assess what the improvement in reliability must be to justify a shorter inspection interval, the Markov approach calculates the reliability and the times between failures.

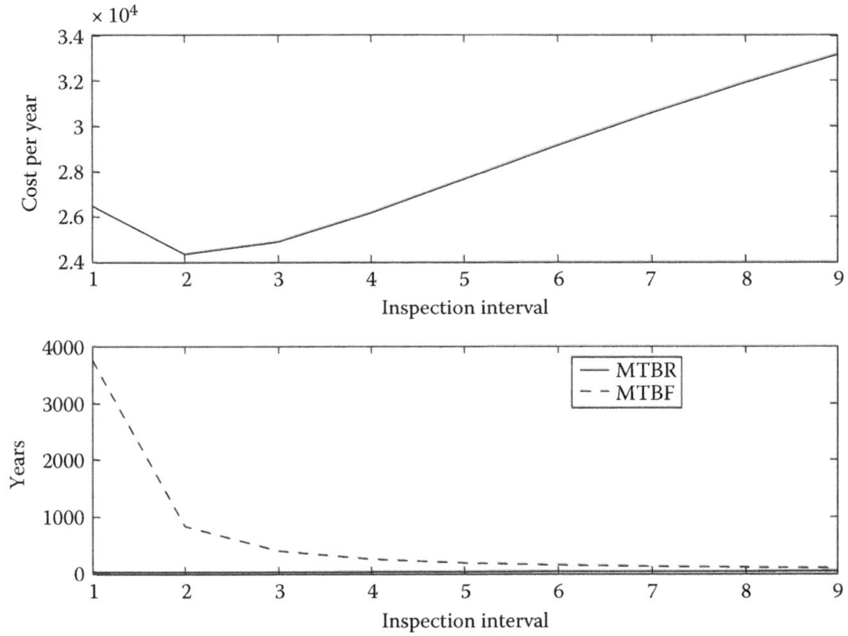

FIGURE 10.25
Results from applying the Markov model to the data from the loose bolts. (Top) Costs of inspection intervals. (Bottom) Mean time between failures. This example contains 10 states. (Adapted from O. Wilhelmsson, *Evaluation of the introduction of RCM for hydro power generators at Vattenfall Vattenkraft*. Master Thesis written at the Royal Institute of Technology KTH, Department of Electrical Engineering, 2004/2005 [265].)

- The LCC approach can be used to analyze each type of inspection separately if there are several types of inspections performed for one piece of equipment. The Markov model handles failures whose maintenance consists of only one type of inspection.
- The LCC approach handles maintenance in the form of inspections. The Markov approach can easily be adjusted to handle scheduled preventive maintenance and other types of maintenance and inspections.
- The data needed in the two different models is different. In the LCC approach, assumptions were made about p, the possibility to detect and prevent a failure when performing inspections at a certain interval of time. In the Markov approach, data about the probabilities after an inspection to repair or return the asset to the same state as before has been fabricated. Both models use approximated costs of failures and times between failures.

10.7 Concluding Remarks

This case study shows an example that the introduction of RCM can be a complex process. The resulting changes in reliability and costs are difficult to calculate. In this case study, the structure and results of the VVK RCM have been studied. One of the main effects of VVK RCM has been a more thorough reasoning when deciding which strategies and tasks to maintain equipment with. A precise task is defined and should be performed at a certain interval of time. Earlier the maintenance plans were not precise and left room for ambiguity.

A pattern in the failure modes analyzed by VVK is the severe consequences resulting from a failure and expensive corrective maintenance compared to preventive maintenance. This has led to preventive maintenance for all failure modes.

The majority of the tasks used to maintain the generator according to VVK RCM are scheduled condition monitoring tasks. Only two continuous condition monitoring tasks have been suggested. When comparing old maintenance plans to plans RCM has resulted in, the tasks were found to be quite similar but the interval between tasks had changed.

To analyze the effects of changed intervals between condition monitoring tasks, an LCC model has been proposed and applied. The model was used to calculate the necessary improvement in the probability to detect and prevent a failure to economically justify a change of inspection interval. The introduction of RCM has resulted in changes of inspection intervals, for example, from every third to every sixth year. The result from this study shows that this change in maintenance is economically justified if the probability to detect and prevent failures does not decrease with more than 2%. The conclusion for this specific case is, under the assumptions given, if the choice is made to increase the inspection interval one should be certain that the probability to detect and prevent the failure only decreases marginally. The change in probability to detect and prevent failures is small because the cost of inspections is small compared to the cost of corrective maintenance and cost due to loss of production.

The models proposed in this case study can be refined and expanded. The Markov model has great flexibility and further states can be introduced representing other maintenance strategies than inspections. The models proposed both handle failures at the component level and address one cause of failure. With some refinement and a higher level of reliability of the input data, these models could be used as an aid in quantitative maintenance planning, for example, with development of RCAM.

Appendix A: Specification of Input Data to RADPOW

There are four types of data that are manipulated in RADPOW: network topology data, customer and power data, component reliability data, and load flow data. These data are used as input for RADPOW in 10 files. The first six files relate to network data (bend, bcomp, nsp, nlp, nnop, and ctype); the following two relate to the power and customer data (ncuspow) and reliability data (crelia); and the last two relate to the load flow data (pflo and pfrx).

A.1 Network Topology Data

The topology of the system is described by the branches of the system. A branch is defined as a set of components in series that start and end with a bus bar. Branches through which the power is permitted to flow in one direction only are defined as unidirectional branches, and those through which power is permitted to flow in both directions are defined as bidirectional branches. Furthermore, branches are identified by a branch number, from 1 ton, where n is the total number of branches in the system.

All components considered in the analysis must be identified. These are identified by a number from 1 to k, where k is the total number of components identified in the system.

The input data consist of:

- Branch numbers, numbers of branch ends, and unidirectional branches
- Component numbers
- Numbers of supply points, load points to be analyzed, and normally open points

A.2 Customer and Power Data

For each load point in the system (Lp), the following customer and power data are specified:

- Total number of customers (tnc)

- Industrial (icp), residential (rep), or commercial (ccp) customer percentages
- Total active power per customer (kW /cust) (tappc)
- Industrial (iapp), residential (rapp), or commercial (capp) active power percentages
- Total reactive power per customer (kvar/cust) (trppc)
- Industrial (irpp), residential (rrpp), or commercial (crpp) reactive power percentages

A.3 Component Reliability Data

For each component, the following reliability data are specified using expected values:

- Permanent failure rate (λ_p)
- Active failure rate (λ_a)
- Temporary failure rate (λ_{te})
- Transient failure rate (λ_{tr})
- Maintenance outage rate (λ_m)
- Repair time (r_r)
- Switching time (r_s)
- Replacement time (r_p)
- Re-closure time (r_c)
- Maintenance outage time (r_m)
- The probability of switching devices being stuck (P)

A.4 Load Flow Data

The most significant restriction in RADPOW_1999 is that the criterion for failure is based on total loss of continuity (TLOC). TLOC means that a load point is interrupted and load disconnected only when all paths between the load point and the sources are disconnected [79]. However, TLOC has been extended to include partial loss of continuity (PLOC). This implies that the failure criterion also includes events that do not fully interrupt the connections between the source and the load point, but are sufficiently serious to require disconnection of some load. The development of a module for load flow analysis has been done [235] and the resulting version of RADPOW is referred to as RADPOW_1999_PF.

A.4.1 Impedance Data

Due to the short length of the lines encountered in distribution systems, their susceptance is neglected. However, their impedance is taken into account to achieve accuracy in the power flow calculations. For each branch in the system, the following input data should be specified:

- Branch number (Brno)
- Sending end (SE)
- Receiving end (RE)
- Resistance p.u. (R(pu))
- Reactance p.u. (X(pu))
- Maximal current p.u. (Max_current(pu))

A.4.2 Load Data

For each load point in the system, the following input data should be specified:

- Active power in p.u. (lapc(pu))
- Reactive power in p.u. (lrpc(pu))
- Load duration curve (ldcu)

In RADPOW_1999_PF, three different load curves have been defined as follows:

- Constant, where $\text{LDC} = 1$
- Sinusoid, where $\text{LDC} = 1 - 0.3 \cdot (1 - \sin(2\pi \cdot 3t/N))$
- Linear, where $\text{LDC} = 1 - (0.4 \cdot t/N)$

and where N is the number of customers and t the time.

For further details about the load flow module, please refer to [235].

A.5 The Flymen System: RADPOW Data

A.5.1 RADPOW Input Data

A.5.1.1 Component Reliability Input Data

Table A.1 presents reliability input data for RADPOW coming from the Flymen system. The input data needed for RADPOW that are not presented in this table have zero values or have no impact on the analysis of the Flymen system.

A.5.1.2 Customer and Power Data

Tables A.2 and A.3 show the customer and power input data for RADPOW that come from the Flymen system. The input data for the reactive power are not presented in the tables since these were not used for the analysis.

A.5.1.3 Output Data from RADPOW

Tables A.4 and A.5 present the results for each load point in the analysis of the Flymen system (FlymenN).

TABLE A.1

Component Reliability Input Data used by RADPOW

Comp.	λ_t	λ_a	λ_{te}	λ_m	λ_r	λ_c	λ_s	P_s
BUSL	0.018	0.018	0.01	1.0	2.0	0.083	1.0	0.0
BUST	0.014	0.014	0.01	1.0	2.0	0.083	1.0	0.0
BUSD	0.0	0.0	0.0	0.0	0.0	0.0	0.0	0.0
BREAK	0.005	0.004	0.02	1.0	12.0	0.083	1.0	0.0
DISCM	0.05	0.05	0.0	1.0	3.0	0.083	1.0	0.0
DISCA	0.051	0.05	0.0	1.0	6.0	0.083	1.0	0.001
LINE1W	0.006	0.006	0.0	1.0	3.0	0.083	1.0	0.0
LINE2W	0.017	0.017	0.0	1.0	3.0	0.083	1.0	0.0
LINE3W	0.029	0.029	0.0	1.0	3.0	0.083	1.0	0.0
LINE4W	0.04	0.04	0.0	1.0	3.0	0.083	1.0	0.0
LINE5W	0.05	0.05	0.0	1.0	3.0	0.083	1.0	0.0
LINE6W	0.06	0.06	0.0	1.0	3.0	0.083	1.0	0.0
LINEIE	0.008	0.008	0.0	1.0	3.0	0.083	1.0	0.0
LINE2E	0.026	0.026	0.0	1.0	3.0	0.083	1.0	0.0
LINE3E	0.04	0.04	0.0	1.0	3.0	0.083	1.0	0.0
LINE4E	0.06	0.06	0.0	1.0	3.0	0.083	1.0	0.0
LINE5E	0.07	0.07	0.0	1.0	3.0	0.083	1.0	0.0
LINE6E	0.09	0.09	0.0	1.0	3.0	0.083	1.0	0.0

TABLE A.2

Customer and Power Input Data for RADPOW

Lp	tnc	icp	rcp	ccp	tappc	iapp	rapp	capp
1	6	0.0	1.0	0.0	0.86	0.0	1.0	0.0
2	4	0.0	1.0	0.0	1.08	0.0	1.0	0.0
3	11	0.0	1.0	0.0	0.63	0.0	1.0	0.0
4	5	0.0	1.0	0.0	1.17	0.0	1.0	0.0
5	13	0.0	1.0	0.0	1.50	0.0	1.0	0.0
6	14	0.0	1.0	0.0	2.43	1.0	0.0	0.0
7	4	1.0	0.0	0.0	5.21	1.0	0.0	0.0
8	4	0.0	1.0	0.0	0.77	0.0	1.0	0.0
9	11	0.0	1.0	0.0	1.68	0.0	1.0	0.0
10	5	0.0	1.0	0.0	0.68	0.0	1.0	0.0
11	7	0.0	1.0	0.0	0.57	0.0	1.0	0.0
12	4	0.0	1.0	0.0	0.22	0.0	1.0	0.0
13	9	0.0	1.0	0.0	0.36	0.0	1.0	0.0
14	2	0.0	1.0	0.0	0.96	0.0	1.0	0.0
15	5	0.0	1.0	0.0	0.50	0.0	1.0	0.0
16	5	0.0	1.0	0.0	1.81	0.0	1.0	0.0
17	7	0.0	1.0	0.0	0.52	0.0	1.0	0.0
18	4	1.0	0.0	0.0	1.14	0.0	1.0	0.0
19	10	0.0	1.0	0.0	1.00	0.0	1.0	0.0
20	2	0.0	1.0	0.0	1.25	0.0	1.0	0.0
21	3	0.0	1.0	0.0	0.45	0.0	1.0	0.0
22	4	0.0	1.0	0.0	1.02	0.0	1.0	0.0
23	4	0.0	1.0	0.0	1.64	0.0	1.0	0.0
24	2	1.0	0.0	0.0	3.99	1.0	0.0	0.0
25	1	0.0	1.0	0.0	0.52	0.0	1.0	0.0
26	6	0.0	1.0	0.0	0.41	0.0	1.0	0.0
27	12	0.0	1.0	0.0	0.71	0.0	1.0	0.0
28	10	0.0	1.0	0.0	0.45	0.0	1.0	0.0
29	9	0.0	1.0	0.0	0.57	0.0	1.0	0.0
30	6	0.0	1.0	0.0	1.18	0.0	1.0	0.0
31	8	0.0	1.0	0.0	0.25	0.0	1.0	0.0
32	3	0.0	1.0	0.0	0.61	0.0	1.0	0.0
33	10	0.0	1.0	0.0	0.64	0.0	1.0	0.0
34	9	0.0	1.0	0.0	0.55	0.0	1.0	0.0
35	5	0.0	1.0	0.0	0.74	0.0	1.0	0.0
36	4	0.0	1.0	0.0	0.35	0.0	1.0	0.0
37	5	0.0	1.0	0.0	0.37	0.0	1.0	0.0
38	15	0.0	1.0	0.0	0.63	0.0	1.0	0.0
39	7	0.0	1.0	0.0	0.58	0.0	1.0	0.0
40	7	0.0	1.0	0.0	0.60	0.0	1.0	0.0

TABLE A.3

Customer and Power Input Data for RADPOW

Lp	tnc	icp	rcp	ccp	tappc	iapp	rapp	capp
41	6	0.0	1.0	0.0	0.64	0.0	1.0	0.0
42	9	0.0	1.0	0.0	0.85	0.0	1.0	0.0
43	5	0.0	1.0	0.0	0.71	0.0	1.0	0.0
44	4	0.0	1.0	0.0	0.64	0.0	1.0	0.0
45	5	0.0	1.0	0.0	0.87	0.0	1.0	0.0
46	4	0.0	1.0	0.0	0.34	0.0	1.0	0.0
47	4	0.0	1.0	0.0	0.65	0.0	1.0	0.0
48	5	0.0	1.0	0.0	0.27	0.0	1.0	0.0
49	11	0.0	1.0	0.0	0.92	0.0	1.0	0.0
50	6	0.0	1.0	0.0	0.66	0.0	1.0	0.0
51	9	0.0	1.0	0.0	1.33	0.0	1.0	0.0
52	4	0.0	1.0	0.0	0.76	0.0	1.0	0.0
53	4	0.0	1.0	0.0	0.66	0.0	1.0	0.0
54	5	0.0	1.0	0.0	0.53	0.0	1.0	0.0
55	2	0.0	1.0	0.0	0.93	0.0	1.0	0.0
56	2	1.0	0.0	0.0	1.30	0.0	1.0	0.0
57	8	0.0	1.0	0.0	0.99	0.0	1.0	0.0
58	43	0.0	1.0	0.0	1.19	0.0	1.0	0.0
59	13	0.0	1.0	0.0	0.78	0.0	1.0	0.0
60	9	0.0	1.0	0.0	0.83	0.0	1.0	0.0
61	6	0.0	1.0	0.0	0.78	0.0	1.0	0.0
62	5	0.0	1.0	0.0	0.83	0.0	1.0	0.0
63	7	0.0	1.0	0.0	1.19	0.0	1.0	0.0
64	4	0.0	1.0	0.0	1.41	0.0	1.0	0.0
65	4	0.0	1.0	0.0	0.53	0.0	1.0	0.0
66	3	0.0	1.0	0.0	0.59	0.0	1.0	0.0
67	4	0.0	1.0	0.0	0.52	0.0	1.0	0.0
68	2	0.0	1.0	0.0	2.06	1.0	0.0	0.0
295	0	0.0	0.0	0.0	0.0	0.0	0.0	0.0
296	0	0.0	0.0	0.0	0.0	0.0	0.0	0.0

TABLE A.4

Results from RADPOW for FlymenNW

Load Point	λ [f/yr]	*U* [h/yr]	*r* [h/f]	*L* [kW]	*LOE* [kWh/yr]
1	4.36	3.72016	0.853248	5.16	19.196
2	4.36	3.67823	0.843631	4.32	15.89
3	4.36	3.72423	0.854182	6.93	25.8089
4	4.36	3.74633	0.859249	5.85	21.916
5	4.36	3.67833	0.843653	19.5	71.7274
6	4.36	3.7244	0.85422	34.02	126.704
7	4.36	3.9844	0.913853	20.84	83.0349
8	4.36	4.0724	0.934036	3.08	12.543
9	4.36	4.0124	0.920275	18.48	74.1492
10	4.36	4.0804	0.935871	3.4	13.8734
11	4.361	3.69977	0.848376	3.99	14.7621
12	4.361	3.6799	0.843821	0.88	3.23832
13	4.361	3.8739	0.888306	3.24	12.5514
14	4.361	3.8999	0.894268	1.92	7.48781
15	4.361	4.0119	0.91995	2.5	10.0298
16	4.361	4.1319	0.947467	9.05	37.3937
17	4.361	4.1099	0.942422	3.64	14.9601
18	4.362	4.3639	1.00044	4.56	19.8994
19	4.362	4.4119	1.01144	10	44.119
20	4.362	4.4879	1.02886	2.5	11.2198
21	4.362	4.4879	1.02886	1.35	6.05867
22	4.362	4.6499	1.066	4.08	18.9716
23	4.362	4.7219	1.08251	6.56	30.9757
24	4.362	4.8559	1.11323	7.98	38.7501
25	4.362	4.8559	1.11323	0.52	2.52507
26	4.362	3.65521	0.837966	2.46	8.99181
27	4.362	3.74934	0.859546	8.52	31.9444
28	4.362	3.94534	0.90448	4.5	17.754
29	4.362	4.01134	0.91961	5.13	20.5782
30	4.362	3.96934	0.909982	7.08	28.1029
31	4.362	4.01734	0.920986	2	8.03468
32	4.362	4.18934	0.960417	1.83	7.6665
33	4.362	4.28334	0.981967	6.4	27.41:34
34	4.362	4.35134	0.997556	4.95	21.5391
35	4.362	3.65543	0.838018	3.7	13.5251
36	4.362	3.68143	0.843978	1.4	5.15401
37	4.362	3.75343	0.860485	1.85	6.94385
38	4.362	3.72943	0.854982	9.45	35.2432
39	4.362	3.67758	0.843095	4.06	14.931
40	4.362	3.65565	0.838068	4.2	15.3538

TABLE A.5

Results from RADPOW for FlymenNE

Load Point	λ [f/yr]	*U* [h/yr]	*r* [h/f]	*L* [kW]	*LOE* [kWh/yr]
41	3.605	3.11209	0.863271	3.84	11.9504
42	3.605	3.14818	0.873283	7.65	24.0836
43	3.605	3.11221	0.863305	3.55	11.0484
44	3.605	3.11228	0.863323	2.56	7.96744
45	3.605	3.14835	0.873328	4.35	13.6953
46	3.605	3.21638	0.892199	1.36	4.37427
47	3.605	3.17854	0.881704	2.60	8.26421
48	3.605	3.24254	0.899457	1.35	4.37743
49	3.606	3.56670	0.989101	10.12	36.095
50	3.606	3.94670	1.094480	3.96	15.6289
51	3.606	3.82270	1.060090	11.97	45.7577
52	3.606	3.62270	1.004630	3.04	11.013
53	3.606	3.79270	1.051770	2.64	10.0127
54	3.606	3.81670	1.058430	2.65	10.1142
55	3.606	3.98070	1.103910	1.86	7.4041
56	3.606	3.85270	1.068410	2.60	10.017
57	3.606	3.92070	1.087270	7.92	31.0519
58	3.605	3.14889	0.873479	51.17	161.129
59	3.606	3.21634	0.891941	10.14	32.6137
60	3.606	3.28234	0.910244	7.47	24.5191
61	3.606	3.27634	0.908580	4.68	15.3333
62	3.606	3.38034	0.937421	4.15	14.0284
63	3.606	3.21843	0.892522	8.33	26.8096
64	3.606	3.15059	0.873707	5.64	17.7693
65	3.606	3.21672	0.892046	2.12	6.81944
66	3.606	3.24472	0.899811	1.77	5.74315
67	3.606	3.11485	0.863797	2.08	6.47889
68	3.606	3.11488	0.863806	4.12	12.8333

Appendix B: Input Data for the RCM Cable Application Study

This appendix summarizes the input data used when implementing the developed failure rate model in the Cable Application Study, that is, for the RCM MATLAB program.

The RCM algorithms implemented require input data from seven files. Five of these files relate to the cable component behavior, and the other two to (i) the reliability input data for the Birka system for RADPOW and (ii) the input data for the cost and benefit analysis. These data are presented in detail below.

B.1 Input Data for the Cable Component Behavior

The input data for the cable component behavior are presented as follows:

- indata_lambda_Svel [86], which are data for modeling the cable component failure rate.
  - Total length for the different cable groups put into operation in the years, 1970–1975
  - Numbers of failures for each cable group during the aging period of 13–18 years after coming into operation
  - The input data are presented in Table 11.2
- indata_lambda
  - The average failure rate for each year is evaluated based on the input data presented above
- indata_wt_KTHSINTEF [82],[115],[119]
  - Assumed maximum water-tree length: lmax = 100% = 5.5 mm
  - Year when water trees reach their maximum length: $t_{lmax} = 13$
- indata_bd_KTH [82],[115]
  - Maximum value for the normalized breakdown voltage, $\max(U_{bd}/U_0) = 10$
  - Minimum value for the normalized breakdown voltage, $\min(U_{bd}/U_0) = 2$

- indata_bdPM_SJNTEF [119]
  - Measurement points for breakdown voltage from the SINTEF study. Normalized for a 24-kV cable with resulting data as presented in Table 11.7.

B.2 Input Data for the Birka System Cable Component

The input data for reliability analysis for the Birka system were defined in Section 10.3. For the RCM Cable Application Study, this data is accomplished with the previously presented component behavior data. The component failure rate model is accomplished with the following system data:

- indata_BirkaLH11
  - Average length for the 11 kV cable, $l_{av} = (259.387/32) \approx 8.10584$ km.
  - Average failure rate for the 11 kV average cable, $\lambda_{LH11} = 58/(18 \cdot 259.387) \cdot l_{av} \approx 0.10069$ f/yr.

B.3 Input Data for the Cost and Benefit Analysis

- indata_Cost
  - Input data for the inflation, interest rate, repair, and related items. Table C summarizes these data.

Appendix C: Fundamentals of Electrical Engineering

C.1 Fundamentals of Electricity

This section introduces the basic laws of electrical circuits namely Kirchhoff's Law and Ohm's Law.

C.1.1 Kirchhoff's Law

Kirchhoff's law is one of the most fundamental laws in electrical circuit analysis and is based on the principle of conservation of charge and conservation of energy. Mathematically it can be written as follows:

$\sum_{j=1}^{n} I_j = 0$, where n is the total number of branches at a node and I is current

$\sum_{j=1}^{n} V_j = 0$, where n is total number of voltages in a closed circuit and V, voltage.

C.1.2 Ohm's Law

Ohm's law gives a relation between the voltage and the current in a conductor. In mathematical terms, it is expressed as follows: $V = IR$, where $V =$ Voltage, $I =$ Current, and $R =$ Proportionality constant called resistance.

Exc 1: Consider a circular conductor with the following materials and properties:

- Aluminum
  - $r = 4\,\text{mm}$ (radius of conductor)
  - $2.82 \cdot 10^{-8}\,\Omega\text{m}$ (specific resistivity of aluminum)
  - $\text{l} = 100\,\text{m}$ (length of conductor)
- Copper
  - $r = 3.33\,\text{mm}$ (radius of conductor)
  - $1.68 \cdot 10^{-8}\,\Omega\text{m}$ (specific resistivity of copper)
  - $\text{l} = 100\,\text{m}$ (length of conductor)

Compute the resistance and conductance of both conductors.

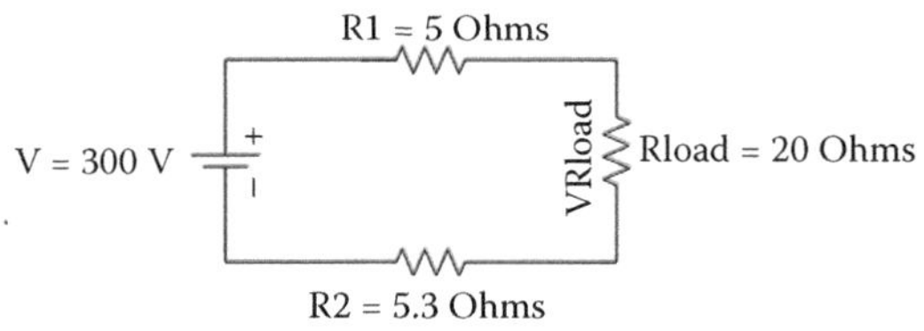

FIGURE C.1
Circuit for Exc 2.

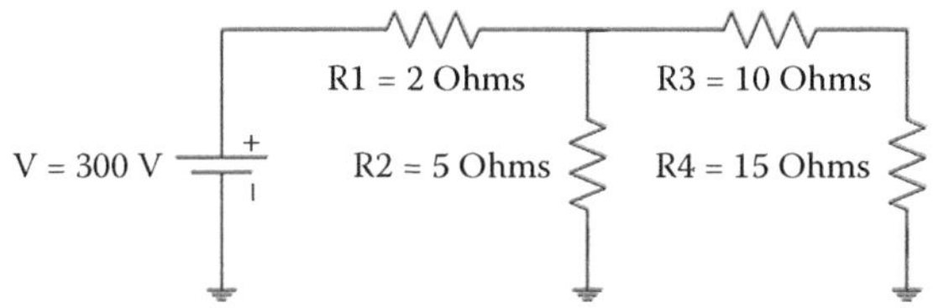

FIGURE C.2
Circuit for Exc 3.

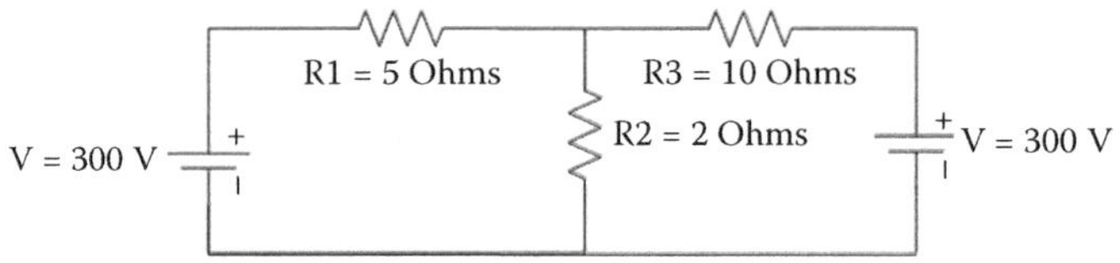

FIGURE C.3
Circuit for Exc 4.

Exc 2: For the circuit drawn below find the current flowing in the circuit and the voltage at the load terminals. Calculate the power dissipated in resistance R1 (Figure C.1). (Use Kirchhoff's voltage law and Ohm's law to solve the circuit.)

Exc 3: Consider the circuit shown below. Solve the circuit to find the current through resistances R1 and R3 (Figure C.2).

Exc 4: Consider the circuit shown below. Find the current and voltage across resistances R1 and R2, using the principle of superposition (Figure C.3).

C.2 AC Power

This section introduces the AC (alternating current circuits). The inductance and capacitance and concept of reactive power are introduced, and root mean square (rms) values. Furthermore, complex power calculations and its significance in the actual system will be explored. The idea for this section is to build a base needed for solving load flow calculations.

C.2.1 Alternating Current and Voltage

In AC circuits, the current and voltage continuously change polarity and the speed at which this change occurs depends on the frequency. The two most common frequencies used in electric power systems are 50 Hz (Europe) and 60 Hz (United States) (Figure C.4).

This oscillating wave means that in one cycle the current and the voltage change polarities, for first half of the cycle the polarity is positive and for second half it is negative. This alternating nature of wave gives rise to the properties of inductance and capacitance in AC electrical circuits.

The waveforms of the voltage and current are typically assumed to be sinusoidal. A sine wave, or sinusoid, is a mathematical curve that describes a smooth repetitive oscillation. It is a continuous wave and is named after the function sine. It's most basic form as a function of time (t) is:

$$y(t) = A\ \sin(2\pi ft + \varphi) = A\ \sin(\omega t + \varphi)$$

where

- A is the amplitude
- f is the ordinary frequency, i.e. the number of oscillations (cycles) that occur each second of time.
- $\omega = 2\pi$ is the angular frequency, the rate of change of the function argument in units of radians per second

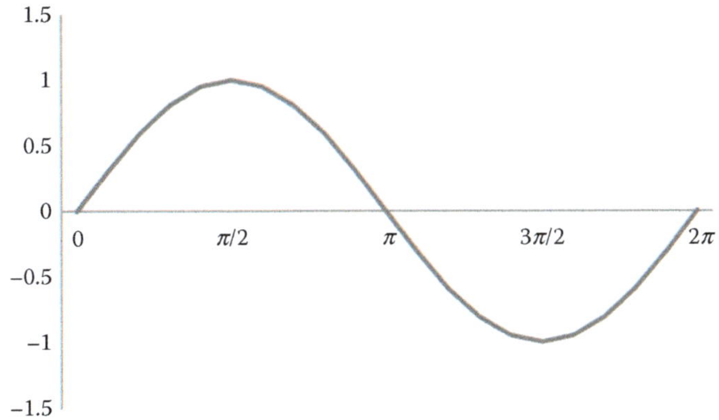

FIGURE C.4
A sinusoid wave oscillation with the period of 2π.

- φ is the phase (in radians) where in its cycle the oscillation is at $t = 0$.

When φ is non-zero, the entire waveform appears to be shifted in time by the amount φ/ω

φ seconds. A negative value represents a delay, and a positive value represents an advance.

The sinus function is an odd and periodic function with the period of 2π which is shown in Figure C4.

C.2.2 Inductance

When an alternating current is passed through a conductor, a magnetic field is induced, this magnetic field produces an effect which opposes the change in the current. A pure inductive circuit produces an effect of phase lag in the current, so that the current lags the voltage in phase by 90 degrees (Figure C.5).

Mathematically, it is written as

$$X_l = \omega L$$

where X_l = inductive reactance;
L = inductance in Henry (H);
$\omega = 2\pi$.

C.2.3 Capacitance

Capacitance is the ability of a body to store an electrical charge. A capacitor essentially resists a change in voltage. A pure capacitive circuit produces an

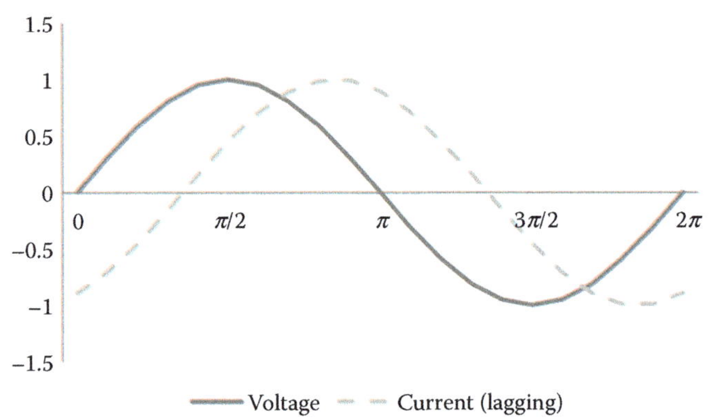

FIGURE C.5
Current lagging voltage by a 90-degree phase difference in purely inductive circuit.

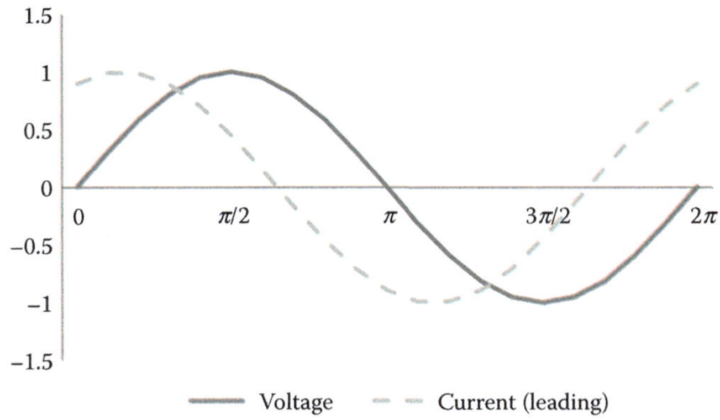

FIGURE C.6
Current leading the voltage by 90 degrees in purely capacitive circuit.

effect of phase lead in the current, so that the current leads the voltage in phase by 90 degrees (Figure C.6).

Mathematically, it is written as

$$X_c = 1/\omega C$$

where X_c = capacitive reactance;
C = capacitance in Faraday (F);
$\omega = 2\pi$.

C.2.4 Impedance

The combination of reactance and resistance that describes the overall behavior of a device in a circuit is called the impedance and is often denoted by Z. Impedance Z is a vector sum of resistance and reactance in a complex plane. Depending on whether the impedance is inductive or capacitive, the current lags or leads the voltage in a circuit. This phenomenon gives rise to the concept of reactive power, which will be dealt with in the next section (Figure C.7).

$$Z = R + jX$$

where Z = impedance;
R = resistance;
X = reactance: either inductive (positive) or capacitive (negative).

The phase difference between current and voltage called angle Φ (Phi) can be calculated as follows:

$$\Phi = \cos^{-1}\left(\frac{R}{Z}\right)$$

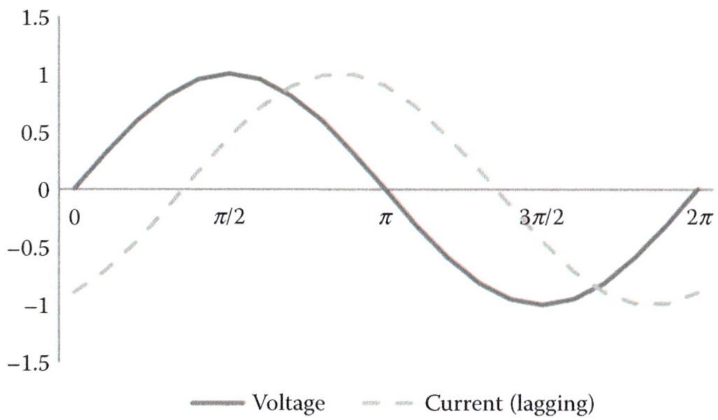

FIGURE C.7
Current lagging the voltage by an angle of 63 degrees.

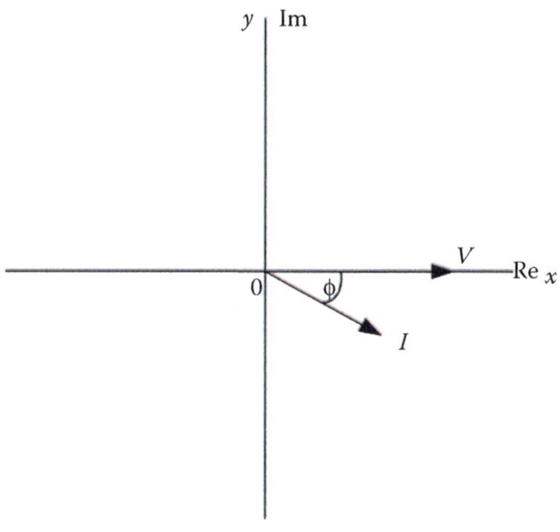

FIGURE C.8
Current lagging voltage shown in the complex plane.

Assume $Z = 5 + j10$, $[Z] = \sqrt{(R^2 + X^2)}$

$$\Phi = 63^\circ$$

Another quantity that is of interest is admittance $Y = 1/Z$ (Figure C.8).

C.2.5 Power

In AC power circuits, there are three different power values that are calculated:

- Active power $P = |V| \cdot |I| \cdot \cos\Phi$ [watts (W)]
- Reactive power $R = |V| \cdot |I| \cdot \sin\Phi$ [var]
- Apparent power $S = |V| \cdot |I| VA$ ($S = V^* \cdot I^*$) [complex power calculation]
  where
  - $|V|$ = rms value of voltage;
  - |I| = rms value of current (I^* is the complex conjugate of the current I);
  - Φ = power factor (angle between voltage and current).

C.3 Generator and Load Modeling

This section introduces the generator and loads modeling for load flow studies.

C.3.1 Generator Modeling

An electric generator is a device used to convert mechanical energy to electrical energy. There are different kinds of generators, for example, DC generator, AC synchronous generator, and AC asynchronous generator. In this section, we will concentrate on the AC synchronous generators as most of the large generators in the electric power system today are AC synchronous generators.

A three-phase synchronous generator produces three-phase voltages with a phase lag of 120° as shown in Figure C.9.

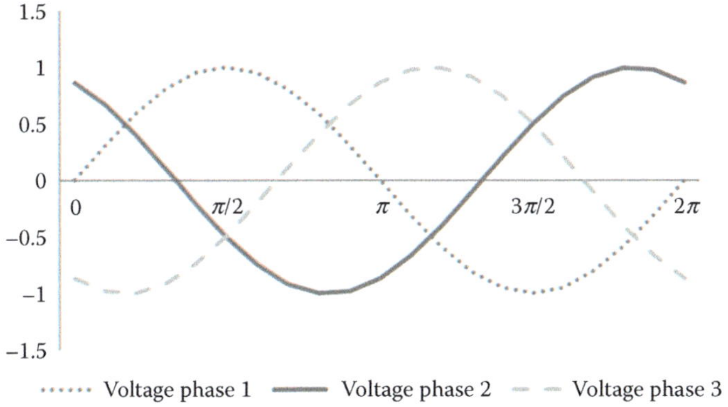

FIGURE C.9
Three-phase voltage output from a generator.

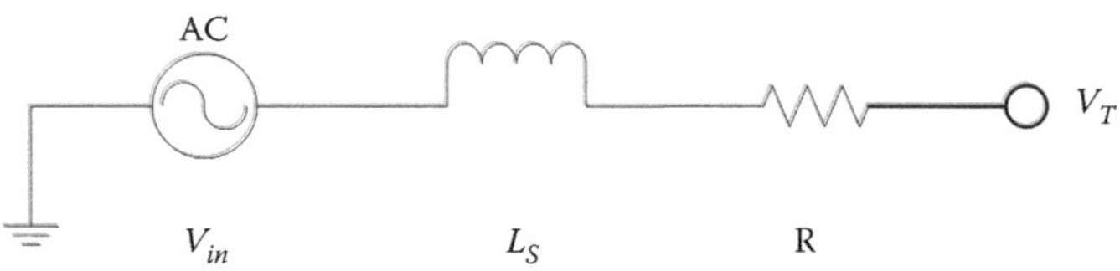

FIGURE C.10
Electrical representation of a generator.

The frequency of voltage, which the generator produces, depends on the number of poles in the generator and the speed at which it rotates and is given by the following equation:

$$f = \frac{n \cdot P}{120}$$

where f = frequency in Hz;
n = speed of rotation in radian per seconds (rpm) [rad/s];
p = number of poles.

A synchronous generator has a set of windings called the stator windings which carry the generated current and are connected to the electric power system. These windings have very little resistance as they are made up of copper in most cases and have short length. However, they have a significant amount of reactance as they are wound in a fashion that creates a lot of inductance in the windings. The generator's synchronous reactance is specified, which is the reactance effective for steady-state balanced operation at synchronous speed. In most cases, the resistance of the generator winding is neglected due to its small value.

Figure C.10 shows a simple single-phase equivalent electrical representation of a three-phase generator. The generator voltage is represented as "V_{in}" as internal voltage of the generator. "V_T" is the terminal voltage of the generator, the voltage supplied by the generator to the connected electric power system. "L_s" and "R" are the internal winding inductance and resistance, respectively.

Note: *A common simplification made is to perform load flow study on single-phase equivalent circuit. This simplifies the calculation without affecting so much on the accuracy as it is assumed that the loads in the system are balanced across all the three phases, which means that the currents in all the three phases are equal. Hence, generator and loads will be represented by their single-phase equivalent circuits.*

C.3.2 Induction Motor Modeling (Constant Power Load)

Induction motors are the most common motor loads in the electric distribution system and are used to convert electrical energy to mechanical energy. Induction motors are robust appliances with simple construction and are available in single-phase and three-phase versions. The windings in a three-phase induction motor are arranged so that when a three-phase voltage is applied to the motor a rotating magnetic field is created. This rotating magnetic field induces current in the rotor, which in turn produces the force on the rotor which makes it rotate. Here we will learn how to model an induction motor for load flow studies with the nameplate rating data for an induction motor.

The electrical rating of an induction motor is given in watts or kilowatts and the mechanical output is measured in terms of torque newton meter. The relation between the electrical power and the mechanical output power can be given by the following equation:

$$P_{out} = \omega \cdot T$$

where P_{out} = output electrical power measured in watts;
ω = mechanical angular frequency ($2\pi \cdot \text{rpm}/60$, where rpm is revolutions per minute);
T = output torque of the machine in newton meters.

For any machine, the input and output powers are related by a factor called efficiency, which depends on the losses in the system. In case of induction motors, there are losses in the electrical system in the form of resistive and inductive losses and in the mechanical system in the form of frictional losses.

$$\eta = \frac{P_{in} - P_{out}}{P_{in}}$$

where η = efficiency
P_{in} = input electrical power;
P_{out} = output electrical power.

Now if we are given the torque and the efficiency of the induction motor, the input power can be easily calculated using the above provided formulae.

For the purpose of load flow study, it is sufficient to model the induction motor as shown below.

Figure C.11 shows single-phase equivalent for a three-phase induction motor. The winding consists of an inductance and resistance; however, the windings may or may not be grounded. In the tutorial, we will learn how to calculate the inductance and resistance values from the nameplate data for an induction motor.

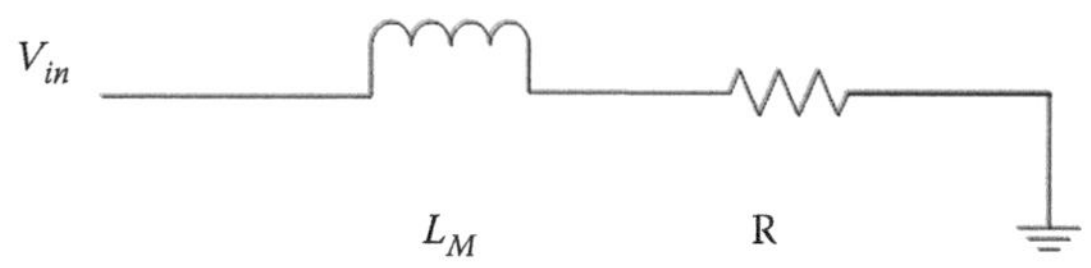

FIGURE C.11
Load flow model for induction motor.

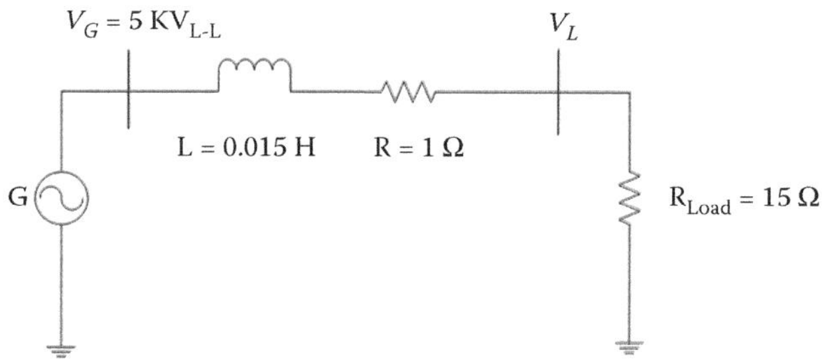

FIGURE C.12
Exc 1 Single-phase equivalent circuit.

Exc 1: A three-phase AC generator is connected to a power system with voltage 5 kV (L–L). The generator bus is connected to a load bus through a transmission line as shown in Figure C.12. Find the current transmitted through the transmission line, the three-phase line to line voltage at the load bus and the internal voltage of the generator.

Consider the following rating for the generator:

Rating: 10 MVA

No. of poles: 2

Synchronous speed: 3000 rpm

Synchronous reactance = 0.02 H (neglect the internal resistance)

Note: *In the above problem, one has to assume that the three-phase system is connected in star "Y" configuration, when calculating the phase to ground voltage from line to line voltage. This is the convention used in electric power engineering.*

Exc 2: A three-phase induction motor is connected in star "Y" configuration to an input voltage of 400 V (line–line) and has an output of 4 kW. The synchronous speed of rotation is 1500 rpm at a frequency of 50 Hz. The motor has an efficiency of 80% and power factor 0.75.

- Find the parameters to model the single-phase equivalent of the induction motor.
- Consider the windings of the same motor are connected in Delta Δ. Now find the output of the motor in kW. (Assume that the efficiency is the same as for "Y" connection.)

Exc 3: Consider a generator rated 500 kW at power factor 0.8 is connected to an electrical distribution system at voltage level 400 V (line–line). The distribution system has two three-phase loads connected in parallel with the following specifications:

Specifications for the generator:

Rating: 500 kW at 0.8 pf

Synchronous reactance: 0.015 H

Internal resistance: 0.1

Rated frequency: 50 Hz
A water pump driven by an induction motor:
Rating: 700 Nm
No. of poles: 4
Rated frequency: 50 Hz
Winding configuration: Y
Efficiency: 86%
Power factor: 0.95
Lighting load (incandescent lamps):
Rating: 50 kW
Connection: Equally distributed load connected in
Power factor: 1

Draw the single-phase equivalent circuit for the above problem showing all the impedances. Calculate the internal voltage of the generator and the current supplied by the generator.

C.4 Transformers and Three-Phase Transmission Lines

C.4.1 Three-Phase Voltage Transformer

A transformer is an equipment used in electric power system to step up (increase) or step down (decrease) the voltage at a given point in the system. Transformer functions in the same way in the electrical system as a gear box function in the mechanical system. Let us understand this analogy in a bit more detail (Figure C.13).

For a gear box, the power transmitted remains the same but the torque and the speed change based on the gear box ratio and the following equation holds true.

$$P_{Mec} = T_1\omega_1 = T_2\omega_2 \text{ (assuming no mechanical losses)}$$

A transformer can be represented in a similar way as in (Figure C.14).

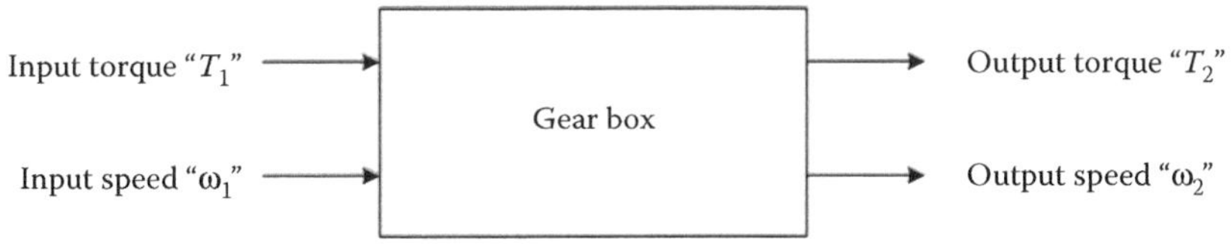

FIGURE C.13
Gearbox representation.

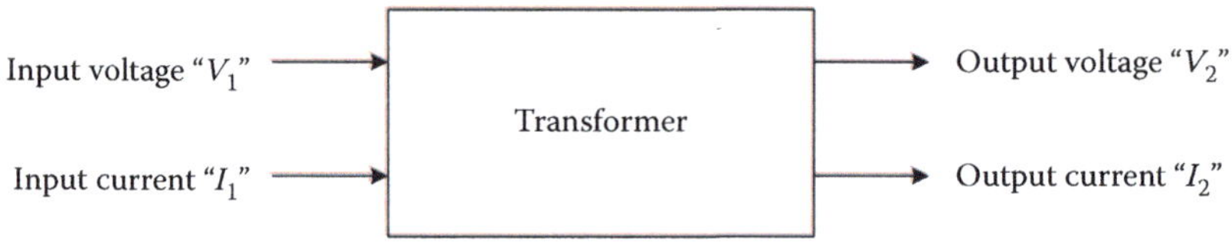

FIGURE C.14
Transformer representation.

As with the gearbox, for a transformer also the input power and the output power are the same (assuming no electrical losses in the transformer).

$$P_{Elec} = V_1 I_1 = V_2 I_2 \text{ (assuming no electrical losses)}$$

where,
V_1 = Primary voltage (voltage on the input side)
V_2 = Secondary voltage (voltage on the output side)
I_1 = Primary current (current on the input side)
I_2 = Secondary current (current on the output side)

A transformer is a static equipment, without any rotating parts, unlike the gearbox. So there are no mechanical losses in the transformer. However, there are electrical losses. To understand where the losses occur we go through the basic construction of a transformer in brief.

The transformer is made up of three main components: the primary winding, an iron core, and the secondary winding. The construction looks like as in (Figure C.15).

Figure C.15 shows a simple transformer construction. As we can see there are two windings, these windings have resistance and inductance, which means they have losses. Also, there is an iron core which is responsible for power transfer from primary to the secondary winding. This iron core consumes power to get magnetized and hence adds to the losses in the transformer. However, it should be noted that the transformer is very efficient electrical equipment with efficiency in the range of 95%–99%.

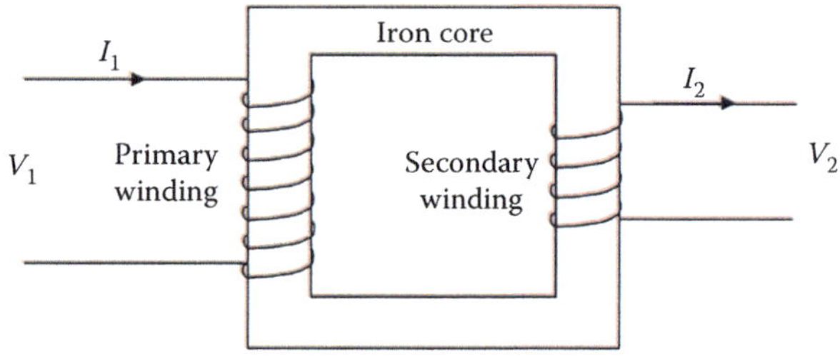

FIGURE C.15
Simple single-phase step down transformer.

Let us consider a 10 MVA, 10/138 kV transformer with an impedance of 5%. Then we know the following:

$S = 10$ MVA

$V_1 = 10$ kV

$V_2 = 138$ kV

$\%Z = 5\%$

In a transformer the impedance is always given in percentage as the actual value of the impedance on the primary and secondary sides is different but their normalized values are the same on both sides. To calculate the % impedance, we first need to calculate the base impedance on both sides

$$Z_{BASE1} = \frac{V_1^2}{S}$$

$$Z_{BASE2} = \frac{V_2^2}{S}$$

Now

$$\%Z = \frac{Z_{ACT(Primary)}}{Z_{BASE1}} = \frac{Z_{ACT(Secondary)}}{Z_{BASE2}}$$

C.4.2 Per Unit System

In electrical engineering, per unit system is often used while performing load flow calculations. In per unit system, all the quantities are expressed as fraction of base unit quantity. This simplifies things as things expressed in per-unit system are the same for all voltage levels.

The first step for per-unit system calculation is the selection of base MVA and base voltage. Base MVA can be selected to any arbitrary value (a general practice is to select base MVA as 100 MVA or 1 MVA). The base voltage selection generally depends on the system under consideration. One way is to select the highest voltage level in the system as the base voltage. The base impedance and the base current are calculated from the base MVA and the base voltage as follows:

$$Z_{BASE} = \frac{V_{BASE}^2}{S_{BASE}}$$

$$I_{BASE} = \frac{S_{BASE}}{\sqrt{3} \cdot V_{BASE}}$$

Exc 1: In a part of a three-phase system, the base power is 100 MVA and the base voltage is 132 kV. Calculate the following:

- The base current
- The base impedance

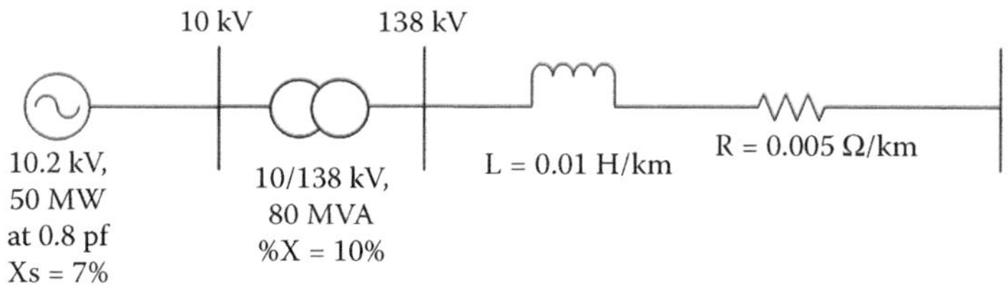

FIGURE C.16
Exc 4 electrical system.

- 210 A expressed in p.u.
- 18 Ω expressed in p.u.

Exc 2: A 20-MVA transformer with 11 kV primary and 66 kV secondary voltage has a reactance of 0.242 per star phase referred to the primary side.

- What is the per-unit reactance on the primary side.
- What is it on the secondary side?

Exc 3: A 50 MVA, 132/33 kV transformer has an impedance of 0.1 p.u. per star phase. What is the impedance in per star phase:

- Referred to the primary side.
- Referred to the secondary side.

Exc 4: The electrical system shown below has a generator, a transformer, and a transmission line of length 100 km. Represent the entire system in per unit system at 100 MVA base power and 138 kV base voltage (Figure C.16).

References

1. ISO, *Standard ISO 55000:2014, Asset Management—Overview, Principles and Terminology*, ISO, 2014. Available at: https://www.iso.org/standard/55088.html.
2. Schlenoff, D. C., What are the 10 greatest inventions of our time? *Scientific American*, November 1, 2013.
3. Elgerd, O., *Electric Energy Systems Theory: An Introduction*. McGraw-Hill, Electrical and Electronic Engineering Series, 1971. 564pp. https://www.amazon.com/Electric-energy-systems-theory-introduction/dp/B0006DY840
4. ASCE Foundation, *Infrastructure Report Card*, 2017. Available at: https://www.infrastructurereportcard.org/wp-content/uploads/2017/01/Energy-Final.pdf.
5. ENTSOE, *10-Year Network Development Plan (TYNDP)*, December 2016.Available at: http://tyndp.entsoe.eu/exec-report/.
6. Nations, U., Transforming our world, the 2030 agenda for sustainable development, resolutions adapted by the un, a/res/70/01. Available at: https://www.infrastructurereportcard.org/wp-content/uploads/2017/01/Energy-Final.pdf, September 25, 2015.
7. *2030 Climate and Energy Goals for a Competitive, Secure And Low Carbon EU Economy*. European Press Release. Available at: http://europa.eu/rapid/press-release_IP-14-54_en.htm, January 22, 2014.
8. *Green Paper: A 2030 Framework for Climate and Energy Policies*. European Commission, Brussels. Available at: http://eur-lex.europa.eu/LexUriServ/LexUriServ.do?uri=COM:2013:0169:FIN:EN:PDF, March 27, 2013.
9. *Wind Energy Scenarios for 2030*. Report by the European Wind Energy Association (EWEA). Available at: https://www.ewea.org/fileadmin/files/library/publications/reports/EWEA-Wind-energy-scenarios-2030.pdf, 2015.
10. *European Energy Industry Investments*. European Parliament, Policy Department, European Union. Available at: http://www.europarl.europa.eu/RegData/etudes/STUD/2017/595356/IPOL_STU(2017)595356_EN.pdf, February 2017.
11. Nowlan, F. S. and Heap, H. F., *Reliability Centered Maintenance*. Technical Report. National Technical Information Service, U.S. Department of Commerce, Springfield, Virginia, 1978.
12. Moubray, J., *Reliability—Centred Maintenance*. Butterworth-Heinemann, Oxford, 1991. Reprint 1995.
13. Bertling, L., *Reliability Centred Maintenance for Electric Power Distribution Systems*, Doctoral Thesis at the Royal Institute of Technology KTH, Department of Electrical Engineering, KTH-EE, ISBN 91-7283-345-9, 2002.
14. Bertling, L., Allan, R. N., and Eriksson, R., A reliability-centred asset maintenance method for assessing the impact of maintenance in power distribution systems, *IEEE Transactions on Power Systems*, **20**(1), 75–82, February 2005.
15. Fischer, K., Besnard, F., and Bertling, L., Reliability-centred maintenance for wind turbines based on statistical analysis and practical experience, *IEEE Transactions on Energy Conversion*, **27**(1), 184–195, 2012.
16. Lloyd, C., *Asset Management Whole-Life Management of Physical Assets*. Thomas Telford, London, ISBN: 978-0-7277-3653-6, February 2010.

17. Lloyd, C., *International Case Studies in Asset Management*. ICE Publishing, ISBN 978-0-7277-5739-5, 2012. 572pp. http://www.icebookshop.com/bookshop_main.asp?ISBN=9780727758187
18. Jardine, A. K. S. and Tsang, A. H. C., *Maintenance, Replacement, and Reliability: Theory and Applications*, Second Edition, CRC Press, Series Mechanical Engineering, March 2017.
19. Billinton, R., Bibliography on the application of probability methods in power system reliability evaluation, 1972, PAS-91.
20. IEEE, Subcommittee on the Application of Probability methods Power System Engineering Committee, bibliography on the application of probability methods in power system reliability evaluation, 1971–1977.
21. Allan, R. N., Billinton, R., and Lee, S. H., Bibliography on the application of probability methods in power system reliability evaluation 1977–1982, *IEEE Transactions on Power Apparatus and Systems*, **PAS-103(2)**, 275–282, 1984. (*Task Force Members of the IEEE Subcommittee on the Application of Probability Methods, Power System Engineering Committee*, IEEE.)
22. Allan, R. N., Billinton, R., Shahidehpour, S. M., and Singh, C., Bibliography on the application of probability methods in power system reliability evaluation 1982–1987, *IEEE Transactions on Power Systems*, **3**(4), 1555–1564, November 1988. *Task Force Members of the IEEE Subcommittee on the Application of Probability Methods*, IEEE.
23. Allan, R. N., Billinton, R., Breipohl, A. M., and Grigg, C. H., Bibliography on the application of probability methods in power system reliability evaluation: 1987–1991, *IEEE Transactions on Power Systems*, **9**(1), 41–49, February 1994. *Task Force Members of the IEEE Subcommittee on the Application of Probability Methods*, IEEE.
24. Allan, R. N., Billinton, R., Breipohl, A., and Grigg, C. H., Bibliography on the application of probability methods in power system reliability evaluation 1992–1996, *IEEE Transactions on Power Systems*, **14**(1), February 1999, *Task Force Members of the IEEE Subcommittee on Reliability, Risk and Probability Applications*, IEEE.
25. Billinton, R., Fotuhi-Firuzabad, M., and Bertling, L., Bibliography on the application of probability methods in power system reliability evaluation 1996–1999, *IEEE Transactions on Power Systems*, **16**(4), 595–602, November 2001.
26. Endrenyi, J. et al., The present status of maintenance strategies and the impact of maintenance on reliability, *IEEE Transactions on Power Systems*, **16**(4), November 2001. *A Report of the IEEE/PES Task Force on Impact of Maintenance Strategy on Reliability of the Reliability, Risk and Probability Applications Subcommittee (RRPA)*.
27. IEEE/PES, *Task Force on Impact of Maintenance Strategy on Reliability Impact of Maintenance Strategy on Reliability*, Technical Report, The Reliability, Risk and Probability Applications Subcommittee, 1999.
28. Meeuwsen, J., Kling, W. L., and Ploem, W. A. G. A., The influence of protection system failures and preventive maintenance on protection systems in distribution systems, *IEEE Transactions on Power Delivery*, **17**(1), 42, 1997.
29. Chantron, J. and Aupied, J., Modeling of the preventive maintenance on the reliability of operating equipment using stochastic Bayesian filtering, in *6th PMAPS, Madeira*, Funchal, Portugal, September 2000.

30. Meeuwsen, J. and Kling, W. L., Effects of preventive maintenance on circuit breakers and protection systems upon substation reliability, *Electric Power Systems Research*, **40**(3), 181–188, 1997.
31. Sanz-Bobi, M., Roman, T. G. S., Zarco, P., and Guillen, G., Oprema: A methodology for preventive maintenance optimization applied to electrical power networks, in *13th PSCC, Trondheim*, Norway, June 1999.
32. Anders, G., Hamoud, G., da Silva, L., A. M. and F. M. L. A., Optimal outage scheduling example of application to a large power system, in *14th PSCC, Sevilla*, Spain, June 2002.
33. CIGRE, WG B5.40 Education, Qualification and Continuing Professional Development of Engineers in Protection and Control. Convener Mladen Kezunovic (US), October 2014.
34. ISO, *Navigating a World in Transition*, Annual Report, 2016.
35. *ISO 55000 Standards for Asset Management*, ISO, 2014. Available at: http://www.assetmanagementstandards.com/.
36. Endrenyi, J., Anders, G., Bertling, L., Kalinowski, B., *Comparison of Two Methods for Evaluating the Effects of Maintenance*, the 8th International Conference on Probabilistic Methods Applied to Power Systems (PMAPS), Ames, Iowa, September 2004.
37. Moubray, J., *Reliability-Centred Maintenance*. 2nd edition. Butterworth-Heinemann, Oxford, 2001.
38. *Maintenance Terminology*. European standard: EN 13306:2010.
39. IEEE, *IEEE Recommended Practice for the Design of Reliable Industrial and Commercial Power Systems, The Gold Book*. IEEE Std 493-2007, February 2007.
40. Romijn, J.-W., *Philosophy of Statistics*. Stanford Encyclopedia, 2014.
41. Li, K.-G., Jiang, H. and Zomaya, A. Y.*Big Data Management and Processing*, Chapman and Hall/CRC, May 25, 2017. 469pp. ISBN 9781498768078.
42. IEC 300: the dependability counterpart of ISO 9000, K. Strandberg, *Annual Reliability and Maintainability Symposium 1991 Proceedings*, pp. 463–467, 1991.
43. Allen, A., *Probability, Statistics, and Queueing Theory*. Academic Press, 2014. 768pp.
44. Endrenyi, J., *Reliability in Electric Power Systems*. John Wiley & Sons US, 1978.
45. Rausand, M. and Hoyland, A., Introduction, in *System Reliability Theory: Models, Statistical Methods, and Applications*, 2nd edition. John Wiley & Sons, Inc., Hoboken, NJ, 1994. http://onlinelibrary.wiley.com/doi/10.1002/9780470316900.ch1/summary
46. VVK RCM generator analysis. Working paper Swedpower, Vattenfall Vattenkraft AB.
47. Blom, G., *Statistics Theory with Applications (Statistikteori med tillämpningar), 2nd edition*. Studentlitteratur, Lund, Sweden, 1984. ISBN 91-44-0559-7 (in Swedish).
48. Raade, L. and Westergren, B., *BETA Mathematics Handbook, 2nd edition*. Studentliteratur, Chartwell-Bratt Ltd, Lund, Sweden, 1990.
49. Wikipedia, *Reliability Engineering According to Wikipedia*. https://en.wikipedia.org/wiki/Reliability_engineering, September 15, 2017.
50. Billinton, R. and Allan, R., *Reliability Evaluation of Engineering Systems, 2nd edition*. Plenum Press, New York, 2002.
51. Allan, R. N., Billinton, R., and Olivera, M. F. D., An efficient algorithm for deducing the minimal cuts and reliability indices of a general network configuration, *IEEE Transactions on Reliability*, **R-25**(4), 226–233, 1976.

52. He, Y. and Bertling, L., *A Comparative Study of Reliability Assessment of Electrical Distribution Systems*. Technical Report A-EES-9705, Department of Electric Power Engineering, KTH, 1997.
53. Ansell, J. I. and Phillips, M. J., *Practical Methods for Reliability Data Analysis*. Oxford University Press, New York, 1993, ISBN 019853664.
54. Billinton, R. and Allan, R., *Reliability Evaluation of Engineering Systems: Concepts and Techniques*. Pitman Advanced Publishing Program, London, 1983. ISBN 0-273-08484-4.
55. Lindquist, T., Bertling, L., and Eriksson, R., *A Method for Age Modeling of Power System Components based on Experiences from the Design Process with the purpose of Maintenance Optimization*, Reliability and Maintainability Symposium (RAMS), IEEE, 2005.
56. Kordonsky, KH. B. and Gertsbakh, I., Multiple Time Scales and the Lifetime Coefficient of Variation: Engineering Applications, *Lifetime Data Analysis*, 2, pp. 139–156. 1997.
57. Duchesne, T. and Lawless, J., Alternative Time Scales and Failure Time Models, *Lifetime Data Analysis*, 6, pp. 157–179, 2000.
58. Bishop, C. M., *Pattern Recognition and Machine Learning*, Springer, 2006.
59. Barreto, G. A. and Andrade, M. G., Bayesian Inference and Markov Chain Monte Carlo Methods Applied to Streamflow Forecasting, *6th PMAPS Conference*, Madeira, Portugal, September 25–28, 2000.
60. Celeux, G., Persoz, M., Wandji, J. N., and Perrot, F., Using Markov Chain Monte Carlo methods to solve full Bayesian modeling of PWR vessel flaw distributions, *Reliability Eng. & Sys. Safety*, 66, pp. 243–252, 1999.
61. CIGRE, *WG 601 of Study Committee C4, Review of the Current Status of Tools and Techniques for Risk Based and Probabilistic Planning in Power Systems*. Convener Poyan Pourbeik (US), March 2010.
62. Olsson, G. and Skärvad, P., *Business Economics (Företagsekonomi 99)*. LiberHermods, Malmö, Sweden, 1993. ISBN 91-23-01704-X (in Swedish).
63. NEPLAN, Computer software. Available at: http://www.neplan.ch/en-company/.
64. Allan, R. et al., *Users Manual for RELNET Reliability Analysis of Power System Networks*. Dept. of Electrical Engineering and Electronics, University of Manschester, 1993. Version 4.1.
65. Kjolle, G. and Holen, A., *Reliability Analysis of Electrical Distribution Systems (Pålitlighetsanalyse av fordelningsnett)*. Technical report, Energiforsyningens Forskningsinstitutt (EFI), 1991. (TR 3853:1-II, Foundation for scientific and industrial research at the Norwegian Institute of Technology, Norway).
66. Brown, R. E., Grupta, S., Christie, R. D., Venkata, S. S., and Fletcher, R., Distribution system reliability assessment using hierarchical Marcov modeling, *IEEE Transactions on Power Delivery*, **11**(4), 1929–1934, 1996.
67. Balu, N., On-line power system security analysis, *IEEE Proceedings*, **8**(2), 262–280, February 1992.
68. EPRIGEN, *Streamlined Reliability Centered Maintenance (SRCM) Implementation Guidelines*. Technical Report, EPRI, California, USA, 1998.
69. Smith, A. M., *Reliability Centred Maintenance*. McGraw-Hill, USA, 1993.
70. Persson, E., *Metoder för kostnadsoptimering av underhållsåtgärder och riskvärdering av felsätt*. Master Thesis, KTH, 2004.

71. von Meier, A., *Electric Power Systems: A Conceptual Introduction.* John Wiley & Sons, Inc., USA, 2006.
72. Ödman, M., The breakthrough for the electrical power (när elstr ommen kom till byn), in *Rabén och Sjögren (in cooperation with the organisation Liv i Sverige),* Kristianstad, Sverige, 1993. (in Swedish).
73. Willis, H. L., *Power Distribution Planning Reference Book.* Marcel Dekker (First book in the Power Engineering Series), New York, USA, 1997.
74. *Voltage Characteristics of Electricity Supplied by Public Distribution Systems.* European Standard EN 50160, European Committee for Electrotechnical Standardization, 2010. 40pp. https://standards.globalspec.com/std/9943573/cenelec-en-50160
75. IEC, 60050-901:2013, International Electrotechnical Vocabulary — Part 901: Standardization, 2013. 47pp.
76. CIGRE, Working Group 38-03. Power System Reliability Analysis—Application guide, 1987.
77. ENTSOE, *Establishing a Guideline on Electricity Transmission System Operating.* Commission Regulation (EU) 2017/1485, August 2, 2017.
78. NERC, North American Electric Reliability Corporation. Available at: http://www.nerc.com.
79. Billinton, R. and Allan, R., *Reliability Evaluation of Power Systems,* 2nd edition. Plenum Press, New York, ISBN 0-306-45259-6, 1996.
80. Wang, L. and Endrenyi, J., Reliability techniques in large electric power systems, *Control and Dynamic Systems,* **42**, 163–243. 1991. https://www.sciencedirect.com/science/article/pii/B9780120127429500094
81. Electric Power Research Institute (EPRI), Composite-System Reliability Evaluation: Phase 1—scoping study, in *EPRI EL-5290, Research Project 2581-1,* 1987.
82. IEEE Guide for Electric Power Distribution Reliability Indices IEEE Std 1366-2012 (Revision of IEEE Std 1366-2003), IEEE Standard, pp. 1–43, 2012.
83. Bollen, M., Wallin, L., Ohnstad, T., and Bertling, L., On operational risk assessment in transmission systems—weather impact and illustrative example, in *Proceedings of 10th PMAPS 2008, June, Puerto Rico,* pp. 1675–1682, IEEE, June 2008.
84. Kundur, P., *Power System Stability and Control.* McGraw-Hill, New York, NY, 1994. 1200pp.
85. Communication with Lars Marketeg, *Responsible for procurement of plans for maintenance at The Swedish National Grid.* 1998.
86. Vainberg, M., Anders, G. J., and Lebow, M., Probabilistic Asset Management and RCM at substations, in *Conference Proceedings, ERA Technology, Report 97-0847,* 1997.
87. Jones, G. R., Laughton, M. A., and Say, M. G., editors. *Electrical Engineer's Reference Book,* 15th edition. Butterworth Heinemann, Oxford, 1993. ISBN 7506 1202 9.
88. Söder, L., *Static Analysis of Power Systems*: Technical Report AEES 9707, Department of Electrical Engineering, KTH, 2001. Course material in 2C1112 for the master program, 1993. ISBN 7506 1202 9.
89. Densley, J., Ageing mechanisms and diagnostics for power cables—An overview, *IEEE Electrical Insulation Magazine,* **17**(1), 14–22, 2001.

90. Holmgren, B., *Dielectric Response, Breakdown Strength and Water Tree Content of Medium Voltage XLPE Cables*. Licentiate Thesis KTH, Stockholm, TRITA-EEA-9705, ISSN 1100-1593, 1997.
91. Bi-cycle is an independent software vendor. More information available at: http://www.bi-cycle.com/.
92. Backlund, F. and Andersson, J.-O., Meeting at Vattenfall (formerly Swedpower), Råcksta, May 16, 2005.
93. Edin, H., *Partial Discharges Studied with Variable Frequency of Applied Voltage*. Doctoral Thesis KTH, Stockholm, TRITA-EEK-0102 , ISSN 1100-1593, 2001.
94. The Lifetime and Usefulness of XLPE Cables, Technical Report, Swedenergy (Former Association of Swedish Electric Power Utilities) (Original title: *PEX-KABLAR livslängd och andvändbarhet*, in Swedish), 1990.
95. Jansson, S., *Availability in Power Supply in the Swedish Electrical Distribution Systems 1995–1997*, Technical Report, Swedenergy AB (Former Association of Swedish Electric Power Utilities), (Original title: *Rapport om leveranssäkerhet i de svenska elnäten 1995 till 1997*, in Swedish), 1999.
96. *DAR User's Guide*, Technical Report, Swedenergy AB (Former Association of Swedish Electric Power Utilities), (Original title: *DAR Användarhandledning*, in Swedish), 1993.
97. Wallnerström, C. J. and Bertling, L., Investigation of the robustness of the Swedish network performance assessment model, *IEEE Transactions on Power Systems*, **23**(2), 773–780, 2008.
98. Elleverantörer, S., *DARWIN Requirements Specification*, Technical Report, Swedenergy AB (Former Association of Swedish Electric Power Utilities), (Original title: *Kravspecifikation för DARWIN*, in Swedish), ISBN 91-7622-137-7, 1980.
99. *Glossary for Operation and Maintenance of the Electric Power Delivery Systems* Technical Report, Swedenergy AB (Former Association of Swedish Electric Power Utilities), (Original title: *Elordlista Drift och underhåll av anläggning för överföring och distribution av elektrisk energi*, in Swedish), 1993.
100. Communication with Sven Jansson, Responsible of the Swedish National Reporting System for disturbance statistics for the electrical distribution systems (Referred as: DAR). Swedenergy AB, 1999.
101. Nitell, P., *Availability analysis for the Stockholm Distribution system (Tillgängligh etsanalys för Stockholms Elnät)*. Master Thesis KTH, Stockholm, MAT-STAT-1998-03, 1998. (in Swedish).
102. Allan, R., Professor of Electrical Energy Systems. The Manchester Centre for Electrical Energy (UMIST), 2002.
103. Nordell, *Statistics for Interruption of Voltage (Driftstörningsstatistik)*. Sorgenfri Tryk, Norway, 1999. (in Swedish).
104. Nordell, *Guidelines for classification of interruption of voltage in the Scandinavian countries (Nordels riktlinjer fr klassificering av driftstörningar och fel preliminär utgåva 10 maj)*. Nordel, Norway, 2000. (in Swedish).
105. Arnborg, S., *Statistics for Interruption of Voltage (Driftstörningsstatistik)*. The Swedish National Grid (Svk) Planning Department, 2000.
106. Statnett, *Annual Statistics for Interruption of Voltage (Driftforrstyrrelser i 33-420 kV nettet—Årsstatistikk)*. Statnett, Norway, 1998. (Transmission system operator in Norway).
107. *Generation Equipment Status Annual Report 1998*. Canadian Electricity Association, 2000.

108. *Forced Outage Performance of Transmission Equipment*. Canadian Electricity Association, 2000.
109. CIGRE, Working Group 06 (Reliability of HV circuit-breakers) of Study Committee 13 (Switching Equipment). Final Report of the Second International Enquiry on High Voltage Circuit-Breaker Failures and Defects in Service, 1994.
110. CIGRE, Working Group 23.02. Report on the Second International Survey on High Voltage Gas Insulated Substations (GIS) Service Experience, 2000.
111. *T-book (3rd edition) Reliability Data of Components in Nordic Nuclear Power Plants*. The ATV Office, Sweden, 1992. 235pp. https://inis.iaea.org/search/searchsinglerecord.aspx?recordsFor=SingleRecord&RN=24057248
112. Bertling, L., RCM-study Birka Nt and KTH-a study of causes of failures based on statistics and practice (Original title: *RCM-studie Birka Nät och KTH—en studie av felorsaker baserad påstatistik och praktik*, in Swedish). Technical Report A-EES-0104, Department of Electrical Engineering, KTH, 2001, p. 64.
113. Bertling, L., Eriksson, R., Allan, R., Gustafsson, L., and hln, M., Survey of causes of failures based on statistics and practice for improvements of preventive maintenance plans, in *Proceeding of the 14th PSCC*, Sevilla, June 2002, pp. 1–7.
114. Kjell Gustafsson, *Responsible for Supply Interruption Statistics*. Ellevio AB (Then Birka Nät), 2001.
115. Gjarde, A., Lundgaard, L., and Faremo, H., *Fault statistics for Mass Impregnated Cable (Feilstatistikk for massekabel, endeavslutninger og skjöter 1991 1995)*. Technical Report, Energiforsyningens Forskningsinstitutt (EFI), Trondheim, Norway, 1997. TR A4550. (in Norwegian).
116. Communication with Mats Åhlén, *Responsible for the Maintenance Planning* of the Stockholm Electrical Distribution System, Ellevio AB (Then Birka Nät), 2001.
117. Bertling, L., Eriksson, R., and Allan, R., Relation between preventive maintenance and reliability for a cost-effective distribution systems, in *Proceedings of IEEE PowerTech '01, 4(208)*, September 2001.
118. CIGRE, Working Group 21.09. Consideration of Ageing factors in Extruded Insulation Cables and Accessories, 1990, in *Published in Electra, No. 140*, February 1990.
119. CIGRE, Working Group 21.05. Diagnostics Methods for HV Cables and Accessories, 1998, in *Published in Electra, No. 176*, February 1998.
120. CIGRE, Working Group 21.05. Diagnostics Methods for HV Cables and Accessories, 1994.
121. Thäring, P., *Watertree Dielectric Spectroscopy*. Licentiate Thesis KTH, Stockholm, TRITA-EEA-9703, ISSN 1100-1593, 1997.
122. KTH, Department of Electrical Engineering, Annual Report 2001, 2002.
123. Werelius, P., Tharning, P., Eriksson, R., Holmgren, B., and Gäfvert, U., Dielectric spectroscopy for diagnosis of water tree deterioration in XLPE cables, *IEEE Transactions on Dielectrics and Electrical Insulation*, **8**(1), 27–42, 2001.
124. Steenis, E. F. and Kreuger, F. H., Water treeing in polyethylene cables, *IEEE Transactions on Electrical Insulation*, **25**(5), 989–1028, 1990.
125. *Diagnostics of XLPE Cable Assets (Diagnostik av PEX-kabelanläggningar Slutrapport från nordiskt samarbetsprojekt 1997–2000)*. ELFORSK, 2000. ELFORSK Report 00:28. (in Swedish).
126. Pilling, J. and Bertini, G., Incorporating cableCURE injection into a cost-effective reliability program, *IEEE Industry Applications Magazine*, **6**(5), 52–56, 2000.

127. Faremo, H., Rehabilitation of XLPE Cables with long Water-trees (Rehabilitering av PEX-kabler med lange ventilerte vanntrae fra ytre halvleder. Felt- og laboratorieforsok.), in *Technical Report, Energiforsyningens Forskningsinstitutt (EFI)*, Trondheim, Norway, 1997. EFI TR A 4512 (in Norwegian), October 1990.
128. *XLPE Cable (PEX-KABEL)*. The Association of Swedish Electric Power Utilities, 1995. (Today a part in Swedenergy AB, in Swedish).
129. Maintenance plans for generators. Inspection plans. Working papers from Swedpower, Vattenfall Vattenkraft AB.
130. Fredriksson, S., Vattenfall service nord ab, telephone interview, February 24, 2005.
131. Maughan, C., Inspection/test impact on maintenance of generators, in *Electrical Insulation Conference and Electrical Manufacturing & Coil Winding Technology Conference, Proceedings 2003*, pp. 443–448, September 2–25, 2003.
132. Bertling, L., On Reliability Assessment of Distribution Systems and Reliability Centred Maintenance. Licentiate Thesis KTH, Department of Electric Power Engineering, Stockholm, TRITA-EES-9902, ISSN 1100-1607, 1999.
133. Investment and Maintenance Policy. Technical Report. Electrical Distribution System Operator E.ON (Formerly Sydkraft), (Original title: *Investerings-och underhållspolicy för SKED Lokalnät*, in Swedish), 1997.
134. Policy for Maintenance of Trees Along Overheadline Feeders. Technical Report. Electrical Distribution System Operator E.ON (Formerly Sydkraft), (Original title: *Policy för ledningsgatuunderhåll av lokalnät*, in Swedish), 1997.
135. Annual Report 2000. Birka Energy AB, 2001.
136. Neimane, V., *Long-term planning studies of the Birka Nät AB distribution network 220–33 kV in the Southern part of Stockholm*. Technical Report, Department of Electric Power Engineering, KTH, 2000.
137. Communication with Kjell Gustafsson, *Responsible for the Distribution Planning of the Stockholm Electrical Distribution System*, Ellevio AB (Then Birka Nät), 2000.
138. Communication with Daniel Terranova, *Distribution Planning department of the Stockholm Electrical Distribution System*, Ellevio AB (Then Birka Nät), 2000.
139. Willis, L. H., *Power Distribution Planning Reference Book*. Marcel Dekker Inc., NY, 1997.
140. Brown, R. E., *Electric Power Distribution Reliability*. Marcel Dekker Inc., NY, 2002.
141. Takács, P., Reliability evaluation of electrical systems with cables are overhead lines. (Originatl title: *Tillförlitlighetsvärdering av elsystem där kablar blandas med luftledningar*. In Swedish), M.Sc. Thesis, Dept. Electrical Engineering, KTH, Stockholm, BEES0206, 2002.
142. Interruption costs for electricity consumers, *Technical Report, Swedenergy AB* (Former Association of Swedish Electric Power Utilities), *(Original title: Avbrottskostnader for elkunder, In Swedish)*, 1994.
143. Bubenko, J. and Skugge, G., Power System Technique for Electrical Distribution System Analysis. Course material for further education course given at KTH School of Electrical Engineering and Computer Science, on contract with STF Ingenjorsutbildning AB (a company for engineering education), 2001.
144. CIGRE, Task Force 38-06-01. Methods to Consider Customer Interruption Costs in Power System Analysis, 2001.
145. Ribrant, J. and Bertling, L., Survey of failures in wind power systems with a focus on Swedish wind power plants, 1997–2005, *IEEE Transactions on Energy Conversion*, **22**(1), 167–173, 2007.

146. Carlstedt, N., Szadkowski, C., and Karlström, C., *Performance of Wind Power Plant*, Annual Report, Technical Reports: Elforskrapporter: 98:4, 99:6, 00:11, 01:16, 02:20, 03:12, 04:19, 05:11. Energiforsk AB (Former Elforsk), (Swedish title: *Driftuppföljning av Vindkraftverk, Årsrapport*), 1997–2004.
147. *Felanalys*, Database of failures for Swedish wind power turbines 1997–2005, used with permission from Vattenfall Power Consultant and the Swedish Energy Authority, 2005.
148. Holttinen, H., Lakso, T., and Marjaniemi, M., *Performance of Wind Power Plants*, Annual Report (Finnish title: *Tuulivoiman Tuotantotilastot Vuosiraportti*). 2000–2005. VTT Processes, Technical Report, VTT Technical Research Centre of Finland LTD, December 2005.
149. Durstewitz, M. (ed.), Wind energy report Germany 2004–2005, Institut für Solare Energieversorgungstechnik (ISET), Kassel, Germany (Original title: Windenergie report Deutschland 2004–2005, in German), 2006.
150. Communication with Nils-Eric Carlstedt at Vattenfall Power Consultant (former swedpower AB), Interview, November 2005.
151. Concerted action on offshore wind energy in Europe, Duwind 2001:006, Delft University Wind Energy Research Institute, 2001.
152. Robb, D., Gearbox design for wind turbines improving but still face challenges, *Windstat Newsletter, Issue: Summer*, **18**(3), 2005, Forlaget Vistoft, Denmark.
153. Faulstich, S., Hahn, B., and Tavner, P. J., Wind turbine downtime and its importance for offshore deployment, *Wind Energy*, **14**, 327–337, 2010.
154. Bertling, L., *RCM-studie Birka Nät och KTH-en studie av felorsaker baserad påstatistik och praktik (RCM study Birka Nät and KTH-a study of failure causes based on statistics and practical experience)*: Royal Inst. Technol., Stockholm, Sweden. Tech. Rep. A-EES-0104, 2001.
155. Fischer, K., Besnard, F., and Bertling, L. A limited-scope reliability centred maintenance analysis of wind turbines, in *Proc. Eur. Wind Energy Conf.*, Brussels, Belgium, pp. 89–93, 2011.
156. Faulstich, S., Hahn, B., Jung, H., and Rafik, K., Suitable failure statistics as a key for improving availability, *Eur. Wind Energy Conf.*, Marseille, France, 2009.
157. *Wind Turbine of Model: V90-1.8MW/2MW*. Manufacturer: Vestas Wind Systems A/S, Randers, Denmark, 2009.
158. *Wind Turbine of Model: V90-1.8MW/2MW*. Manufacturer: Vestas Wind Systems A/S, Randers, Denmark, 2010.
159. Endrenyi, J., Anders, G. J., and Silva, A. M. L. D., Probabilistic evaluation of the effect of maintenance on reliability—An application, *IEEE Transactions on Power Systems* **13**(2), 576–583, 1998.
160. Balschuweit, R., Stress and damage analysis of wind turbines using the example of Vestas V90-2MW, (German title: *Beanspruchungs-und Schadensanalyse von Windenergieanlagen am Beispiel der Vestas V90-2MW*), MSc thesis, Eng., Sci., Beuth University of Applied Sciences Berlin, Berlin, 2009.
161. Musial, W., Butterfield, F., and McNiff, B., Improving wind turbine gearbox reliability, *Eur. Wind Energy Conf.*, Milan, Italy, 2007. http://www.townoflenox.com/Public_Documents/LenoxMA_Wind/may200741548.pdf
162. *Bearing Failures and their Causes, Product Information 401*. SKF, Gothenburg, Sweden, 1994.

163. Bellens, R. and Chemweno, P., *Business case for condition based maintenance of wind turbine gearboxes*. M.S. thesis, Faculty Eng., Katholieke Universiteit Leuven, Leuven, Belgium, 2010.
164. Ruo, P. and Olver, A. V., Hydrogen in lubricated contact, in *Artificial Intelligence for Industrial Applications: 2nd PhD Conference Paperback*, TEWISS, Hannover, Germany, 2008. 58pp.
165. Pasaribu, R., *Bearing Failures and their Causes, Product Information 401*. SKF Engineering and Research Centre, Nieuwegein, the Netherlands, 2011.
166. Alewine, K. and Chen, W., Wind turbine generator failure modes analysis and occurrence, in *Windpower 2010*, Dallas, TX, 2010.
167. Carlson, O., Helmer, M., and Liljegren, C., Analys av havererade asynkrongeneratorer (analysis of failed asynchronous generators), in *Chalmers Univ. Technol.*, Gothenburg, Sweden, 2001.
168. Manwell, J. F., McGowan, J. G., and Rogers, A. L., *Wind Energy Explained: Theory, Design and Application*. 2nd edition. John Wiley & Sons, Ltd, Chichester, UK, 2009.
169. Nilsson, K., Torque estimation of double fed induction generator using a dynamic model and measured data. MS thesis, Dept. Ind. Electr. Eng. Automat., Lund University, Lund, Sweden, 2010.
170. Polinder, H., Lendenmann, H., Chin, R., and Arshad, W. M., Fault tolerant generator systems for increasing availability of wind turbines, in *Eur. Wind Energy Conf. Exhib.*, Marseille, France, 2009.
171. Roed, C., Jeppesen, O. M., Jensen, K. L., and Ahler, C., Hydraulic system and method for operating a brake of a wind turbine, US patent 20110014048a1, January 1, 2011.
172. Nilsson, J. and Bertling, L., Maintenance management of wind power systems using condition monitoring systems—life cycle cost analysis for two case studies in the nordic system, *IEEE Transactions on Energy Conversion*, **22**(1), 223–229, IEEE, March 2007.
173. Ribrant, J., *Reliability performance and maintenance, a survey of failures in wind power systems*. Master's thesis, KTH, Stockholm, Sweden, XR-EE-EEK 2006:009, 2005–2006.
174. Puglia, G., Banglore, P., and Tjernberg, L. B. Cost efficient maintenance strategies for wind power systems using lcc, in *PMAPS 2014*, IEEE, Durham, July 2014.
175. *EN13306, Maintenance terminology*. Swedish Standard Institute, 2010.
176. Lindquist, T. and Bertling, L., Hazard rate estimation for high-voltage contacts using infrared thermography, in 2008 Annual Reliability and Maintainability Symposium, IEEE, 2008, pp. 231–237.
177. Gill, P., *Electrical Power Equipment Maintenance and Testing*. Marcel Dekker Inc., New York, 1998. 600pp. https://www.abebooks.com/book-search/title/electrical-power-equipment-maintenance-testing/author/paul-gill/
178. Korendo, Z. and Florowski, M. Thermography-based diagnostics of power equipment, *Power Engineering Journal*, 33–42, 2001.
179. Lindquist, T. M. and Bertling, L., A method for calculating disconnector contact availability as a function of thermography inspection intervals and load current, in *Proc. CIGRE Symposium on System Development and Asset Management under Restructuring*, Osaka, Japan, November 2007.
180. IEC, *694:1996, Common clauses for high-voltage switchgear and controlgear standards.*
181. VAST, *[Thermography in substations]*. Vattenfall (in Swedish), 1989.

182. Ascher, H. and Feingold, H., *Repairable Systems—Modeling, Inference, Misconceptions and Their Causes*. CRC Press/Marcel Dekker, New York, 1984.
183. Nachlas, J. A., *Reliability Engineering*. CRC Press, Boca Raton, 2005.
184. Mathworks, T., *Statistics Toolbox*: version 4, The Mathworks, 2002. https://www.mathworks.com/products/statistics.html
185. Franzén, A. and Bertling, L., *State of the art—life time modeling and management of transformers*. Royal Institute of Technology KTH, Department of Electrical Engineering, TRITA-EE 2007:041, August 2007.
186. Carlson, A., *Power transformer design fundamentals*. ABB Transformers, 2000.
187. Dietrich, W., An international survey on failures in large power transformers in service, *Electra*, (88):21–48, 1983.
188. Aubin, J. et al., Thermal aspects of transformers. Lifetime evaluation of transformers, in *CIGRE Report 96*, pp. 49–55, August 1995.
189. Gabrielsson, A., *Livslängdsfördelningar för krafttransformatorer: tillämpning av bayesiansk metodik. Elforsk report 06:46*, June 2006.
190. Auvinen, O., *Wood pole condition management and transformer failure rate assessment in Finnish electricity distribution utilities*: VTT Processes, Technical Report, VTT Technical Research Centre of Finland LTD, June 2006.
191. Wikipedia, Commonwealth of independent states. Wikipedia, the free encyclopedia, March 2007.
192. Drobyshevski, A., Sokolov, V., and Tsurpal, S., Reliability problems with large power transformers and shunt reactors. Typical failure modes and failure causes, in *CIGRE SC A2 Colloquium on "Transformer Reliability and Transients"*, Moscow, Russia, pp. 20–23, June 2005.
193. Vainberg, M., Kurtz, C., and Ford, G., Managing aged transformers. Utility develops repair/refurbish/replace strategy using innovative risk-based methodologies, *Transmission and Distribution World*, **57**, 30–43, 2005.
194. de Klerk, P., Minhas, M. S. A., and Reynders, J., Failures in power system transformers and appropriate monitoring techniques, in *High Voltage Engineering Symposium*, in Conference Publication No. 467, pp. 22–27, August 1999.
195. Wilson, A., Jarman, P. N., and Lapworth, J. A., Life assessment of 275 and 400kV transmission transformers, in *CIGRE, Paper 12–210*, 1998.
196. Averweg, J., Minutes of the round table discussion at Edison Electric Institute Transmission Subject Area Subcommittee. Transformer Aging, April 2005.
197. Hammer, C., Weck, K. H., Virayavanich, S., and Seiler, A., Reliability of on-load tap changers with special consideration of experience with delta connected transformer windings and tropical environmental conditions, in *CIGRE, Paper 12-103*, 1996.
198. Lapworth, J. A., Transformer reliability surveys, *Electra*, (227), 10–14, August 2006.
199. Lapworth, J. A., Transformer reliability surveys, *Electra*, 2006.
200. *Power transformer—Loading guide for oil-immersed power transformer*. IEC Standards 60076-7, 1st edition, December 2005.
201. Borsi, H., Zickler, U., Schwan, M., Machkine, A., Schnettler, A., Zhang, X., and Gockenbach, E., Determination of the actual condition of the electrical components in distribution systems, in *XIVth Internal Symposium on High Voltage Engineering*, Tsinghua University, Beijing, China, August 25–29, 2005. https://www.ieee.org/conferences_events/conferences/conferencedetails/index.html?Conf_ID=10652

202. Lipstein, G. R., Lokhanin, A. K., Morozova, T. I., Kassihin, S. D., and Lizunov, S. D., Service experience and reasons of bushings failures of EHV transformers and shunt reactors, in *CIGRE, Paper 12-105*, 1996.
203. *Maintenance Terminology*. European standard: EN 13306:2010.
204. Lager, U., Underhållsinstruktion, krafttransformator., e.on elnät sverige ab, September 2006.
205. Svensson, M., Underhållsinstruktion, krafttransformator. felsökning, e.on elnät sverige ab, September 2006.
206. Lorentzon, M., Underhållsinstruktion, underhållsintervall, e.on elnät sverige ab, September 2006.
207. Svensson, M., Underhållsinstruktion, isolerolja, e.on elnät sverige ab, September 2006.
208. Fofana, I., Wasserberg, V., Borsi, H., Gockenbach, E., Drying of transformer insulation using zeolite, *IEEE Electrical Insulation Magazine*, **20**(1), 20–30, 2004.
209. How does your transformer work? Regeneration of transformer oil, Technical Report ABB Power Technologies AB, Power Transformers Installation and Service. (Original title: *Hur mår din transformator? Regenerering av transformatorolja*, in Swedish), April 2005.
210. Svensson, M., Bengtsson, C., and Tapper, M., Isolation Maintenance of Power Transformers. (Original title: *Isolationsunderhåll av krafttransformatorer*, in Swedish), Energiforsk (Former Elforsk), Technical Report: Elforsk report 01:18, April 2001.
211. Svensson, M. and Bro, C., Isolation Maintenance of Power Transformers. Phase 2. (Original title: *Isolationsunderhåll av krafttransformatorer. Fas 2*, in Swedish), Energiforsk (Former Elforsk), Technical Report: Elforsk report 03:39, December 2003.
212. Lager, U., Maintenance instructions, winding clutches, Technical Report. E.ON. (Original title: *Underhållsinstruktion, lindningskopplare*, in Swedish), September 2006.
213. Lager, U., Maintenance instructions, winding clutches, ASAE Type HD, Technical Report. E.ON. (Original title: *Underhållsinstruktion, lindningskopplare*, in Swedish, ASEA typ HD), September 2006.
214. Communication with Lars Pettersson, ABB Transformers, Ludvika, 2006.
215. Öfverbeck, P. and Hällgren, B., Some practical aspects on maintenance interval optimization, in *12th INTER-RAM Conference for the Electric Power Industry*, Baltimore, pp. 9–12, April 1985.
216. Jolicoeur, A., Foata, M., and Rajotte, C., On-load tap changer reliability and maintenance strategy, in *CIGRE International Council on Large Electric Systems*, Paris, France, August 28–September 1, 2006, No. A2-102.
217. Storesund, J., *Handbook for Life Time Analysis of Energy Plants*. Second Edition, Technical Report 2004:885, Energiforsk AB. (Former Vårmeforsk). (Original title: *Handbok för livslångdsarbete för energianlåggningar*, in Swedish), 2004.
218. Egan, D. M. and Chen, Q., A Bayesian method for transformer life estimation using Perk's hazard function, *IEEE Transaction on Power Systems*, **21**(4), 1954–1965, 2006.
219. Li, X., Swift, G., Weekes, T., and Molinski, T., Risk assessment using transformer loss of life data, *IEEE Electrical Insulation Magazine*, **20**(2), 27–31, 2004.

220. Ramu, T. S. and Pradhan, M. K., On the estimation of elapsed life of oil-immersed power transformers, *IEEE Transactions on Power Delivery*, **20**(3), 1962–1969, 2005.
221. Gockenbach, E. and Zhang, X., Assessment of the actual condition of the electrical components in medium-voltage networks, *IEEE Transactions on Reliability*, **55**(2), 361–368, 2006.
222. Wasserberg, V., Borsi, H., Zhang, X., and Gockenbach, E., Estimation of the lifetime of the electrical components in distribution networks, *IEEE Transactions on Power Delivery*, **22**(1), 515–522, 2007.
223. McCalley, J., Zhang, Z., and Jiang, Y., *Bayesian Analysis of Power Transformer Failure Rate Based on Condition Monitoring Information*, Iowa State University, 2004. http://home.eng.iastate.edu/~jdm/WebConfPapers/BayesianXfmrNAPS.pdf
224. Singh, C. and Jirutitijaroen, P., The effect of transformer maintenance parameters on reliability and costs: A probabilistic model, *Electric Power Systems Research*, (72), 213–214, 2004.
225. Franzén, A. and Karlsson, S., *Failure Modes and Effects Analysis of Transformers*: Royal Institute of Technology KTH, Department of Electrical Engineering, TRITA-EE 2007:040, January 2007.
226. Communication with David Ribbenfjärd, KTH Department of Electrical Engineering (he presented his PhD thesis on Electromagnetic transformer modelling including the ferromagnetic core in 2010), 2006.
227. Carlson, A., *Power transformer design fundamentals*, Technical Report, ABB Transformers, 2000.
228. Facilities Instructions, standards and Techniques, volume 3-31, Transformer Diagnostics, Unites States Department of the Interior Bureau of Reclamation, June 2003.
229. Sanghi, R., Chemistry behind the life of a transformer, in *Resonance*, **8**(6), 17–23, 2003.
230. Lundgaard, L. E., Hansen, W., Linhjell, D. and Painter, T. J., Aging of oil-impregnated paper in power transformers, *IEEE Transactions on Power Delivery*, **19**(1), 230–239, 2004.
231. *V90-1.8MW & 2MW-Built on Experience*. Vestas Wind Systems A/S, Randers, Denmark, 2007.
232. Harlow, J. H., editor, *Electric Power Transformer Engineering*, 2004.
233. ABB, *On-Load Tap-Changer Type UC*, Components Technical Guide, 3rd edition, Mars 1996.
234. IEC, *Standards 60076-7 Power Transformer—Loading Guide for Oil-Immersed Power Transformer, 1st edition*. Technical Guide, 3rd edition, December 2005.
235. Rumbaugh, J. et al., Power flow assessment for reliability evaluation in RAD-POW. Diploma Thesis, KTH, Stockholm, B-EES-0005, 2000.
236. Gabrielsson, A., *Livslängdsfördelningar för krafttransformatorer: tillämpn- ing av bayesiansk metodik*. Technical Report 06:46, Elforsk, June 2006.
237. Sundermann, U. and Petersson, A. N. F. L., Life assessment: Ranking of power transformer using condition based evaluation, a new approach, in *CIGRE Paris Conference Paper*, pp. 12–204, 1998.
238. J. Persson, Walldén, K., and Petersson, A. N. L. F. L., Condition based evaluation of net transformers—Experience from a new ranking procedure, in *CIGRE Paris Conference Paper*, pp. 12–108, 2002.

239. Petersson, L. and Fantana, N., Condition-based evaluation: A new platform for power equipment life management: ABB review, *The Corporate Technical Journal of the ABB Group*, (4), 45–54, 2000.
240. Banglore, P. and Bertling, L., An artificial neural network approach for early fault detection of gearbox bearings, *IEEE Transactions on Smart Grid*, **6**(2), IEEE, 2015.
241. Bangalore, P. and Bertling, L., An approach for self evolving neural network based algorithm for fault prognosis in wind turbine, *Proc. IEEE PowerTech Conf.*, Grenoble, France, pp. 1–6, June 2013.
242. Haykin, S. S., *Neural Networks and Learning Machines*, 3rd edition, Pearson, New York, NY, 2009.
243. Hagan, M. T. and Menhaj, M. B., Training feedforward networks with the Marquardt algorithm, *IEEE Transactions on Neural Networks*, **5**(6), 989–993, 1994.
244. Niu, G., Singh, S., Holland, S. W., and Pecht, M., Health monitoring of electronic products based on Mahalanobis distance and Weibull decision metrics, *Microelectronics Reliability*, **51**(2), 279–284.
245. Wang, Y., Miao, Q., Ma, E. W. M., Tsui, K. L., and Pecht, M. G., Online anomaly detection for hard disk drives based on Mahalanobis distance, *IEEE Transactions on Reliability*, **62**(1), 136–145.
246. Oyague, F., Gearbox reliability collaborative (GRC) description and loading, in *Nat. Renew. Energy Lab., Washington, DC, USA, Tech. Rep. NREL/TP-5000-47773*, November 2011.
247. Garcia, M. C., Sanz-Bobi, M. A., and del Pico, J., Simap: Intelligent system for predictive maintenance: Application to the health condition monitoring of a windturbine gearbox, *Computers in Industry*, **57**(6), 552–568.
248. Zaher, A., McArthur, S. D. J., Infield, D. G., and Patel, Y., Online wind turbine fault detection through automated scada data analysis, *Wind Energy*, **12**(6), 574–593.
249. Hilber, P., Miranda, V., Matos, M. A., and Bertling, L. Multiobjective optimization applied to maintenance policy for electrical networks, *IEEE Transactions on Power Systems*, **22**(4), 1675–1682, 2007.
250. Hilber, P. and Bertling, L., A method for extracting reliability importance indices from reliability simulations of electrical networks, in *Proceedings of the 15th Power Systems Computation Conference*, Liege, Belgium, 2005.
251. Keeney, R. L. and Raiffa, H., *Decisions with Multiple Objectives*, Cambridge University Press, Cambridge, UK, 1993.
252. Bertling, L., Eriksson, R., Allan, R. N., Gustafsson, L. A., and Hlén, M. A., Survey of causes of failures based on statistics and practice for improvements of preventive maintenance plans, in *Proceedings of the 14th Power Systems Computation Conference*, Sevilla, Spain, 2002.
253. Besnard, F., Fischer, K., and Tjernberg, L. B. A model for the optimization of the maintenance support organization for offshore wind farms, *IEEE Transactions on Sustainable Energy*, **4**(2), 443–450, 2013.
254. Rademakers, L. W. M. M., Braam, H., Obdam, T. S., Frohböse, P., and Kruse, N. Tools for estimating operation and maintenance costs of offshore wind farms: State of the art, in *Proceedings of the European Wind Energy Conference 2008*, Brussels, Belgium, March 31–April 3, 2008.
255. Gross, D., Shortle, J. F., Thompson, J. M., and Harris, C., *Fundamentals of Queuing Theory*, 4th edition. Hoboken, NJ,Wiley, ISBN 978-0-471-79127-0, 2004.

256. Nilsson, J., *Maintenance Management of Wind Power Systems Cost Effect Analysis of Condition Monitoring Systems,* Master's thesis, Royal Institute of Technology, Stockholm, Sweden, 2010.
257. Broehl, J., *Servicing a Turbine*: Wind Power Special Report—Supplement to Wind Power Monthly, March 2010.
258. *Offshore Wind in Europe.* 2010 Market Report KPMG/German Offshore Wind Energy Foundation, 2010. http://www.southbaltic-offshore.eu/news/imgs-media/Offshore-Wind-in-Europe-2010-Market-Report.pdf
259. *Annual Availability Report Horns Rev 2009–2011.* Vattenfall Tech. Rep., 2012.
260. Nilsson, J., Wojciechowski, A., Stromberg, A-B., Patriksson, M., and Bertling, L. An opportunistic maintenance optimization model for shaft seals in feed-water pump systems in nuclear power plants, in *2009 IEEE Bucharest PowerTech,* Bucharest, IEEE, July 2009, pp. 1–8.
261. Andréasson, N., *Optimization of opportunistic replacement activities in deterministic and stochastic multi-component systems.* Licentiate thesis Chalmers University of Technology and Göteborg University, Göteborg, Department of Mathematical Sciences, ISSN 0347-2809, 2004.
262. *Forsmark Power Plant and PCI Information AB, "Facility Knowledge Introduction Forsmark".* Forsmarks Kraftgrupp AB, 1987–1988.
263. Backlund, F., *Managing the Introduction of Reliability-Centred Maintenance, RCM.* Doctoral thesis Luleå University of Technology, *ISSN:1402 – 1544,* 2003.
264. Reliability-centred maintenance (Original title: *Pålitlighetsstyrt vedlikehold (rcm),* in Norweigan). Energy Supply Joint Organization ENFO (Energiforsyningens Fellesorganisasjon), ISBN 8243602682, Norway, 1997.
265. Wilhelmsson, O., *Evaluation of the introduction of RCM for hydro power generators at Vattenfall Vattenkraft.* Master Thesis written at the Royal Institute of Technology KTH, Department of Electrical Engineering, Report X-ETS/EEK-0517, 2005.
266. Systembeskrivning rcm generator system working paper, Vattenfall.
267. Patrick, D. and Fardo, S., *Rotating Electrical Machines and Power Systems.* 2nd edition. The Fairmont Press 1997. 399pp.
268. Mörtsell, M., Interview, Vattenfall service nord ab, March 4, 2005.

Index

D

G

P

* 9 7 8 1 4 9 8 7 0 8 6 7 8 *
#0003 - 220318 - C2 - 229/152/32 [34] - CB - 9781498708678